Earthquake Hazard and Seismic Risk Reduction

Advances in Natural and Technological Hazards Research

VOLUME 12

The titles published in this series are listed at the end of this volume.

Earthquake Hazard and Seismic Risk Reduction

Edited by

SERGUEI BALASSANIAN

Armenian National Survey for Seismic Protection, Yerevan, Republic of Armenia

ARMANDO CISTERNAS

Institut de Physique du Globe de Strasbourg, France

and

MIKAEL MELKUMYAN

Armenian National Survey for Seismic Protection, Yerevan, Republic of Armenia

KLUWER ACADEMIC PUBLISHERS

DORDRECHT / BOSTON / LONDON

A C.I.P. Catalogue record for this book is available from the Library of Congress.

ISBN 0-7923-6390-6

Published by Kluwer Academic Publishers,
P.O. Box 17, 3300 AA Dordrecht, The Netherlands.

Sold and distributed in North, Central and South America
by Kluwer Academic Publishers,
101 Philip Drive, Norwell, MA 02061, U.S.A.

In all other countries, sold and distributed
by Kluwer Academic Publishers,
P.O. Box 322, 3300 AH Dordrecht, The Netherlands.

Printed on acid-free paper

Printed in the Netherlands.

TABLE OF CONTENTS

PREFACE

In 1998 Armenia was commemorating the tenth anniversary of the catastrophic Spitak earthquake. The Second International Conference on "Earthquake Hazard and Seismic Risk Reduction" sponsored by the Government of the Republic of Armenia and United Nation's International Decade for Natural Disaster Reduction (UN/IDNDR) was held in dedication to that event between 14-21 September (later referred to as Yerevan Conference). The Yerevan Conference has been organized by the National Survey for Seismic Protection (NSSP) of the Republic of Armenia.

All level's decision-makers (from the ministers to the local authorities), politicians, scientists, leaders of the executive and legislative powers, psychologists, leading businessmen, representatives from the private sector and the media as well as from the International Organizations have been invited by the Armenian NSSP to take part in joint discussion of the Seismic Risk Reduction Problem for the first time in the history of such forums. Armenian NSSP's such initiative has been triggered by the experience of the Spitak earthquake and other disasters. They showed that it will be possible to reduce the risks, posed by the natural disaster, only through the common efforts of all the community in co-operation with the International institutions.

The Yerevan Conference has brought together more than 400 participants from 43 countries and summed up the achievements and experiences of different countries in disaster reduction since 1988, provided a deep scientific evaluation of Earthquake Hazard, Earthquake Engineering and Seismic Risk Reduction problems as well as discussed possible models and strategies for disaster reduction in the XXI century.

The Yerevan Conference has served for the UN as the IDNDR Regional Conference for the countries of the Commonwealth Independent States (CIS) and Central and Eastern Europe and played crucial role in the regional consolidation of the achievements of disaster reduction during the decade and the projection of the risks that the region will be faced in the forth coming millennium. It is noted in the Yerevan Declaration adopted at the Yerevan conference that with the increasing scale of disasters that experts project for the XXI century, scientists, local, regional and national authorities and religious community leaders must come together in order to devise a strategy for the prevention of the impacts of hazards on the population, vital infrastructure and property. Particular emphasis must be placed on the protection of vital social and economic infrastructure because these are intimately linked to the ability of a country to function appropriately, for the guarantee of business continuity and hence economic growth, and the potential of a country to prosper and develop. The subsequent stressed that disaster reduction should be a major priority in regional, national and international development and civil protection strategies.

While there have been notable achievements in natural disaster reduction during the Decade, and while awareness of risk from natural disasters has increased considerably at the global scale, there remain areas where loss of life and the impoverishment of large communities continue to increase at an alarming rate. The participants anticipate that the severity and frequency of natural and environmental disasters, and their impact on society, will intensify in near future. The participants reaffirmed the central importance of disaster reduction as an essential element of government policy. Recent events and disasters have again demonstrated the need for permanent sustained strategies to reduce disaster risk in order to save lives and protect economic and social assets.

Within the framework of the Yerevan Conference, the V Conference on Cooperation of Central, Eastern and South- Eastern European Countries on the Protection Against Natural and other disasters (The "Magdeburg Process") was held from 15-17 September. The V Magdeburg Conference brought together all the Ministers and Heads of Departments for Civil Protection from the countries of the CIS, and many representatives from Central and Eastern Europe, including Hungary, Slovenia, Germany, Croatia and Austria. The V Magdeburg Conference provided the necessary political backdrop for the discussions on disaster prevention, reduction and response taking place on the 10th Anniversary of the Spitak Earthquake. The V Magdeburg Conference held rescue exercises in Gyumri, Armenia as well as discussions on the future of the IDNDR. The V Magdeburg Conference was organized by the Emergency Management Administration of the Republic of Armenia.

Another important component of the Yerevan Conference was the organization and hosting of the RADIUS Initiative mid-term Workshop for the evaluation of progress in the RADIUS case Studies and for discussion on urban seismic risk reduction practices. The RADIUS Initiative was lunched by the IDNDR Secretariat, aiming to prepare earthquake disaster scenarios and the risk management plans in 9 selected cities worldwide and develop manuals for seismic risk assessment in urban areas. The progress of the RADIUS Case Studies was reported during the session as well as the achievements of similar effects in order cities. It was stressed in the session that the scientific knowledge should be applied in an appropriate way, involving various sectors of the community such as the decision makers, press, public and private sectors.

The Yerevan Conference noted the need for the exchange and transfer of up-to-date scientific technology and the strengthening of international cooperation in the field of disaster reduction, in particular in the field of seismic risk reduction. A forum for this cooperation, in the form of an international coordinated framework, was proposed as a desirable initiative for the improvement of scientific and technical cooperation in the field of disaster reduction.

The participants reiterated the need to move away from a disaster response oriented approach towards disaster prevention and reduction methodologies. The participants stated their appreciation for the important work that the IDNDR has carried out over the past decade in order to reduce the vulnerability of communities around the world to natural disasters, through the promotion and coordination of Disaster Reduction worldwide.

The participants expressed their belief that effective disaster reduction depends upon a multi- sectoral and interdisciplinary collaboration among all concerned actors, as successfully demonstrated during the Yerevan Conference.

Strategies for seismic risk reduction in 21st century must focus more on "Preparedness and Prevention" rather than only the " Recovery" approach.

One of the most important elements for the implementation of "Preparedness and Prevention" is the development of long-term Government policies in the field of disaster risk reduction.

The importance of consolidating the final phase of the Decade on a regional scale was recognized as appropriate since countries of the same region share a common history, types of risk and often have coordinated strategies for disaster reduction. The participants noted in particular that the Yerevan Conference, which is the IDNDR Regional Conference for the Countries of the CIS and Central and Eastern Europe has been a very useful forum for the exchange of disaster reduction information and the coordination of future disaster reduction strategies in the region.

Special attention in the Yerevan Conference has been paid to the scientific aspect of the Seismic Risk Reduction Problem. Three sections of that direction, i.e. Disaster Reduction, Earthquake Hazard, Earthquake Engineering were specially efficient. Presented Conference Proceedings are dedicated to these exactly significant components of the Seismic Risk Reduction demanding periodically a deep scientific analysis and evaluation.

ACKNOWLEDGMENTS

International Organizing Committee of the Second International Conference on "Earthquake Hazard and Seismic Risk Reduction" consisting of:

Balassanian, Serguei – Chairman
National Survey for Seismic Protection, Armenia

Weber, Scott
IDNDR Secretariat

Eisenberg, Jakob
Russian National Committee on Earthquake Engineering, Russia

Chuanzen, Zhu
Asian Seismological Commission, Institute of Geophysics, SSB, China

Cisternas, Armando
Institut de Physique du Globe, France

Der-Kiureghian, Armen
University of California at Berkeley, USA and American University of Armenia

Giardini, Domenico
Institute of Geophysics, ETH, Switzerland

Hamada, Kazuo
Association for the Development of Earthquake Prediction, ERC, Japan

Kelley, James
University of California at Berkeley, EEC, USA

Nikolaev, Alexey
United Institute of Physics of the Earth, Russia

Simpson, David
IRIS CONSORTIUM, USA

Sobolev, Gennadi
European Seismological Commission and United Institute of Physics of the Earth, Russia

Tucker, Brian
GEOHAZARDS INTERNATIONAL, USA

Zschau, Jochen
GEOFORSCHUNGSZENTRUM, Potsdam, Germany

is expressing its deep gratitude to the Conference sponsors: the Government of the Republic of Armenia and UN/IDNDR and also all the co-sponsoring organizations:

American University of Armenia
Amyot Exco Armenia, Auditors & Business Advisers, Yerevan, Armenia

Armenia Red Cross, Armenia
Armenian – American Exploration Company, Armenia
Armenian Association of Builders
Armenian General Benevolent Union, Armenia
Armimpexbank, Armenia
Centre on Earthquake Engineering and Natural Disaster Reduction Ministry on Construction, Russia
EMERCOM Russia
EPPO, Greece
GEOFORSCHUNGSZENTRUM, Potsdam, Germany
GEOHAZARDS International, USA
GEO-PROCESSORS PTY. LTD., Australia
Government of Germany
Government of Slovenia
Hotel "Dvin", Armenia
HYDRA Engineering Co. for Industry & Trade, Aleppo, Syria
IASPEI Commission for IDNDR
Pacific Earthquake Research Center, University of California, Berkeley, USA
State Emergency Management Administration of Armenia
UNITED BANK, Armenia

SEISMIC RISK REDUCTION STRATEGY IN THE XXI CENTURY

S. BALASSANIAN
Armenian National Survey for Seismic Protection (NSSP)

1. Introduction

The Spitak destructive earthquake (M=7.0) occurred in December, 1988 in Armenia has shook entire world by enormous scale of destructions and casualties, when 25.000 people died, over 20.000 were injured, 515.000 became homeless, all the large settlements of the Northern part of Armenia was practically destroyed, which is not turning out well to be recovered till now. For the first time in disasters practice all the countries of the world have took sincere and active part to support suffered country. Spitak earthquake of 1988 has indicated, that the states, even with opposite political order are ready to render humanitarian assistance to any country in case of large-scale disaster. Successful interaction of different countries during Spitak tragedy became a trigger in 1988 for many of international projects' and programs' development directed to disaster reduction. One of such programs has became the UN-IDNDR.

Large-scale international assistance rendered to any country suffered from the strong earthquake or some other destructive event has became a norm of international relations after the 1988.

At the same time, the idea of seismic risk reduction as a main way for disaster prevention has became more attractive at the end of the XX century both at the national and international levels.

2. The State of the Art

During the XX Century, both at the national and international levels the main attention was being paid to the Recovery in the field of protection of the population against disasters generally and seismic disaster in particular. For that reason, there have been Disaster Management Plans and Forces created at the national levels and also corresponding funds provided into the state budgets for Recovery purposes in case of various disasters. The same steps have been undertaken also at the international, particularly, UN level, since it has turned out that many of, in particular, developing countries being aware about the possible hazards in their own counties have never been just ready for the Recovery but even for Disaster Management in case of disaster.

Coming from the international experience, never depreciating Disaster Management and Recovery significance, it could be concluded that such policy unfortunately has never solved the problem of the population protection against

S. Balassanian et al. (eds.), Earthquake Hazard and Seismic Risk Reduction, 1–11.

disasters. Neither Disaster Management nor the Recovery can make irretrievable human and huge economical and social losses.

In connection with that, the idea of Preparedness priority over the Recovery and, in accordance with that, the Risk Reduction Strategy including Disaster Management as one of its important elements became more and more attractive at the end of the XX Century.

Armenia has became one of the designers and conductors of the Seismic Risk Reduction Strategy after the destructive Spitak earthquake of 1988. Some countries' experience, particularly USA, are indicating at the end of the XX Century that only well organized Preparedness can minimize or even exclude at all the Recovery need. It means not the Preparedness for the Recovery, but the Preparedness for the Recovery exclusion.

For the Seismic Risk Reduction in any country, irrespective of its development degree, it is following steps are required:

- seismic hazard assessment (SHA);
- seismic risk assessment (SRA);
- seismic codes and standards design;
- seismic strengthening and upgrading of existing buildings;
- seismoresistant construction;
- education of the population;
- current seismic hazard assessment (CSHA);
- early warning and notification (EWAN) system design;
- emergency response and rescue operations (Disaster Management);
- disaster relief and people rehabilitation;
- insurance;
- state disaster low and regulations design;
- responsible governmental institution's establishment for the Seismic Risk Reduction.

The countries passed different ways for the indicated problems' overcoming in the XX Century. Developed countries such as USA, Japan, etc have progressed further than any others. But in spite of that, some certain problems still are remaining unsolved for them also. The majors, irrespective of that countries development degree, are:

- proper seismic hazard assessment;
- establishment of the responsible governmental institution for the Seismic Risk Reduction Policy design and implementation.

The Cobe earthquake (Japan, 1995, M=7.2) shows that the strong earthquakes being significantly different one from the other, irrespective of all the most contemporary and conservative Japanese codes and standards, are always bearing an element of suddenness which relates to the SHA and automatically spreads upon the SRA. In effect, there is always a possibility of victims and destructions as it happened in 1995 in Japan.

The key issue of the Preparedness is the proper selection of the institution, that will elaborate the policy of seismic risk reduction and consecutively implement it.

That problem in many countries is assigned to the Coordination Councils which are called for coordinating the activities of different institutions involved into various aspects of the seismic risk reduction.

Many countries' experience is showing that the Coordination Councils are being faced known difficulties while implementing the Seismic Risk Reduction National Plans:

- members of the Coordination Councils are representing interests of different institutions that pursue various objectives and goals, frequently in direct conflict with each other;
- many of the Coordination Council members are scientists who are representing various institutions and not carrying direct legal responsibility for the accuracy and timeliness of decision making;
- decision making process by the Coordination Council is often being delegated to the actually non competent Government official who is obliged to make subjective decision relating to professional issues.

3. Seismic Hazard Assessment

The problem of SHA is complicated by the fact that the always existing respective percentage of uncertainty of the long-term SHA must be compensated by the monitoring of the CSHA.

The CSHA idea being designed and implemented in the Armenian NSSP (Balassanian, [1,2,3,4]) and having the same goal as the earthquake prediction is principally differing from it. The difference is that the CSHA is supposing the possibility of intermediate scientific results' use from the earthquake prediction research by the governmental institution responsible for the seismic protection of the population. It is necessary for the Early Warning of the government and taking the adequate preventive measures by decision makers directed to the population protection against possible strong earthquake. Thus, elaborated CSHA strategy (Balassanian, 1991 [ ]) in contrast to the earthquake prediction ordinary strategy, is considering the both sides of the prediction problem i.e. the scientific one and the social as well.

Successes of the scientific side of the prediction problem are still mainly relating to the empiric knowledge accumulation about pre-, co- and post-seismic processes occurred in the seismically active zones.

Sufficiency of the empiric knowledge about some strong earthquake scenarios within some seismically active zones is resulted to the series of successful predictions in China (from 1974) [5], Russia (from 1966) [6], Armenia (from 1991) [2,7,8] etc.

The obtained results in the earthquake prediction research field are proving the fact, that it could be considered today that strong earthquakes ($M \geq 5.5$) can be predicted in principal. But it is always should be added: there under the certain physico-geological conditions; under those scenarios which well studied in the given region; in case of well organized multiparameter monitoring of the medium in the real or quasireal time; by means of dense observation networks with data acquisition in the Unified Center; based on the Expert Data Analysis System being repeatedly tested by the seismic events that took place in the given seismoactive zone previously.

Failures in the earthquake prediction scientific part are connected with the following main reasons: empiric knowledge deficiency about the various scenarios of the earthquake preparation in different physico-geological mediums; unclear nature of the being observed many physical, physico-chemical, chemical phenomena accompanying the process of the strong earthquake preparation and occurrence; known uncertainty of the strong earthquakes' origin and mechanism.

Successes in the social part of the earthquake prediction problem are mainly relating with such countries as China and Armenia. Only here, both sides of the prediction problem, scientific and social, are being under the permanent attention thanks to established Seismological Surveys with the high State status. In spite of that, all the scientific earthquake prediction researches intermediate results are being used for solution of the social part of the problem. It is being achieved by the Government's and, if it is necessary, also the population's attention concentration upon the Current Seismic Hazard. This strategy is coming from the fact that the more unexpected is an earthquake the higher is number of casualties and an economical losses.

Thus, both the Chinese State Seismological Bureau and the Armenian National Survey for Seismic Protection are using actually the same strategy, i.e. the Current Seismic Hazard Assessment. Both institutions have designed the clear mechanism for prediction information's transmission from the scientists to the government for making the adequate decisions directed to the protection of the population, i.e. there an effective relation Scientists - Government - Population is acting.

In whole, the main failures in the social aspect for many of developed countries are caused by lack of real links between the scientists taking care of earthquake prediction (working in universities and other institutions) and the government making decisions, i.e. lack of the Scientists - Government - Population link in case of existing prediction.

For developing countries it is firstly a lack of the State Policy in the field of seismic protection of the population and the State Institution responsible for State Policy design and implementation.

Actuality of the Current Seismic Hazard Assessment, irrespective of the fact, that how much optimistically or pessimistically scientists are disposed regarding to the earthquake prediction, is arising continuously and will become in our opinion one of the major problem in the Seismic Risk Reduction Strategy of the XXI Century.

The destructive earthquake in Cobe (Japan, 1995) has displayed that the attempt to pass over the problem of earthquake prediction by means of long-term seismic hazard assessment and then the adequate seismically resistant construction can not be considered as a successful one. In the long-term seismic hazard assessment there is always uncertainty exists, related with data deficiency about the past, present and future development of deformation process in the seismically active zones, which should be compensated by the Current Seismic Hazard Assessment. That problem for the developing countries is becoming more actual, since here has not even provided the initial condition, i.e. the adequateness of the existing buildings', and constructions' seismoresistance to the seismic hazard which known for this moment. Current Seismic Hazard Assessment, for example in Armenia, will be extremely actual as long as all the constructions and inhabited buildings of the country would be possible to strengthen and reconstruct. Since many decades will be required for that matters, then NSSP is implementing large-scale activities on the Current Seismic Hazard Assessment technologies design with the simultaneous design of the new methods of buildings and structures strengthening and also participation in the new building codes design.

Current Seismic Hazard Assessment in Armenia is intending to provide preventive and operative actions of the NSSP, Government and Local Governmental bodies adequate to the seismic hazard evaluated by the NSSP's Expert Council, based on the intermediate earthquake prediction research results' use.

Coming from the above stated it is getting clear that the Early Warning and Notification, which is actual for any country irrespective of its development level, is coming out from the Current Seismic Hazard Assessment .

The general preconditions for the successful CSHA are:

- necessary geological geophysical data about the geological medium patterns (data base) and long (centuries-old) statistics of seismic events (representative and homogeneous earthquakes catalogue);
- favourable geological, geophysical medium being investigated where seismotectonic signs of strong seismic events are being well pronounced and also current seismic processes are clearly reflected;
- well organised, unified multiparameter observation networks allowing to ran geological medium's monitoring in the real time with data transmission onto the Unified Centre of data acquisition, processing and analysis;
- efficient methods of data processing and analysis;
- highly qualified experts in different scientific fields who are able to carry out the complex data analysis jointly;
- international co-operation of the scientists.

Possibility of current seismic hazard assessment should be based upon the following principal physical preconditions:

1. Preparation and occurrence of the seismic event with $M \geq 5.5$ should be accompanied by the strong changes of physical, physico-chemical, chemical parameters of deformed medium.
2. Strong changes of physico-chemical parameters of deformed medium should be accompanied by formation of temporary sources of various geophysical fields that could be pronounced in all (solid, liquid, gaseous) or at least one phase of the lithosphere.
3. Temporary sources of the various geophysical fields being formed at the macro- and micro- hierarchical levels of the lithosphere during the elastic deformations accumulation within the geological medium are getting disintegrated in case of its discharging.
4. Temporary sources are getting formed in the lithosphere being deformed (primary seismogenous sources), provocating the secondary sources in the atmosphere and biosphere in the stages which are close to the strong seismic event ($M \geq 5.5$).
5. Temporary sources are the system with changeable parameters that creating various non-linear dynamic physical fields being confidently recorded on the observation surfaces.

Current Seismic Hazard Assessment carried out particularly in the NSSP is based upon the "Expert" System (Balassanian, 1999 [9]) operating by means of being continuously replenished multifactor retrospective analysis of all the seismic events occurred in Armenia and adjoining countries.

As a result of CSHA the territory of Armenia is being daily tested (based on retrospective scenarios of earthquake preparation) on the probability of seismic event occurrence with $M \geq 5.5$ during each current time (Δt) interval (where 1 hour$\leq \Delta t \leq$10 days). Seismic hazard assessment and even earthquake prediction are losing its applied sense if there is no effective direct connection between the institution assessing and predicting the hazard and the decision-makers in the Government.

That problem had been solved in Armenia in 1991 when the National Survey for Seismic Protection was established. NSSP is co-ordinating all the ministries and state institutions in the field of Seismic Hazard Assessment and Seismic Risk Reduction.

Received information in case of increased current seismic hazard confirmed by the NSSP's Expert Council along with recommendations on adequate actions is being transferred to the Government by means of direct communication between the President of the NSSP and the Prime-Minister of the Republic.

The adequate actions, in any case, having preventive nature, are:

- concentration of the government's attention on the zone of high current seismic hazard;
- concentration of the local governmental body's attention;
- more frequent observations in the stations of the National Observation Network, particularly in the zone of increased seismic hazard;
- additional concentration of the mobile stations in the zone of possible earthquake;
- inspection of buildings and constructions for high seismic risk areas determination by the mobile groups of the NSSP's experts;
- high trim declaration for the NSSP's Special Rescue Brigade;
- Armenian NSSP's and Armenian Emergency Management Administration's (EMI) high trim joint actions plan correction in case of possible earthquake in certain area;
- Armenian Government's, NSSP's, EMI's, Local Governmental Bodies' action plans correction in case of possible earthquake in certain area;
- carrying out of the extraordinary training of the population for proper actions before, during and after an earthquake.

Examples of the stage by stage Current Seismic Hazard Assessment Strategy implementation in Armenia are presented in the presentations of the NSSP's experts in the Second International Conference on Earthquake Hazard and Seismic Risk Reduction.

4. Armenian National Survey for Seismic Protection - Is One of the Probable Future Models for the Seismic Risk Reduction Management.

Coming from tragic lessons of the 1988 Spitak catastrophe the National Survey for Seismic Protection (NSSP) has been established under the Government of the Republic of Armenia according to the project presented to the Parliament and Government (Balassanian, [10]).

The main responsibility of the Armenian NSSP is the Seismic Risk Reduction in Armenia.

Distinctive peculiarities of the NSSP are:

- high governmental status and direct responsibility for the seismic risk reduction in Armenia;
- direct subordination of the NSSP's President to the Prime-Minister of Armenia;
- unification of the separated earlier groups and institutions carrying out various problems into unified NSSP's structure;
- special NSSP's structure consisting of Specialised Centres connected to each other by adjoining tasks and responsible for all the elements of the seismic protection system (from the seismic hazard and risk assessment to the seismoresistance of buildings and structures, from the population training to the rapid response);
- administrative structure based upon the combination of horizontally interacting Centres with its clear vertical subordination to the NSSP's President;
- territorial principal of management; NSSP consists of four territorial departments;
- the common goal is the Seismic Risk Reduction which is consolidating activities of all the Centres and Departments;
- internationalisation of the NSSP's Program implementation based upon the interlinked international programs and agreements with the leading Centres world-wide on all the aspects of the Seismic Risk Reduction;
- training of the NSSP's staff in various scientific and engineering International Centres aimed to master high technologies and its wide use in Armenian NSSP;
- design of the modern technical base consisting of various high tech equipment combination which could be adapted to Armenian conditions in a best manner;
- subordination of all the observation sites in the territory of Armenia to the NSSP, unified into the common National Observation Network consisting of three-level stations, with are incorporated into the Global International Networks (IRIS, GPS, READINESS etc.); Regional Network; Local networks, controlling densely populated areas (Yerevan city etc.) and Life Lines (Armenian NPP and etc.);
- round-time-clock data recording, processing and analysis, i.e. whole data concentration from all the observation sites in the real and quasi-real time into the Unified Centre for Seismic Hazard Assessment (UCSHA);
- using of all kinds available communication means (from satellite to telephone) amongst observation sites and UCSHA;
- unique National Data Bank including all the seismological, geophysical, geological, geochemical and other information about the lithosphere state in the territory of Armenia and adjoining areas. The special part of its is the National Earthquakes Catalogue containing data about 18.000 seismic events beginning from the 18 Century B.C.
- decision making about the current seismic hazard by the President of the NSSP on the NSSP's Expert Council with the following information's and

recommendation's transmission to the Government regarding adequate actions.

Indicated peculiarities of the NSSP allowed to solve the following problems of high importance within unprecedently short period of time:

- to elaborate and confirm by the Government and the President of Armenia the long-term Seismic Risk Reduction State Program for Armenia:
- to assess probabilistic seismic hazard at the scale of 1:1000000 for the Crimea-Caucasus-Copet-Dag region (carried out in the frame of GSHAP Program on INTAS GRANT N...) [11].
- to assess probabilistic seismic hazard at the scale of 1:500000 for the territory of Armenia the task is carried out jointly with the Swiss Seismological Survey [12,13,14];
- to assess seismic risk for the territory of Armenia and Yerevan city [10,15];
- to take part jointly with the Ministry of Urban-Planning of Armenia and the other Armenian Institutions in the new Seismoresistant Building Codes and Standards design [16];
- to elaborate, taste and use in practice the new unique methods of seismoresistance strengthening of the existing buildings without tenants evacuation along with the Ministry of Urban-Planning assistance (by support of the World Bank) [17,18,19];
- to elaborate and implement stage by stage the Program of the population training and awareness increase as well as the Governmental institutions' preparation to act before, during and after an earthquake [20];
- to establish the Rapid Response Forces including seismologists, earthquake engineers, psychologists, instructors for the population training, NSSP's Special Rescue Brigade staff, equipped by equipment and uniform received as a gift from the Swiss Disaster Relief [21];
- to design unique National Observation Network integrated into the Global World-wide, Regional and Local networks consisting of more than 150 multiparameter observation stations for the monitoring of the lithosphere, atmosphere and biosphere in the real time [22];
- to develop current seismic hazard assessment technology including the new methods of observation, data processing and analysis [9];
- to design the Early Warning Program [23];
- to develop Population Seismic Protection State Low Project for Armenia;
- to design the Seismic Risk Reduction Regional Program for Caucasus.

In the frame of the various International Programs Armenian NSSP has manifested with the initiative on the establishment of the Regional System for the Population's Protection against strong earthquakes unifying the efforts of all the Caucasus countries, Russia, Turkey and Iran.

In 1998 the NSSP has certified as a finalist by the United Nations Sasakawa Disaster Prevention Award in appreciation for its distinguished contribution to disaster prevention, mitigation and preparedness and furthering the goals of the IDNDR (International Decade for National Disaster Reduction) in Geneva.

5. Conclusion

1. In the area of protection of the population in the XXI Century against seismic and other disasters, the Recovery Strategy and relating with it Disaster Management should be replaced by the strategy of Preparedness priority over the Recovery.
2. Preparedness strategy is a strategy of Seismic Risk Reduction in which Disaster Management is one of the important elements.
3. For the successful Seismic Risk Reduction it is necessary to make a proper Seismic Hazard Assessment and have an effective Risk Management Model.
4. The proper Seismic Hazard Assessment is based on the long-term and Current Seismic Hazard Assessment.
5. Current Seismic Hazard Assessment allows to compensate uncertainty of the long-term Seismic Hazard Assessment at any current moment of time.
6. One of the possible Models for Seismic Risk Reduction Management could become Armenian NSSP's model.
7. The main task of the Preparedness for the developed countries could become Current Seismic Hazard Assessment and for the developing countries - State Policy elaboration concerning the Seismic Risk Reduction and the Responsible Governmental Institution Establishment, which would be able to elaborate that State Policy and its stage by stage implementation.
8. Seismic Risk Reduction Policy on the national level should be in well correspondence with the International Policy of Seismic Risk Reduction at the UN level.
9. International Policy of Seismic Risk Reduction at the UN level can be elaborated and implemented in the frame of the International Program like IDNDR.
10. The most important aspect of the International Policy is to assist the countries on elaboration and implementation of the Seismic Risk Reduction National Policy.

6. References

1. Balassanian S. (1994) "Problem of Earthquake Prediction and the Ways of its Solution", *Proceeding of the XXIV General Assembly of the European Seismological Commission*, Greece, p. **1029**.
2. Balassanian S., Arakelian A., Martirossian A., Harutunian V., Avanessian A., Mouradian S., Petrossian H., Sahakian A., Kalinin S., Arakelian S., Babayan H., Pribilsky G., Guemilian B. (1995) "Experience of current seismic hazard on the territory of Armenia", *Proceeding of the second international conference on seismology and earthquake engineering*, Iran.
3. Balassanian S. (1998) "Current Seismic Hazard assessment Strategy" 2nd International Conference on "Earthquake Hazard and Seismic Risk Reduction", *Abstract volume, Yerevan, Armenia, September 15-21, 1998,* p. **13**.
4. Balassanian S. (1999) "Earthquake Prediction or Current Seismic Hazard Assessment" - *European Geophysical Society XXIV General Assembly, the Hague, 19-23 April, 1999, Abstract Volume,* p. **889**.

5. Ulomov W., Mavashhev B. (1971). "Tashkent earthquake precursor. - Tashkent earthquake of 26 April, 1966". *Tashkent, Academy of sciencies*, 1971, pp. **188-192**.
6. Zhu Fung-ming (1976) "An outline of prediction and disaster prevention of the Haicheng earthquake of magnitude 7.3 (in Japanese)", *Proceedings of the Lectures by the Seismological Delegation on the People's Republic of China, Seismol. Soc. Japan, 15-26* (Available in English in Proceedings of the Lectures by the People's Republic of China, J.P. Muller (ed.), Spec. Publ. 43-32, Jet Propulsion Laboratory, California Inst. of Tech., Pasadena, 1976, 11-19).
7. Balassanian S. (1997) "Earthquake Prediction Research in Armenia", *The 29th General Assembly of IASPEI, abstract volume*, Greece, p. **138**.
8. Balassanian S., Mouradian A., Sahakian A., Kalinin S., Babayan M., Pogossian A. (1997) "The investigation of electromagnetic precursors to earthquakes in Armenia", *Annali di Geofisika, Italy*, vol. XL, N2, pp. **209-226**.
9. Balassanian S. (1999) "The Anomalies Daily Dynamics of Local Geophysical and Geochemical Fields (ADF) Effect Study in the Connection with Earthquake Preparation and Occurrence", *The Physics and Chemistry of the Earth*", vol. 24, No8, pp.**741-752**, 1999.
10. Balassanian S., Manukian A. 91994) "Seismic Risk on the Territory of the City of Yerevan, Armenia", *Issues in Urban Earthquake Risk, Kluwer Academic Publishers*, Netherlands, pp. **167-182**.
11. Balassanian S. (jointly with Giardini D., Chelidze T., Pustovintko B., Achirow T., Ghafory-Astiani M., Erdik M., Ulomov V., Kondorskaya N., Trifonov V., Molchan G., Mayer-Rosa D., Stucchi M., Griintal G.) (1997) "Seismic Hazard Assessment for the Caucasus Test Area", *The 29th General Assembly of IASPEI, abstract volume*, Greece, p. **385**.
12. Balassanian S., Nazaretian S., Avanessian A., Arakelian A., Igoumnov V., Badalian M., Martirossian A., Ambartsumian V. (1995) "The new seismic zonation map for the territory of Armenia", *Proceedings of the XXI General Assembly of the International Union of Geodesy and Geophysics*, Boulder, Colorado, USA, p. **B347**.
13. Balassanian S., Nazaretian S., Avanessian A., Arakelian A., Igumnov V., Badalian M., Martirosyan A., Ambartsumian V., Tovmassian A. (1997) "The New Seismic Zonation Map for the Territory of Armenia", *Natural Hazards, ed. by Schenk V., Kluwer Academic Publishers*, the Netherlands, vol. 15, N 2-3, pp. **231-249**.
14. Balassanian et all (1999) "Seismic Hazard Assessment in Armenia", *Kluwer Academic Publishers*, NH 18. pp **227-236**.
15. Balassanian S., Manukian A. (1995) "Seismic risk assessment for the territory of Armenia", *Proceeding of the Fifth International Conference on Seismic Zonation, OUEST EDITIONS Press Academicues*, vol. II, pp. **1289-1296**.
16. "Earthquake Engineering. Seismic Cod. CPHA II-02 94", Yerevan, 1994.
17. Balassanian S., Melkoumian M., Azarian A. (1997) "Seismic Risk Assessment for the Territory of Armenia and Strategy of its Mitigation", *The 29th General Assembly of IASPEI, abstract volume*, Greece, p. **7**.
18. Melkumian M. (1996). "Dynamic tests of the 9-storey R/C full-scale building with an additional isolated upper floor acting as a vibrator damper". *Proceedings of the*

third European Conference on Structural Dynamics, Eurodyn '96, Florence, Italy, 5-8 June 1996, vol. 1., pp. **557-560**.

19. Melkumian M., Beybutian L., Vartanian G., Azarian A., Kazarian V., Sarkisian G., Nersesian T. (1998). "Testing of the 4-storey Full-Scale Base Isolated Apartment Building in the City of Spitak". *Report of EEC NSSP*, pp. **1-23**.
20. Balassanian S. (1998) "Lesson of the Spitak Earthquake an the Armenian NSSP's Activity on Seismic Risk Reduction". *Second meeting of Asian Seismological Commission and Symposium on Earthquake Hazard Assessment and Earth's Interior Related Topics, abstract volume*, Hyderabad, India, December 1-3, 1998, p. **28**.
21. Balassanian S. (1998) "National Survey for Seismic Protection (NSSP) as One of the Possible Models for Seismic Risk Reduction abstract volume, p. **319**, Yerevan, Armenia, September 15-21, 1998.
22. Balassanian S., Babayan H., Muradian S., Karakhanian A., Arakelian A. (1995) "Seismological and earthquake prediction observation network of the NSSP", *Proceeding of scientific meeting on the seismic protection*, Venice, Italy, pp. **77-83**.
23. Balassanian S.(jointly with Martirosyan A., Arzoumanian V.) (1999) "Project of Creation of Earthquake Early Warning System Around the Capital of Armenia". *International workshop on recent earthquakes and disaster prevention management.* 10-12 March 1999, Ankara, Turkey, pp. **293-300**.

BUILDING LOCAL CAPACITY FOR EARTHQUAKE LOSS REDUCTION

SHIRLEY MATTINGLY
Former Regional Director, Federal Emergency Management Agency, USA
1965 Vistazo West, Tiburon, CA 94920, USA

Abstract

In 1988 the catastrophic Spitak earthquake rocked Armenia and shocked the world with its devastation. Six years later, the Northridge earthquake ripped through Los Angeles and instantly put to the test the lessons learned from Spitak and other disasters. This paper explores what has been learned and put into practice in building local capacity for earthquake loss reduction since 1988, when the Spitak tragedy influenced disaster management thinking.

Since the Spitak earthquake, there have been rapid technological advances allowing us to build more earthquake resistant structures, strengthen existing buildings, and harden critical lifelines against earthquake effects. Our understanding of social and economic aspects of disaster impacts, and what factors strengthen response and recovery capabilities, has advanced. We have learned that effective disaster management and loss reduction depends on networks of individuals and organizations united by a common goal of protecting lives, livelihoods, investments, and the environment. These lessons were tested in the 1994 Northridge earthquake. This paper describes disaster management advances born of necessity in Northridge and improved upon in practice in subsequent disasters.

1. Learning from Earthquakes

In the 1994 Northridge earthquake, lifelines and buildings in Los Angeles and neighboring communities shook convulsively—some failing, others resisting. Responders swung into action, hoping to save lives and minimize suffering. State and local disaster management systems and networks, developed and strengthened through years of practice in California, were launched.

The quake's physical, social and economic impacts were far-reaching: businesses collapsed, "ghost towns" of abandoned apartment houses were created, and even modern steel moment resisting frame structures built in accordance with code suffered damage, far from the quake's epicenter. New, innovative strategies were needed to meet the needs of victims. But the quake demonstrated that a community can build resilience to disaster before the event by strengthening its physical environment and its systems and networks to resist nature's forces.

S. Balassanian et al. (eds.), Earthquake Hazard and Seismic Risk Reduction, 13–20.

Northridge demonstrated that rapid and effective response is, in large part, a function of quality pre-disaster planning, of practice, and of strong inter-agency and inter-governmental coordination. Rapid response is facilitated by shared, standardized procedures. After Northridge, more than any other disaster I have had a part in managing, I realized that all the hard work—planning, training, exercising, building an organization, strengthening relationships with outside agencies, and promoting an earthquake mitigation agenda--had really paid off. Perhaps these lessons have application elsewhere.

Experiences such as preparing for, responding to, and helping people to recover and rebuild from a major earthquake teach researchers and practitioners alike practical lessons that can be applied in the next damaging quake. Many of the lessons of Spitak and Northridge are universally applicable regardless of the specific hazards a city faces, its building stock, or its political system. Human lives and social and economic systems are at risk from disaster, and local capacity to reduce disaster losses must be strengthened.

Being involved in a disaster as a responder, a public official, a researcher, or a victim is a hard teacher. Being involved in earthquakes and floods and riots and fatal fires, I have seen people suffer, I have talked with families who have lost their homes and husbands who have lost their families and children who have nightmares and cannot sleep peacefully for years, because of the horror they have experienced.

I have seen the response to a disaster jeopardized by failures in command and control, by breakdowns in communications systems, and by inadequate coordination between responding agencies or between elected officials unwilling to set aside their differences or allow their political adversaries any possible political advantage.

In 1980, when we began to develop a disaster management system and organization for the City of Los Angeles, the only preparation for working in disasters and disaster management was a one-week training course. When the Mayor said, let us figure out appropriate public policy for dealing with short-term earthquake predictions, let us plan in advance what will be our response to a prediction, no one had ever done it before. And when the 1985 Mexico City earthquake occurred and we saw its horrendous impacts, making hundreds of thousands of people instantly homeless, many of them our brothers or sisters or cousins, we knew we had to act. We felt compelled to help by immediately sending a train load of backhoes and tractors and dump trucks and other heavy equipment to remove rubble. Moreover, we also felt compelled to begin developing our own plan for recovery and reconstruction plan before *our* big earthquake, before a quake with the effects of Mexico City or Spitak occurred in Los Angeles. But no one had done that before, either. No one had prepared such a plan. There was no model. So we just put together a team of people and put our minds to work and figured it out as we went along.

Now, after working to reduce disasters' toll year after year, many of the lessons previously learned the hard way have been recorded in disaster literature as accepted theory and practice. Now it is possible to learn much about disaster management and loss reduction techniques through case studies, university courses, and specialized training, rather than having to be a responder or a disaster victim or have your friend crushed in a collapsed building.

2. Challenges and Issues in Seismically Vulnerable Cities

Local leaders in seismically vulnerable cities face numerous challenges, with the earthquake peril being only one of many serious threats to the well-being and security of their citizens. However, new tools and techniques are constantly being developed to help communities meet their challenges.

Clearly, in regard to the earthquake menace, we all face a very big job. Human society—all over the world—continues to place itself at risk to earthquake effects, often even when it knows better.

All cities have unique circumstances and special challenges. But they also share some common challenges and common difficulties:

- Inadequate budgets even to meet basic needs
- Inadequate public awareness and concern about the earthquake hazard and risk, and
- Inadequate political will to make the hard decisions
- To protect the environment and the residents
- To limit growth in hazardous areas
- To enact and enforce strong seismic code provisions and controls over construction quality.

We can learn from each other and then help the decision-makers in our communities apply what we learn in ways that make sense in each community's own social, economic, political and natural environment.

3. Strategies for Strengthening Local Capacity

Some of the tools and strategies used successfully in California to confront the earthquake challenge can be adapted to other earthquake-prone cities' political and socioeconomic realities and their formal and informal power structures.

3.1. KNOWLEDGE TRANSFER

Earthquakes are not a problem that can be managed within one discipline or through the efforts of one group of professionals. Understanding earthquakes and reducing earthquake risk require collaboration and communication among professionals, public officials and the public. There are different worlds inhabited by earth scientists, social scientists, engineers, elected officials, and the public. They must come together and exchange information and concerns. Our cities' leaders may not understand the earthquake threat and what to do about it. Elected leaders may not <u>want</u> to know, because they already face enough other challenges and problems.

Knowledge transfer requires multidisciplinary collaboration, learning to speak someone else's language, in a sense, and openly sharing information. Getting people with diverse backgrounds and interests together in the same room is often not easy. But it can be done and it needs to be done more than once.

One example from California experience is the collaboration between the City of Los Angeles and a consortium of universities, known collectively as the Southern

California Earthquake Center, or SCEC, working together to unite science, engineering, and public policy. SCEC's scientists represent all the disciplines in the earth sciences. Before SCEC brought them together, in many cases the scientists did not really talk to each other, let alone talk to engineers and bureaucrats. But now they have united around a common goal of coordinating research on the area's earthquake hazards and applying earth science to earthquake hazard reduction.

The partnership with the City of Los Angeles took years to develop, but gradually it became mutually beneficial. For instance, the City contracted with SCEC to provide site-specific earth science data for City seismic retrofit construction projects, and the scientists have worked with City planning and building officials in exploring options for implementing seismic zonation.

In late 1998 SCEC teamed up with a U.S. national laboratory to organize a workshop for city and state governmental agencies and utilities. The purpose was to get user input on science and technology based tools being developed for disaster planning training and response and recovery management. In this way and others, SCEC is transferring knowledge and bridging gaps that we did not even know existed a few years ago.

A second example of effectively sharing knowledge and bridging gaps is the California Clearinghouse. The Clearinghouse was created by the State of California's Office of Emergency Services to provide a mechanism for coordination of post-earthquake research. Launched into action in the hours following the Northridge quake, the Clearinghouse helped bring immediate focus to scientific, engineering, and social science research on the quake effects. It brought together investigators representing dozens of institutions and agencies in information sharing sessions at the end of every day for several weeks following the quake. Following Northridge, a core group of individuals continued to meet, to plan, to exercise its plans, and to formalize responsibilities assigned among over a dozen participant organizations. Therefore, after the next major California quake, the Clearinghouse should function even more effectively.

This kind of effort can be easily replicated in any large city or state or country by someone, or preferably a few people, deciding that it is a priority, and taking on the task.

3.2 EFFECTIVELY COMMUNICATING HAZARD AND RISK INFORMATION

The Southern California Earthquake Center, described above, seeks to understand and explain where the area's active faults are, how often they slip, and what are the characteristics of the earthquakes they can be expected to produce. SCEC regards communicating this earthquake hazard information to engineers, insurers and financial institutions, local government, and the public, as an integral part of its mission. It's a great asset for Los Angeles, but it is not unique. The partnering between researchers or engineers and government can be replicated wherever the will exists.

The media play an enormous role in communicating risk, in raising awareness of local and distant earthquakes and efforts to reduce risk. Having an active, ongoing dialogue with your local media will help ensure that they accurately portray hazard and risk information. Moreover, they can provide tremendous assistance in disseminating appropriate information in the moments, hours and days after an earthquake strikes. They are a very important partner to have.

3.3 WELL-COORDINATED RESPONSE

An effective and efficient response to a disaster can save lives and reduce losses. It is based on clear assignments of responsibilities, lines of command, mechanisms for coordination and communication, and training and practicing response operations. It must overcome competing interests, overlapping jurisdictions, and diverse responding organizations each with their own way of doing business. There are local agencies responding and federal agencies responding and private sector responders and voluntary and private non-profit agencies responding.

Effective response relies on skillful coordination among people, agencies and institutions with diverse interests and objectives. It is crucial that the roles and responsibilities and capabilities of various organizations are clearly defined and understood. Achieving effective response involves an endless cycle of planning, preparing, training staff, exercising or testing plans, and coordinating the plans and response of the multiplicity of response organizations. Hospitals and utilities—providers of electricity and water and other essential services—have special responsibilities to the community to be well prepared to respond and to restore services.

Who should be in charge depends on what makes sense in a particular environment. In the U.S. it is a small agency that has the responsibility to lead and coordinate the efforts of all federal agencies, including the massive Department of Defense, in response to a disaster. In some countries the military is in charge. It is whatever makes sense. But it must be decided, and preferably it should be formalized in national and/or local law.

3.4 INFORMATION MANAGEMENT

When an earthquake occurs, proficient response relies on strong communications systems contributing to an accurate and timely information flow. These systems include hardware, software, and people systems. Public officials need prompt information assessing the situation and estimating resource needs in order to direct response personnel and equipment to where they can be used most effectively. Reliable information assessing the extent and location of damage is crucial to assist in early restoration of services, planning for recovery, and determining how best to build in mitigation to protect against future losses.

Meeting this challenge in the U.S. are two new tools. One is a standardized earthquake loss estimation model called HAZUS developed by the Federal Emergency Management Agency (FEMA). HAZUS is a software program that uses mathematical formulas and information about local geology, building stock, economic data, and the expected size and location of potential earthquakes to estimate earthquake losses. The model can produce estimates of casualties, numbers of persons displaced, buildings damaged, damage to transportation systems and disruptions in utility systems. Its development has involved a coalition of partners representing federal, state and local government and the private sector.

HAZUS was pilot tested in Boston, Massachusetts, and Portland, Oregon, and was calibrated against the Northridge earthquake. The HAZUS software and default database were made available free to local governments in the U.S.; all they require is a computer and the software for one of two Geographic Information Systems in order to map and display the information. Work is underway to create modules to estimate losses for hurricanes and floods as well as earthquakes.

Loss estimation methodologies like HAZUS are potentially a very useful tool for planning and for presenting critical information to assist in decision-making in both pre- and post-earthquake environments. They can help in planning response, relief, recovery and reconstruction, and help increase awareness and motivate action to reduce potential earthquake losses through heightened understanding.

A second new tool for information management is also a major advance in managing overall response. The State of California has developed the Standardized Emergency Management System in response to legislation rising out of the devastating Oakland-Berkeley Hills fire in 1991. Its intent is to improve coordination and mobilization of resources and flow of information between different levels of government in a disaster. Because emergency response involves a broad range of disciplines, a system was needed with certain attributes: (1) to operate within a clear and consistent organizational system, (2) to utilize a common terminology, and (3) to facilitate priority setting, interagency cooperation, and the efficient flow of resources and information.

Because this is so important, State legislation in California mandates the adoption and use by all state agencies and local governments of this standard system. A part of SEMS is a new data collection and standardized incident reporting system, known as RIMS, or Response Information Management System, which allows the state to quickly gather and share information on situation assessments, deployments, requests for mutual aid assistance, etc., in standard formats and standard language.

SEMS is based on well-practiced principles and components of emergency management including ICS, the Incident Command System, and the mutual aid system, and it is adaptable and transportable to other environments to improve management of disaster response.

3.5 PROMOTING LOSS REDUCTION

Scientists, engineers, local officials, administrators, and managers in private and voluntary organizations all share responsibility for making risk reduction happen in their community. Scientists and engineers, for example, should collaborate in assisting business people to understand technical information regarding earthquake hazard, risk, the response of structures, and design options. Representatives of the insurance industry, local financial institutions, and international organizations and lenders are also members of the community to involve, as they can be a source for incentives and play a role in controlling losses. After the earthquake, they are major players in financing repairs and reconstruction. In order to get and keep earthquake mitigation on the public policy agenda as an issue of importance to a city or community, it is necessary to team up, pool resources, and be persistent.

City officials must determine out who is to be responsible for what, and create mechanisms for coordinated action both prior and subsequent to the earthquake. In some cities (and countries), a seismic safety commission, board or committee provides the framework for bringing together diverse interests and pursuing loss reduction goals.

3.6 BUILDING DISASTER RESISTANT COMMUNITIES

In the years since the Spitak earthquake, governmental and non-governmental organizations in the U.S. have begun focusing more attention on incorporating local community interests into disaster management and disaster loss reduction agendas. In order to create an informed and involved public that demands reduction of their vulnerability to earthquakes and other disasters, we must create opportunities for the public's personal involvement in local preparedness activities and loss reduction decisions. Through personal involvement, members of a community can come to recognize that preparedness and mitigation are in their own self-interest.

In the U.S., this community-based approach to disaster loss reduction is known as building "disaster resistant communities." This approach strives to generate new and sustainable public-private alliances in an atmosphere of mutual benefit. National and local media are involved, providing publicity and helping to build public awareness and support. Regional and national organizations provide technical assistance, training and supplies, and in some cases the federal government provides "seed money" matched by local businesses. The theme is for communities to become, through their own efforts, more resistant to disaster and more resilient, better able to manage recovery and reconstruction, after a disaster has occurred.

Experience has shown us that preparedness and mitigation need to become an integral part of everyday life. If residents and businesses in a community place low value and low priority on earthquake protection and preparedness, it is likely their governmental and academic leaders and elected officials will, too. Of course, it is our job to get communities to take on the task and set their own goals. It requires leadership.

At this time, similar initiatives are being pursued in hazard-prone cities around the world. In the state of Maharashtra, India, a broad coalition of people from the community, the universities, and government have been working together for months to turn their cities and communities from potential earthquake victims to disaster-resistant communities. They have developed a mitigation strategy they are calling Partnerships for Building Safer Communities. Why? They have experienced earthquake's devastation. The 1993 Latur earthquake killed 10,000 people abruptly and unexpectedly. The World Bank, in an admirable decision, determined that this mitigation and preparedness planning process would be required and supported by the Bank as a condition of the reconstruction loan.

Like the people of Maharashtra, many people in cities around the world are learning about reducing earthquake risk the hard way. They are learning how difficult it can be to change the culture of an organization, of a governmental structure, of a community. But culture must change from one that reactively responds to disasters, to one that proactively helps families and neighbors avoid becoming disaster victims.

In California, in some communities, we also have this experience of learning as we went along. Our experience has taught us many lessons. We've learned we need:

- Consistent leadership and commitment
- To make the connection between scientists and decision-makers
- To increase the open exchange of information about hazard, vulnerability and risk among engineers, building owners, and building code officials, and
- Training and realistic exercises to test response and recovery plans and capability.

4. Conclusion

In all societies in today's world, a core business of government is to protect its citizens and their rights, and to contribute to the public health, safety, and well-being. We find ourselves in an era when government alone cannot provide all the solutions to all our challenges. So, we need to draw on all our resources—public, private, academic and non-profit institutions, and individuals and families—to advance society's common interests.

In most if not all societies, government is where we put the hard problems, the most important things that relate to the welfare of the people. Earthquake preparedness and mitigation competes with other needs related to the welfare of people. But in some cities and countries at high risk from earthquake, the problem is little recognized and is not even on the table for consideration. This is a big challenge if we are trying to strengthen disaster management capability at the local level and reduce future disaster losses.

If I had to distill what I've learned all down to one piece of advice, I would say, simply, do not go it alone. Build a team. Establish a network. Reach out. Involve others. Pool your resources—your knowledge, your contacts, your experience in other endeavors. You will be amazed what you can accomplish when you simply team up with other people with different skills, backgrounds, and even interests. There is always common ground.

The challenges facing cities at risk from major earthquakes are so great that they require concerted and coordinated efforts on the part of diverse interests and forces within each city and among different cities. It is a shared responsibility; everyone must band together with unity of purpose to reduce losses from future earthquakes.

5. References

Nathe, Sarah K. (1997), California's Post-Earthquake Information Clearinghouse, *Western States Seismic Policy Council Newsletter,* Spring 1997, pp. 3-4.

DISASTER ASSISTANCE IN COLD WEATHER CONDITIONS: AN OVERVIEW OF ISSUES AND OPTIONS

CHARLES KELLY
Disaster Management Consultant
Ste. 211, 7758 Wisconsin Ave.
Bethesda MD., 20814, USA,
Email: 72734.2412@Compuserve.com

Abstract

Disasters in cold weather pose extreme and demanding conditions on victims and assistance providers, as evidenced by the Spitak earthquake. This type of disaster situation, where the weather alone can pose an immediate and significant threat to the lives of victims and helpers, is a major risk for large areas of the world during at least part of the year. The paper identifies issues concerning disasters during cold weather and suggests options to consider in addressing these issues. These options include: (1) Increase Indigenous Self-Support Capacities, (2) Integrate Local and External Assistance (3) Better Define External Assistance Support Requirements, (4) Consider Changes in Basic Support Systems, and (5) Document Experience and Lessons Learned. The paper is only an initial scoping of issues and options and concludes that a more structures consideration of the relationship between cold weather and disaster response is needed to avoid unnecessary suffering when major disasters occur in temperate climates.

1. Introduction

The Spitak earthquake occurred in the winter of 1988 and resulted in over 25,000 deaths. The earthquake also left the survivors without shelter and normal life-line services. Although assistance arrived quickly, the disaster occurred during cold winter weather. This increased the general suffering and hardship of those lucky enough to have survived. This hardship and suffering continued for many months, if not years, as the housing and infrastructure were slowly rebuilt to provide a normal level protection against the often harsh weather in the uplands of Armenia.

The winter conditions also had an impact on those who came to assist in recovery efforts. Some were ill-equipped to live in cold weather conditions in the absence of locally available shelter and other support. Some of this outside assistance may even have had to rely on the disaster victims for support.

The Spitak earthquake occurred ten years ago. It is fortunate a disaster of similar magnitude has not affected a temperate climate area since. But the outlook is not rosy.

Recent earthquake relief operations in Iran (1997) and Afghanistan (1998) were hampered by winter conditions. There are the vast regions of Russia and neighboring states, which have had severe earthquakes in heavily populated areas. One need only think of the impact of a severe earthquake in Almaty in December to understand the

S. Balassanian et al. (eds.), Earthquake Hazard and Seismic Risk Reduction, 21–30.

challenge posed by providing housing and basic life-line services -heating, water, electricity, gas--to a city of over 1.5 million for at least four months of cold weather.

But the problem extends beyond Eurasia. There are many population centers in the western hemisphere, as well as in New Zealand and Australia, which can be affected by seismic events, with the impacts exacerbated by harsh weather. And, as the 1998 ice storm disaster in Canada demonstrated, any disaster which damages a critical life-line in winter can result in serious suffering and hardship for victims and relief personnel exposed to a normally hostile environment.

The objective of this paper is to identify issues concerning disasters which take place during cold weather conditions and to identify options to consider in addressing these issues. The larger goal of the paper is to contribute to improving the understanding of how to deal with disasters in cold weather conditions, thus promoting more efficient and effective disaster management.

This paper is only an initial scoping of issues and options. Cold weather affects a large part of the land mass and considerable portion of the earth's population. Dealing with disasters in cold weather conditions, conditions which may affect a locale for six months of the year or more, is not simple. This paper is a first step towards defining, refining and improving disaster response operations which take place in the cold.

Disaster operations in cold weather conditions do not appear to be extensively documented or, to date, extensively researched. For instance, an on-going review of humanitarian assistance operations in Bosnia is only addressing cold weather issues, such as support for expatriate personnel, winterization and the effects cold weather on the victim needs and delivery of assistance, in a peripheral manner (Fawcett). A literature search by the Natural Hazards Research and Applications Information Center in Boulder, Colorado identified only 23 items, most concerning the impact of winter weather disasters, rather than the management of disaster in cold weather conditions.

On the positive side, the recent Canadian ice disaster has produced some documentation on disasters in cold weather (See Emergency Preparedness Canada 1998; Eckles; or "IBM Correspondent"). There are a few articles, reports and books which provide insight into the impacts of cold weather on disaster operations (see: Watson and Vespa; Department of Humanitarian Affairs) or are useful as general guidance in planning cold weather operations (see Schofield and Mason, Eade and Williams).

But, these items are exceptions. They highlight the limited formal sources of information on dealing with disasters in cold weather conditions, when compared to what is available on disasters in warm weather situations.

2. Cold Weather and Disaster Management

This paper uses a working definition for cold weather conditions as when daily average temperature is such that sleeping outside without a heavy cover is not possible without a immediate risk to health and well being.

The core problem posed by cold weather conditions is that the normal environment poses a direct threat to human well being. In other words, shelter, heating and other infrastructure, special clothing and increased calorie intake may all be required for basic survival. When people loose access to some or all of these resources due to a disaster, the environment returns as a direct threat to victims and those who come to their assistance.

The extreme cold weather situation is, of course, deep winter at high latitude. But cold weather conditions can occur in areas which also experience very warm weather, such as the steppe of Central Asia. They can also develop where temperatures drop to close to, but not below the freezing point, making unprotected outside living a risk.

Two other aspects of cold weather conditions are important. First, the colder and more common cold weather conditions are, the greater the support requirements are for a normal life. Second, the colder and more common cold weather is, the more difficult the outside working environment becomes.

In other words, cold weather makes life difficult and it makes the already difficult process of disaster assistance more difficult. If these difficulties can be reduced, a more effective and responsive disaster response can be expected. Issues with an impact on making cold weather operations less difficult Options

2.1. SOCIAL AND POLITICAL CONTEXT

Every disaster takes place in a social and political context. For cold weather operations, a critical concern is whether the context presents a latent or active conflict. In the case of the latter--war--one or more parties may actively be trying to destroy the support infrastructures and control access to food and other resources.

Of course, humanitarian assistance in war is a challenge in any environment. But, cold weather conditions require the delivery of a greater range and volume of assistance (e.g., heating, water, shelter, clothing) than a warm weather disaster.

In turn, greater resources are needed to move this assistance in more difficult operating conditions (e.g., trucks to move heating oil over mountain passes in winter). These increased requirements make the relief operation more vulnerable to attack and, at the same time, under pressure to operate larger and more complex relief operations over longer periods of time.

Options for dealing with the technical and logistical aspects of cold weather assistance in war were well developed during the Bosnia conflict. Unfortunately, little institutional memory remains in most humanitarian assistance organizations on cold weather operations (Neumann, Fawcett). Few of the lessons have been formally documented. More critically, the people involved in providing cold weather assistance in Bosnia are now scattered (Neumann, Floret).

Latent conflict, whether in the form of oppressive political systems or as the result of social and political change, is also an important factor in cold weather operations. The most common challenge is in assisting people who are in need of basic services to survive when the normal support infrastructure has collapsed.

The best broad example of this situation is the need to assist the disadvantaged (e.g., elderly, disabled, single parent families) following the post-1990 change in social and political systems in Europe and Central Asia. Although these changes did not fit the classic parameters of a disaster, the resulting hardship (e.g., due to the cutting-off of natural gas or reduction in coal supplies as a result of economic problems), led to a catastrophic exposure of many people to the impacts of cold weather conditions.

Again, ways were developed to deal with these disasters, from the provision of stoves and fuel here in Armenia, to delivering coal to hospitals in Tajikistan. But, with few exceptions, the methods and means developed to address these disasters are not easily accessible or well documented.

Although social and economic disruption on the scale of the last 10 years is unlikely in the near future, the prospect for latent conflict in one or more countries which experience cold weather is real. As with active conflict situations, lessons learned from latent conflict need to be defined and documented. Without this information there is no assurance that the mistakes of the past will not be repeated the next time the gas or electricity is cut off in the middle of winter.

2.2. TIMING

Time is critical to any disaster response: assistance delivered too late is a waste of effort. The challenge for cold weather operations is twofold:

- First, to accomplishing a timely delivery of large volumes of assistance. In cold weather victims need more support than for the same disaster in warm conditions.
- Second, to operate faster under more difficult conditions than in warm weather conditions.

The delay in reaching the required delivery levels for external assistance defines how long victims need to survive on their own resources. In cold weather conditions even 48 hours may be too long to wait for external assistance to provide temporary shelter and heating.

Obvious options to address the assistance delay problem include reducing disaster damage risk through mitigation, and increasing the level of preparedness on the part of potential victims to cope with cold weather. A third option, of maintaining a standing large scale logistics capacity is logical, but probably not possible in this period of tight budgets.

A second timing issue involves post-disaster recovery. For instance, an earthquake in September in Yerevan will damage infrastructure, but weather conditions are not be too extreme. However, by November cold weather conditions will set-in, with a resulting need for a higher level of support and assistance, particularly for traditionally vulnerable groups.

Thus, timing and cold weather conditions also need to be considered in planning post-disaster assistance. Most types of building construction can continue in cold weather. However, a disaster in summer may provide only two to three months to re-establish (or build a new) heating infrastructure for an urban area. And this is clearly not just a problem for cities, as demonstrated by the electrical outages caused by this year's ice storm in Canada.

Dealing with these timing issues requires a good planning of the recover effort. But, no matter how good the planning is, the second part of the reconstruction process, project financing, could be a major problem.

Simply put, if it takes 6 months to plan and finance the rehabilitation of a major heating infrastructure system, then whenever a disaster occurs, the victims will spend at least one cold weather season without normal heating.

2.3. INDIGENOUS SELF-ASSISTANCE

A much more simple issue is the level of local resources which could be available after a disaster. As suggested above, external assistance will not arrive immediately. Thus, the level of self-assistance is a critical element in reducing immediate disaster impact and speeding recovery.

Options for promoting self-assistance capacity involve the normal disaster management tasks of planning, preparedness, training and mitigation. Yet, the challenge posed by cold weather goes beyond the normal.

An earthquake can destroy buildings and gas and electrical infrastructure. Self-assistance may need to involve quickly building new shelter out of the rubble, manufacturing stoves and organizing fuel supplies. These tasks are not too complicated individually. But, to be accomplished quickly and at the same time requires an organizational structure much more elaborate than likely to exist just after a major earthquake.

In addition, these types of tasks are not normally performed by most people, particularly in an urban setting. Getting people to shift occupations and become specialists in basic cold weather survival is a major undertaking.

Certainly, training and general preparedness are important. But, the difficulties created by a disaster which takes people abruptly out of a protected environment and requires their survival in cold weather conditions, needs to be thought through and understood before a disaster.

In the end, it may be that the realistic level of self-assistance in cold weather conditions is much less than the actual level of local resources which can be made available. The burden then falls on the external assistance effort to provide more and quicker post-disaster assistance.

2.4. EXTERNAL SUPPORT REQUIREMENTS

A second relatively straightforward issue is the level of external support which might be needed in cold weather. As suggested above, external assistance needs are governed by the level of local self-assistance and the timing in which the assistance is needed. In short, the process of assistance in cold weather is not much different than in any other type of disaster situation.

Differences do exist, as also mentioned above, in the level of assistance which is needed: cold weather conditions will require a larger volume, and probably a broader range, of assistance than if the disaster took place in warm weather. Logistics requirements are increased and include the need for additional transport capacity, the means to operate safely in cold weather conditions, and to maintain vehicles under more difficult than normal field conditions.

A second part of this issue concerns the mandate of those providing the outside assistance. For instance, following a major disaster in Eurasia, NATO would be quickly involved in providing large scale assistance. Clearly, NATO is able to work in the most harsh conditions (Piazzi).

However, it is not clear how long this assistance channel will operate. Will or can external military organizations continue to furnish logistics and basic service support for three, six or twelve months, providing enough time for the full re-establishment of local support capacities? Or would these large capacity (but costly) organizations seek to withdraw when the immediate rescue work is completed (i.e., weeks after the disaster has begun), with the burden of external assistance falling on less well endowed non governmental assistance organizations (Faller)?

This is not to question the motives of military organizations. They will certainly have a key role in responding to any major disaster, and particularly those in cold weather conditions, because of the depth of their resource base.

Yet, the questions remains as to how a city of 1 million will be supplied with heat, water and other basic services when major damage to shelter and infrastructure has occurred and the prospects are for another four months of very cold weather. Will external assistance organizations have the means and capacities to stay for the duration of the disaster?

2.5. SUPPORT NEEDS FOR OUTSIDERS

A disaster in Miami may require a hat, couple of T-shirts, tropical weight pants and some sun screen. After a hurricane, ties are optional.

Going to a disaster in Spitak in December requires a lot more. Not only do you need several layers of winter clothing, but also a sleeping bag, a way to heat water, some basic food supplies which require minimum cooking and, in extreme conditions, a weather proof shelter. On the other hand, keeping something frozen is not a problem.

Providing for an expatriate working in cold weather conditions is more costly and demanding than for the same person to work in a warm environment. This is not to suggest that expatriate assistance managers are a waste of money in a cold weather situation. Rather, the support requirements need to be taken into account before sending someone off to a disaster which is covered in 50 cm of snow.

These support requirements extend beyond the immediate emergent phase of the disaster. Expatriates living in a disaster area for several months need adequate shelter, appropriate food, heating and a way to clean themselves and their clothes. All this could mean providing assistance to an expatriate which places them in an clearly better situation than the local population. And, the cost of this assistance package, even if very basic, may exceed the per capita assistance received by the disaster victims. This can pose ethical and political dilemmas for the external assistance organization, and the expatriate.

The Bosnia disaster provided a number of models of how expatriates can operate in cold weather conditions, as have operations conducted by the International Federation of Red Cross and Red Crescent in Central Asia. Bosnia has also highlighted different views as to the appropriate level of support to be provided (i.e., how much greater support should be than that available to the local population), and how different levels of support affected the capacities and output of the expatriate personnel (Fawcett).

Overall, the experiences of expatriate assistance personnel in cold weather conditions are not well documented, but the topic needs serious consideration. Most external organizations would not send staff to the field with so little support that they require help from the disaster victims. But defining the appropriate level of support for expatriates working in cold weather conditions should be define before the next major disaster in cold weather conditions, not witty

Disasters in cold weather conditions most likely will result in an increased density of populations. This can have positive, and possibly negative, impacts. The increase in density takes place as people use fewer rooms to conserve heat and seek protection from the weather.

For example, victims may cover windows in the least damaged room following an earthquake and move most of their activities into this room. This is a logical way to conserve limited resources needed to counter cold weather conditions.

This increase in density can also reduce the volume of immediate external assistance needed. For instance, only one heater need be allocated per family if they are only using one room for most activities. Thus, increasing density is an effective initial step in a post-disaster response.

However, an increase in density carries several risks. The first is disease transmission. Having many people living in a confined space for most of each day can lead to the spread of disease more easily than if the same people lived in a more disbursed setting. This risk is probably minimal if increasing assistance leads to the creation of more living space of a short period of time. A long term concentration probably poses a more significant risk to communicable disease transmission, although this presumption needs formal confirmation.

The second health problem which may develop from crowding involves dysfunctional behavior. Immediately after a disaster one can expect social and personal tensions to decrease. Tensions may then increase the longer people live together with little privacy and in close proximity. Dysfunctional behavior, such as excessive drinking, or physical abuse, may then increase.

The severity of this problem probably depends on the proximity of people in normal conditions. People who expect more personal space could be more affected than those who tend to normally live in closer association with family and neighbors. As with disease transmission, dysfunctional behavior related to disaster in cold weather conditions needs further study.

A third risk from increase density comes from the possibility of additional disaster events. These can range from a secondary after shock to shelling during a war. The concentration of people increases the possible level of death and injuries which could occur from a new event. In other words, concentrating people makes for good targets.

As a result, the logical tendency to concentrate populations to facilitate basic services after a disaster needs to be moderated by longer term considerations for the risk posed by further disaster events. In cold weather conditions, this may require a delicate balancing between quickly reducing hardship and increasing the likelihood of additional injuries and loss of life.

2.6. TECHNICAL ISSUES

An exhaustive discussion of technical issues relating to cold weather conditions is beyond the scope of this paper. Four points are worth noting as a starting point in considering technical issues.

First, as noted, military organizations have considerable experience, resources and expertise in providing technical support for cold weather operations. These resources and skills are of invaluable assistance.

However, two cautions are warranted. The typical disaster victim does not have the training and skills to use military resources. These resources may need to be accompanied with personnel to ensure the resources are used correctly, or at all. This, of course, increases the logistics and support requirements.

As well, the physical and psychological profile of a typical soldier is different from the general population. Thus, military resources for cold weather operations may be inappropriate or ill adapted for use by a non-military population. A particular care is needed in cold weather conditions to ensure that resources originally intended for one use (soldiers) are appropriate for a typical disaster victim, a 60 year old woman.

Second, military and off-the-shelf resources useful in cold weather conditions can be expensive. Thus, providing sleeping bags or pre-fabricated temporary shelters (see Emergency Preparedness Canada 1996) may be logical from a humanitarian perspective, but prohibitively expensive for thousands of disaster victims. Using the newest and best resources needs to be tempered with doing the greatest good for all the victims.

A third point involves technical specifications: what may be technically appropriate in normal conditions is less than adequate for a disaster in cold weather conditions (Neumann). For instance, only one type of material used as a blanket may perform adequately when exposed to water, although several types perform well when not wet. But, most disasters in cold weather are also disaster in wet conditions. Thus, cold weather conditions require resource specifications need to meet a "most extreme" threshold.

Finally, adaptability is a basic technical issue. Disaster victims adapt assistance to meet specific needs. For example, reinforced plastic sheeting or glass can be used to fix windows. But, the former can also serve to cover holes in the roof and partition rooms. The selection and specifications of material assistance need to consider multiple use and adaptability, to get the greatest impact under the limitations posed by winter conditions.

A considerable wealth of experience on technical specifications and adaptation options was developed during the Bosnia disaster, as well as during the disruption following the break-up of the Soviet Union. Unfortunately, these experiences appear to be poorly documented and, more critically, not incorporated into most assistance planning resource materials (e.g., procurement manuals) (Neumann). Unless this experience is documented, it is likely the next major disaster response in cold weather conditions will have to relearn many of the technical lessons of the recent past.

3. Conclusions

This paper has provided a brief overview on disaster assistance in cold weather conditions. Five recommendations are provided as a starting point for efforts to improve cold weather operations.

> **Increase Indigenous Self-Support Capacities**: Because of the operational difficulty and volume of assistance needed after a disaster in cold weather conditions, the victims are going to have to provide a significant level of their own recovery assistance well into the post-disaster period. This capacity for

self-assistance needs to go beyond search and rescue to include the provision of shelter, heating and other basic support services.

Integrate Local and External Assistance: External assistance will play a critical role in responding to a disaster in cold weather. Yet, because of the volume of assistance needed, the difficulty of providing the assistance, and the need to assist quickly, this assistance needs to be well integrated into self-assistance efforts. Since pre-assistance assessments will be difficult in cold weather conditions, some level of assistance needs to be triggered automatically on word of a disaster to move critical assistance to those exposed to cold weather.

Better Define External Assistance Support Requirements: The expectations, requirements and resource needs to support external assistance operations, that is personnel, logistics and facilities, needs to be better defined. Like with external assistance, an ad hoc approach to placing personnel in the field reduces the effectiveness and timeliness of assistance operations. It can place an unnecessary burden of support on the local population. External assistance organizations should not get involved in cold weather operations if they just make the post-disaster burdens greater.

Consider Changes in Basic Support Systems: Although only implied in the paper, a disaster may provide an opportunity to make the basic support infrastructure more disaster resistant. For instance, options for rebuilding centralized heating systems or strengthening electrical systems in the reconstruction period need to be considered before a disaster. Systemic changes are probably impractical before a disaster. But the damage caused by a disaster can provide an opportunity to make systems more resistant, and reduce future vulnerability to cold weather conditions.

Document Experience and Lessons Learned: As suggested throughout the paper, there is a clear need to document current knowledge and lessons learned on providing disaster assistance in cold weather. A lot of experience has been gained since the Spitak earthquake. Unless this experience is recorded and available, everything will start over from close to zero when the next major disaster occurs in cold weather conditions. This learning-what-you-should-already-known will be an embarrassment to assistance organizations. But, more importantly, it will mean less effective help, and thus greater than necessary hardship, for the disaster victims.

In conclusion, this paper has raised a number of issues and options on responding to disaster in cold weather conditions. This discussion has been brief and intended to start a process of defining and improving the way in which disasters in cold weather conditions are managed. Other issues and options exist and need to be defined and considered.

Cold weather has a significant impact on the effects of a disaster and assistance operations. Improving the capacity to manage disasters in cold weather will make for a better disaster response, and a better response is to the ultimate benefit of the disaster victim.

4. References

Babayan, T. (1998), personal communication.
Department of Humanitarian Affairs (1994), *Georgia: DHA activities in 1993 (Case Report)*, Geneva.
Eades, D., and Williams, S. (1995), *The Oxfam Handbook of Development and Relief*, Oxfam (UK and Ireland), Oxford.
Eckles, J. L. (1998), When it Rains - It Freezes, *Disaster Recovery*, Spring.
Emergency Preparedness Canada (1998), *Emergency Preparedness Digest* (issue on "Lessons Learned from Ice Storm '98"), Public Works and Government Services, Canada.
-- (1996), Canadian shelters have responders covered, *Emergency Preparedness Digest*, Public Works and Government Services, Canada.
Faller, M. (1998), personal communication.
Fawcett, J. (1998), personal communication.
Floret, J.-P. (1998), personal communication.
Gordenker, L. and Weiss, T. G. (19910, *Soldiers, Peacekeepers and Disasters*, St. Martin's Press, New York.
IBM Correspondent (1998), D/R Plans Key to Business Survival During January Ice Storm, *Disaster Recovery*, Spring.
Kelly, C. (1996), Limitations on the use of military resources for foreign disaster assistance, *Disaster Prevention and Management: An International Journal*, **5** 1.
Neumann, W. (1998), personal communication.
Piazzi, S. (1998), personal communication.
Schofield, E.C. and. Mason , J. B. (1996), Setting and Evaluating the Energy Content of Emergency Rations, *Disasters,* **20** 3.
Watson, F. and Vespa, J. (1995), The Impact of Reduced and Uncertain Food Supply in Three Besieged Cities of Bosnia-Hercegovina, *Disasters*, **19** 3.
Weiss, T. G. and Minear, L. (1993), *Humanitarianism Across Borders: Sustaining Civilians in Times of War*, Lynne Reinner Publishers, Boulder.

ACTUAL STATE OF TECHNICAL COOPERATION ON EARTHQUAKE DISASTER PREVENTION
CASE STUDY: JICA EARTHQUAKE DISASTER PREVENTION PROJECT IN MEXICO

H. KITAJIMA
ETRA Environmental Technology, Research & Associates
Kanagawa Science Park West 4f 3-2-1 Sakado Takatsu-Ku Kawasaki
Japan 213-0012

The government of Japan is requested technical cooperation and grant aid in disaster prevention area from various countries world widely, where the experiences and expertise have been tried to accumulate to different environments. Especially Latin America (Mexico), South-East Asia (Indonesia) and Near-Middle East (Turkey) are three strong points for the technical transfer in this area.

The main executing organization of those cooperation is the Japan International Cooperation Agency(hereinafter referred to as JICA).

In order to introduce and discuss the actual state of technical cooperation in this field, JICA EARTHQUAKE DISASTER PREVENTION PROJECT IN MEXICO is used as a case study in this paper. This project has been carried out during 1990-1997 by the Japanese and Mexican team in "Centro Nacional de Prevencion de Desastres " (hereinafter CENAPRED), MEXICO.

INTRODUCTION & BACKGROUND

Japan's Official Development Assistance (ODA) has been standing as one of the top donors in the Development Assistance Committee (DAC) members of the Organization for Economic Cooperation and Development (OECD). There is no cooperative system among the members.

INTRODUCTION

The amount of Japan's ODA disbursement has been the biggest scale 1991-1996, nevertheless the percentage of technical cooperation is less than two-third of DAC's average. To achieve the effective and efficient development assistance, it is necessary and important to increase the ratio of grant aid and technical cooperation in Japan's ODA,at least, to the DAC's average level. The ODA disbursement of major DAC countries is shown as follows <No.1>.

S. Balassanian et al. (eds.), Earthquake Hazard and Seismic Risk Reduction, 31–40.

No.1 OFFICIAL DEVELOPMENT ASSISTANCE (ODA)

BACKGROUND

Japan's ODA categories are shown in <No.2>. The disbursement of contributions, subscriptions and loans to multilateral donor organizations are more than that of the bilateral grant including grant aid and technical cooperation. Recently there are combined forms of technical cooperation and grant aid in the project-type technical cooperation.

JICA's technical cooperation includes (1)TRAINING IN JAPAN, (2)DISPATCH OF EXPERTS, (3)PROVISION OF EQUIPMENT, **(4) PROJECT-TYPE TECHNICAL COOPERATION,** (5)DEVELOPMENT STUDY. And the project-type technical cooperation is based on three forms of technical cooperation, they are (1) TECHNICAL TRAINING, (2) DISPATCH OF EXPERTS, (3) PROVISION OF EQUIPMENT.

No.2 CATEGORIES OF JAPAN'S ODA

CASE STUDY: JICA EARTHQUAKE DISASTER PREVENTION PROJECT IN MEXICO

This project in Mexico is a typical case of project-type technical cooperation with grant aid for the construction of a center. To contribute to the safety of the population of Mexican, and eventually Central America and the Caribbean Region is the final goal of this project.

Usually a JICA's PROJECT-TYPE TECHNICAL COOPERATION□has three kinds of main functions. They are (1) STUDY, (2) TRAINING and (3) DISSEMINATION .

THE EARTHQUAKE DISASTER PREVENTION PROJECT IN MEXICO

The chronological review of the project is as shown in Table <No.3>.

In September,1985, a catastrophic earthquake occurred in Mexico and caused significant damages to the population and urban infrastructure. It was three years before that of SPITTAK, ARMENIA.

With this experience, the government of MEXICO requested technical cooperation and grant aid to JAPAN. Based on the request, Japanese survey teams were dispatched several times. In March, 1990,The National Disaster Prevention Center (or CENAPRED) was established by the Japan's grant aid. And also the both sides signed the master plan of the earthquake disaster prevention project for five years.

Before the commencement of the project, almost 5-years were needed. Finally it was 7-years project, 1990-1997. The former 5-years, 1990-1995, was the original project and the latter 2-years,1995-1997, was follow-up project. The author was dispatched to this project as a long term expert, in September, 1992 to September, 1993, for one year.

No.3 CHRONOLOGICAL REVIEW OF THE PROJECT

LONG-TERM RESEARCH PLAN OF THE PROJECT & RESULTS

(1) GENERAL

Just after the center (CENAPRFED) has been established, the project was commenced according to the tentative schedule at April,1990. But in the beginning of the project, because of the lack of researcher posts in CENAPRED and so on, it was hard to run the project effectively. The research areas were SEISMIC STRUCTURES and MATERIALS, BUILDING CODES, STRONG MOTION OBSERVATION and ESTIMATION.

The contents of the research plan is as shown in <No.4>.

No.4 RESEARCH PLAN OF THE PROJECT

(2) LONG-TERM RESEARCH PLAN

To promote the project, both Japanese and Mexican team have discussed and agreed to collaborate according to the long-term research plan in 1992. After this, surrounding conditions for reseat in CENAPRED were improved. The scheduled long-term research plan and the objectives of research is shown in this table<No.5>.

No.5 LONG-TERM RESEARCH PLAN

(3) RESULTS

The Japanese evaluation team was dispatched in November 1994. And they reached the understanding that the Follow-up Project for 2-years should be needed. Also in November, 1996 the Japanese evaluation team was dispatched again. (cf. reference). In March, 1997, the JICA's Earthquake Disaster Prevention Project in Mexico has been achieved.

ANALYSIS ON THE PROJECT RESULTS

According to the evaluation reports for the projects submitted in December, 1994 and November, 1996, this project has been completed successfully and achieved fruitful result. During the original project, activities of research area have been carried out satisfactory and the follow-up era, activities of training, dissemination area were active.

GENERAL

In the beginning of the project, the lack of researcher posts was the most serious problem. But in 1993, it was improved to be able to employ sufficient number of researchers. Input to the project from Japan side is shown as follows in <No.6>. Financial and personnel input from Mexican side was increased, in particular during follow-up project years. The knowledge and technologies of the technical transference by Japan are suitable for the Mexican side. Financial sustainability should be the most important factor for the activities of this center after the project.

No.6: INPUT FROM JAPAN

PROJECT ACHIEVEMENT

The activities of the project are divided in three areas, they are (1) RESEARCH, (2) TRAINING and (3) DISSEMINATION area. The activities of the both projects are shown in <No.7>. During the original project, main activities were carried out in research area, and the activities of training and dissemination area were considerably behind schedule. In the follow-up project, these activities were increased.

(1) RESEARCH

- SEISMIC INSTRUMENTATION AREA. The most important research carried out is on seismic safety of low-cost housing in Mexico and Latin America. And this is particularly focussed on housing of confined masonry type, which is the most common construction materials used.
- SEISMIC TESTING AREA. The main activities carried on installation, improvement and operation of seismic observation network.

(2) TRAINING

In order to strengthen the technical training activities for earthquake disaster prevention, the seminars for DRO (the directors responsible for works and works supervising) on this subject in Mexico were held in CENAPRED several times.

(3) DISSEMINATION

Hundreds of events (symposiums, congresses, workshops, seminars, etc) were carried out in CENAPRED. In 1994, 20 titles of "Cuadernos de Investigacion" and so on were produced and published. It is an intention of the center to give strong push on the dissemination of technical aspects.

IMPACT OF THE PROJECT

The research area has made meaningful achievements. It is necessary to elaborate implementation plans of Research, Training and Dissemination for the practical application of these achievements.

No.7 ACTIVITIES OF THE PROJECT

PROBLEMS OF TECHNICAL TRANSFER

A model of diffusion mechanism is shown in <No.8>. The transfer of technologies by technical cooperation is one of the most important transferring pattern, particularly for the developing countries.

GENERAL

The words used in <No.8> are as follows. PROVIDER□is/are donor(s). PROVIDER (=ACCEPTOR) is a counterpart of a project. CHANNEL is an international technical cooperation between donors and developing countries. CHANNEL is (and are) the diffusion processes in receiving countries. If appropriate technologies were selected, the technical level of an acceptor could be improved and the acceptor would become a second provider.

The cycle of the technical diffusion were functioned properly, the effective technical transfer would be expected.

PROBLEMS TO BE SOLVED

For an execution of effective transfer, this cycle should be rotated smoothly. For this purpose, it is necessary to select an appropriate technology that is acclimate to the new technical environment in developing countries.

As shown in the case study, compared with research area, it is difficult to achieve training and dissemination area by a 5-10 years project. Significant time is needed to reach results of practical application.

FURTHER PLANS

To select an appropriate technology, it is needed and important to set up the way how to evaluate and specify the technical environment in a developing country.

No.8 THE DIFFUSION MECHANISM OF TECHNOLOGIES

SUMMARY

To contribute to the safety of the population is the final goal of technical cooperation in this area. For this purpose, it is the most important to diffuse technologies of disaster prevention to ordinary engineers and people in developing countries.

SUMMARIZING THE RESULTS

(1) INTERACTION OF RESEARCH AND PRACTICE

Interaction among research, training and dissemination activities should be increased. In the future, the research staff should be more involved in training and dissemination activities. It is necessary to establish a system to fairly evaluate not only academic activities, but also practical activities which include training and dissemination by the research staff.

(2) THE CHOOSE OF COUNTERPART

Not only researchers, but also the practical engineers or ordinary people must be chosen as counterparts of JICA's projects. For the necessity and importance of training and dissemination was indicated in a case-study and a diffusion mechanism.

(3) SUPPORT OF SURROUNDING CONDITIONS

To specify a technical environment in developing countries, it is necessary to improve related codes, social systems and so on.

CONCLUSION

It is the most important to set up the technical environment to verify acceptable technologies for a country. Appropriate technology could be selected easily for the specified environment. Then effective and efficient technical transfer will be expected in this area

Reference

1. Joint Evaluation Repot on the Japanese Technical Cooperation for the Earthquake Disaster Prevention Project in the United States of MEXICO NOVEMBER 1994 & NOVEMBER 1996

2. DAC PRINCIPLES FOR EFFECTIVE AID Development Assistance Manual 1992 O .E .C. D.

<№1 Official Development Assistance>

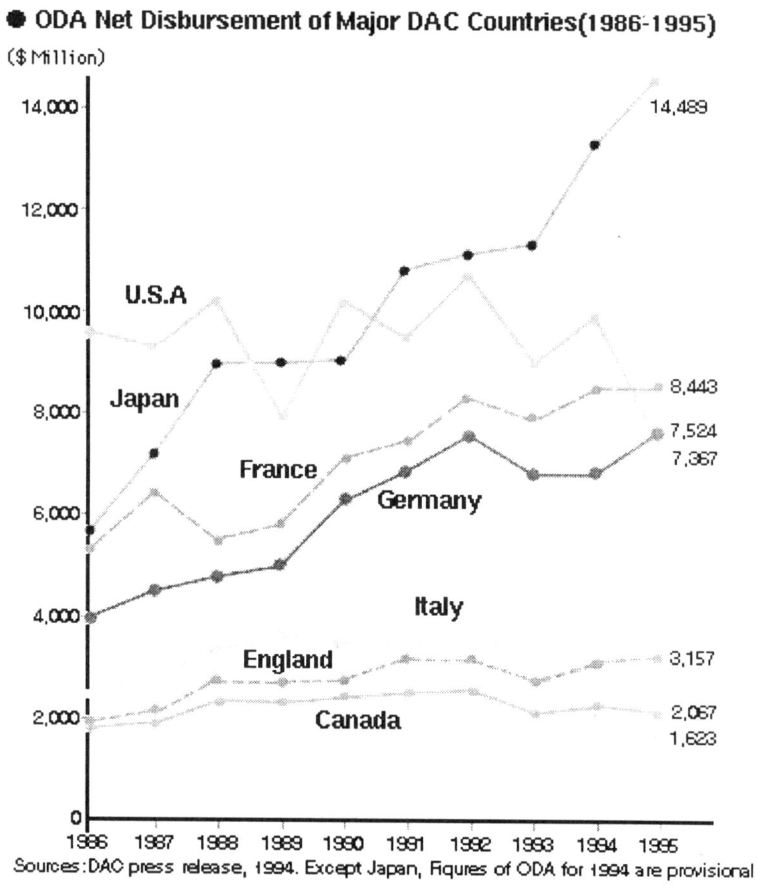

<Unit: US$ 1 million>

ODA NET DISBURSEMENT OF DAC COUNTRIES(1986-1995)

JAPAN'S OFFICIAL DEVELOPMENT ASSISTANCE□□□□□

1. US$14,500,000,000 (1995) TOP FOR 5-YEARS
2. PERCENTAGE OF TECHNICAL COOPERATION & GRANT AID
47% (JAPAN) 77% (Average of □□□)
3. EFFECTIVE AND EFFICIENT DEVELOPMENT ASSISTANCE

<№2 Japan's ODA Categories>

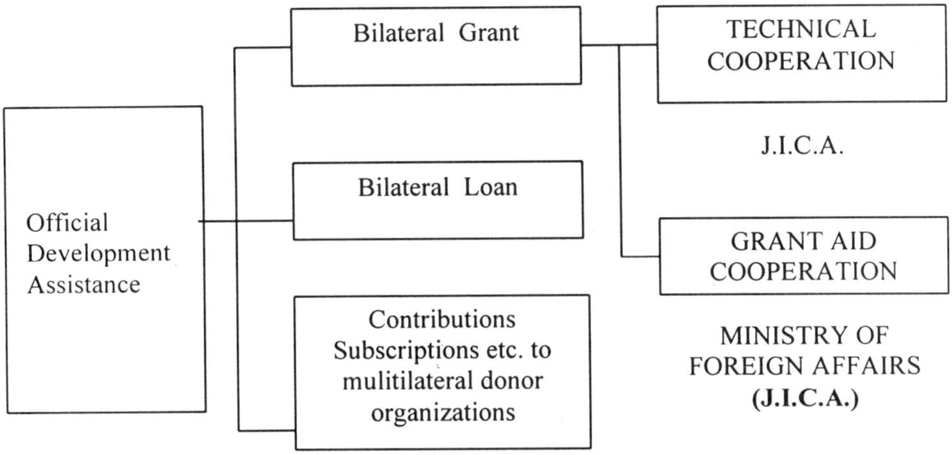

<N3 Chronological Review of the Project>

THE EARTHQUAKE DISASTER PREVENTION PROJECT IN MEXICO

Sep./1985	The earthquake in mexico: 19 th September 1985 7:18 AM (M=8.1) 36 hours later (M=7.5)
Dec./1986	Mexico Requested Technical Cooperation And Grant Aid
Jul./1987	The Japanese Preliminary Survey Team (For Feasibility Study)
Dec./1987	The Japanese Preparatory Survey Team
Mar./1988	The Japanese Survey Team of Basic Design Study for the Center
Feb./1989	The Japanese Long-Term Survey Team (I) <5-years>
Oct./1989	The Japanese Long-Term Survey Team (II)
Mar./1990	A Technical Cooperation Agreement Signed
Mar./1990	The National Disaster Prevention Center Established
Apr./1990	The Earthquake Disaster Prevention Project in Mexico (1990-1995)
Apr./1991	
Aug./1992	The Long-Term-Research Plan (1992-1994) For the Project
1993	Sep. 1992 Study & Investigation On Sep. 1993 Construction Materials <5-years Project>
Nov./1994	The Japanese Evalution Team for the Project (1990-1995)
Mar./1995	The Earthquake Disaster Prevention Project Achieved
Apr./1995	The Japanese Evaluation team for the Follow up Project (1995-1997)
Oct./1996	The Japanese Evaluation Team (for 1995-1997) <2-years Project>
Mar./1997	The Follow up Project Achieved

<N4 Research Plan of the Project>

(I) SEISMIC STRUCTURES and MATERIALS & (II) BUILDING CODES

1. STATIC LOADING TEST AND ANALYSIS OF CONFINED MASONRY WALL AND BUILDINGS
2. NONLINEAR EARTHQUAKE RESPONSE ANALYSIS OF CONFINED MASONRY BUILDINGS
3. STATISTICAL EVALUATION OF STRUCTURAL PROPERTIES OF CONSTRUCTION MATERIALS, PARTICULARLY THOSE RELATED TO CONFINED MASONRY CONSTRUCTION
4. SURVEY ON THE CURRENT STATE OF LOW-COST HOUSING SUPPLIED BY PUBLIC AGENCIES
5. DESIGN OF STANDARD EXAMPLES FOR LOW-COST HOUSING
6. UP-GRADING OF EXISTING LOADING AND DATA-ACQUISITION SYSTEMS
7. DEVELOPMENT OF A COMPUTER-CONTROLLED LOADING SYSTEM
8. COMPARISON OF MATERIAL TESTING METHODS IN MEXICO, THE U.S.A. AND JAPAN, AND MANUAL-MAKING FOR MATERIAL TESTING
9. DEVELOPMENT OF A GENERAL NONLINEAR EARTHQUAKE RESPONSE ANALYZING SOFTWARE FOR FRAME-WALL STRUCTURES

(III) STRONG MOTION OBSERVATION ESTIMATION

1. COLLECTION OF SEISMIC DATA AND IMPROVEMENT OF THE OBSERVATION NETWORK
2. DEVELOPMENT OF AN EARLY INTENSITY ESTIMATION SYSTEM
3. DEPLOYMENT OF A BROAD-BAND HIGH-SENSITIVITY OBSERVATION SYSTEM
4. DEVELOPMENT OF A STRONG MOTION DATABASE AND PROCESSING SYSTEM
5. ANALYSES OF OBSERVED STRONG MOTIONS
6. INVESTIGATION INTO EARTHQUAKE SOURCE PROCESSES FOR STRONG MOTION EVALUATION
7. UNDERGROUND STRUCTURES DETECTION
8. DEVELOPMENT OF A GROUND INFORMATION SYSTEM
9. MICRO-ZONING

<N6 Input from Japan>

I. GRANT AID

1. CENTER BUILDING & EQUIPMENT

Y1,246,000,000

Ⅱ.PROJECT-TYPE TECHNICAL COOPERATION

1. DISPATCHED LONG-TERM EXPERTS

22 experts

2. DISPATCHED SHORT-TERM EXPERTS

85 experts

3. PROVISION OF EQUIPMENT

Y217,000,000

4. TRAINING IN JAPAN

23 personnel

<N7 Input from Japan>

	The Original Project <Apr. 1990→Mar. 1994>		The follow-up project 1995→1997	
1. Research	• Meaningful achievement • Needs of Practical Study • Connection with Practice	(1)		(2)
2. Training	• D.R.O. (Directors Responsible works and works supervisors)		Seminars	
3. Dissemination (Diffusion)	• Symposium • Seminars • Workshops		Publications	

(1) : Joint evalution report on the project 1990-1995 (Dec., 1994), JICA

(2) : Joint evalution report on the project 1995-1997 (Dec., 1996), JICA

RADIUS INITIATIVE FOR IDNDR
A PRACTICAL APPROACH TO REDUCE URBAN SEISMIC RISK

KENJI OKAZAKI
IDNDR Secretariat, OCHA, United Nations
Palais des Nations, CH -1211 Geneva 10, Switzerland

1. Introduction

A Strong Earthquake Hits Tashkent, Uzbekistan, at 5:00 am
Strong tremors hit the city of Tashkent, capital of Uzbekistan, at 5:00 am in early morning. It is estimated that the epicenter is located just beneath the city with a magnitude of 6.1 (in the Richter scale). The intensity is 7 and 8 (MSK) in most of the parts, and reaching 9 in some areas of the city. The shaking lasted for about half a minute.

Traditional (adobe) houses located near the central part of the city are devastated. Some of modern reinforced concrete buildings are damaged considerably, having cracks in walls and bearing elements. Self-built homes located in the old town collapse, killing many inhabitants. Water supply pipelines are damaged throughout the city. As there is no emergency cut-off device in water supply system, large quantity of water is lost due to breakage of pipes. Some of sewage pipes are ruptured, causing serious pollution in the city. Damages to power supply system keep the city without electricity, immediately after the earthquake.

Survived and non-injured city-dwellers are searching in ruins of destroyed houses for their relatives and friends, trying to free victims from the destroyed structures. Some of people trapped are still alive. But the death toll is frightening. It is estimated that several thousands of people are homeless. Some older houses catch fire as a result of sparkling short-circuits or from explosion of broken gas pipes. Hospitals are overcrowded with injured patients. There is a deficit of medicaments in many of medical establishments, and a great deal of medical equipment has been damaged - - - - - - - - .

The above was edited from the earthquake damage scenario prepared by the City of Tashkent, as a case study of the RADIUS (**R**isk **A**ssessment Tools for **Di**agnosis of **U**rban Areas against **S**eismic Disasters) Initiative. This kind of scenario is very useful to visualize the estimated effect of an earthquake, and to raise the public awareness and take actions to reduce the seismic risk in urban areas, which are growing rapidly all over the world. Local people could easily understand how their city, buildings and infrastructure, could be destroyed, and how many people could be killed by probable earthquakes.

S. Balassanian et al. (eds.), Earthquake Hazard and Seismic Risk Reduction, 41–48.

2. RADIUS - A world-wide initiative to reduce the destructive impact of earthquakes in urban areas

The world is rapidly being urbanized with almost half of its population living in cities. Cities where population and all kinds of human activities are concentrated, are more and more vulnerable to disasters, particularly to earthquakes, which could not be predicted precisely. Once an earthquake takes place in a big city, the damage would be tremendous both in human and economic terms. Even an intermediate earthquake could cause a destructive damage to a city as in the case of the 1995 Kobe Earthquake in Japan.

However, there is a tendency to look at disasters only from a humanitarian angle, bringing us into the position of giving priority to the response to disasters. There is also an unfounded tendency to consider that the investment to strengthen the existing infrastructure before disasters will cost much more than the cost of response after the disasters. It is actually just the reverse. It is clear that from the economical view point, preparedness pays off in the long term. Besides, the response activities never save human lives which have already been lost. In order to reduce the impact of the disasters, it is therefore essential to concentrate our efforts on prevention and preparedness.

The Secretariat of the International Decade for Natural Disaster Reduction (IDNDR 1990-2000), United Nations, therefore launched the RADIUS Initiative in 1996 with assistance of the Government of Japan, to reduce seismic disasters in urban areas, particularly in developing countries. RADIUS aims to aid vulnerable communities reduce physical, economic and social damage caused by earthquakes, and to raise public awareness in earthquake prone cities. In collaboration with 9 selected cities around the world, the initiative has developed common tools for seismic risk assessment and management in the urban areas.

The objectives of the RADIUS Initiative are four fold:

1) Develop seismic damage scenarios and risk management plans for the nine case study cities selected worldwide
2) Develop practical tools for the seismic damage assessment and management, which could be applied to any earthquake prone city in the world.
3) Conduct a comparative study to understand urban seismic risk around the world
4) Promote information exchange for the seismic risk mitigation at city level

The total cost is approximately $ 2.5 million. The Japanese Government has contributed approx. $2 million while several international organizations such as UN University (UNU), UN Centre for Regional Development (UNCRD), and Japan International Cooperation Agency (JICA) have cooperated to hold a seminar and symposium.

3. Case Studies

The objectives of the case studies are two fold;

1) Develop ***an earthquake damage scenario*** which describes the consequence of a possible earthquake

2) Prepare ***a risk management plan*** and propose **an action plan** for earthquake disaster mitigation

An earthquake damage scenario describes physical damage to buildings and infrastructure (roads, water supply, electric power supply, etc.) and human losses as well as the effects to the urban functions and activities during and after a probable earthquake. The result may be shown in visual forms by Geographical Information System (GIS).

A risk management plan has been prepared, based on the earthquake damage scenario. It contains the following aspects.

- urban development plan to mitigate seismic disasters.
- improvement plan for the existing urban structures such as reinforcement of vulnerable buildings and infrastructures, securing of open spaces and emergency roads and designation of areas for evacuation
- emergency activities such as life saving, fire fighting, emergency transportation, and assistance to the suffering people.
- individual counter measures for important facilities
- dissemination of information to and training of the public and private sectors

Finally, a practical "Action Plan" has been proposed. It prioritizes necessary actions so that they can be implemented soon after the project. The scenario and action plan are disseminated to relevant organizations and the public.

The goals of the case studies are;

1) to raise the awareness of decision makers and the public for seismic risk, involving mass media
2) to transfer appropriate technologies to the cities from the advanced world as well as the scientific world
3) to set up a local infrastructure for a sustainable plan for earthquake disaster mitigation
4) to promote multidisciplinary collaboration among the local government, scientists and the public
5) to promote worldwide interaction with other earthquake prone cities

Nine cities were selected from 58 cities, which had applied for the case studies, under consultation with the STC (Scientific and Technical Committee for IDNDR) subcommittee for RADIUS (Members are Dr. Tsuneo Katayama (Chair), Mr. Robert Hamilton, Prof. Mustafa Erdik). The case studies were carried out from February 1998 until July 1999 with financial assistance from the IDNDR Secretariat ($50,000 to a full case study city) and technical assistance from internationally renowned institutes in this field.

Three international institutes offered technical assistance to these cities in each region while three regional advisory committees visited the cities to provide technical advice and to raise public awareness of the seismic risk there together with the international institutes.

<u>9 case study cities</u>
Addis Ababa (Ethiopia), Antofagasta (Chile), Bandung (Indonesia), Guayaquil (Ecuador), Izmir (Turkey), Skopje (TFYR Macedonia), Tashkent (Uzbekistan), Tijuana (Mexico), Zigong (China)

<u>Three international institutes</u>
- Asia (Bandung, Tashkent, Zigong)
Center for Disaster-Mitigation Engineering (INCEDE) and OYO Group, Japan
- Europe, Middle East and Africa (Addis Ababa, Izmir, Skopje)
Bureau de Recherches Géologiques et Minières (BRGM), France
- Latin America (Antofagasta, Guayaquil, Tijuana)
GeoHazards International (GHI), USA

<u>Regional Advisory Committee Members</u> (in alphabetical order)
- Asia
Dr. Anand S. Arya (India), Dr. Jack Rynn (Australia), Dr. Tsunehisa Tsugawa (Japan)
- Europe, Middle East and Africa
Dr. Mohamed Belazougui (Algeria), Dr. Victor Davidovici (France), Prof. Dr. Rainer Flesch (Austria)
- Latin America
Dr. Andrew Maskrey (Peru), Ms. Shirley Mattingly (USA), Prof. Carlos E. Ventura (Canada)

Two training seminars were held mainly for the RADIUS case study cities in 1998. The JICA (Japan International Cooperation Agency) Seminar on "Seismology and Earthquake Engineering," was held for the technical experts with assistance of IISEE, BRI, Japanese Ministry of Construction, in support of the RADIUS initiative in Tsukuba, Japan, from 11 May to 19 June 1998. The seminar was attended by 17 scientific/technical experts from the 9 RADIUS case study cities and other cities pre-selected for the RADIUS case studies.

The RADIUS training seminar for city government officials was held from 22 to 30 June, 1998, in Tokyo and Fukui, Japan, with 18 participants from 13 cities, including the RADIUS case study cities. It was co-organized by UNU (United Nations University, UNCRD (UN Centre for Regional Development), and the IDNDR Secretariat.

All the case study cities held the Earthquake Scenario Workshop from October 1998 to March 1999 at the end of the first stage of the case study. The common objectives of the Workshop are;

- to present the damage estimates to the city and ask for feedback from the participants
- to estimate the impact of the estimated damage on the city activities
- to produce ideas of actions that could reduce the impact of an earthquake on the city
- to discuss the conditions needed to institutionalize the risk management activities

The Action Plan Workshop was held in most of the case study cities at the end of the second stage in order to develop the risk management plan, based on the evaluation of the earthquake damage scenarios and propose Action Plan for immediate actions.

4. Development of practical tools

A technical manual for the urban seismic risk assessment has been developed, based on the analysis of the case studies. By applying the manual, the result will show how buildings and infrastructure in a city could be damaged and how many human loss would be estimated. It will be available as a CD-ROM. It should be noted that the result will be so rough that it must be only the first step for the seismic risk assessment of the city.

In order to facilitate any earthquake prone city to conduct a RADIUS type project, guidelines have been also developed, based on the experiences of the 9 case studies. The emphasis were made on:

- How to involve decision makers, relevant organizations/institutions, communities, private sectors, mass media, and scientists at a multidisciplinary way
- How to transfer the scientific data into decision making information in a practical way
- How to disseminate information and educate the people, particularly through the mass media
- How to prepare a risk management plan as well as an action plan
- What would be the next step

These tools are useful to decision makers and government officials, who are responsible for disaster prevention and disaster preparedness in the cities;

1) to decide priorities for urban planning to mitigate seismic disasters
2) to prepare an improvement plan for the existing urban structures such as reinforcement of vulnerable buildings and infrastructures, securing of open spaces and emergency roads and designation of areas for evacuation
3) to prepare for emergency activities such as life saving, fire fighting, emergency transportation, and assistance to the suffering people

The tools are useful to communities, NGOs, and citizens;

1) to understand the vulnerability of the area where they live
2) to understand how to behave in case of an earthquake disaster
3) to participate in preparing a plan for disaster prevention

They are also useful to semipublic and private companies, particularly to those who maintain urban infrastructure such as electricity and telephone, to understand the necessity of prevention and preparedness so that they could minimize the damage on their business.

5. Comparative Study on "Understanding Urban Seismic Risk Around the World"

More than 70 cities participated as "Member Cities" in the comparative study on "Understanding Urban Seismic Risk Around the World," which started in June 1998 and was completed in August 1999. The study developed a better understanding of the various aspects contributing to the seismic risk of a city, underlining the common earthquake risk problems in different urban areas of the world, and identifying solutions and risk management practices that were successful and could be duplicated.

GeoHazards International (GHI) conducted this study. The report on the study is being finalized.

There were several benefits by joining the study. First, the participating cities could gain a better understanding of the characteristics of their seismic risk as well as the factors that contribute to this risk, such as the city's vulnerability, economic exposure, emergency response and recovery capabilities. This information could then be used by the city to generate public awareness and support the city's risk management efforts. Second, the study helped the cities recognize and prioritize projects they could use for their disaster management. Third, the study also offered the opportunity for the cities to establish partnerships with other cities facing similar problems.

6. Information Exchange

Associate cities

More than 30 cities, which have carried out a seismic risk assessment or are in the process of doing so with independent resources, joined RADIUS as "Associate Cities" for information exchange and international cooperation. The Associate Cities are expected to offer their valuable experience to other cities mainly through the RADIUS home page while they can obtain useful information from the other cities.

RADIUS Homepage

The IDNDR Secretariat launched the RADIUS website, providing a clear, user-friendly and up-to-date access to all the information available concerning the RADIUS Initiative. It provides a fully interactive medium to exchange information on the experience of RADIUS. The midterm reports from the 9 case study cities, the three international institutes, and the associate cities are presented in the home page. The results of RADIUS, including the developed tools, will be also presented there.

The address of the RADIUS home page is; **http://www. geohaz.org/radius**

RADIUS Symposium in Tijuana, Mexico, in October 1999

International IDNDR Symposium on "The RADIUS Initiative - Towards Earthquake Safe Cities" will be held from 11 to 14 October 1999 in Tijuana, Mexico. The Symposium is an important closing event for RADIUS to present and discuss the results of the case studies, developed tools, comparative study on the urban seismic risk, and reports of similar efforts. It will bring together many city government officers and experts involved in RADIUS worldwide. The Symposium will be co-organized by the City of Tijuana, UN Centre for Regional Development (UNCRD), UN University (UNU), and the IDNDR Secretariat, in close cooperation with several international organizations such as International Association for Earthquake Engineering (IAEE), International Association of Seismology and Physics of the Earth's Interior (IASPEI), and World Seismic Safety Initiative (WSSI).

7. Conclusion

The main goal of RADIUS is to raise the public awareness, particularly at city level. The RADIUS case studies with assistance from the United Nations involved decision makers, communities, and mass media in the cities. Its activities were broadly covered by newspapers, televisions, radios, etc. It also created active partnership between scientists and local people. The local scientists closely cooperated with the local governments to transfer scientific data into decision making information. Thus, the RADIUS case studies are of great success.

RADIUS does not draw a closed circle but draws an open circle. While all the results will be open to anybody interested, any kind of information and contribution are welcome. It is expected that further actions will be taken to reduce the impact of probable earthquakes in as many cities as necessary, following RADIUS by learning its experiences and by utilizing the practical tools to be developed by the end of IDNDR.

Appendix

Cities concerned

58 cities that applied for RADIUS case studies

- **Asia** (27 cities)
Almaty (Kazakhstan), Amman (Jordan), Ashgabat (Turkmenistan), Bandung (Indonesia), Baoji (China), Bishkek (Kyrghistan), Calcutta (India), Damascus (Syria), Daqing (China), Dushanbe (Tajikistan), Hefei (China), Istanbul (Turkey), Izmir (Turkey), Kathmandu (Nepal), Mandalay (Myanmar), Metropolitan Manila (Philippines), Mumbai (India), Shiraz (Iran), Tabriz (Iran), Tangshan (China), Tashkent (Uzbekistan), Tbilisi (Georgia), Tehran (Iran), Urumqi (China), Yangon (Myanmar), Yerevan (Armenia), Zigong (China)

- **Europe and Africa** (12 cities)
Accra (Ghana), Addis Ababa (Ethiopia), Algiers (Algeria), Belgrade (Yugoslavia), Bucharest (Romania), Conakry (Guinea), Dodoma (Tanzania), Giza (Egypt), Petropavlovsk-Kamchatsky (Russian Federation), Skopje (TFYR Macedonia), Sofia (Bulgaria), Tirana (Albania),

- **Latin America** (19 cities)
Ambato (Ecuador), Antofagasta (Chile), Cali (Colombia), Cumana (Venezuela), Guayaquil (Ecuador), Kingston (Jamaica), La Paz (Bolivia), Lima (Peru), Manizales (Colombia), Medellin (Colombia), Pasto (Colombia), Pereira (Colombia), Popayan (Colombia), Quito (Ecuador), San Juan (Argentina), Santiago (Chile), Santo Domingo (Dominican Rep.), Tijuana (Mexico), Toluca (Mexico)

73 Member Cities

Accra (Ghana), Addis Ababa (Ethiopia), Algiers (Algeria), Almaty (Kazakhstan), Ambato (Ecuador), Antofagasta (Chile),Athens (Greece), Bandung (Indonesia), Baoji (China), Beijing (China), Bogota (Colombia), Bucharest (Romania), Cairns (Australia), Caracas (Venezuela), Colima (Mexico), Delhi (India), Dhaka (Bangladesh), Gilgit (Pakistan), Giza (Egypt), Guadalajara (Mexico), Guatemala City (Guatemala), Guayaquil (Ecuador), Gyumri (Armenia), Huaraz (Peru), Irkutsk (Russia), Izmir (Turkey), Jakarta (Indonesia), Kampala (Uganda), Kathmandu (Nepal), Khartoum (Sudan), Kingston (Jamaica), La Paz (Bolivia), Lima (Peru), Lisbon (Portugal), Manizales (Colombia), Metro Manila (Philippines), Mumbai (India), Newcastle (Australia), Pasto (Colombia), Pereira (Colombia), Pimpri (India), Popayan (Colombia), Potenza (Italy), Quito (Ecuador), Rome (Italy), St. George's (Grenada), San Jose (Cost Rica), San Juan (Argentina), San Salvador (El Salvador), Santiago (Chile), Santiago (Dominican Republic), Santo Domingo (Dominican Republic), Seattle (USA), Seoul (Republic of Korea), Shiraz (Iran), Skopje (TFYR of Macedonia), Sochi (Russia), Sofia

(Bulgaria), Spitak (Armenia), Tabriz (Iran), Tai'an (China), Tashkent (Uzbekistan), Tbilisi (Georgia), Tehran (Iran), Tijuana (Mexico), Tirana (Albania), Tokyo (Japan), Tuscan Region (Italy), Ulaanbaatar (Mongolia), Urumqi (China), Vladivankaz (Russia), Yerevan (Armenia), Zigong (China)

35 Associate Cities

Algiers (Algeria), Baoji (China), Beijing(China), Bogota (Colombia), Cairns (Australia), Calcutta (India), Dalian (China), Damascus (Syria), Gyumri (Armenia), Hefei (China), Istanbul (Turkey), Jabalpur (India), Kathmandu (Nepal), Khartoum (Sudan), Lima (Peru), Manizales (Colombia), Mumbai (India), Newcastle (Australia), Pereira (Colombia), Pimpri (India), Quito (Ecuador), St. George's (Grenada), San Juan (Argentina), Shiraz (Iran), Sochi (Russia), Spitak (Armenia), Suva (Fiji), Tai'an (China), Tangshan (China), Tehran (Iran), Tianjin (China), Tuscan Region (Italy), Ulaanbaatar (Mongolia), Urumqi (China), Yerevan (Armenia)

URBAN SEISMIC RISK MITIGATION IN ASIA: EXAMPLES FROM RADIUS CASE STUDIES

R. K. SHAW[1], F. KANEKO[2], S. SEGAWA[2] and J. SUN[3]
1: OYO International, Tokyo, Japan (rajib-shaw@oyonet.oyo.co.jp)
2: OYO Corporation, Tokyo, Japan
3: OYO International, Singapore

1. Introduction

As a part of the International Decade for Natural Disaster Reduction (IDNDR: 1990-2000), numerous projects have been launched on natural disaster reduction. A good level of awareness has been reached at the national, regional and international level for disaster mitigation and management. A significant shift of focus from the post-disaster rehabilitation and reconstruction to pre-disaster preparedness and planning has been observed. IDNDR is creditable for this very important achievement. As a part of these activities, a project called RADIUS (**R**isk **A**ssessment Tools for **Di**agnosis of **U**rban Areas against **S**eismic Disaster) has been launched with the financial help of Japanese Government. The aim of this project is to reduce the seismic risk in the cities of developing countries. A unique methodology has been incorporated in this project in order to ensure people participation in the risk reduction activities. In total, nine cities have been selected for case studies, from a total number of fifty-eight cities worldwide. These cities are geographically well-distributed: three being in Asia, three in Middle East-Europe-Africa, and three in Latin America. In the present paper, the achievements of the RADIUS project in Asian cities will be discussed. Three cities are selected in the Asian region from three different geographic areas. Cities are located in the tropical area in Indonesia (Bandung), in the vast country of China (Zigong) and in the land-locked country of Uzbekistan (Tashkent).

2. Why earthquake and why developing countries

In spite of significant progress of science in the 20th century, human beings are still helpless in front of natural events. Earthquake being one of the most unpredictable natural phenomenon, it always remains in a dilemma. With the help of modern data acquisition system, advent of remote sensing technology, we are now being able to forecast some of the natural phenomena like flood, cyclone volcanic eruption; and has been able to improve the early-warning system. However, in case of earthquake, all these techniques usually do not work, and people become helpless in front of the devastating events, being either victims or just a passive witness. Earthquake prone

S. Balassanian et al. (eds.), Earthquake Hazard and Seismic Risk Reduction, 49–67.

areas have been identified by intense geological and geophysical survey, location of active fault systems, and from the historic seismicity. However, there are still many areas, which are regarded relative non-earthquake prone areas, but still having disastrous events. This has been demonstrated and pointed out repeatedly by several big earthquakes in the unexpected areas (e.g., Latur Earthquake of India, 1993, Northridge Earthquake of USA, 1994 and Kobe Earthquake of Japan, 1995). Although search and rescue is an important component, the most rational way to reduce the impact of earthquakes is through proper mitigation and preparedness efforts. It is now established that earthquake disaster management requires a joint and collective effort from different parts of the society. This concept is once again solidified in the present project.

In the developing countries, the population is gradually increasing, and migration of people is occurring from the rural to the urban areas. By the year 2000, approximately 3 billion people will live in the urban areas (Tucker et al., 1994, *Figure 1*), out of which a significant amount will be living in the earthquake threatened cities. This rapid urbanization in the developing countries is often associated with the unplanned and unrestricted aerial growth of the city, which in turn makes it more vulnerable, since the soil condition and other geological features are often ignored in this process of development. The physical vulnerability of the city is enhanced by the concentration of population in the slum areas, which is mainly dominated by the low-income group. Most of these settlements are usually illegal, and therefore the building conditions are also poor. For very natural reason, these groups of people have their other priority area: to survive in the day-to-day life. The lack of proper education and awareness among these groups of people contributes to the increasing risk of a city. Thus, it is very much needed to focus on the cities in the developing countries. In these cities, the risk is often under-estimated or not properly understood. Therefore, the present project focuses on the awareness raising in the decision-makers, government officers and most importantly the common people.

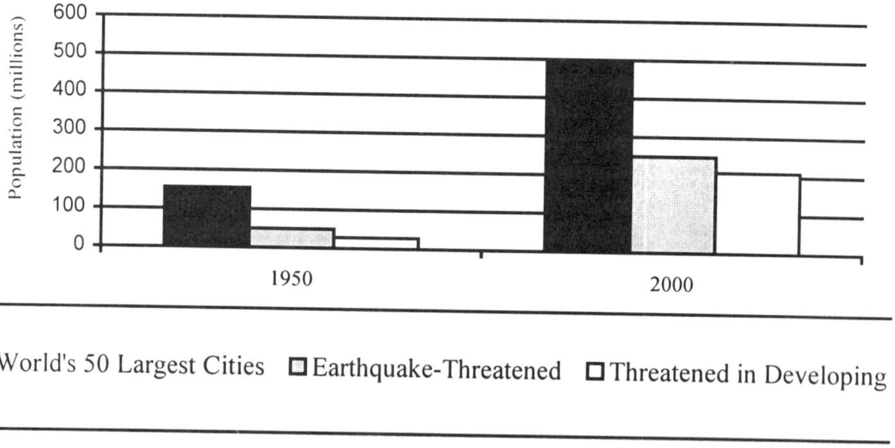

Figure 1. Comparison of earthquake threatened city and its percentage in the developing countries (Source: Tucket et al., 1994)

3. Why these three cities have been chosen

Around twenty-seven cities applied from the Asian region, out of which seven cities were pre-selected based on some criteria, determined by the IDNDR Secretariat. A team of experts visited these seven pre-selected cities to know the existing condition and policies of the earthquake risk reduction. Based on their report and critical reviewing, three cities were finally chosen for case studies. All these three cities are very important in their respective country as well as in the regional basis, although the infrastructure and local condition is quite different from one city to the other. Bandung is a 'tropical resort' city with a cluster of universities and research institutes. The city is situated on the ancient lake deposits, and is surrounded by the mountain ranges. It is a rapidly growing city, the largest in the Western Java Province, with a very high population growth rate, and one of the most important center of the business and trading in this region.

Tashkent, in contrast, is the capital city of Ubekistan, and is one of the most strategic city in the central Asian region in connection to education, culture, trading and business. The city is located in the eastern part of Uzbekistan, which a plain crossed by several rivers. High-rise mountains surround the city in the north and northeast. Tashkent region itself contributes more than one fifth of the total GDP of the country. This city had a very glorious past, being one of the main city on the 'Silk-road', and ancient capital of Amir Timur.

Zigong, on the other hand, is a city in south China, located in the Sichuan province within the mountain ranges. The city is one of the major growing industrial hub, with several industries located within the city. The key sectors are mechanical engineering, chemical engineering, and salt making. The city also produces building materials, textile, electronic products, foodstuffs, and other light industry products. Relicts of Dinosaur fossils and ancient salt producing well (more than 1000 m deep) are other major characteristics. Table 1 summarizes the demographic features of these three cities.

TABLE 1. *Basic demographic data of case study cities*

City	Area (Sq. km)	Status	Population (million)	Annual Growth	GDP contribution
BANDUNG	168	Provincial capital	2.06	3.48%	9.13% (regional GDP)
TASHKENT	326	Country capital	2.08	2.00%	21% (country's GDP)
ZIGONG	817	Industrial city	3.13	0.74%	7.6% (regional GDP)

3.1 BANDUNG

The city of Bandung does not have any historic record of earthquakes, since the city is very new one, only having a history of 120 years. However, there is a very distinct fault called Lembang Fault, which is running E-W, and is in near vicinity to the city. The seismicity along this fault has historic records, however the present condition is unknown. The seismicity caused by the subduction zone activities in the southern part of Java island will also affect Bandung. Moreover, the Semangko fault system, which ends in the Sunda Strait, originating from the south Sumatra region, can also contribute to the seismic activity to the city (Sengara et al., 1998). The major risk of the city of Bandung is its building types, most of which does not conform with the national building codes. Among the decision-makers and the government officers in the city, there is a pre-concept that Bandung is relatively safe with respect to earthquake, which leads to flexible rules, and violation of norms in building construction. Most of the lifeline facilities are also not aware of the seismic risk of the city. This lack of awareness is also exemplified in the existing plan of the municipality. There is a coordination body in the municipality for disaster activities, which mainly focuses on the rescue and relief operation. The personnel come from the Office of the Civil Defense, which has other responsibilities regarding the law and order condition of the city. There is no focus on the earthquake-related disaster in the future planning, although the disaster itself is considered in a limited way. Therefore, considering the increasing risk of the city, the RADIUS project is considered to be very timely and suitable. In Bandung, the major objectives of the project are: a) to raise awareness of the decision-makers and common citizens, b) to improve the local preparedness, in order to reduce loss and casualties during earthquake disaster, c) to set up a local infrastructure to execute the action plan, and d) to promote multidisciplinary collaboration and coordination among the local government agencies and scientific communities. Following scope of work has been identified to achieve the above objectives: a) to assess the local public awareness and preparedness toward earthquake, b) to assess the preparedness and safety of various critical facilities, infrastructures, lifelines for seismic disaster, and c) to prepare an action plan to improve the capability of the assessed sectors in emergency planning and to promote inter-agency co-operation.

3.2 TASHKENT

Tashkent had experienced damaging earthquake in 1966, while Uzbekistan was a part of the USSR. The memory of the earthquake is still fresh in the common people and in the decision-makers. Tashkent is located within the Karzhantau fault system, which is responsible for more than 20% of strong earthquakes in this region (Rashidov et al., 1999). The fault having total length of 130 km, and part of this segment right below the city was activated during the 1966 earthquake. Since the epicenter was located just beneath the city, it had severe impact on the housing and buildings. Immediately after the earthquake, people from Moscow and other parts of USSR rushed to this area with aid and relief materials and constructed new buildings. The methodology they used to reconstruct the buildings, were also used in Armenia, where 95% of similar types building were destroyed in the Spitak Earthquake of 1988 (King et al. 1999). The similar effects have also been observed in Leninakan and Sakhalin region (1995 Sakhalin earthquake). In spite of these recent experiences similar construction practices are continued in the city and making it more vulnerable day by day. In contrast to Bandung, since there has been past damaging earthquakes, the city emergency department is aware of the risk of the city, and has special drill and training program, which they conduct every year in different critical facilities. Several scientific research programs are currently going in the institutes of Academy of Sciences. Different facilities have their own manual, which describes what to do during the earthquake. Ministry of Emergency Situations, created after 1966 earthquake is aimed to organize and coordinate programs related to emergency management, education and research. The level of public awareness regarding the earthquake risk of the city is also very high. However the lacking part is the inter-agency co-ordination and to form an effective planning and command structure to cope with the disaster situation. Therefore, in the present project, emphasis has been given to prepare the necessary and organized set-up to use the existing data in an effective way. Another important aspect is to create interest and awareness among the government officers in the municipality to work together for the seismic risk planning and management. It is also needed to have a good and workable collaboration between the decision-makers and the academicians to make the full use of the research results in actual implementation program.

3.3 ZIGONG

Similar to Tashkent, the Zigong city has also experienced several past earthquakes (1954 with M 5.0, 1959 with M 5.0, 1965 with M 4.6 and 1985 with M 5.0). In the 1985 earthquake, more than 400 houses were destroyed, 2,500,000 square meter of buildings were damaged and structures and 280 people were direct victims (2 persons killed), which gave rise to a total economic loss of more than 8 million US$ (in 1985 value exchange rate). The administrative department for earthquake disaster prevention and mitigation is the Zigong Seismological Bureau, which has been set up in 1971. Zigong Seismological Bureau usually co-ordinates with the provincial Seismological Bureau (in the Sichuan Province) to carry out related seismological work. In 1987 the local government invited the seismic scientists to make investigation on preliminary seismic microzonation in Zigong. On this basis, the government approved and issued "The Plan for Earthquake Resistance and Disaster Relief in Zigong" and "The Seismic Countermeasures of Zigong Municipality". Since 1990s, the government has put the seismic countermeasures into "The Ninth Five Years Plan of the National Economy and Social Development of the Zigong City" and "2010-Year Developing Plan". Therefore, it is found that different activities are on-going to reduce the earthquake impact on the city. However, to boost these activities, and to use the outcome in an effective way, the RADIUS project was very much appropriate in Zigong. The project aims to the following main aspects: (1) to enhance the understanding of seismic risk by decision makers and the public, (2) to transfer appropriate technologies to the cities, (3) to improve the local infrastructure for a sustainable plan for earthquake disaster mitigation, and (4) to promote multidisciplinary collaboration among the local government as well as between government officers and scientists.

4. Organization and Methodology

IDNDR Secretariat is in-charge of the overall administrative control of the project, including fund and policy decision. The direct coordination of the case study is made by the international institute, which in case of Asia is represented jointly by International Center for Disaster Mitigation Engineering (INCEDE) and OYO Corporation. Their role is to coordinate all the activities during the project period, and to provide technical assistance to the case study city. At the city level, a steering committee is responsible for the administrative and monetary matters. It consists of representative of the city government, local educational institute/university and the international institute. Besides, there are several working groups, designated for specific tasks at different levels, with the participation of related stakeholders. These activities are monitored by two advisory committees: one at the regional level, consisting of renowned international experts in the related field; and the other at the local level with the participation of important decision makers, government officers, academicians and eminent personalities.

The total project duration is of 18 months (February 1998 to July 1999). At the beginning, a kick-off meeting has been organized to initiate the project, with the participation of decision-makers, high ranking government officers, community representatives and educational specialists. The first half of the project was assigned to evaluate the seismic risk of the city. This was done in several stages: data collection, hazard assessment, vulnerability analysis and damage estimation. After that, an earthquake scenario is prepared. Intensive interview procedures have been conducted with different lifelines, utility systems and infrastructure organizations. Special emphasis has been given on the community participation and disaster education and awareness raising at different levels. After the completion of the scenario, it is presented in a workshop, and intensive discussion has been made.

In the second half of the project, based on the earthquake scenario, earthquake impact reduction plan or an action plan has been formulated. Here, emphasis has been given on the availability of resources, priority in the national plan and cooperation among different stakeholders. The action plan has been presented in a workshop, where diverse people from the community presented their opinion through intensive group discussion sessions.

5. Earthquake Scenarios

The earthquake scenario depicts the condition of the city, including the human casualty, damages to buildings, lifelines and infrastructures, considering that a certain hypothetical earthquake occurs in the city at a certain time. This scenario has two different aspects: first, the technical procedures including hazard analysis and damage assessment quantifying the percentage of building collapsed, lifeline damages etc. The aspect of scenario is the non-technical part of the soft part, which includes the social, economic and psychological aspects. A combination of both these approaches are highly required to make a workable and valid scenario, which can be easily understandable by the citizen. This approach has been first incorporated in Quito, Ecuador in the Quito Risk Management Project (Quito, 1994), has been revised in Kathmandu, Nepal (Kathmandu, 1999) in the Kathmandu Valley Earthquake Risk Management Project, and has been applied in the present RADIUS project. At first, based on the past earthquake records, a hypothetical scenario earthquake has been chosen for each city, in consultation with the technical community and the government decision-makers. Table 2 shows the characterization of the scenario earthquakes.

Ground classification is the second step in hazard analysis. Deep geological structures were studied in and around the city, and geological profiles were made. Peak ground acceleration (PGA) and intensity distributions were calculated and the collateral hazards were determined. Inventory of the buildings, lifelines, infrastructures were prepared, and the vulnerability functions were decided, modified from ATC materials and examples from similar earthquake damage examples. Based on these vulnerability functions, damage calculations were made and presented in the maps using GIS software, wherever available.

TABLE 2. *Scenario Earthquake in Case Study Cities*

City	Earthquake	Date and Time
Bandung	200 year/ 60% probability (maximum 0.238g at Surface)	March 2 / 14:30 hrs.
Tashkent	M = 6.1 / D = 10 km (same as 1966 earthquake)	April 26 / 05:20 hrs
Zigong	M = 5.5 and M = 6.0 (max. and exceeding max. estimated)	Thursday Spring / 17:00 hrs.

A parallel process includes the interview procedures, where different stakeholder organizations have been interviewed by the working group representatives. At first, a detailed questionnaire has been prepared, and sent to the organizations for initial reply. Based on this, a detail discussion and interview has been made, where the top ranking decision-makers of individual organizations and the technical officers took part. Some of the major topics of interview include earthquake preparedness, emergency drills, earthquake risk assessment, earthquake recovery, major earthquake impact (if any), vulnerable points, responsible department, and damage estimation from the scenario earthquake. The interview results and damage estimation output were compiled to prepare the final scenario, which was described in the common people language, and written by the media people. The scenario was described in time frame, i.e., how the city behaved immediately after the earthquake, and how it recovered in time. This sequential scenario describes the condition of city based on the existing relief and rescue capabilities and inter-agency coordination of different lifeline and utility systems. This scenario was presented in a workshop in each city. Several damage maps were produced for different utility, lifeline and infrastructure system.

5.1 BANDUNG

Following is an example of the sequential event describing the situation of the city of Bandung, in case of the scenario earthquake (Earthquake Scenario, Bandung, 1999a).

First ten seconds...

Tuesday, March 2nd 1999, about 1430 WIB, citizen's activities are reaching the peak. Some main roads are crowded and there are traffic jams on some places. It is a usual view when most of employees of government office and morning-class-students ended their activities. Meanwhile afternoon-class-students just started their activities. Factories just applied new employee's turns......

All of sudden, great tremors occur everywhere all around the city. Those tired people are startled by an extreme natural power. No one of them is in alert condition when the disaster happens. For ten seconds, people can do nothing but hanging out. They do not even have any preparation or training to face earthquake.

Panic occurs everywhere in people's residencies, offices, and other public facilities. People throng out of the house to safe themselves. However, just in seconds, masonry-wall housed unequipped with anti –quake construction starting to cracks down and collapse over the dweller inside. Painful shrieks heard everywhere when people the house crashed by surrounding things fallen over them.

School's buildings made without perfect construction's calculation also ruined. Some students and teachers in the building can not flee and trapped below the ruins. .....

On the highways, quake shakes the drivers. Fast moving vehicles are braked suddenly causing collisions among them. Some drivers lose control on the vehicles and hit against trees and road pillars. Some telephone and electric pillars, traffic controller, and signboards fall over the vehicles. Injured people suffering....

Panic also happen in hospitals and maternity clinics. Doctors and nurses trying to guide unable-to-walk patients to open yard, while less-serious-affected patients walk out of the building by themselves. Nurses on pediatrics department take just-born babies to outside of the building. Hospital's condition becomes so hectic. Collision among peoples happens in narrow hospital's corridors. Meanwhile open yard can not take the entire patient in........

This city, Bandung, has come into a mess. Smokes swell everywhere. Prayers to God The Almighty uttered by secured people mixed by saddening shrieks of injured and crashed peoples. They are being in panic and frustration so that they do not even think of finding helps. In fact, they do not know where to find helps.

First few minutes...

When the tremor stops, trapped people run to the exits. The possibility of following earthquakes brings new panic. People rush to the stairs and exits while there is no light as well as the elevator is not functioning. The number of victims increases due to people falling down and being trampled on.

Ruined or damaged houses start to catch on fire. Smokes emerge from multi stories buildings and factories. Secured people trying to rescue injured and dead people from the ruins...

Worried parents rush to schools to find their children. Some uses the telephone to confirm their condition. As the results, the telephone line is congested.

In Hospitals, technician and important facilities operator hurried to set up emergency equipments. Generator turned on to supply power to vital instrument for patient's life.

Employees rushed to home to check their family. Again, some uses telephone and can hardly be connected due to the line's congestion.

On highways, people get out from their vehicles. They can not continue riding because collapsed pillars scattered on the roads. Victims of traffic accidents start to be evacuated.

Landslide happens in some places especially in the north area of Bandung. Houses built on slopes falling down brought by landslide.....

First few hours...

Five seismic stations around Bandung report the earthquake's scale. Experts determining epicenter and checking the possibility of following quake happen in term of 3 to 10 days. Special team from Geology research and Development Center who are involved in Quick-response team come to the Head of the district to give information on how high the earthquake's level, its epicenter, and possibility of following quake.

It is not an easy task for the team to explain the scientific information to the local government especially when the explanation should be understandable by common peoples. On the other side, Head of local government should be convinced about what had happened exactly and the possibility of the post disaster...

Disaster traumatic peoples can not accept the explanation from experts and Head of district in full senses. Public figures even blame geological team who is assumed being unready to anticipate the disaster. They have assumption that experts should be able to predict the disasters by fully equipped and sophisticate seismic means located surrounding Bandung. They make assumptions there should not be any victims.

People's knowledge of earthquake is very low because the disaster has never been thought. In their mind, earthquake as well as other disaster such as floods or landslide can be predicted days before. With that kind of pre-concept, people do not trust expert's explanations...

Meanwhile, all radio stations, which are undamaged, broadcast these disasters. Assistance demand from the society starts to flow and calls through telephones to radio operator get busy. The society asking for assistance from the radio stations to check certain places and to broadcast it widely....

A Day after the Disaster

The community is still possessed with panic and fright. The worrying feeling kept in mind by the Bandungee because the aftershocks happened a few times. They are afraid that bigger earthquake will occur.

Most of the society live in emergency tents while the rest seek for shelter in somewhere else due to the size of tents. None of them dare to return for their houses even though it still can be lived in.

Meanwhile the evacuation efforts keep on performed. Groans of victims are still heard from ruins of collapsing buildings. Military troop sends searcher dogs to find victims trapped inside those ruins. Rescue team having trouble to evacuate victims from ruins of high building due to insufficient or mal-functioning of tools needed to lift up concrete...

The above scenario shows that the city becomes a very unusual condition, partly due to lack of awareness of people, and partly due to inefficiency of decision-makers and government officers. Apart from this descriptive scenario, several damage maps were produced for buildings, lifelines (water supply, sewage system, power supply, telecommunication network) and infrastructures (road network, bridges) and critical buildings. *Figure 2* shows the damage maps of road network in Bandung. Thus, the above scenario has been produced through the collective efforts of the technical and non-technical people and involvement of media communities.

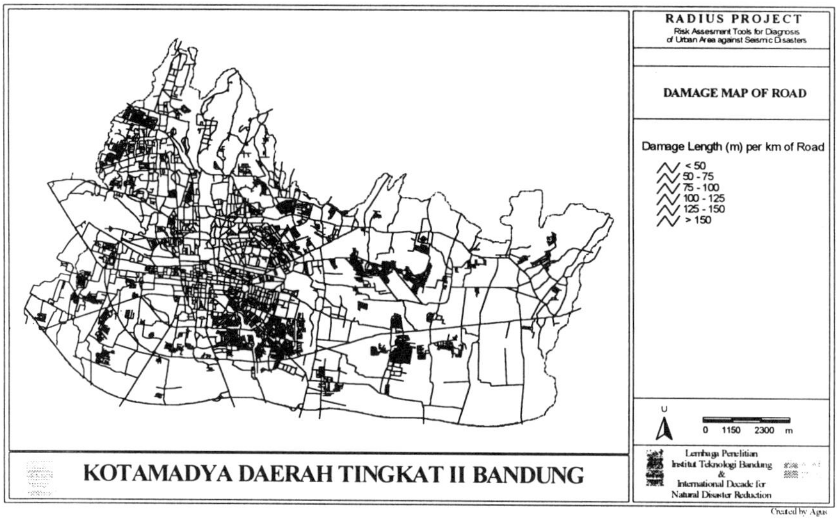

Figure 2 Damage map of road network in Bandung

5.2 TASHKENT

The city of Tashkent also prepared similar earthquake scenario, which has been written in common people language. Following is the fragmented part (Tashkent, 1999).

It is an early spring morning, just after 5:00 am... The city is awakening little by little... But most people are still in bed sleeping peacefully. The usual run of things at the usual time. There is no foreboding of evil... . Nobody knows that all of a sudden the time allotted to seismic risk management and disaster preparedness is over. There comes a time to check how well it has been done. The timer is on...

Seconds go...

...Strong shaking is felt throughout the city. Sleeping people are awaken. They are frightened and try to run outdoors.... For inhabitants of tall buildings there is no possibility to run from their homes because stairways and elevators become most dangerous under continued shaking. Their only hope for these crawling seconds is that their homes are strong enough to withstand this terrible shaking. But buildings behave differently… Individual and adobe homes located in the epicentral zone are devastated. Some of modern reinforced concrete buildings are damaged considerably, having cracks in walls and bearing elements. Self-built adobe homes located in the old part of Shaykhantahur and Sabir-Rakhimov districts collapse killing their inhabitants... Water supply pipelines are damaged throughout the city... Some of transit sewer pipes are ruptured polluting the city.... Damages to power supply system immediately remain the city without electricity... The shaking continues for about half a minute gradually dying down. For these seconds the earthquake ends up a disaster. On its way of development to future prosperity the young independent state is struck in the heart. The capital has suffered very seriously.

This damage scenario is supported by damage maps of the buildings and lifelines. Figure 3 shows the damages to residential buildings for Tashkent city.

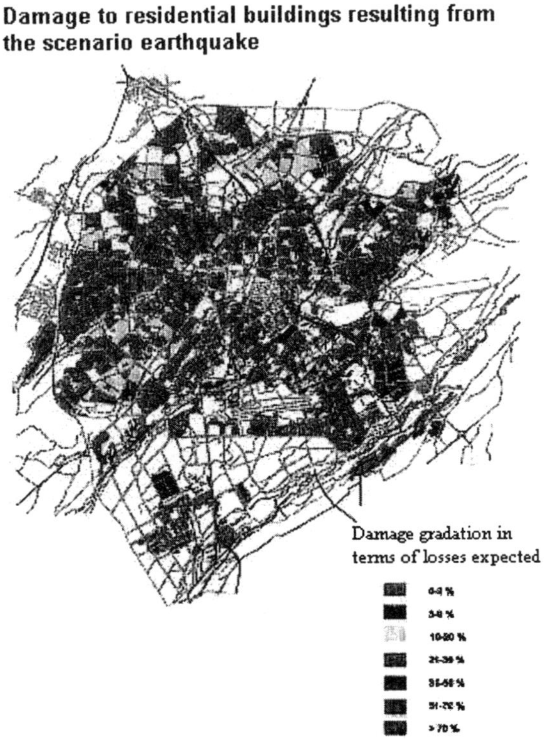

Figure 3. Damage to residential buildings in Tashkent, resulted from the Scenario Earthquake

5.3 ZIGONG

The Zigong city also produced similar scenario. Following is the fragmented part of the scenario (Zigong, 1999).

The first few seconds...

Strong shaking is felt throughout the city. It is violent because the earthquake is shallow and directly below the city. Residents of tall residential buildings were terrified but did not try to run out. They know the shaking makes it not possible to run (crawl) out in time, and their buildings are designed against earthquakes. They take shelter under the desk, table or bed to avoid injury from falling objects...

Adobe houses suffer heavier damage. The tiles on the roof fall off completely and injured many people. Some walls collapse and a few houses are reduced to rubbles...People in the public buildings (shopping center, cinema, government office, etc.) are also scared. Some try to run out but it is not possible due to the shaking. These buildings were designed and built with higher anti-earthquake requirement than normal residential buildings, and perform well.

On the street, a new driver panics and his car skids to the reverse side of the road. In a desperate effort to avoid the vehicle, an incoming bus swifts and hits a tree. No one is killed immediately but many are injured. The road is blocked by the damaged bus. The natural gas bad on top of the bus is punched, and gas is leaking. ...

The shaking lasts 15 seconds. It must feel like a year for those experiencing it. The shaking is finished, but the drama has just begun.

The first few minutes...

When the shaking stops, people quickly run out of the building. Fortunately no stampede is resulted in the schools, shopping centers and cinemas, as the people do not see any sign of collapsing. In a few minutes, all are in the streets or open ground, and very shocked and confused. Water begins to flood the street because of the broken water pipe. Fire breaks out in a high-rise building and a few adobe houses. The smoke makes it like a battle field and adds to the apprehension. Passengers flee the crashed bus. The odor of the leaked natural gas is chilling.

The first hour...

The mayor survives unhurt, and starts to gather information. The facilities in the Seismological Bureau of Zigong is still functional, and the staff-on-duty determine the epicenter and magnitude of the earthquake based on records from the local seismic network operated by the Bureau, but he cannot report it to the Mayor by phone because too many people try to call the Mayor and the Bureau. People try to contact their families and others by phone but nobody can dial through because all are trying to dial out. Telecommunication is down although the facility is undamaged. Worried for the safety of their children, parents rush to schools. Residents near the collapsed building try to search and rescue any survivors but it is impossible with bare hands and simple tools.

The Armed Police in Zigong is mobilized by now. It is the main unit, under the China Seismic Disaster Reduction Law, to act in the emergency. Its brigades are dispatched to designated districts. The traffic has come to complete standstill in the street (it is bad enough even without earthquake), so they have to run to the sites. Other emergency groups also start to gather, but not as fast.

The first night...

Most people are able to make contact with the people they want, and through the radio and TV, have learned the situation in the city. The Seismological Bureau of Sichuan Province confirms the earthquake parameter determination by Zigong Seismological Bureau, but unsure whether there will be aftershock. A group of seismologists rush to Zigong to survey the damage. Traffic police take control of the roads and streets, and traffic is now possible, except the more damaged area. More accidents occur at site of damaged road. At the collapsed buildings and houses, the armed police and others try to find and rescue survivors, but to no avail. The collapsed building causes short circuit and the blocks surrounding it has no electricity. They work in poor lighting. Natural Gas company has turn off the gas supply to those areas as precaution against fire. Fire fighters and volunteers are still trying to control the blaze. They are helpless in front of the 10-story building on fire because their water cannon is not powerful enough.

Hospitals are crowded with injured people. Many facilities are damaged by the shaking and doctors have difficulty in treating the patients. Some off-duty doctors come back to hospital to help, but there are just too many patients to deal with. It is fortunate they still have electricity and water supply. Most part of the city has electricity supply. Many people dare not return to the their houses and apartments, not knowing whether the houses and buildings are safe, and whether aftershock is coming. Civil engineers try to inspect buildings for safety assessment but the darkness makes it difficult.

The first week...

Most people return home following the inspection of their buildings by engineers. Some buildings and houses are no longer safe. Some homeless people have to live in temporary shelters but most of them stay with relatives and friends, partly because of shortage of tents. Damage scenario and casualty figure are gradually clear. 23 people died and 116 people are seriously injured. 67,900 people are homeless.

The first month...

At the end of the first month, life returns almost normal, except some people live in temporary shelters. The economic loss has been estimated to be 800,000,000 RMB (about 88 million US$). More people suffer from diseases, but the situation is not out of control. Damaged substations and water pipes have been repaired. Collapsed houses and buildings are being cleared. Lightly damaged buildings are being retrofitted. Insurance company starts to process various claims....

Similar to other cities, Zigong also produced damage maps to buildings, lifelines and infrastructures. *Figure 4* shows the building damages from a potential M 6.0 earthquake in Zigong city.

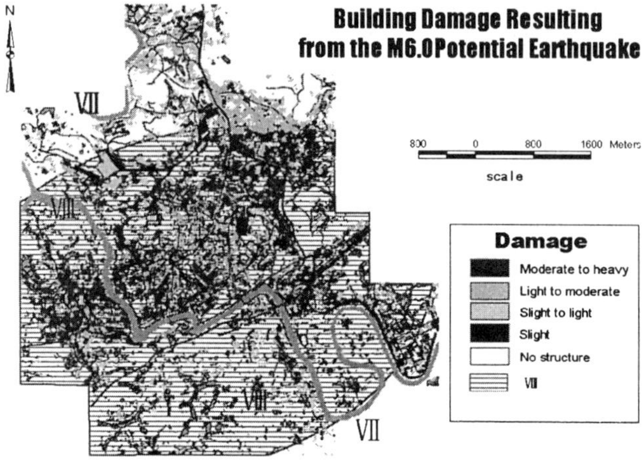

Figure 4. Damages of buildings resulting from the Scenario Earthquake in Zigong

From the above three scenario, the difference in awareness and preparedness levels I n these three cities are quite evident. In Bandung, everybody is in panic, since this is the first time people are experiencing the earthquake. However, the Zigong city looks better prepared and the level of public awareness is high, which is evidenced in people's behavior during the earthquake, as well as the response and rescue team's ability to restore the situation.

6. Action Planning

Action planning, in principal, includes the following phases: problems and opportunities, goals and priorities, options and tradeoffs, resources and constraints, project teams and tasks, and implementation and monitoring (Hamdi and Goethert, 1997). The process begins with identifying the problems, which is done in the preparation of the earthquake scenario, and identification of the vulnerable elements in the city. Analyses at different stages are necessary in order to make conformity with the available resources and the priority of the city. The overall aim of the risk management plan is to assist the high level decision-makers of the city and concerned agencies for policy making regarding the present infrastructure and existing elements, as well as the future development. Also, it aims to help the people to mitigate earthquake risk through community participation and disaster education. For this, different priority areas have been chosen in different cities. However, in common, the emphasis has been given on the following main aspects: improvement of emergency response planning and capability, improvement of public awareness of earthquake risk related issues, improvement of seismic performance of buildings and infrastructures including lifelines and critical buildings, improvement of school buildings and safety measures of the school children.

In the process of formation, first the current status of the actions and responsibilities related to the seismic disaster are listed and reviewed. Both physical and social aspects of planning have been stressed. In the physical aspects, interdisciplinary approaches have been incorporated including several stakeholder organizations. Intensive interview sessions were carried out with different related organizations, and the results were compiled to prepare the draft version of the action plan. In the social planning, education, capacity building and community participation have been stressed, where emphasis have been given on the basic issues like health, education, poverty etc. The results of both these exercises have been combined together to produce the integrated earthquake risk management plan. This integrated plan was then presented in the workshop, and the interdependence and the interactions among different agencies have been studied. Group discussions were made to reach consensus regarding the proposed plan. The suggestions from these discussion sessions were incorporated in the final version of the plan. An implementation strategy has been formulated for each case study city depending on the local condition, which is in accordance to their national and local development policy. Emphasis has been given on the social and practical acceptance of the plan and the technically sound approach for the mitigation efforts. Special caution has been taken so that the plan becomes suitable for its implementation.

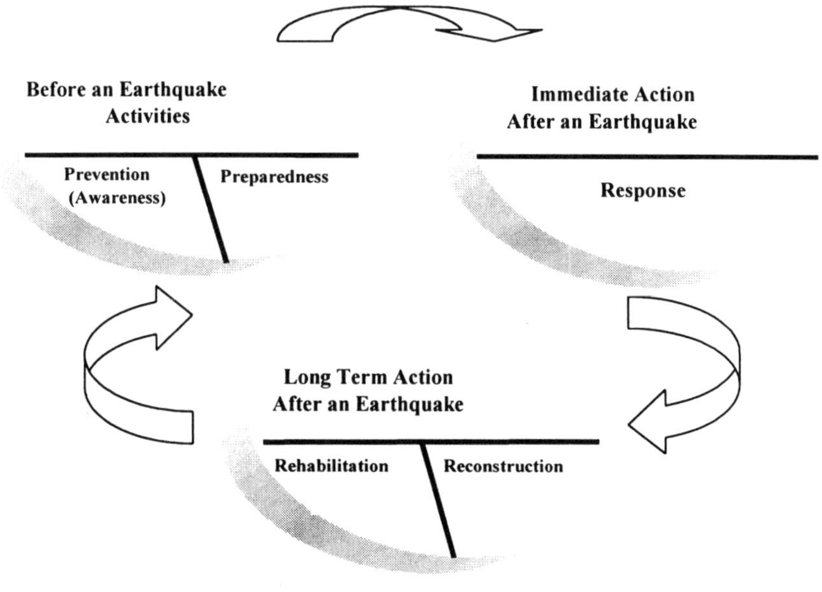

Figure 5. P2R3 Disaster Cycle
(Source: Bandung Earthquake Disaster Mitigation Action Plan, 1999)

As an example, let us discus briefly the Action Plan of Bandung (Bandung 1999b). The basic concept of the Bandung Action Plan is based on the P2R3 Disaster Cycle, which is shown above (*Figure 5*). As a part of activities in this cycle, different actions have been proposed. These actions include some long-term actions before an earthquake (which include prevention and preparedness), immediate action after an earthquake (including emergency response and relief), and long-term actions after an earthquake (rehabilitation, reconstruction and urban planning). The following few lines describe the roles of different organizations in these activities and their mutual interdependence.

Prevention: The main goal of prevention activities is to build public awareness that will be expected for supporting the program successfully. Awareness activities for decision-maker, school student, local community will be specially emphasized. Educational and Cultural Department collaborate with appropriate agencies responsible for public education campaign for earthquake hazard awareness.

Preparedness: Preparedness activities are aimed to develop strategic planning of all the member groups to reduce the earthquake impact through implementing seismic building code, strengthening existing structures, planning and training for emergency response action and conducting earthquake research, etc. The Geological Research and Development Center (GRDC), Department Mining and Energy has been preparing and being responsible for seismological study and isoseismal mapping. The building Monitoring Division of the Bandung Municipality is responsible for monitoring building and strengthening the building regulation. Public Works Department is responsible for assessing vulnerability of critical building and recommending the retrofit of the most vulnerable ones. Assessing vulnerability of existing lifelines, infrastructures, housing and retrofitting are also needed. The emergency operation center of disaster "SATLAK" is responsible for coordinating all agencies involved in emergency response activities through legislative and regulatory arrangements. Currently SATLAK's standard operating procedure has been established to explain the responsibilities of each organization involved in emergency response activities. Improving and socializing the existing SATLAK's standard operating procedure and training emergency response personnel are needed to mitigate earthquake disaster.

The Response Action: These activities include both response and relief actions. The objectives of the first part is to make the situation under control, such as: coordinating emergency response efforts of all involved agencies, fire suppressions, investigating and containing hazardous materials spills, etc. The objective of the second part is to help the victims caused by an earthquake in order to get some better conditions, such as: providing emergency medical treatment, collecting and distributing relief supplies, etc. SATLAK is responsible for coordinating these activities. Regional Drinking Water Company responsible for provision of potable water. Social department responsible for collecting and distributing relief supplies (such as food, blanket, etc.) and work together with Indonesian Red Cross, Program Planning Division, Local Planning Division (Bappeda) and other agencies which are indirectly involved. The Meteorological and Geophysical Agency (MGA) and Geological Research and Development Center (GRDC) are responsible for monitoring aftershocks, other necessary information collection, and then send it to "the emergency operation Center" (SATLAK) as a center for information during emergency action. The Metro Police Department, District Military Command, Civil Defense, SAR, Emergency Room, The Fire Brigade, Volunteer Rescue Association or other agencies receive those results to implement depend on their function and task.

Rehabilitation/Recovery Action: It aims to help the community to recover from the effects of earthquake, focused on temporary arrangements and rapidly executed projects to put vital buildings, lifelines, and infrastructures back into services, providing temporary housing and school, with an expected duration of three months for temporary measures. These activities are the responsibilities of SATKORLAK PB. Social Department is responsible for providing temporary housing for homeless families. Recovery of lifelines is the responsibility of the related individual agencies. The rehabilitation and recovery of damaged infrastructures such as bridges and roads should be carried out by the Public Works Division. Rehabilitation is also needed for the victims of earthquake. Social Department and Educational and Cultural Department are responsible for this activity.

Reconstruction Action: It comprises the long term process of rebuilding all aspect of community following an earthquake (e.g. rebuilding houses and building, financing for rebuilding, psychological counseling, etc), and lasted for about 60 days of planning, and two or three years of execution. The coordinating Minister for Economic Affairs and the State Minister for Development Planning (currently the Chairman of the National Development Planning Agency, BAPPENAS), in cooperation with other government agencies, especially the Minister of Public Work (MPW) are responsible for those action. Special emphasis will be given to the use of seismic resistant construction in the reconstruction program.

From the above description it is quite evident that earthquake risk mitigation is a networking of activities among different related organizations. The understanding of this inter-dependence is an important aspect, and the related stakeholders should be well aware of it. A flow chart and a correlation matrix has been prepared (see Table 3) to show the mutual dependence of each organization in the city of Bandung before, during and after earthquake. This has been discussed in different sessions of the Working Groups and during the Group discussions in the Action Plan Workshop. A consensus has been reached among different organizations regarding the responsibilities.

7. Conclusion

From the above discussion, it can be seen that the current project incorporates a unique methodology for mitigating of the seismic risk in the city. The most important part of this project is the involvement of different diverse organizations and participation of communities at the grass root levels in mitigation efforts. The understanding of risk at the top most decision makers as well as the common people are both combined to make the overall process. Through its learning stage it has been found that several invisible aspects, especially social and cultural features are deeply related to the risk of the city, and therefore should be taken into consideration in the future disaster management initiatives. The active participation of the common people and different communities are regarded as most effective means for disaster mitigation. The involvement of the community and the awareness and preparedness in the city level has been very much evident in the recent examples of Columbia earthquake of January, 1999. A comparison of two cities in Armenia and Pereira (Columbia, 1999) shows that this preparedness aspects have been counted for the minimum casualty of only 44 people in Pereira, whereas the death toll was 543 for Armenia (as of preliminary report of January 1999). The utility system and financial system was running almost normally in Pereira, whereas the a very few percent of the same systems were working in Armenia. This is to be noted that Armenia did not have any historic earthquake, which is a very similar to the situation in Bandung. Therefore, it is very much needed that the experiences and lessons learned through this project should be disseminated in the national and the regional level.

In any of the time-bound earthquake disaster mitigation project, another vital point is the continuation of the activities after the completion of the project. Since the earthquake-mitigation efforts are non-ending, and effects of past earthquakes are often forgotten, it is very important to make the efforts sustainable. For the present project also, the first priority is the implementation of action plan in the case study cities. In some cities, implementation process has already been started. In Bandung, school project has been initiated, which has two main parts: one, the vulnerability assessment of the schools and other, disaster education for the school children. In Zigong, the new Disaster Management Center is now being constructed. It is quite obvious that in the developing countries, there are different other priority areas and basic needs, e.g., food, education, poverty. Therefore, it is often difficult for the national or local government to focus on earthquake-related issues, where a big earthquake may come once in 50 years or 100 years. In many cases, 'disaster' itself is regarded as a temporary issue, only for a certain period of time, during and after the damaging event. More emphasis should be given on the long-term efforts to make the total process more continuous, and meaningful. The national and international development agencies should consider these issues in a more committed way, and to incorporate disaster issues in their development planning, to make it more sustainable. It is important that the decision-makers should understand the need of preparedness efforts. Through the present project a certain level of awareness has been achieved both at the community level and at the government decision-makers level. However, there are many areas, which needs future actions. More similar projects are necessary, which should focus the grass-root level activities with the proper community participation and using local and affordable technologies.

8. References

Bandung (1999a) *The Earthqake Scenario*, Bandung, Indonesia

Bandung (1999b) Bandung Earthquake disaster Mitigation Action Plan, Bandung, Indonesia

Columbia (1999) Columbia Earthquake: preliminary report. OYO Corporation

Hamdi N. and Geothert R. (1996) *Action Planning for Cities*, John Willey and Sons, UK

Kathmandu (1999) *The Kathmandu Valley Earthquake Risk Management Project*, NSET-Nepal and GHI.

King S. A., Khalturin V. I., and Tucker B. (1999) *Seismic hazard and building vulnerability in Post-Soviet Central Asian Republic*, NATO ASI Series, Kluwer Academic Publisher

Quito (1994) *The Quito, Ecuador Earthquake Risk Management Project*, Geohazards International Publication

Rashidov T., Plotnikova L., and Khakimov S. (1999) Seismic Hazard and Building Vulnerability in Uzbekistan. In *Seismic hazard and building vulnerability in Post-Soviet Central Asian Republic*, NATO ASI Series, Kluwer Academic Publisher, pp. 147-180

Sengara W., Surahman A., and Pribadi K. S. (1998) Seismic hazard and countermeasures in Bandung-Indonesia. In *IAEES Training Seminar on Earthquake Engineering*, Tsukuba, Japan

Tashkent (1999) The 2nd RADIUS semiannual report, Tashkent, Uzbekistan

Tucker B.E., Trumbull J. G., and Wyss S. J. (1994) Some remarks concerning worldwide urban earthquake hazard and earthquake hazard mitigation. In *Issues in Urban Earthquake Risk* eds. Tucker et al., NATO ASI Series, Kluwer Academic Publisher, pp. 1-10

Zigong (1999) The 2nd RADIUS semiannual report, Zigong, China

THE KATHMANDU VALLEY EARTHQUAKE RISK MANAGEMENT PROJECT (KVERMP): PROJECT MOTIVATION AND DESCRIPTION

A.M. DIXIT*, L. DWELLEY-SAMANT, M. NAKARMI*,**
S.B. PRADHANANG*, B. TUCKER**
**National Society for Earthquake Technology – Nepal (NSET-Nepal), Kha-2-731, Mahadevsthan Baneshwar, Kathmandu, Nepal, email: nset@mos.com.np*
***GeoHazards International, 200 Town and Country Village, Palo Alto, CA, 94309 USA, email: info@geohaz.org*

1. Abstract

This paper introduces the Kathmandu Valley Earthquake Risk Management Project (KVERMP), being implemented by the National Society for Earthquake Technology – Nepal (NSET-Nepal) and GeoHazards International (GHI) as part of the Asian Urban Disaster Mitigation Program (AUDMP). The motivation for this project, selected project activities, and the progress of the project as of August 1998 are described below. The project activities discussed are the simplified earthquake scenario and action plan, the school seismic safety program, and public awareness and institution building activities.

2. Introduction

The Kathmandu Valley Earthquake Risk Management Project (KVERMP) is an 18-month project which started on Sept. 1, 1997. The project is being implemented by the National Society for Earthquake Technology – Nepal (NSET-Nepal) and GeoHazards International (GHI), and is part of the Asian Urban Disaster Mitigation Program (AUDMP), funded by USAID/OFDA.

KVERMP has four stated objectives: 1) to evaluate Kathmandu Valley's earthquake risk and prescribe an action plan for managing that risk; 2) to reduce the public schools' earthquake vulnerability; 3) to raise awareness among the public, government officials, the international community resident in Kathmandu Valley, and international organizations about Kathmandu Valley's earthquake risk; and 4) to build local institutions that can sustain the work launched in this project.

This paper describes the motivation for this project, selected project activities, and the progress of the project as of August 1998.

S. Balassanian et al. (eds.), Earthquake Hazard and Seismic Risk Reduction, 69–74.

3. Project Motivation

Kathmandu Valley, located in the Himalayan nation of Nepal, has a long history of destructive earthquakes. Most recently, a 1934 earthquake produced intensities of IX-X MMI in Kathmandu Valley, and destroyed 20% and damaged 40% of the valley's building stock. Many of the valley's treasured cultural monuments were also destroyed[1]. Three earthquakes produced intensities of IX-X in Kathmandu Valley in the 19th Century: in 1810, 1833, and 1866. The seismic record of the region extends back to 1255, and indicates that such major earthquakes affect the valley approximately twice a century.

Like many urban areas in developing countries, Kathmandu Valley's risk has increased significantly since the last major earthquake. The valley has a burgeoning population of almost 1 1/2 million people, uncontrolled development, and a construction practice that has actually degraded over this century. Nepal is among the poorest and least developed countries in the world. It has a gross domestic product (GDP) of US$ 2.9 billion, 60% of which is derived from agriculture, or a per capita GDP of US$ 145. Approximately 14% of the GDP (US$ 400 million) is derived from foreign development aid. A weak economy and abundant poverty affect earthquake risk management in many ways: a lack of government funds to support earthquake hazard mitigation programs, inexpensive and poorly constructed dwellings that often fail even in the absence of earthquakes, a tendency in the general population to ignore the earthquake hazard due to more immediate needs, and a lack of awareness about the earthquake hazard. Kathmandu Valley has an urban growth rate of 6.5% and one of the highest urban densities in the world. Nepal currently has no official building code[2], and practically all construction is built without the input of an engineer and without considering seismic forces.

Despite this threat, there is no institution within Kathmandu Valley to assess earthquake risks or promote an earthquake risk management program. The technical information about earthquake risk in Kathmandu Valley is incomplete and scattered among several governmental agencies. However, a more important contributor to the region's lack of earthquake preparedness is that the technical information that is available has not been synthesized, has not been applied to the infrastructure of modern day Kathmandu Valley, and has not been presented in a form that the public and government officials can digest.

4. Project Description and Progress

KVERMP includes a wide variety of activities aimed at beginning a self-sustaining earthquake risk management program for Kathmandu Valley. Described below are the simplified earthquake scenario and action plan, the school seismic safety program, and public awareness and institution building activities.

---

[1] Kathmandu Valley contains seven UNESCO World Heritage Sites, some of which were rebuilt after being destroyed in 1934.

[2] The United Nations Development Program sponsored the development of a building code which was approved by the national Parliament in 1998. The building code still needs to be legalized and enforced.

4.1 SIMPLIFIED EARTHQUAKE SCENARIO AND ACTION PLAN

This component of the project has three parts: a simple loss estimation study, a written scenario which translates the findings and consequences of the loss estimation study into layman's language, and an action plan which presents the activities which will follow the completion of this project.

A simple loss estimation study was conducted for a repeat of the 1934 earthquake in modern day Kathmandu Valley. Although the next major earthquake to affect the valley will not be identical to the 1934 event, Kathmandu Valley's soft-surface geology makes it likely that the relative pattern of shaking will be approximately the same for all distant earthquakes. The location and vulnerability of Kathmandu Valley's infrastructure was determined through interviews with over thirty institutions. The information collected in these interviews was combined with previously conducted studies, and then a loss estimation study was conducted using ATC-13 and other earthquake loss models. Loss estimates were conducted for the road system, the water system, the electricity system, the telephone system, and typical structures. In addition, possible death and injury figures were determined by looking at statistics from previous comparable earthquakes from around the world. All of this information was compiled on maps which showed the Modified Mercalli Intensities experienced in 1934 and liquefaction susceptibility as determined in a previous study. These maps, although very simple, have been extremely effective in communicating with non-specialists in Kathmandu Valley, and have helped many decision makers to visualize the potential effects of Kathmandu Valley's next earthquake. The project team has made an effort to emphasize the point that this loss estimation study provides an understanding of the order of magnitude of damage to expect, as opposed to an accurate prediction of future events. Some of the results of the loss estimation study will be published in the Earthquake Risk Management Action Plan, to be complete in January of 1999 (see description below).

A scenario document that explains the results of the earthquake loss estimation study in layman's terms is currently being written. This document includes a description of possible damages to various vital systems in Kathmandu, and an explanation of the repercussions of this damage on life in Kathmandu Valley. It also presents the story of one common man, "Bhaicha", for up to one year after the scenario earthquake, and illustrates how this imaginary character's life is impacted. Issues discussed include the chaos after the earthquake, lack of adequate emergency medical care, living in mass shelters, rebuilding using earthquake-resistant techniques, and many others. This document will be published in large quantities and will be widely distributed along with information on how to prepare for an earthquake and where to get more information about earthquake safety. It is expected that this document will provide emotional understanding of the earthquake phenomenon to complement the technical information in the loss estimation study.

The loss estimation study and the scenario provide motivation for the Earthquake Risk Management Action Plan, currently being developed by the project team. The purpose of this plan is to provide a feasible "road map" for effectively managing Kathmandu Valley's earthquake risk after the completion of this project. This plan aims to prioritize and coordinate the earthquake risk management activities of all institutions in Kathmandu Valley. NSET-Nepal will act as an umbrella organization to promote, assist, coordinate, and monitor all of these actions. In Nepal, NGOs do not typically conduct this type of work, although non-governmental organizations in other parts of the world have been successful in undertaking such coordination roles. Therefore, NSET-Nepal has the difficult task of simultaneously establishing the concept of a multi-disciplinary earthquake risk management plan and the concept of an NGO in a coordination role.

The success of the action plan largely depends on the willingness of Kathmandu Valley institutions to participate in earthquake risk reduction activities. The loss estimation study has so far proved successful in raising the interest of national institutions, local institutions, and international groups active in Kathmandu Valley to implement risk reducing strategies. The project team has been soliciting and suggesting earthquake risk reduction projects for a wide variety of institutions throughout the course of the project through workshops, letters, and face-to-face meetings. In particular, low-cost, locally fundable activities such as education programs, emergency planning, and non-structural building improvements have wide appeal. The project team is now in the process of collecting all of the actions which have strong support and seem possible to implement, and therefore should be considered for the action plan.

The Action Plan will include a long-term strategy for how to create a sustainable earthquake risk management program for Kathmandu Valley, a short and focused list of highest priority feasible actions to start in the short-term that NSET-Nepal will actively aid and promote, and a longer list of other important endorsed actions that need to be focused on in future years. In the process of developing the action plan, openness and transparency are being emphasized in order to reduce suspicion of the process. The Action Plan will be finalized during a workshop in October of 1998. NSET-Nepal will monitor the plan's effectiveness and update the plan's contents on an annual basis.

4.2 SCHOOL SEISMIC SAFETY

The project includes a vulnerability assessment of Kathmandu Valley's public schools as an example of how to conduct earthquake risk management projects in Nepal. The purpose of this assessment is not to identify individual schools as vulnerable, but to quantify the risk faced by the entire system. In addition to beginning the process of reducing Kathmandu Valley's public schools' risk, this component of the project has broader applications such as determining appropriate methods and costs of retrofits for typical Nepali construction, and examining such issues as acceptable levels of risk for Nepali culture.

First, the project team created a questionnaire that can be filled out by school headmasters. This questionnaire includes topics such as size of buildings, density of students, year(s) of construction, whether or not an engineer was involved in the building design or construction, etc. Additionally, simple questions are asked about structural characteristics, which are presented through illustrations and descriptions. The project is conducting 15 seminars with school headmasters to teach them about earthquake risk, about the necessity of planning for earthquakes in their school, and how to fill out the project questionnaire. One pilot seminar was held in November of 1997 to test the draft survey form. Changes were made to the survey based on comments from the headmasters who participated, primarily relating to ease of use.

The methodology for assessing the schools' vulnerability based on the survey responses was developed by analyzing ten representative school buildings in detail. This analysis will be used to make generalizations about the level of risk of all the schools based on the survey responses. Following the completion of the surveys by the headmasters, a structural engineer will visit approximately 10% of all schools in the survey sample to assess the reliability of the data collected. Investigations will be made into potential methods for retrofitting these existing buildings, and possible costs of retrofitting. Detailed retrofits designs and cost estimates will be made for three typical school buildings by a local engineer. The project will try to raise local and international resources to complete these retrofits, testing the level of community interest and the ability of Kathmandu Valley communities to improve the earthquake resistance of their schools.

There has been a very high level of cooperation and interest from the Kathmandu Valley schools and school officials in this project. In fact, the project has already resulted in some unanticipated benefits, such as collecting statistics on schools, not related to structural safety, that were previously unknown to the school authorities.

4.3 PUBLIC AWARENESS AND INSTITUTION BUILDING

Prior to this project, public awareness about earthquakes was very low. The publicity of this project has greatly increased that awareness. The project has appeared frequently in newspaper stories, has been featured on television, and has been presented in a wide variety of other forums.

January 15 will be designated as Kathmandu Valley Earthquake Day, in recognition of the occurrence of the last earthquake tragedy to strike the valley on January 15, 1934. The focus of this day will be to raise public awareness at all levels of society. School children in the valley will receive a lesson about earthquake safety prior to the event, and on January 15th will practice earthquake drills, receive informational pamphlets about earthquakes, and watch demonstrations on earthquake safety techniques. Television shows relating information about earthquakes and earthquake safety will be aired around this date. Public figures will be present, and there will be a press conference to officially announce the Earthquake Risk Management Action Plan. The event is expected to attract many residents and will involve many Kathmandu Valley institutions. It is hoped that Earthquake Day will be established as an annual event.

An important component of this project is to institutionalize the earthquake risk management processes started in this project. In order to reduce Kathmandu Valley's earthquake risk, the processes started by this project need to continue for many years. The project's institutionalization efforts have focused on two areas: first, establishing NSET-Nepal as a neutral seismic safety advocate for Nepal; second, to incorporate earthquake risk management and other disaster management activities into local government. This project has given NSET-Nepal an opportunity to establish an office, train its staff, gain experience in earthquake risk management, and develop a positive reputation through its actions. It has also provided an opportunity, through the development of the Earthquake Risk Management Action Plan and other activities, for

NSET-Nepal to plan its long-term strategy in tackling Kathmandu Valley's earthquake risk. Progress has been made in establishing local government earthquake risk management institutions, as well. The Kathmandu Metropolis created a Disaster Management Unit as part of the city government, which has been included in project activities and was trained in organizing disaster management activities by a KVERMP consultant. Other municipalities in the valley have also considered establishing Disaster Management Units, and are working with NSET-Nepal to get the process started. NSET-Nepal has also been active in educating ward-level officials (a ward is a subset of a municipality, the legal equivalent of a neighborhood), and at this time two wards have created their own Disaster Management Units made up of neighborhood residents.

5. Conclusions

Kathmandu Valley faces an immense an growing risk from earthquakes. KVERMP, is beginning the process of managing that risk responsibly. Through the loss estimation study and the scenario, KVERMP is raising awareness about earthquake hazard and risk among Kathmandu Valley's leaders. Through the Earthquake Risk Management Action Plan, KVERMP is outlining a long-term process with short-term achievable results to manage this risk. Through the school seismic safety program, KVERMP is providing an example of how to conduct earthquake risk management projects in Nepal. Through Earthquake Day, KVERMP is raising awareness among the general population, in order to build a broad base of support for earthquake risk management activities. Last, through institutionalizing multiple organizations to manage earthquake risk, KVERMP is planting the seeds for a locally driven and sustainable risk management process.

The experience gained in this project is already being applied in nine other cities world-wide through the UN's RADIUS project.

6. References

R. Yadav, et al. "Status of Seismic Hazard and Risk Management in Kathmandu Valley." Issues in Urban Earthquake Risk. Ed. Brian E. Tucker, et al. Dordrecht, The Netherlands: Kluwer Academic, 1994. 183-198.

GeoHazards International, National Society for Earthquake Technology - Nepal. "Kathmandu Valley Earthquake Risk Management Project: Project Proposal". Submitted to Asian Disaster Preparedness Center, June 1997.

THE ROLE OF THE SUBJECTIVE FACTORS IN THE SEISMIC RISK

A. TARVERDYAN
Department of Natural Calamities, Armenia

The territory of the Republic of Armenia is located in a seismo-active zone, which risk is assessed by the objective and subjective factors.

The objective factors are ardent. The power of the earthquakes in the region reach up to seven magnitude; the frequency of the strong earthquakes is 35-40 years.

The subjective factors are numerous: the existence of industrial, energetic, water-economic, civil and other dangerous objects, the density of the population and numerous hearths of geological dangerous events.

The most dangerous role from the earthquake subjective factors plays for the Republic of Armenia the location of reservoirs and tailing dams in the highest points of the mountainous rivers.

Armenia is ranked to the number of countries with the scarcity of water, where the average annual water consumption per capita is 1.2-1.5 thousand m water.

The water resources total to 6.2-6.5 mln. m^3 annually, including subterranean waters of 0.5 mln. m^3.

Taking into consideration the scarcity of the water for the satisfaction of agricultural, industrial, energetic, communal and nature-protection needs of Armenia, there have been constructed 80 reservoirs with 0.3-525 mln.m volume of water, about 1.4mld.m^3 total water volume. The construction of reservoirs was implemented primarily in 1950- 1985 years and during 40- 45 years of their exploitation the complex of hydrotechnical constructions was physically and morally used up and doesn't satisfy to the needs of contemporary norms.

The observations of the experts from the World Bank and local experts showed that the significant part of the dams does not satisfy to the new norms of the seismic marks in the country, and the water output constructions of the maximum water draining are calculated in accordance with the limited water output conditions, which is also unacceptable.

Based on the current needs to raise the security of the reservoirs exploitations and to ensure the safety of the population, the reconstruction works of four large reservoirs of Armenia are included in the 1st stage of the Irrigation Program of the World Bank, the studies are conducted for 22 reservoirs for works to be implemented in the 2nd stage of the Program, for the remaining reservoirs subprograms are being prepared.

In addition to the above mentioned, there are 7 reservoirs in Armenia with the incomplete constructions, which are not passed for exploitation, and the existence of their temporary constructions are potentially dangerous for the population of the country.

S. Balassanian et al. (eds.), Earthquake Hazard and Seismic Risk Reduction, 75–77.

For the safety of the population and security of the social-economic enterprises the following measures are undertaken:

1.The territories and objects of reservoirs overflow are assessed in the case of its reservoirs damage.

2.Activities are elaborated to provide with alarming, control and measuring equipment.

3.Jointly with interested Ministries, annual inspection monitoring is organized and prophylactic activities are undertaken to improve their technical situation.

4.The reservoirs database is created, which is periodically refreshed with the inspection control data.

Despite of the undertaken measures, during the last 5 years two cases of reservoir damages took place on the territory of Armenia, from which the first caused human losses.

Armenia is famous for its mining enterprises and with tailing dams collecting their waste materials, which is also one of the subjective important factors of seismic risk.

Nowadays on the territory of the Republic there are operating and conserved 12 tailing dams with 300 mln.m^3 total volume.

The tailing dams are complex systems of hydrotechnical constructions with 30-100 m. height and 2-170 mln.m^3 volume of washed dams.

Three tailings dams from twelve now are in exploitation, and the remaining from the technological view point are conserved, but they continue to operate as water output and resisting hydrotechnical constructions, which are potentially dangerous for the people of nearby populated areas.

At present the tailing dams of Dastakert and Geghanush in the region of Sunik are in the serious state of wrecking.

The Emergency Management Administration under the Government of the Republic of Armenia jointly with interested Ministries organizes annual control and undertakes prophylactic measures to raise the reliability of the tailing dams' exploitation. But taking into consideration the long period of their operation, they are in technically unsatisfactory situation and require large capital investments for basic reconstruction.

The Government of RA realizing the importance of reservoirs and tailing dams safety issues, many times was referring to their discussion and elaborated certain preventive measures to increase the reliability of their exploitation during emergency situation.

Conclusion

The damages of complex hydrotechnical constructions of reservoirs and tailing dams, caused by earthquakes and other reasons, bring to significant material losses among population and many human losses. In order to prevent the above mentioned consequences it is necessary to:

1. Recalculate the stability of water reservoirs and tailing dams for the established on the territory of Armenia new seismic 9 marks instead of 8.

2. Recalculate high flood water output constructions in accordance with the world norms requirements instead of factual limited calculated water outputs .

3. Reconstruct and provide with the measuring, controlling equipment and mechanize the communication and alarming systems.

4. Increase the responsibilities and technical readiness of the exploiting staff .

MAJOR AND GREAT EARTHQUAKES IN THE HIMALAYAN REGION: AN OVERVIEW

HARSH K. GUPTA
National Geophysical Research Institute, Hyderabad-500007, India

Abstract

The Himalayan portion of the Alpide belt is seismically one of the most active intra-continental region anywhere in the world. Between 1897 and 1952 there was a phase of very high seismic activity when 14 major earthquakes (M≥7.5) including 5 great earthquakes of M≥8 occurred. No major earthquake has occurred since 1952. Re-estimation of the magnitude of the earthquakes in the early part of the century and comparison of various catalogues has confirmed that quiescence of a major earthquake since 1952 is real. There have been efforts to estimate the repeat time of earthquakes in the vicinity of the Himalayan region. However, most of these were statistical and related approaches using earthquake catalogues and information of the historical earthquakes. An effort to date repeat time on the Chedrang Fault, the local of the 1897 Shillong Plateau earthquake of M 8.7 has revealed that in the past at least 4 comparable earthquakes had occurred in this region with a repeat time of about 400-600 years. In another interesting study it has been shown that if the 1905 Kangra earthquake which had claimed 30,000 lives was to repeat today, between 88,000 and 3,44,000 human lives would be lost spending upon the time of the day when the earthquake occurs. This requires adequate preparation and emergency planning to reduce the seismic risk.

Key Words: Alpide Belt, earthquake risk, Himalaya, major earthquakes, repeat time.

1. Introduction

The Himalayan frontal arc constitutes the central part of the Alpide seismic belt. As far as earthquakes are concerned, this is one of the most active intra-continental regions in the world. The northward movement of the Indian plate caused continental collision and created the Himalayan mountain range. The current seismic activity is the result of the continental collision between the Indian and the Eurasian plates. The instrumental recording in the region started towards the end of the 19th Century with the installation of Milne seismograph at Alipore, Calcutta in 1998 and later at Colaba, Bombay and Madras. Table 1 gives earthquakes of magnitude $\geq 7^1/_2$ in the Himalayan region since

S. Balassanian et al. (eds.), Earthquake Hazard and Seismic Risk Reduction, 79–85.

1897. Figure I depicts earthquakes of magnitude > 7 and the ones that claimed human lives at Himalayan frontal arc. This has been updated from Chandra (1992).

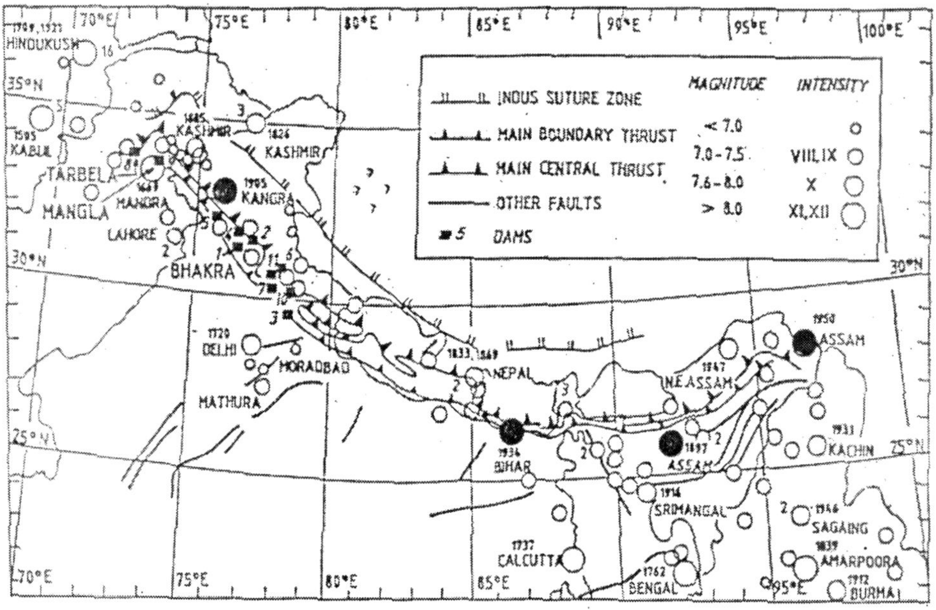

Figure 1. Earthquakes of magnitude greater and equal to 7 as well as earthquakes that claimed human lives in the vicinity of the Himalayan Frontal Arc (updated from Chandra, 1992)

TABLE 1. *Earthquake of M ≥7¹/₂ in the Himalayan region since 1897*

Date	Latitude (N)	Longitude (E)	Location	Magnitude (M)
Junt 12, 1897	25.9	91.8	Assam	8.7
April 4, 1905	33.0	76.0	Kangra Valley	8.6
Dec. 12, 1908	26.5	97.0	N. Burma	7.5
May 23, 1912	21.0	97.0	Burma	8.0
July 8, 1918	24.5	91.0	Assam	7.6
Jan. 27, 1931	25.6	96.0	N. Burma	7.6
Jan. 12, 1934	26.5	86.5	Bihar-Nepal	8.4
May 30, 1935	29.5	66.7	Quetta	7.6
Sept 29, 1947	28.5	94.0	India-Burma	7.9
July 29, 1947	28.5	94.0	N.E. Assam	7.9
Aug. 15, 1950	28.5	96.7	Assam	8.7
Nov. 18, 1951	31.1	91.4	North of India	8.0
Aug. 17, 1952	30.5	91.5	North of India	7.5

2. Quiescence of Major Earthquakes

The earthquake catalogue for Himalaya and North-East India regions (latitude 20°N to 38°N and longitude 75°E to 100°E) comprising of portions of Himalaya and Arakan Yoma fold belt for the period 1997 to present shows that there have been a total of 14 major earthquakes (M≥7.5) during the period 1897 to 1952. This includes 5 great earthquakes exceeding magnitude 8, i.e Shillong Plateau earthquake of 1897, Kangra earthquake of 1905, Bihar-Nepal earthquake of 1934 and India-China earthquake of 1950 and its aftershock of M 8 of November 1951. No major earthquake of magnitude 7.5 or larger has occurred since 1952. There was an apprehension that possibly during the first half of the current century the earthquake magnitudes were over-estimated in the Himalayan region. To verify the same, we took up a detailed study, where we re-determined the magnitudes of 37 earthquakes (M≥7) in the Himalayan region for the period 1903-1985 (Gupta et al 1995). For these earthquakes, the micro films of Gottingen Observatory seismograms were used and the magnitude values were estimated using P-, S- and surface waves from long- and medium-period instruments. The redetermined magnitudes of the Himalayan earthquakes were within the error limits of observations of the earlier magnitude values and confirmed that indeed no rnajor earthquake of M≥7.5 has occurred since 1952. In Table 2 the distribution of the earthquakes in different magnitude ranges for the periods 1897 to 1952 and 19S3 through 1999 is given. (updated from Satyabala and Gupta, 1996).

TABLE 2. *Number of earthquakes in different magnitude ranges during the period 1897- 1999 in the Himalayan region*

Magnitude	1897-1952	1953-1999
M≥7.5	14	0
7.5>M≥7.0	11	7
7.0>M≥6.5	19	20

An examination of Table 2 shows that while major earthquakes of M > 7.5 have not occurred since 1952, there are comparable number of earthquakes in 6.5 to 7.4 magnitude ranges, indicating that indeed there is a quiescence of major earthquakes in the Himalayan region. This quiescence was further confirmed by comparing several other catalogues. For example, those of Abe (1981, 1984 and 1994) and Pacheco and Sykes (1992).

3. Return Period of Great Earthquakes

There has been a sincere effort to estimate the repeat time of earthquakes in the vicinity of the Himalayan region. One paper which needs to be mentioned is that of Seeber and Armbruster (1981) where from a detailed study of spatio temporal distribution of earthquakes in the Himalayan region they have estimated that the entire Himalayan arc ruptures in about 180 to 240 years and the repeat time of a typical magnitude 8 earthquake with a rupture length of 300 km is 200 to 270 years. However, such efforts are hampered by the fact that our earthquake catalogues for the Himalayan region are not very reliable and in any case the instrumental data is available only for the period of last 100 years, i.e. from 1898 till present. Over the past few decades a sincere effort in different parts of the world has been made to identify, characterize and date the deformed sedimentary structures caused by the past violent earthquakes and thereby estimating the time of occurrence of past earthquakes (Sieh 1978, Yeats et al 1997 and McCalpin 1996). Sukhija et al (1999) have reported the result of paleo liquefaction evidence on the periodicity of large pre-historic earthquake in Shillong Plateau. They conducted very systematic studies of collecting samples in several trenches opened in Krishnai and Dudhani rivers close to the Chedrang fault in Shillong Plateau. Chedrang fault is known to have been associated with the great 1897 Shillong Plateau earthquake. They found very well preserved liquefaction and deformed syndepositional features at 10 selected sites in the alluvial deposits along these rivers. Using C^{14} dating method they have provided evidence of events which occurred in the past. In addition to 1897 event they found two more events that occurred during 1450 to l650 and 700 to 1050 A.D. The third event pre-dates 600 A.D. They infer a return period of 400 to 600 years for large earthquakes in the Shillong Plateau region.

This finding indicates that earthquakes of the type of 1897 in the Shillong Plateau may repeat only after around 400 to 600 years. However, what is required is to make similar estimates at several other sites where great earthquakes are known to have occurred such as the 1934 Bihar-Nepal earthquake and the 1905 Kangra earthquake.

4. A Medium Term Earthquake Forecast in the North- East India

There is a global effort to establish precursor patterns that precede major' earthquakes. An effort in this direction was made by Gupta and Singh (1986, 1989) in the North-East India region. Encouraged by the discovery of precursory swarm and quiescence preceding the Cachar earthquake of December 30, 1984, Gupta and Singh (1986, 1989) carried out an indepth study of all the earthquakes of magnitude $\geq 7^1/_2$ since 1987 and several smaller magnitude earthquakes that occurred after 1962. In their study they discovered that the main shock magnitude (Mm) has correspondence with the magnitude of the largest events (Mp) in the swarm and the time interval (Tp) between the onset of swarm and the occurrence of main shock in days. Following are the relations found by them:

$Mm = 1.37\ Mp - 1.41$ and

$Mm = 3 \log Tp - 3.27$

Gupta and Singh (1986) observed that it is important to recognize swarm and quiescence before the occurrence of the main shock. They discovered one such region in the vicinity of Indo-Burma border and concluded that – "(1) Moderate magnitude to great earthquakes in the North-East India region are found to be preceded, generally, by well defined earthquake swarms and quiescence periods, (2) On the basis of an earthquake swarm and quiescence period, an area bound by 21°N and $25^1/_2$°N latitude and 93°E longitude is identified to be the site of a possible future earthquake of M $8\pm^1/_2$ with a focal depth of 100±40 km. This earthquake should occur any time from now onwards. Should it not occur till the end of 1990, this forecast could be considered as a false alarm."

The occurrence of August 6, 1988 earthquake with focal parameters mentioned in Table 3 has made this medium term forecast come true. This success encourages one to make similar investigations elsewhere in the Himalayan Frontal Arc for concentrating hazard-related investigations in a few critical areas.

TABLE 3. *Forecast of August 6, 1988 earthquake*

Earthquake Parameters	Forecast (Gupta and Singh 1986)	Occurrence NEIS (Preliminary Determination)
Epicenter	21°N to 25$^1/_2$°N 93°E to 96°E	25.149°N 95.127°E
Magnitude	8±$^1/_2$	7.3
Depth	100±40 km	90.5 km
Time	February 1986 to December, 1986	August 6, 1988 (00.36.26.9G.C.T.)

5. Future E: How Damaging Will They Be?

There has been a phenomenal increase in the population density in the foothills of the Himalaya during the last several decades. Therefore, the number of people likely to be affected by great earthquakes has increased considerably as the construction continues to be of poor quality. Arya (1992) has made an estimate of damage scenario in the Kangra region if the 1905 great earthquake was to repeat today. He presumed that the distribution of earthquake intensity will be similar to the 1905 earthquake putting areas of 500 sq. km., 2,400 sq. km., 5000 sq. km. and 26,000 sq. km. under intensities X, IX, VIII, and VII respectively on the Modified Mercalli Intensity Scale. Keeping in view the building material and construction, it is estimated that a total of about 1,4S,400 houses will collapse completely and another 2,67,800 houses will suffer severe damage. Arya (1992) estimates that depending upon the time of the day when the earthquake occurs, the loss of human lives would vary between 88,000 and 3,44,000 (Table 4). He has underlined the importance of retrofitting to important buildings to reduce the hazard.

TABLE 4. *Estimates of human lives likely to be lost if the Kangra earthquake of 1905, which claimed 20"000 lives, was to occur today (Arya, 1992).*

Time of Occurrence	Deaths in Collapsed Houses	Deaths in Part-Collapsed Houses	Total Potential Deaths
Midnight (Sleeping)	40%	20%	3,44,00
Morning (Awake and Sleeping)	20%	10%	1,77,000
Noon Time (out working)	10%	5%	88,000

6. Concluding Remarks

From the foregoing it is obvious that major and great earthquakes have occurred in the Himalayan region and will continue to occur. However, the entire region has become much more vulnerable due to increase in population and vulnerability of structures. This is obvious from the kind of damage which the Himalayan regions suffered during the 1991 Uttarkashi earthquake. The Uttarkashi earthquake of October 20, 1991, although of only magnitude 6.8 claimed an estimated 2000 human lives and became the most significant earthquake in the year 1991. Occurrence of a bigger earthquake will create much more problem. What is necessary is to educate the public on what needs to be done during and after an earthquake as well as teach them methods of simple retrofitting of non-engineered structures so that damage due to earthquakes should be reduced considerably. This all needs to be done in a short time frame so that people are prepared to reduce the hazard of a major earthquake.

7. References

1. Abe, K., 1981, Magnitudes of large shallow earthquakes from 1904 to 1980, Phys. Earth Planet. Interiors, 27, pp. 72-92.
2. Abe, K., 1984, Complements to magnitudes of large shallow earthquakes from 1904 to 1980, Phys. Earth Planet. Interiors, 34, 17-23.
3. Abe, K., 1994, Instrumental~magnitudes of historical earthquakes, Bull.Seism.Soc.Am., 84, 415-425.
4. Arya, A.S., 1992, Possible effects of a major earthquake in Kangra region of Himachal Pradesh, Current Science, Special Issue, 62, No. 1 k 2, pp. 25 l-256.
5. Chandra, U., 1992, Seismotectonics of Himalaya, Current Science, Special Issue, 62, No. 1 A 2, pp. 40-71.
6. Gupta, H.K., Khanal, K.N., Upadhyay, S.K., Sarkar, D., Rastogi, 8.K. and Duda, S.J., 1995, Verification of magnitudes of Himalayan region earthquakes of 1903-1985 from Goettingen Observatory, Tectonophysics, 244, pp. 267-284.
7. Gupta, H.K. and Singh, H.N., 1986, Seismicity of the north-east India Region, Part II: Earthquake swarms precursory to moderate magnitude to great earthquakes, J. Geol. Soc. India, 28, No. 5, pp. 367-406.
8. Gupta, H.K. and Singh, H.N., 1989, Earthquake swarms precursory to moderate to great earthquakes in the northeast India region, Tectonophysics, 167, pp. 285-298.
9. McCalpin, J.P. (Ed.), 1996, Paleoseismology, Academic Press, London, S88 pp.
10. Pacheco, J.F. and Sykcs, L.R. (1992), Seismic moment catalog of large shallow earthquakes, 1900-1989, Bull.Seism. Soc.Am, 82, 1306-1347.
11. Satyabala, S.P. and Gupta, H.K., 1996, I s the quiescence of major earthquakes ($M \geq 7.5$) since 1952 in the Himalaya and Northeast India real?, Bull.of Seis. Soc. Am., 86, No. 6, pp. 1983-1986.
12. Seeber, L. and Armbruster, John G., 1981, Great detachment earthquakes along the Himalayan arc and long-term forecasting, AGU publication, Earthquake Prediction, An international review, Maurice Ewing Series 4, pp. 259-277.

13. Sieh, K.E., 1978, Prehistoric large earthquakes produced by slip on the Andreas fault at Pallett Creek, California, J. Geophys.Res., 83, pp. 3907-3939.
14. Sukhija, B.S., Rao, M.N., Reddy, D.V., Nagabhushanam, P., Hussain, S., Chadha, R.K., Gupta, H.K., 1999, Paleoliquefaction evidence and periodicity of large prehistoric earthquakes in Shillong Plateau, India, Earth and Planetary Science Letters, 167, pp. 269-282.
15. Yeats, R.S., Sieh, K.E., Allen, C.R., 1997, The geology of earthquakes, Oxford University Press, New York 568 pp.

SEISMICITY OF THE SPITAK EARTHQUAKE SOURCE ZONE VICINITY

R.E.TATEVOSSIAN(1), S.S.AREFIEV(1), H.HAESSLER(2)
(1) United Institute of Physics of the Earth, RAS
Moscow, 123810, ul.B.Gruzinskaya, 10.
(2) Institut de Physique du Globe
5, rue Rene Descartes, 67084, Strasbourg, France

1. Introduction

Spitak earthquake of December 7, 1988 (M=6.9), is an outstanding seismic event of the region. Its epicentral area was considered incapable to produce an earthquake with M≥6.0 according to Seismic Zoning Map of the USSR Territory [1]. Immediately after the earthquake epicentral expedition was organized [2]. There are several publications on Spitak earthquake source zone [3-10]. Seismic history of the area has been discussed [11]; two paleoearthquakes with M≈7 are described there. But the study of seismic history still is incomplete because these publications did not used regional instrumental catalogues. Our study is devoted to that very problem: study of seismic regime of the Spitak earthquake source vicinity based on the regional scale instrumental catalogue.

2. Method

Seismicity analysis includes separation of spatial distribution of earthquakes into dispersed and concentrated components, based on the cluster analysis approaches using method of averages [12]. Cluster analysis solution is presented by the coordinates of cluster centers with the list of earthquakes belonging to each cluster. The procedure is iterative, it starts with distribution of trial cluster centers. Then each earthquake is ascribed to the nearest cluster center. Average coordinates (center) of the events belonging to the given center have to be determined; this center will be accepted as trial center for the next step. Before going to the next iteration step distances between all cluster centers are checked: if two clusters are closer to each other than some fixed distance these clusters have to be merged into one. This fixed distance is called separation coefficient (DR). For each DR value we have a set of cluster solutions. Iteration is stopped when all initial clusters are merged into one. As integrative descriptor of the solution average cluster radius (R_{av}) is used. The character of dependence of R_{av} versus DR is used to make a choice of optimal cluster solution from successive iterative steps. After the choosing of cluster solution clusters belonging to the concentrated and dispersed components of the seismicity have to be identified. This identification is based on the critical number of events K in cluster: clusters with larger

S. Balassanian et al. (eds.), Earthquake Hazard and Seismic Risk Reduction, 87–97.

number of events, than K, are identified as concentrated component. The K value is accepted based on the character of distribution of number of earthquakes in clusters.

3. Initial data

The basic data is the catalogue of the Caucasian Regional Centre (CRC). It includes 4267 earthquakes with magnitudes more than 1.9 starting from 1962 up to the mainshock (07.12.88) in spatial frames 40.3°N – 41.7°N and 43°E - 45°E. Epicentral map of this catalogue is presented in Figure 1,a. Thin line shows boundaries of Javakhety highland according to the regionalization in [13], hatched contour indicates its geological frames. Epicentral area is within the solid rectangle. Surface ruptures (see near town of Spitak) are shown according to [14, 15]. Most of the earthquakes, especially at the beginning of the catalogue are given in it with rounding of coordinates to 0.1°. This creates false grids in epicentral map. To avoid this we add random error with maximum value 0.025° (which is less than mean error value in epicentre location. Corresponding map is shown in Figure 1,b – it gives some idea concerning possible distortions of the real seismicity distribution created by the data format in the catalogue.

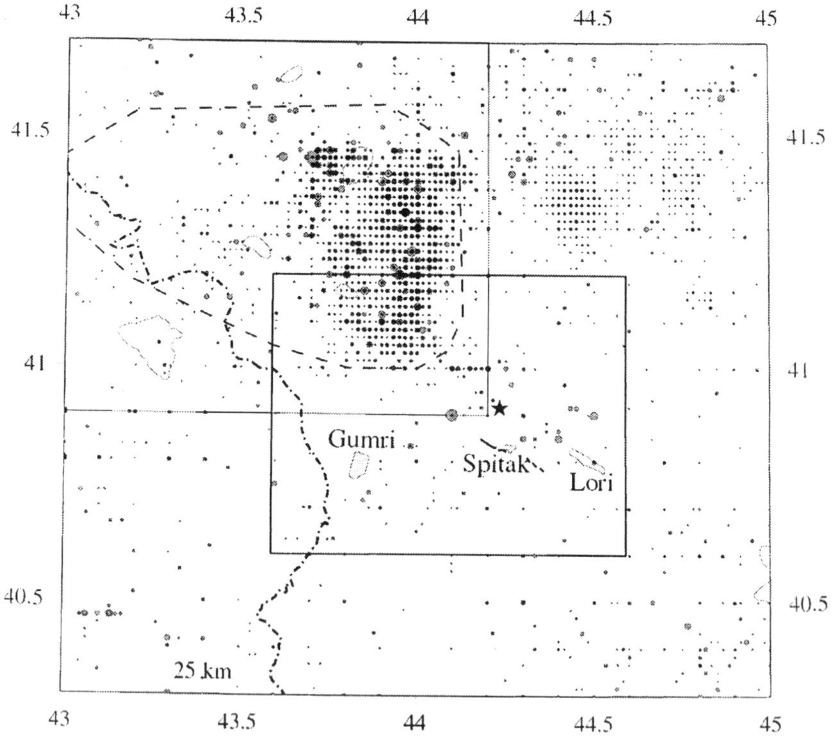

Figure 1,a. Epicentral map of the CRC catalogue. Asterisk is for mainshock. Symbol size correlates with earthquake magnitude. Explanations of indicated sub-regions see in the text.

Because there were no aftershocks registered to the south from surface rupture trace [3, 4], as well as relatively strong earthquakes in historical [16, 17] and recent times [18], we have checked if the events mentioned there in CRC catalogue could be erroneously included explosions (this is region of quarries). Unfortunately, we could not apply spectral methods of discrimination of earthquakes, as their records were not available. But we used some non-direct considerations. First of all, 38 from 50 events

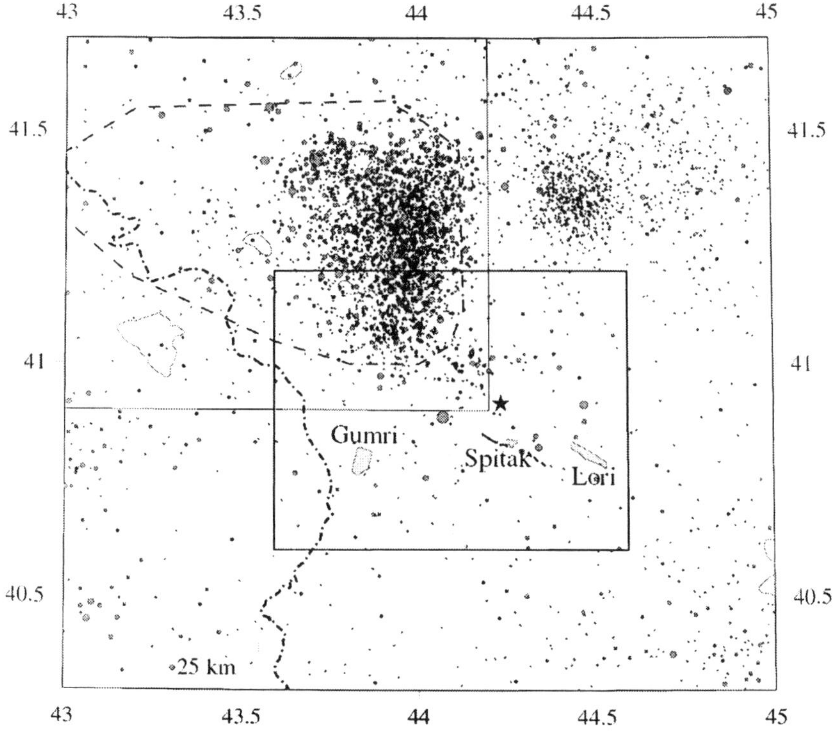

Figure 1,b. The same as in 1a, but with random error added to the coordinates.

in this occurred in daytime (during working hours). To all of them 5 km depth is ascribed. Then, all of them have the same magnitude 2.2, except one event with magnitude 2.5 (formally, b-value for this set is 1.53). So, there are grounds to suspect inclusion of records of fake quakes in the catalogue. Most probably these are events with M≈2.2.

Representavity of data in catalogue differs strongly in space and time. According to corresponding annuals Earthquakes in the USSR [18] in 1962 earthquakes with M≥2.2 were recorded without omissions in Javakhety highland, for Armenian highland this threshold value was 3.3; in 1988, the corresponding thresholds were 1.7 and 2.2. So, to avoid distortions related with representavity level and possible presence of explosions in the catalogue, events with magnitude less than 3.0 were cut out from the catalogue for the further analysis of seismicity. For Javakhety highland this threshold value was fixed at M=2.5.

4. Results

Epicentral map of earthquakes with M≥3.0 is shown in Figure 2. The preliminary analysis of initial data suggests that this map is representative for the whole observation time interval and spatial frames. The boundary between northern and southern wings of surface ruptures is very well pronounced here. The very active spot of epicentres, which was marked in Figures 1a,b to the east from Javakhety highland, disappeared. The most active zone adjacent to the epicentral area is Javakhety highland.

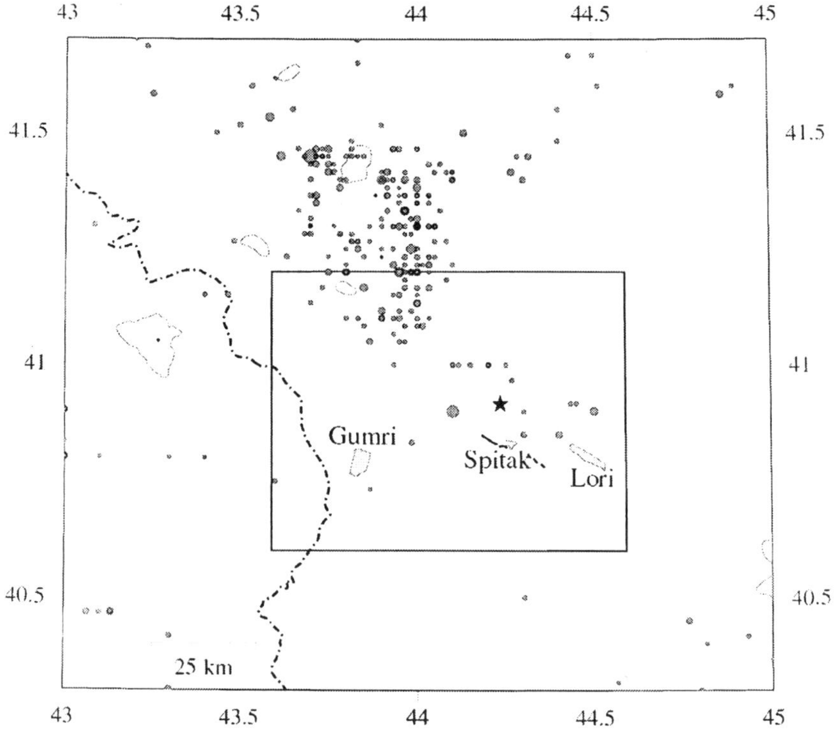

Figure 2. Epicentres of earthquakes with M≥3.0.

Some information on recurrence time of large earthquakes (M≈7) in the epicentral area could be derived from following facts. There are four moderate earthquakes (M = 4-5.5) in epicentral area in the instrumental catalogue. There is also one more moderate shock in 1916 [16]. During all the rest period of observations (27 years of instrumental catalogue + episodic data from historical catalogue covering approximately last 150 years) the zone was practically aseismic. Meanwhile, there are two large paleoearthquakes with M = 6.5-7.5 reported in [11]. So, the epicentral area is characterized by sporadic seismicity, with very large recurrence time. Duration of instrumental catalogue (even being prolonged with historical data) is nearly 3 orders smaller than recurrence time of large events. This leaves no chances to find out any statistically meaningful regularity in time domain. Taking into account that aftershocks of Spitak earthquake occurred in Javakhety highland (Figure 3) a possibility of some seismotectonic relationships between these regions were supposed. To reveal this possible interaction, an analysis of seismicity of Javakhety highland was carried out.

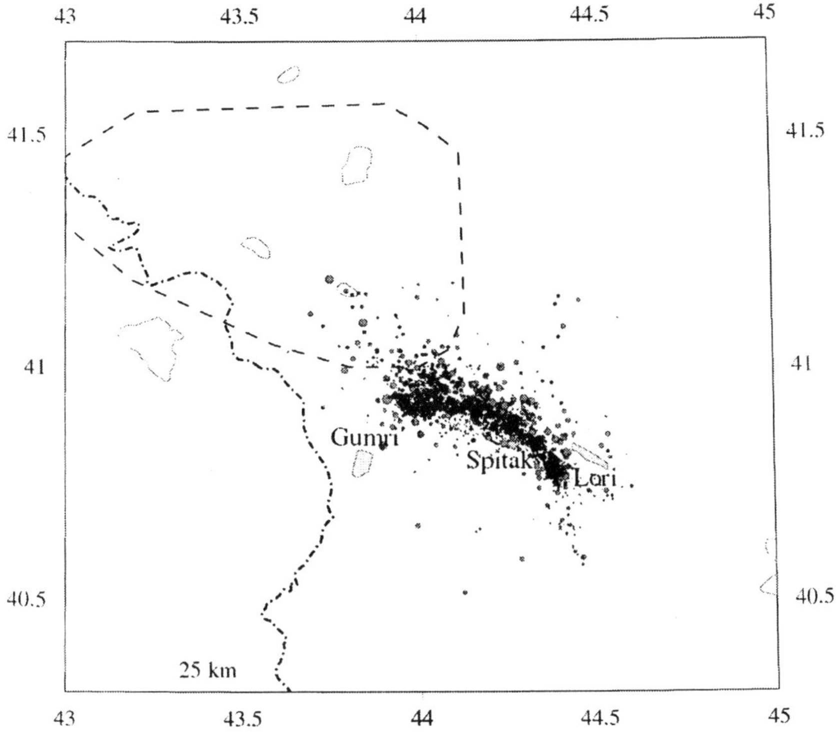

Figure 3. Epicentre map Spitak earthquake aftershocks.

The epicentre map of earthquakes with M≥4.2 is shown in Figure 4. Fractionality value (b-value) is mapped (details see below). Though, Javakhety highland is seismically very active, there are no really large earthquakes known there. The largest recorded magnitude in Javakhety highland is 5.7, to the largest historical event (in December 31, 1899) magnitude 6.3 is ascribed, so we will consider event above magnitude threshold 4.2 as large earthquake in Javakhety highland.

Above described method of cluster analysis was applied to the catalogue in spatial frames of Javakhety highland. Average radius (R_{av}) versus separation coefficient (DR) is plotted in Figure 5. Initial value of separation coefficient is 2.5 km; it increases with 2.5 km step. There is a flat segment on graph for separation coefficient interval 15 – 20 km. R_{av} for this DR interval is equal 6 km. As a cluster solution was chosen the one for DR = 15 km. This solution consists from 16 clusters. Number of earthquakes in clusters is presented in Figure 6. The smallest cluster (R_{av} = 6.9 km) contains maximum events (193). Conventionally, three groups of clusters could be distinguished: 1) with 1 – 25 events; 2) with 96 – 126 events, and 3) from 189 – 193 events. According to this

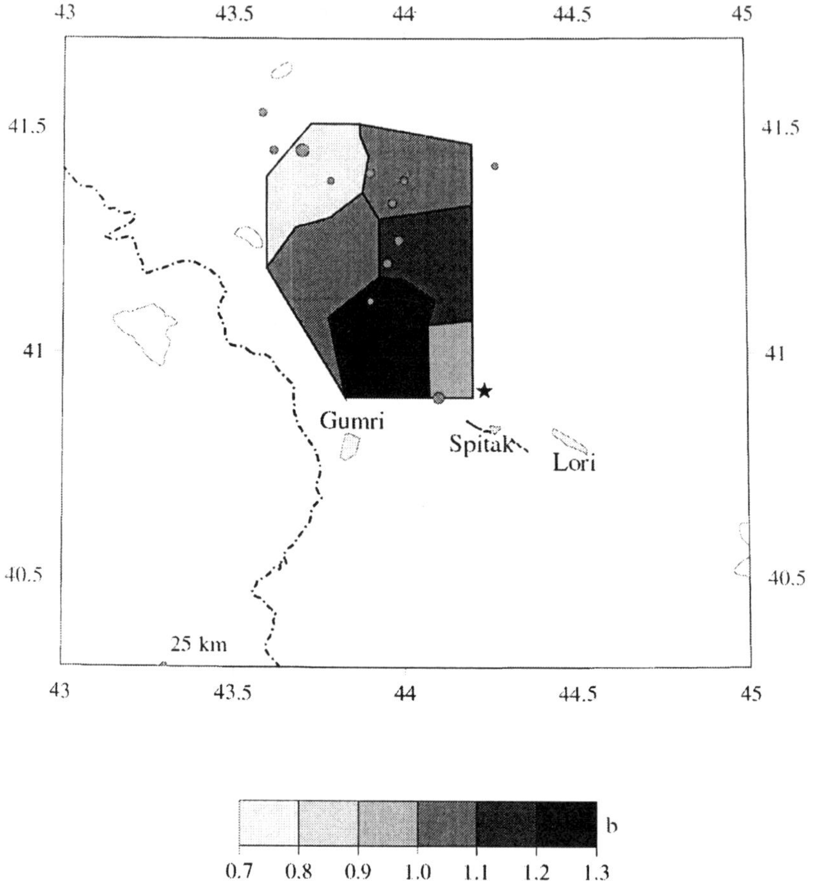

Figure 4. Epicentre map of earthquakes with M≥4.2. b-values of clusters are mapped (see text).

distribution we identify as dispersed seismicity clusters containing 1 – 25 earthquakes. Then only 6 clusters from 16 will compound concentrated component and they will include 724 earthquakes from 789 (which means that only 8% of events belongs to the dispersed component). For comparison, it could be mentioned that in the Caucasus the dispersed component can comprise 30% [19]. Dispersed component occupies more than the half of the territory of Javakhety. Epicentre maps for the whole catalogue and different components of seismicity is shown in Figure 7.

b-value for the dispersed component is 0.90, for concentrated one is 1.03. Fractionality (b-value) of each cluster of concentrated component was calculated, map is presented in Figure 4. Southeastern part of Javakhety highland, which borders with Spitak earthquake epicentral zone is characterized with b-value 0.90. Adjacent to it are two clusters with very high fractionality (1.21 and 1.08), farther to the northwest there are clusters with "normal" fractionality (0.99 and 1.01), the lowest b-value is at northwestern extremity of the concentrated component – 0.81. It is interesting that in zones of highest fractionality there were no earthquakes with M≥5.0 not in time-

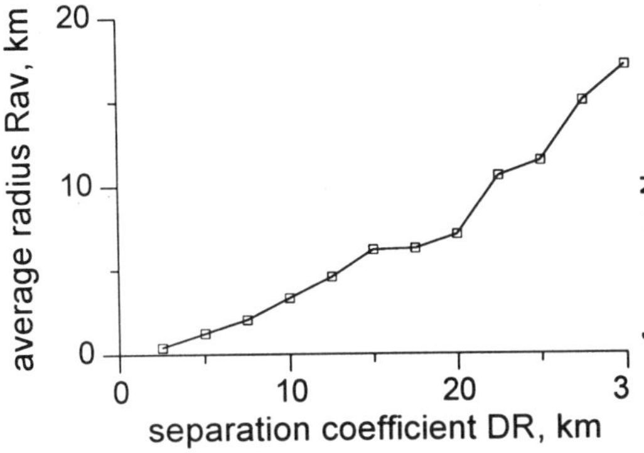

Figure 5. Average radius of clusters R_{av} versus separation coefficient DR.

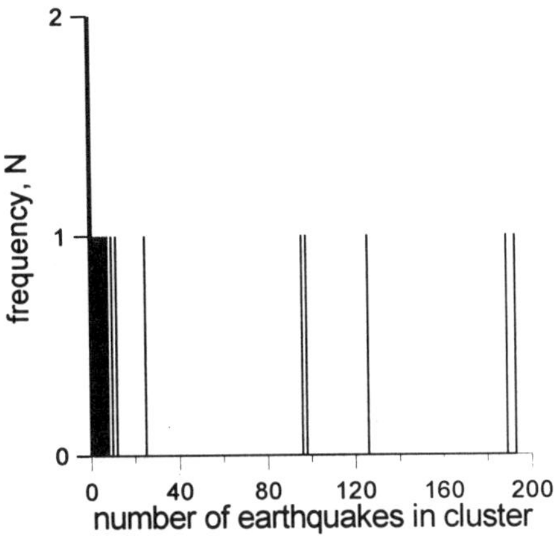

Figure 6. Distribution of clusters by number of events in it for separation coefficient DR=15 km.

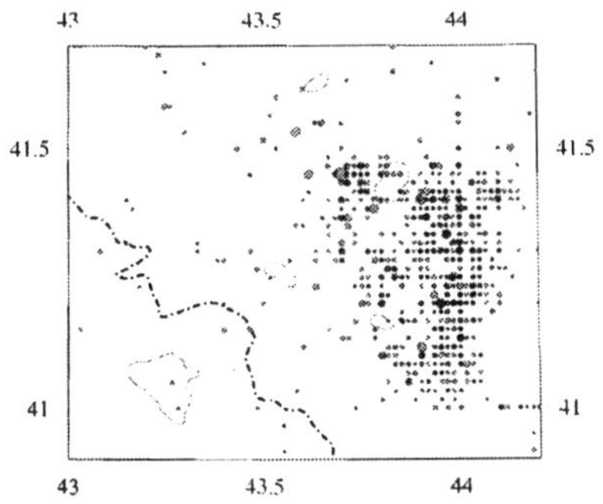

Figure 7a. Epicentre map of Javakhety highland (M≥2.5)

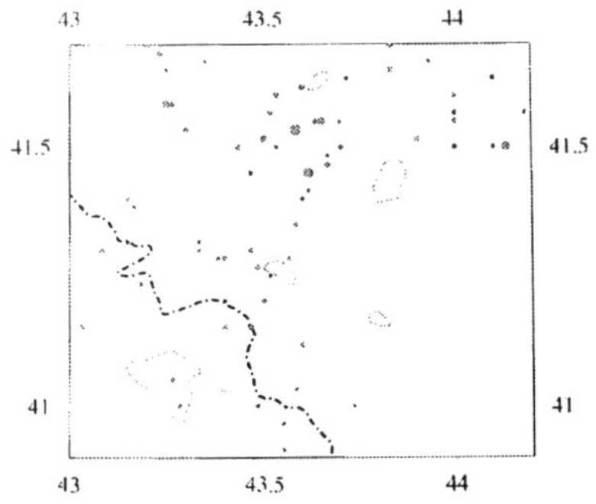

Figure 7b. Dispersed component

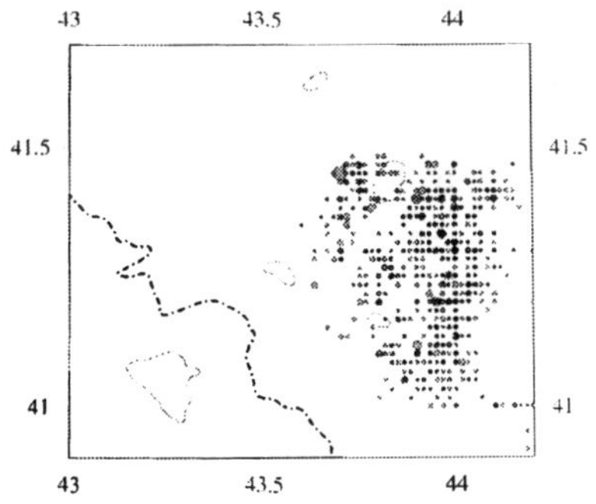

Figure 7c. Concentrated component

interval 1962-1988 (for which interval b-value was calculated), neither earlier or later. So, there are grounds to consider the regionalization based on cluster analysis of spatial distribution of earthquakes reflects some principal seismotectonic features. The Spitak epicentral zone is separated from main part of Javakhety highland with high fractionality belt.

To reveal the possible seismotectonic relationships between Javakhety highland and Spitak zone activity and fractionality time-variations in Javakhety were studied. These variations are plotted in Figure 8. Only concentrated component was used in calculations. Each b-value on graph is calculated for 100 event samples with overlapping of 50 events. Depending on the seismic activity the time duration of each sample set is different: from 3 to 19 months. There is a short-term quiescence before Spitak earthquake marked in activity (Figure 8a), which observed after Paravani earthquake of 1986. This relative quiescence accompanies with decreasing of fractionality (Figure 8b), i.e. with deficit of weak shocks. This time-behaviour of parameters before large earthquake is not typical [20]. If we exclude from the analysis the Paravani cluster, which is the farthermost one from the Spitak zone, then the fractionality time behaviour will display well-known bay-shape form (Figures 8c). At the same time, duration of quiescence will increase up to the 5 years (Figures 8d). So exclusion of distant Paravani zone leads to often observed time variations of seismic parameters before large shock. Calculation of presented time-variations of activity and fractionality includes several steps (cluster analysis, sample set volumes, overlapping interval, etc.), so the result depends on a certain amount of adjustable parameters. It was decided to check the results with the simplest approach – direct comparison of origin times of large earthquakes in Javakhety and the Spitak zones (Figure 9). Bold lines indicate earthquakes in Javakhety highland, hatched – in the Spitak zone. 14 earthquakes from 17 occurred in groups from 2 to 16 months long. Total duration of groups is 4 years while the whole time interval is 27 years. All three isolated shocks are from Javakhety highland (Paravani zone). No any earthquake in the Spitak zone occurred out of the group with Javakhety earthquakes; they can start the group or be within it, they can have largest or smallest magnitude in a group.

Grouping of large earthquakes in Javakhety highland and Spitak zone together with time-variations of some seismicity parameters indicate a certain synchronization of seismicity in both regions. Earthquake sequence in groups is such, that it is impossible to identify activity in which region provokes the other. Some indirect evidence of more independence of Javakhety highland could be associated with the fact that some earthquakes there happened isolately, while all large earthquakes in Spitak zone are in groups with Javakhety earthquakes. Very possible, that the provoking source is outside of both zones and related with more global tectonic processes, for which both Javakhety highland and Spitak zone are peripheral regions [21].

5. Conclusions

Though seismicity of Spitak zone and Javakhety highland differs significantly by largest recorded magnitude, recurrence times, etc. there is a certain synchronisation in seismicity between them. Most probably that the source which provokes activity in these regions is governed by global seismotectonic interaction.

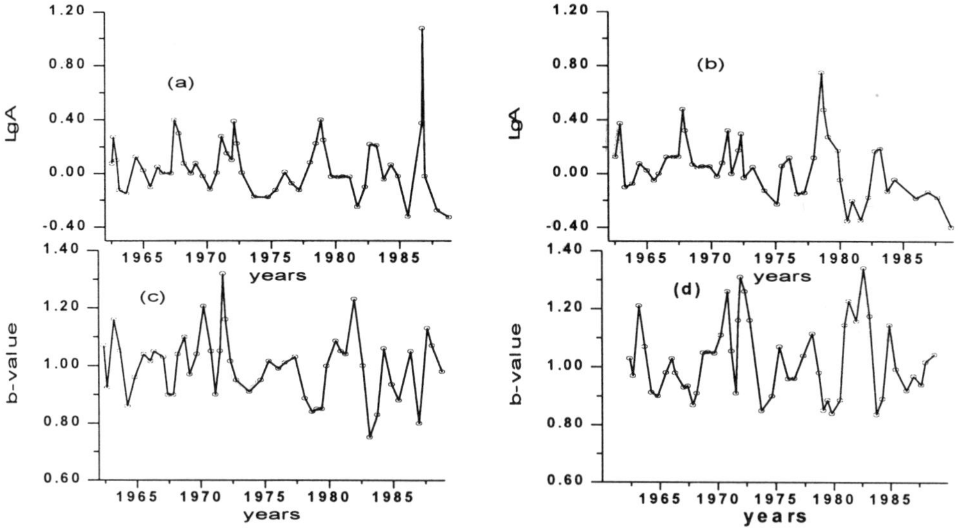

Figure 8. Activity time-variations for the whole region (a), the same, excluding Paravani region (b); fractionality time-variations for the whole region (c), the same, excluding Paravani region (d)

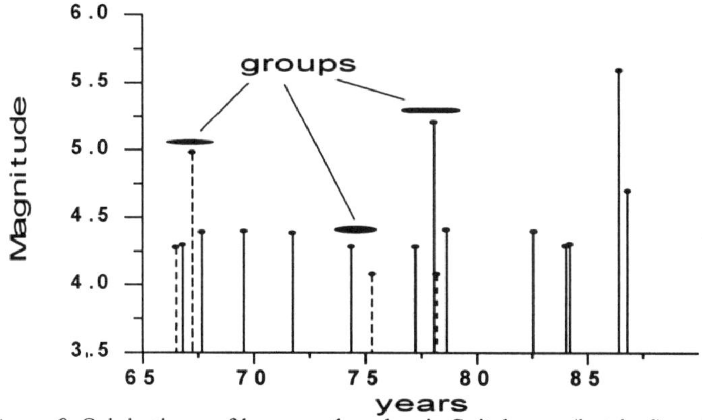

Figure 9. Origin times of large earthquakes in Spitak area (hatched) and in Javakhety (solid bars).

Acknowledgments

This work was partly supported by the RFBR grant 99-05-65418 (Russia).

References

1. Seysmicheskoe rayonirovanie territorii SSSR. (1980). Moskva. (*in Russian*).

2. Arefiev S.S., Aptekman J.Ya., Afimina T.V., et al. (1991) Katalog aftershokov Spitakskogo zemletryaseniya 7 dekabrya 1988 g. *Fizika Zemli*, **11**, 60-73. (*in Russian*).

3. Dorbath L., Dorbath C., Rivera L., et al. (1992) Geometry, segmentation and stress regime of the Spitak (Armenia) earthquake from the analysis of the aftershock sequence. *Geophys. Journal Internat.*, **108**, 309-328.

4. Arefiev S.S., Tatevossian R.E. (1991) Struktura I seismicheskiy rezhim ochagovoi zony Spitakskogo zemletryaseniya. *Fizika Zemli*, **11**, 74-85. (*in Russian*).

5. Dorbath C., Dorbath L., Aptekman J.Ya. (1991) Geometriya ochaga Spitakskogo zemletryaseniya po dannym detalnogo izucheniya aftershokov. *Fizika Zemli*, **11**, 86-95. (*in Russian*).

6. Aptekman J.Ya., Lander A.V., Dorbath C., Dorbath L. (1991) Protsessy v ochagovoi zone Spitakskogo zemletryaseniya po dannym o mekhanizmakh ochagov ego aftershokov. *Fizika Zemli*, **11**, 96-105. (*in Russian*).

7. Graizer V.M., Erteleva O.O., Cistersnas A. et al. (1991) Modelirovanie ochaga Spitakskogo zemletryaseniya 7 dekabrya 1988 g. *Fizika Zemli*, **12**, 46-55. (*in Russian*).

8. Duda S., Fastoff S., Kaizer D. (1991) Spitakskoe zemletryasenie 7 dekabrya 1988 g.: parametry ochaga po shirokopolosnym zapisyam Tsentralnoi seismologicheskoi observatorii FRG. . *Fizika Zemli*, **11**, 27-38. (*in Russian*).

9. Shebalin N.V., Dorbath C., Tatevossian R.E. (1991) Razmery, forma i polozhenie ochaga Spitakskogo zemletryaseniya: aftershoki i makroseismika. *Fizika Zemli*, **11**, 18-26. (*in Russian*).

10. Treusov A.V., Arefiev S.S., Borissoff B.A. (1993) Tomograficheskoe issledovanie zony Spitakskogo zemletryaseniya. *Fizika Zemli*, **5**, 23-27. (*in Russian*).

11. Albarede Ph., Bogachkin B.M., Oleynik A.E. (1990) O doistoricheskom silnom zemletryasenii v rayone g.Spitaka (Severnaya Armeniya). *Dokl. AN SSSR*, 3**13**, 397-400. (*in Russian*).

12. Mandel I.D. (1988) *Klasterniy analiz*. Moskva.

13. Stroenie, napryazhenno-deformirovannoe sostoyanie i usloviya seismichnosti litosfery Malogo Kavkaza. (1983). Tbilisi. (*in Russia*).

14. Rogozhin E.A., Rybakov L.N., Borissoff B.A. (1990) Seismodislokatsii Spitakskogo zemletryaseniya. *Dokl. AN SSSR*, **311**, 435-439. (*in Russian*).

15. Rogozhin E.A., Philipp H. (1991) Geologo-tektonocheskoe izuchenie ochagovoi zony Spitakskogo zemletryaseniya. *Fizika Zemli*, **11**, 3-19. (*in Russian*).

16. Kondorskaya N.V., Shebalin N.V. (1982) *New catalog of Strong earthquakes in the USSR from ancient times through 1977*. Boulder.

17. Nikonov A.A. (1991) Razrushitelnye zemletryaseniya v epitsentralnoi zone Spitakskogo zemletryaseniya 1988 g. v proshlom. *Fizika Zemli*, **12**, 3-16. (*in Russian*).

18. Zemletryaseniya v SSSR. (Annuals since 1962). Moskva.

19. Shebalin N.V., Arefiev S.S., Vasiliev V.Yu., Tatevossian R.E. (1991) Ot seismichnosti ploshchadei k structure seismichnosti. *Fizika Zemli*, **9**, 20-28. (*in Russian*).

20. Arefiev S.S., Tatevossian R.E., Shebalin N.V. (1989). Some features of the seismicity prior to strong aftershocks. ESC proceedings, Sofia. 524-532
21. Philip H., Cisternas A., Gvishiani A., Gorshkov A. (1989) The Caucasus: an actual example of the initial stages of continental collision. Tectonophysics, **161**, 1-21.

SOUTHERN JAVAKHET EARTHQUAKE SWARM. STRESS TENSOR ORIENTATION DERIVED FROM FOCAL MECHANISM DATA

A. K. TOVMASYAN
National Survey for Seismic Protection
Davidashen, 4-th massif, Yerevan, Armenia
ashkhen @nssp.r.am

Abstract

The earthquake swarm which occurred in 1996, in the period from May 29 to June 9, was recorded by both analog and digital stations and more than 120 small earthquakes (M< 2.5) were recorded by Armenian and Georgian networks during that time. Focal mechanisms for 19 relatively strong events were determined using first polarities of direct P waves. The variety of methods was used to analyze focal mechanism solutions. Based on obtained results one may conclude that these earthquakes are not related to well known and mapped active faults. Using Gepart and Forsuth's improved method the regional stress tensor was determined.

We have analyzed in detail a sequence of earthquakes in Javakhet Upland and find some peculiarities of that. A detailed study of the earthquake catalogue showed that this swarm as some of preceding ones, was followed by relatively strong earthquake in Iran and Turkey

1. Introduction

Investigation of stress field and determination of regional stress directions are the important problems when assessing seismic hazard. In solving these problems, by means of seismological methods based on focal mechanism's solution due to various reasons, only the data obtained from strong earthquakes were used. Today is already well known that weak seismicity study is very important, especially of weak earthquakes focal mechanisms. Weak local events should reflect the small scale tectonics and their mechanisms have the potential to inform us even at an early stage of the rock fracturing and the fault development. Hence, it contains a lot of useful information of strong earthquake preparation process.

This paper attempts to use the Gephart's method for stress tensor determination based on focal mechanism data for the Javakhet (southern Georgia and northern Armenia) 1996 earthquake swarm. These earthquakes were with 1.5 <M<4.2 .

Earthquake swarm is a type of earthquake sequence characterized by a long series of large and small shocks, with no one outstanding, principal event (Lee and Stewart 1981). Swarms are sequences of earthquakes that are clustered in space and time and are not associated with an identifiable mainshock (Torne Lay Terry C. Wallace1995). They often start and end gradually (Christopher H. Scholz1994). Swarms most commonly occur in volcanic regions because earthquakes associated with volcanic activity often occur in swarms. But swarms also occur in many non- volcanic regions (Bullen & Bolt,

S. Balassanian et al. (eds.), Earthquake Hazard and Seismic Risk Reduction, 99–107.

1993), or in regions of geologically young volcanism thought it is not a contemporary one. Among these regions is Javakhet volcanic massif, which is the most seismically active zone in the Lesser Caucasus where earthquakes occur permanently.

The Javakhet Upland is situated at a very particular point: in the collision zone (figure 1a) between Arabian and Eurasian plates and directly north of the zone where two pairs of faults with a different type of faulting, joint together (figure 1b).

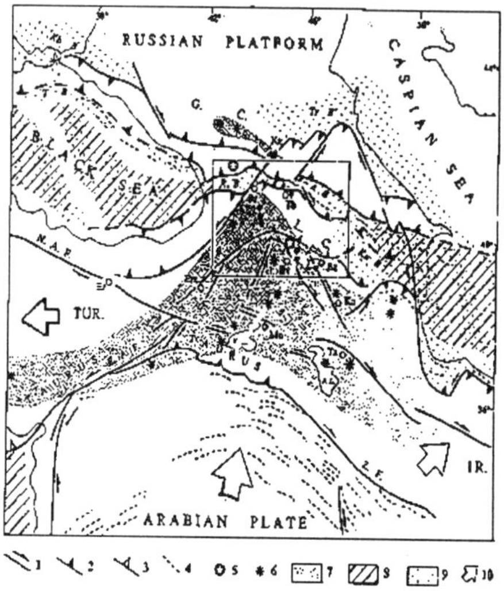

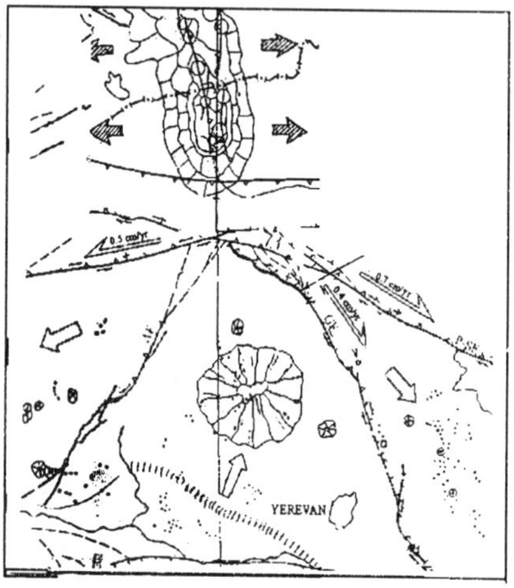

Fugure 1a Active tectonics in the front of the Arabian collision and the Caucasus.Key:1 major strike –slip fault; 2,major trust fault; 3,subduction trench; 4,recent folding;5, epicentral zone of the Spitak (December 7,1988) and the Racha-Dzava (April 27,1991) earthquakes;6, recent volcanic cones;7;plio-cuanternary volcanic zonez;8, oceanic and intermediate crust; main neogenic and quaternary basins; and 10, relative motions of blocks with respects to Eurasia.(from Rebai et al., 1993).

Figure 1b First pair is composed by the Zheltorechensk-Sarikamish (Z-SF) and Pambak-Sevan (PS-F) active faults. The second one consists of the Akhurian (AF) and Garni (GF) active faults. Pambak-Sevan) and Garni (GF) faults are characterized by right - lateral strike slip and reverse faulting , Zheltorechensk-Sarikamish and Akhurian are characterized by left-lateral strike slip faulting. The horizontal slip rate is 0.5-0.7 cm/yr for Z-SF and P-SF and a little lower for the next AF and GF pair which is 0.3-0.4 cm/yr respectively (from NPP project 1996)

2. Southern Javakhet earthquake swarm on 1996

In 1996, for the period from May 29 to June 9, a swarm of earthquakes has occurred. The swarm was recorded by both analog stations of Armenian and Georgian regional networks and two networks of the telemetric digital stations of the National Survey for Seismic Protection (NSSP) of the Republic of Armenia. The telemetric networks with the digital recording greatly improve ability to locate and study earthquakes in this region (figure 2a,b).

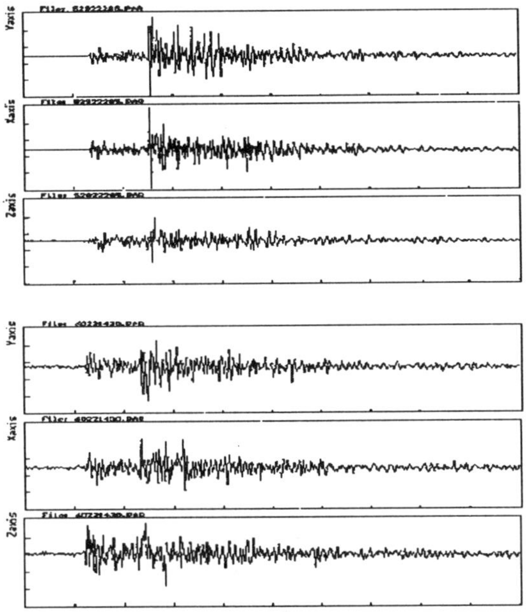

Figure 2. Two examples of earthquakes recorded on short period three component seismographs .Short- period three- component seismograms of the earthquake swarm on 29 May – 9 June 1996 in Southern Javaketi. Up on the Z, X(NS), Y(EW) corresponds to upward, northsouth and eastwest ground motion, respectively. The onset of S can be seen on both horizontal-component seismograms. The vertical-component short- period records display clear P.

During this period the "Bavra" seismic station which was the closest one and the "Stepanavan" seismic station which was the more sensitive one (magnifecation =20820) have recorded about 150 weak shocks. From total number of detected events, twenty seven were large enough to be located. The largest event (M=4.2), on June 9, was located at a depth of 39 km.

The earthquake swarm has started with one small shock which was followed by a stronger one of magnitude M=3.4 on May 29, 1996 . Three shocks were recorded on May 30, and then there was a pause lasting for two days. Four earthquakes approximately with the same magnitudes occurred on June 4 with weaker shocks in the intervals between them the total number of which being 14. Two strongest events of the swarm have occurred 10 days later, on June 9. The magnitudes were M=4.2 and M=4.0 with the focus depths of 39 and 23 correspondingly. The hypocenters were determined for 27 stronger shocks. As seen in figure 3 seven earthquakes had magnitude higher or equal to 3. Some of these earthquakes were accompanied by a hum and felt by population of the adjacent regions.

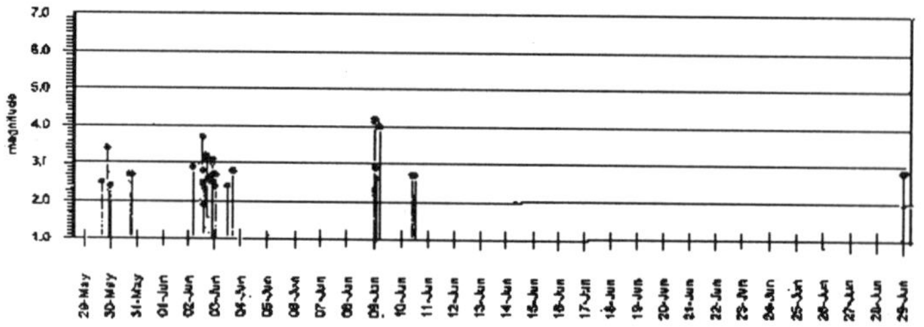

Fugure 3. Event magnitude as a function of time for the Javakhet Upland swarm. The largest quake of the swarm , M=4.2 struck the area on June 9 .Note ,there is not clear mainshock..

Figure 4 shows the distribution of earthquake epicenters in the southern Javakhet area. The locations were determined by HYPO71 program using seismograms of above-mentioned networks.

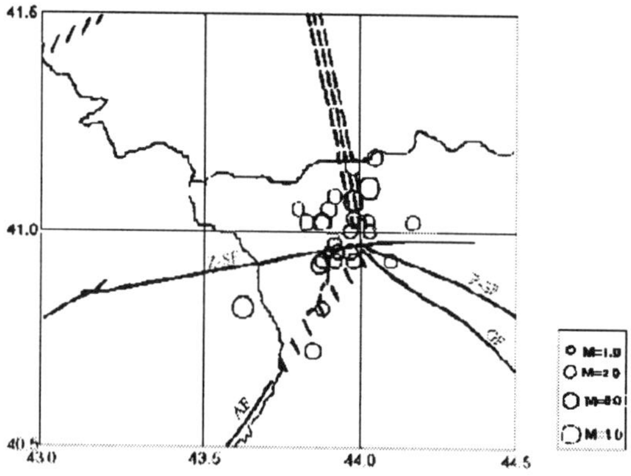

Figure 4. Map of the distribution of the earthquake epicenters determined by the two regional and telemetric networks. The source location for all events. Open circles indicate epicenters. The earthquakes are located chiefly in the southern part of Javakhet Upland.

According to this map spatially the swarm's epicenters are located within the intersection zone of the Pambak-Sevan, Zheltorechenck-Sarikamysh, Garni and Achouryan faults. Two concentrations are observed in the map: one is located in the southern margin of the Javakhet Upland, and the second one is in the intersection zone of the Zheltorechenck- Sarikamysh and Akhurian faults.

3. Focal mechanism and stress tensor

Focal mechanism information was needed to study the stress field in the territory under study, to determine the stress tensor. In addition to the NSSP data first motions were obtained from the stations operating in Georgia. However, focal mechanism determination was not available for many of the smaller events. Only for the 19 best recorded events focal mechanisms were determined using direct P – wave first motions.

As seen in Figure 5 pure or nearly pure normal faulting is observed and this type of faulting is dominant, but even a case of trust (event N 3) and strike slip faulting (events N2, N12) is also observed.

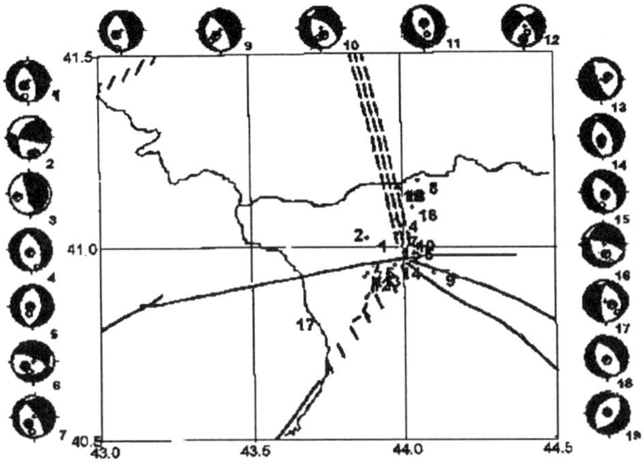

Figure 5. Focal mechanisms (upper hemisphere projection) of the 19 best recorded earthquakes

Based on the obtained results according to Zoback's approach [M.L.Zoback, 1992] the stress regime [Table 1] and Shmax direction were determined in each earthquake focus According to the obtained results the stress regime is normal faulting in the 13 focuses, strike slip in two and trust in one focus.

TABLE 1. *Focal mechanism solutions, stress regime*

y m d	hm	Fi	al	H	M	azP	plP	azT	plT	stkl	dpl	slipl	stk2	dp2	slip2	SR
960529	2229	4102	4389	10	3.4	37	66	264	16	18	32	-57	160	63	-109	NF
960530	0039	4105	4380	11	2.4	341	17	238	36	25	51	15	286	78	140	SS
960530	1801	4094	4392	21	2.2	77	36	247	54	191	10	118	343	81	85	TF
960602	1313	4107	4398	21	3.7	23	778	258	6	359	39	-75	159	52	-102	NF
960602	1316	4096	4390	24	1.9	72	82	278	6	12	39	-84	185	52	-94	NF
960602	1318	4101	4403	23	2.4	59	55	183	22	235	33	-14	114	72	-62	NF
960602	1319	4104	4399	13	2.5	37	56	251	30	19	21	-41	148	76	-101	NF
960602	1430	4117	4406	5	3.1	49	63	262	22	16	25	-58	161	69	-104	NF
960602	1542	4094	4410	12	2.3	40	60	250	27	9	22	-53	149	73	-103	NF
960602	1621	4102	4402	26	3.2	338	58	234	7	355	46	-45	119	60	-126	NF
960602	2053	4094	4389	23	2.5	174	75	66	5	171	42	-69	323	51	-108	NF
960602	2143	4092	4388	23	3.1	2	33	93	2	143	66	-157	43	69	-26	SS
960603	0042	4082	4388	17	2.4	247	60	68	30	159	15	-88	338	75	-90	NF
960603	0056	4095	4397	20	2.7	14	72	253	90	1	38	-65	150	56	-108	NF
960603	1208	4100	4398	15	2.4	338	78	243	2	345	45	-73	141	48	-106	NF
960609	0216	4110	4404	23	4.2	348	51	218	28	153	29	-27	107	77	-116	NF
960609	0539	4082	4362	39	4.0	299	60	84	23	145	25	-129	6	71	-74	NF
960613	2307	4099	4348	3	2.5	2	78	241	6	342	40	-74	142	52	-103	NF
960629	0122	4115	4399	15	2.8	126	86	306	4	36	41	-90	216	49	-90	NF

The comparison of two nodal planes (one of them is supposed to be a rupture in the earthquake focus) to the fault tectonics shows, that the striking of one of the nodal planes fits well to the fault striking only in three foci, which are located in the southern margin of the Javakhet Upland [Figure 6]

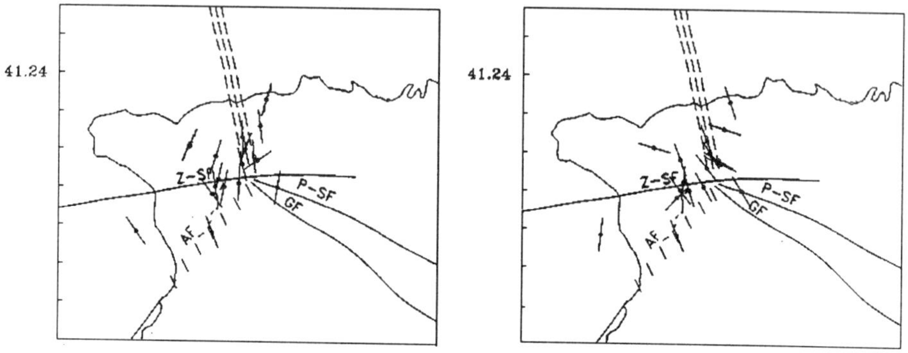

Figure 6. Maps showing orientations of the first (N1) and the second (N2) nodal planes. It is clear that the nodal planes are not consistent with mapped active faults.

In all the other cases the nodal planes are directed across fault striking. It seems probable that these weak shocks are related to secondary faults that do not reach a day surface. It evidences once more that it is impossible to confine foci to faults, which is often done in seismology practice, based only on spatial distribution of foci without detailed analyses of mechanisms.

It is well known that groupings of P and T axes from a number of fault plane solutions are generally good indicators of regional maximum and minimum compression directions and the maps of P, T and Shmax axes directions were drown up. Figure7, 8.

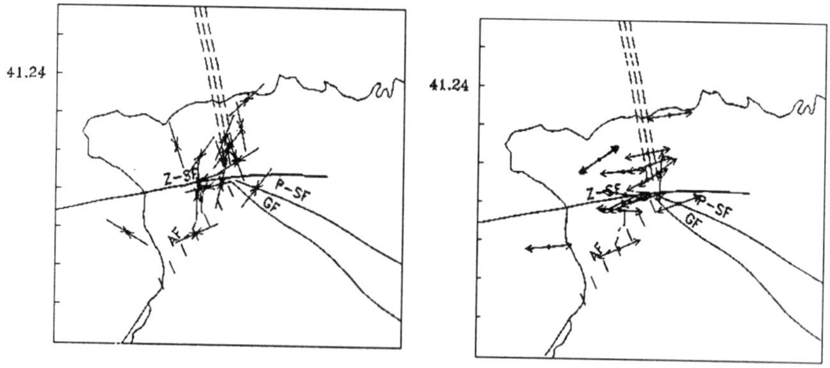

Figure 7. Map showing orientations of P and T and Shmax ploted on a base of active faults

The study of maximum P and minimum T compression axes shows that despite the scattering of P axes directions, two dominating directions are revealed: the first one is a near meridional direction (in 8 foci of 17) coming from S158SN - S168SN in 4 foci .the other direction is S182SE-S203S and only in one case P is directed from the North to the South. In three foci P axes are of a near-latitudinal direction. T axes are of a near latitudinal direction predominantly with AZ=68°-88°. Despite of the fact that a weak seismicity reflects usually specific features of local tectonic structures, in the given case the dominating directions of P and T stresses do not differ from regional ones.

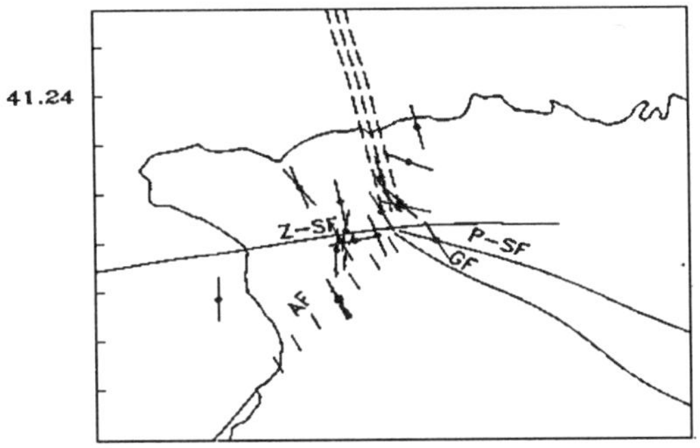

Figure 8. Orientations of Shmax

In 10 focuses Shmax of near-meridional direction was observed.

But on the other hand it is also well known that P and T axes for an individual earthquake may differ significantly from the actual stress orientations producing the slip [McKenzie, 1969]. For that reason we have used the Gephart method based on focal mechanism using, to reveal the directions of regional maximum and minimum compressive stresses. We used approximate method using P and T axes. It is clear that "if only one type of focal mechanism is observed, then the direction of the principal stresses would be poorly constraint"(modifide by Gillard and Wyss, 1995).

The stress tensor obtained is of normal style with a PHI=60.1, a measure of relative stress magnitudes R=0.60. The average approximate (minimum rotation) misfit between this model and all the data is about 5 degrees and all events have misfits of less than19°. The maximum principal stress ($\sigma 1$) strikes 9° and is nearly vertical (plunge=60°) the minimum principal stress ($\sigma 2$) is a nearly horizontal (plunge=14°) and strikes 253° when the intermediate stress ($\sigma 3$) is also nearly horizontal (plunge=26°) and strikes 155°.

This method also enables an objective selection of the preferred nodal plane from each focal mechanism and as part of analysis we identify one of the nodal planes from each focal mechanism as a more reasonable fault plane than the other for a given stress model [Figure 9]

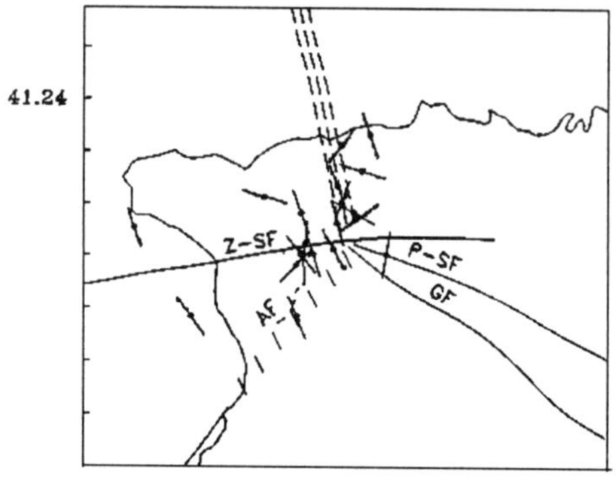

Figure 9. Map showing the directions of identified nodal planes from each focal mechanism as a more reasonable fault plane than the other for a given stress model. Comparisons of nodal planes with active tectonics shows that only in two cases nodal plane direction and direction of mapped known fault are the same.

A detailed study of the earthquake catalogue, for the instrumental period from 1962 till 1999, shows that the small earthquake swarms are a common phenomenon in the southern part of the Javakhet Upland. In some cases they occurred independently in the other ones they are related to strong earthquakes. They are different according to duration because they may last from several hours to several days, to their number- usually five to six events, and to magnitudes.

But this swarm as more preceding ones, was followed by relatively strong earthquake in Iran and Turkey. It is understandable that we mean only the territory near their border, restricted by Van lake and the towns of Khoy and Maku, and not the entire territories of Iran and Turkey. Many events in catalogue where the Javekhet Upland's earthquakes were followed by strong one in mentioned above territory, were found. Based on this fact it is possible to suggest that the Javakhet Upland seismicity in some way is related to the seismicity of Iran and Turkey (playing the role of trigger?). This relationship is not very clear but it is a real fact.

4. Conclusons

1. The study of NSSP catalog of seismisity for 1962-1996 time period shows that in most cases Javakhet earthquake swarms were followed by relatively strong events located in Turkey- Iran border region.
2. The spatial distribution of hypocentres seems to be consistent with mapped active faults but the comparison of chosen nodal planes and the fault tectonics shows that they are not consistent with any of the mapped active faults. Based on this one may conclude they are related to secondary, small and blind fault.
3. The local stress tensor regime from earthquake swarm is normal regime.
4. From the orientation of the stress tensor on one hand and from the orientation of P axes and Shmax on the other hand one can conclude that maximum compressive stress in the observed area is horizontal and near-meridional.

5. References

1. Bullen, K.E., Bruce A.Bolt(1993) An Intrudaction to the Theory of Seismology, Cambridge University Press. Printed in the Great Britain.
2. Christopher H. Scholz (1994) The Mechanics of Earthquakes and Faulting, Cambridge University Press. Printed in the USA
3. Gepart J. W. and Forsyth D. W.(1984) An Improved Method for Determining the Regional Stress Tenzor Using Earthquake Focal Mechanism Data: Aplication To the San Fernando Earthquake Sequence, Journal of Geophisical research, Vol.89, No. B11, Pages 9305-9320,October 10,
4. Lee, W.K., Stewart, S. W.(1981) Principles and Applications of Microearthquake Networks, Academic Press, New York.
5. McKenzie, D. E., (1969) The relation between fault plain solutions for earthquakes and the directions of the principal stresses, Bull. seismol. Soc. Am., 59, 591-601.
6. Thorne, L., Terry C.W. (1995) Modern Global Seismology, Academic Press, San Diego.
7. Zoback M. L. (1992) First and Second –Order Patterns of Stress in the Lithosphere: The World Stress Map Project, Journal of Geophisical research, Vol.97, No.B8, Pages 11, 703-11, 728, July30, 1992.

THE DIGITAL ACCELEROGRAPH NETWORK IN THE CAUCASUS

SMIT P., ARZOUMANIAN V., JAVAKHISHVILI Z., AREFIEV S., MAYER-ROSA D., BALASSANIAN S., CHELIDZE T.
Imperial College of Science, Technology and Medicine, London, UK; National Survey for Seismic Protection of Republic of Armenia, Yerevan, Armenia; Georgian Academy of Sciences, Institute of Geophysics, Tbilisi, Georgia; Russian Academy of Sciences, Unified Institute of Physics of the Earth, Moscow, Russia; Swiss Seismological Service, Zurich, Switzerland

Abstract

The paper describes the permanent digital accelerograph network in the Caucasus area operated by the National Survey for Seismic Protection of the Republic of Armenia, the Georgian Academy of Sciences and the Swiss Seismological Service and granted by the Swiss Disaster Relief. The network, which became operational in 1990 and since then gradually extended, consists today of 12 digital free-field stations as well as 6 digital strong-motion instruments for aftershock studies and is complemented with 16 analogue strong-motion stations and 2 structure-related arrays. The acceleration time histories recorded by this network and the routine operation and data processing procedures are outlined. The results on empirical scaling of response spectra and peak acceleration of earthquake induced ground motion in the Caucasus and adjacent area are summarised. These models provide the basis for reliable empirical estimates of ground motion parameters and seismic hazard assessment in the Caucasus.

Key words: accelerograph network, attenuation, Caucasus

1. Introduction

The seismic hazard is the result of the strong ground motion, especially near to the epicentre. From an engineering point of view, the recordings of acceleration time histories in the near-field are therefore an indispensable tool because they provide the input for seismic hazard and risk analysis and design of buildings and structures under seismic actions. Such an important database providing strong-motion parameters and recordings which are typical and useful for the Caucasus has to be collected with a network of accelerographs.

Several months after the Spitak earthquake all temporary accelerograph stations, which were installed immediately after the earthquake, were removed. No permanent accelerograph stations were in operation after that time due to political, economical and technical problems.

In 1989, in the frame of prevention measures for future strong earthquakes in the Armenian upland the Swiss Disaster Relief (SDR) decided to provide substantial financial support for the installation, operation and maintenance of a digital strong-

S. Balassanian et al. (eds.), Earthquake Hazard and Seismic Risk Reduction, 109–118.

motion network in the epicentral area of the Spitak earthquake. The main goal of this network is to collect strong-motion records typical for the area, which will serve for seismological, engineering and future planning purposes.

2. The Accelerograph Network

In June 1990, in the frame of a long term agreement between the Institute Physics of the Earth of the former USSR Academy of Sciences (IPE), the Swiss Seismological Service (SED) and SDR, a network with 9 digital strong-motion stations was installed in the epicentral area of the Spitak earthquake. All sites were chosen on the basis of the seismic activity in the Armenian upland and the Javakheti area (southern Georgia) and accessibility of each site during the whole year. All instruments are installed in existing seismograph stations or in private houses. The sensors are well fixed to the seismometer pedestal or the concrete foundation of the building.

In addition to this network 6 digital strong-motion instruments are provided as stand-by equipment for rapid epicentral observation expeditions organised by IPE. In 1994 a new agreement was set up between SDR, SED, the National Survey for the Seismic Protection of the Republic of Armenia (NSSP), the Institute of Geophysics of the Georgian Academy of Sciences (IG) and the Unified Institute Physics of the Earth of the Russian Academy of Sciences (UIPE, former IPE).

Between 1996 and 1998 the digital strong-motion network was subsequently improved and extended and consists today of 12 digital strong-motion stations. The annual costs for service, maintenance and data compilation for the permanent network and the stand-by equipment is covered by SDR until end of 1998 and contracted to the local partners (NSSP, IG and UIPE). SED provides the project-management, repairment and upgrading of the instruments and quality control of the data and associated parameters.

Since 1993 another 16 analogue strong-motion stations were gradually installed throughout Armenia by NSSP. Furthermore there are other organisations, such as the operators of the nuclear power plant "Metsamor" in Armenia and the "Ingurin" dam in Georgia (highest arch dam of the world), operating accelerograph networks of smaller scale for specific purposes. The location of all permanent strong-motion stations in the Caucasus and adjacent area are shown in Figure 1.

Between 1992 and 1995 the political situation in the Caucasus made it extremely difficult to operate and maintain the digital strong-motion network. Thanks to the robust design and careful selection of the sites all instruments "survived" this difficult period with little or even no damages. Since 1996 all instruments were gradually upgraded and equipped with large external batteries and chargers, providing an autonomy of two weeks, in order to overcome the instability and frequent breakdown of the public power supply.

Since 1990 IPE/UIPE organised several aftershock studies and operated temporary strong-motion networks (using the stand-by instruments) in the epicentral area of the earthquakes. The aftershock data collected with these temporary networks complemented the database (Table 1) from the main shock especially in regions where no strong-motion data were recorded during the main shock.

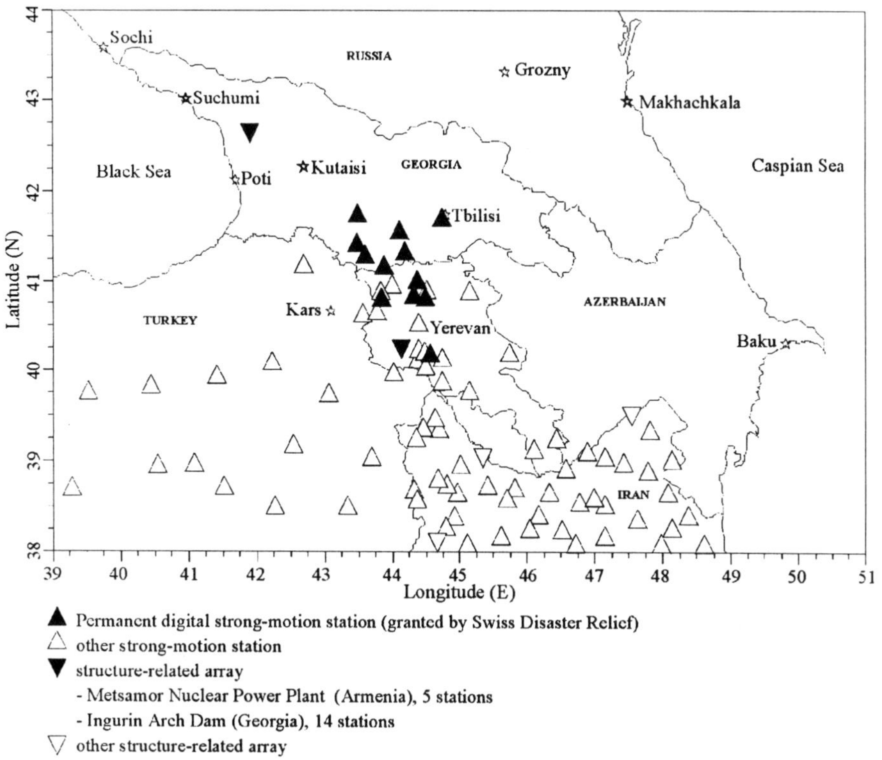

Figure 1. Location of permanent strong-motion stations and structure-related accelerograph arrays in the Caucasus and adjacent area.

TABLE 1. *Acceleration time histories recorded with permanent and temporary networks.*

Earthquake	Observation period	Number of records	Magnitude range	Distance Range
Caucasus	6.1990 – ...	62	2.0 - 6.4	0km - 203km
Manjil, Iran 20.6.1990, 7.7Ms	8.1990 – 9.1990	96	2.0 - 5.0	1km - 41km
Racha, Georgia 29.4.1991, 7.0Ms (Arefiev et al., 1993)	4.1991 – 8.1991	299	1.5 - 7.1	1km - 178km
Erzincan, Turkey 13.3.1992, 6.8mb (Wepf, Smit, 1993)	3.1992	12	2.8 - 6.1	25km – 83km
Susamyr, Kyrgystan 19.8.1992, 7.4Ms	8.1992 – 9.1992	68	2.0 - 4.2	4km – 57km
Shikotan, southern Kuril Islands (Russia) 4.10.1994, 8.2Ms	10.1994 - 2.1995	125	4.3 - 6.8	17km – 171km
Sachalin, Russia 27.5.1995, 7.6Ms	6.1995	7	3.4 - 5.0	7km – 25km

3. Acceleration Time Histories

Since the installation of the first digital strong-motion station in the Caucasus area 469 acceleration time histories from 278 earthquakes were recorded. In addition 130 system- and sensor-tests were processed, inspected and archived using the software package SM1_PACK (Smit, 1996).

A summary of all time histories recorded in the Caucasus area are regularly published in strong-motion bulletins of the Swiss Seismological Service (Arefiev et al., 1991a; Arefiev et al., 1991b; Balassanian et al., 1997). All time histories and parameters are freely available as soon as the corresponding bulletin is published.

4. Attenuation Of Ground Motion

Attenuation relationships play a vital role in earthquake hazard analysis and seismic design of structures. In recent years a large number of investigations were carried out in this field and many regional relationships between earthquake, magnitude, source-station distance, local site conditions and ground motion parameters such us peak acceleration, spectral acceleration or duration of ground motion have been derived. Most of these studies used acceleration time histories to calibrate the empirical attenuation relationships.

Based on the acceleration time histories recorded between June 1990 and September 1998 with the permanent and temporary digital strong-motion network in the Caucasus and adjacent area, 84 corrected horizontal acceleration time histories and response spectra from 26 earthquakes with magnitudes between 4.0 and 7.1 were selected and compiled into a new data-set. All time histories were recorded at sites where the local geology is classified as “alluvium”. The attenuation relations derived in this study are therefore only valid for the prediction of the ground motion at “alluvium” sites.

The relationship adopted in this study, suggested by Ambraseys and Bommer (1991), is given by:

$$\log Y = a + b \cdot M + c \cdot \log R + d \cdot R + p \cdot \varepsilon \qquad (1)$$

and

$$R = \sqrt{D^2 + h^2}$$

where Y is the maximum absolute acceleration response (larger horizontal), SA, for 5% critical damping in [cm/sec^2], M is the surface-wave magnitude and D is the hypocentral-distance in [km]. p is 0 for 50-percentile values and 1 for 84-percentile. The attenuation model was derived by constraining the coefficient for geometric attenuation (c) to –1.0, which corresponds to a spherical model of energy spreading.

The calculation of the correlation coefficients (a, b, c, d, h) and the residual root mean square ε of (1) was performed with the well known Joyner and Boore (1981) two step regression model. This method allows a de-coupling of the determination of the magnitude dependence from the determination of the distance dependence of the attenuation of ground motion. Using the larger horizontal component for spectra of the selected acceleration time histories, the following values were obtained for the coefficients of (1) at different frequencies (Table 2).

Empirical attenuation models with peak ground acceleration as dependent parameter still play an important rule in different seismic hazard and earthquake engineering studies. However from different point of views it is obvious, that peak ground acceleration is poorly representing the character of the ground motion.

TABLE 2. *Coefficients of the attenuation relationship for maximum absolute acceleration response (larger horizontal) SA for 5% critical damping.*

Frequency	Larger spectral horizontal ground motion on alluvium					
[Hz]	a	B	c	D	H	ε
1.0	-1.733	0.804	-1.0	-0.00237	1.0	0.38
1.5	-1.673	0.846	-1.0	-0.00301	0.0	0.40
2.0	-1.150	0.778	-1.0	-0.00273	6.0	0.40
2.5	-0.890	0.768	-1.0	-0.00399	8.5	0.39
3.0	-0.739	0.767	-1.0	-0.00492	6.5	0.39
4.0	0.052	0.632	-1.0	-0.00437	9.0	0.34
5.0	0.390	0.570	-1.0	-0.00341	6.5	0.31
6.0	0.841	0.472	-1.0	-0.00167	5.5	0.30
7.0	1.111	0.419	-1.0	-0.00152	4.5	0.27
8.0	1.433	0.355	-1.0	-0.00139	4.5	0.28
9.0	1.706	0.303	-1.0	-0.00117	5.5	0.28
10.0	1.955	0.254	-1.0	-0.00095	6.0	0.28
11.0	1.840	0.279	-1.0	-0.00158	5.0	0.28
12.0	1.834	0.275	-1.0	-0.00145	4.5	0.29
13.0	1.843	0.266	-1.0	-0.00109	4.0	0.29
14.0	1.690	0.298	-1.0	-0.00154	4.0	0.29
15.0	1.587	0.318	-1.0	-0.00186	4.0	0.30
16.0	1.591	0.309	-1.0	-0.00149	2.0	0.30
17.0	1.533	0.319	-1.0	-0.00151	0.0	0.30
18.0	1.454	0.332	-1.0	-0.00149	0.0	0.29
19.0	1.341	0.351	-1.0	-0.00159	0.0	0.29
20.0	1.301	0.358	-1.0	-0.00185	0.0	0.29

The resulting equation for larger horizontal values of peak horizontal acceleration is:

$$\log pha = 0.72 + 0.44 \cdot M - \log R - 0.00231 \cdot R + 0.28 \cdot p$$

$$R = \sqrt{D^2 + 4.5^2}$$

where *pha* is the peak horizontal acceleration in [cm/sec^2], M is the surface-wave magnitude and D is the hypocentral-distance in [km]. p is 0 for 50-percentile values and 1 for 84-percentile.

It is important to bear in mind that the models represent a best fit of the selected data-set, and therefore represent mean values about which there is a considerable scatter. In the case of the attenuation model for the larger horizontal value of peak horizontal acceleration the predicted mean plus one standard deviation is equal to 1.91 times the mean value. The scatter of the *pha*-models is the same as similar models for Europe (Ambraseys et al., 1996) and Western North-America (Joyner and Boore, 1982).

The comparison of the attenuation relationships for peak horizontal acceleration with similar relations for other areas shows a good agreement with the models from Western North-America.

5. Conclusion

The digital strong-motion network in the Caucasus area, which was installed in 1990, may record the strong ground motion from earthquakes in the Armenian upland and Javakheti area. Events as small as magnitude 2.5-3.0 are expected to be recorded at the nearest portion of this network. Thus the accelerograph network is recording both major and lesser seismic events and is benefiting both engineers and seismologists in their seismic hazard reduction programs. The 669 acceleration time histories and all future time histories, which were or will be recorded by the permanent network and several temporary networks, are contributing towards answering many questions such as spectral attenuation of strong ground motion, site characteristics of strong ground shaking.

The empirical attenuation relationships for maximum absolute acceleration spectra at 5% critical damping and peak acceleration both for larger horizontal ground motion on alluvium in the Caucasus and adjacent area are considered to be valid for hypocentral-distances between 4km and 230km and surface-wave magnitudes between 4 and 7. The predicted ground motion values are in good accordance with corresponding models from Western North-America. Due to the complex structure of the Caucasus the scatter and absorption of ground motion is somewhat higher than in European areas and hence the ground motion in the Caucasus attenuates faster.

The variability of the models may be improved by increasing the volume of input data, especially with ground motion data in the near-field of the earthquake sources. Furthermore all seismological- and geological-parameters (local site conditions) have to be completed and uniformly measured and the number of variables of the attenuation models should be increased, considering the source effects especially in the near-field and the influence of the local site to strong ground motion. However, such an extension of the independent parameters of the attenuation models requires a fundamental knowledge of the source mechanism of the earthquakes and the dynamic behaviour of the local sites of all accelerograph stations.

6. Acknowledgement

Despite of all political and economical problems the operation of the permanent and temporary digital strong-motion network in the Caucasus and adjacent area and the data exchange between all partners involved in this project is extremely successful. This is only possible thanks to the motivation and commitment of all participating persons and organisations and the substantial financial support provided by the Swiss Disaster Relief.

7. References

Ambraseys, N.; Bommer, J. (1991) The attenuation of ground motions in Europe. *Journ. Earthq. Struct. Dyn. 20*: 1179-1202

Ambraseys, N.; Simpson, K.; Bommer, J. (1996) Prediction of horizontal response spectra in Europe. *Earthq. Eng. Struct. Dyn. 25*: 371-400

Arefiev, S.; Mayer-Rosa, D.; Parini, I.; Romanov, A.; Smit, P. (1991a) The Racha (Georgia, USSR) earthquake of 29.April 1991: Strong-Motion data of selected aftershocks 3.May 1991-30.June 1991. *Publication Series of the Swiss Seismological Service* 103, Zurich

Arefiev, S.; Mayer-Rosa, D.; Parini, I.; Pletnev, K.; Romanov, A.; Smit, P. (1991b) Spitak (Armenia, USSR) 1988 earthquake region: Strong-Motion data of selected earthquakes June 1990-April 1991. *Publication Series of the Swiss Seismological Service* 104, Zurich

Arefiev, S.; Pletnev, K.; Tatevossyan, R.; Borissof, B.; Aptekman, J.; Vassiliev, V.; Delitsyn, L.; Romanov, A.; Osher, B.; Parini, I.; Afimina, T.; Shilova, N.; Shumilina, L.; Javakhishvili, Z.; Cisternas, A.; Haessler, H.; Rivera, L.; Dorbath, L.; King, G.; Fuenzalida, A.; Owen, T.; Mc Cormack, D.; Baker, C.; Langer, Ch.; Mayer-Rosa, D.; Smit, P. (1993) The Racha 1991 earthquake: Results of the field seismological observation. *Physika Semlja* 3: 12-23, Moscow

Balassanian, S.; Melkumian, M.; Arzoumanian, V.; Nazaretian, S.; Oganesian, A.; Gevorgian, E.; Arefiev, S.; Smit, P.; Mayer-Rosa, D. (1997) Spitak Area Strong Motion Network. Strong Motion Bulletin June 1991-February 1997. *Publication Series of the Swiss Seismological Service* 110, Zurich

Joyner, W.; Boore, D. (1991) Peak horizontal acceleration and velocity from strong motion records including records from the 1979 Imperial Valley, California earthquake. *Bull. Seism. Soc. Am.* 71: 2011-2038

Joyner, W.; Boore, D. (1982) Estimation of response-spectral values as functions of magnitude, distance and site conditions. *U.S. Geol. Survey, Open File Report* 82-881

Smit, P. (1996) Datenerfassung und Bestimmung der Abminderung der Bodenbewegung bei Erdbeben in der Schweiz. *PhD thesis* 11396 Swiss Federal Institute of Technology, Zurich

Wepf, D.; Smit, P. (1993) Das Erdbeben in der Türkei vom 13. März 1992. *Schweiz. Ing. Arch.* 4: 43-49, Zurich

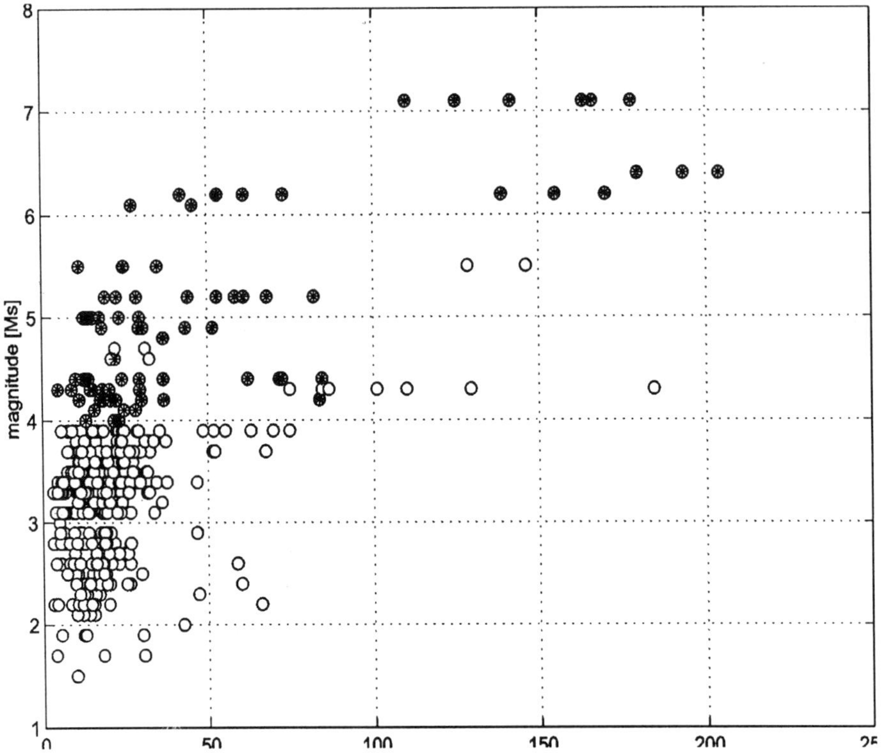

Figure 2. Distribution of the dataset with respect to magnitude and distance. Data marked with a star were used for the calibration of the attenuation models.

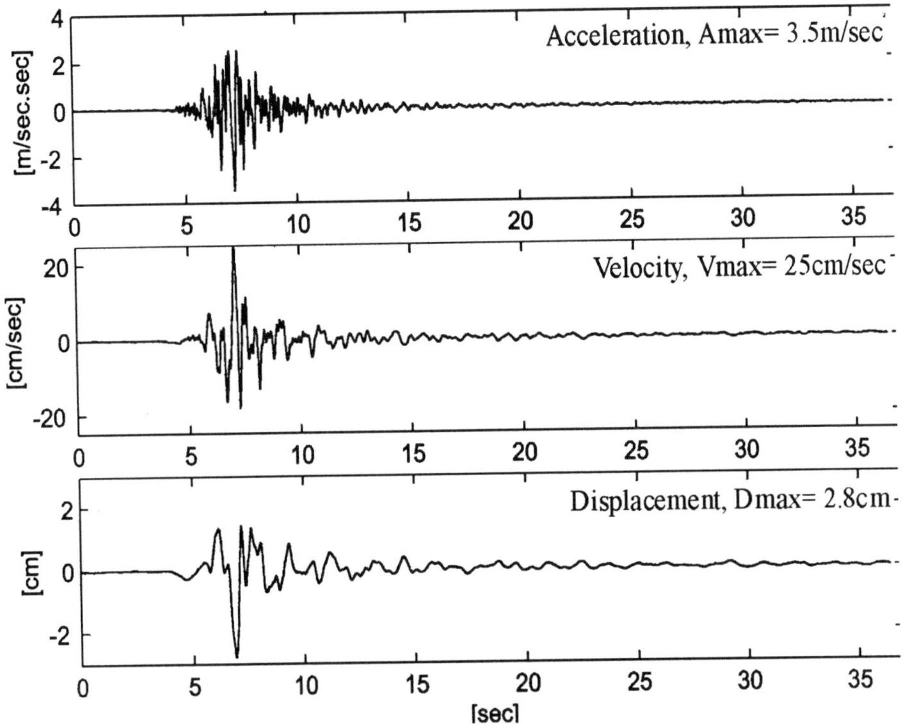

Figure 3. Corrected acceleration-, velocity and displacement-time histories of NS-component recorded in Ambrolauri (Georgia) during a 5.3Ms aftershock (3.5.1991 20:19) of the Dzhava-Racha earthquake (Δ= 9km).

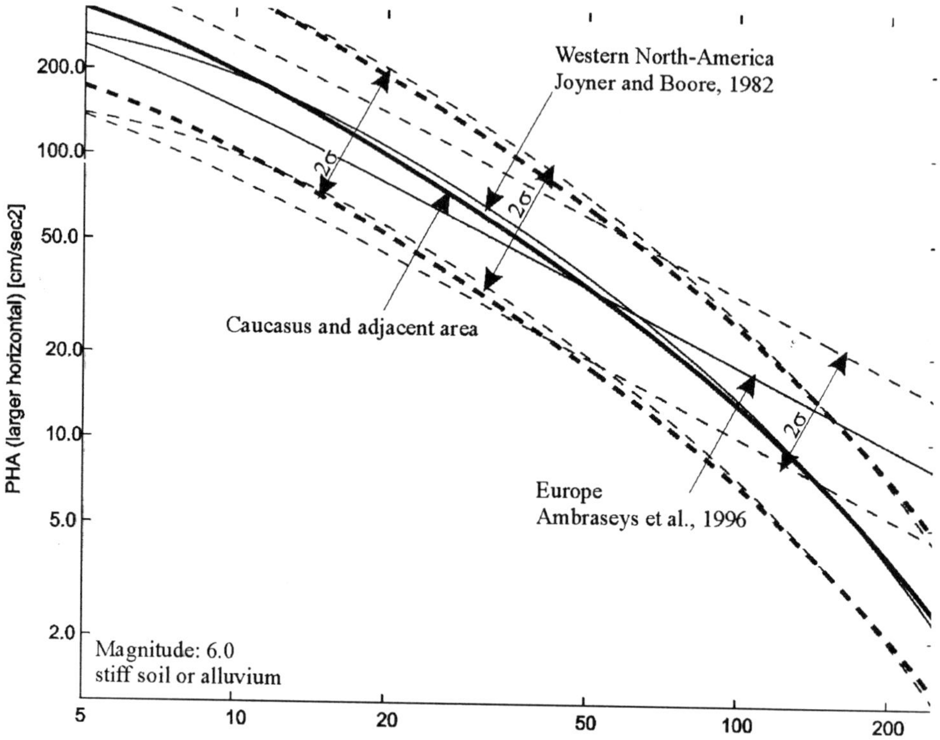

Figure 4. Comparison of European and Western North-American models for peak horizontal acceleration with the corresponding model for the Caucasus and adjacent area. The solid lines represent the mean-values whereas the dotted lines mark the upper and lower bound (36%- and 84%-percentile) of the models.

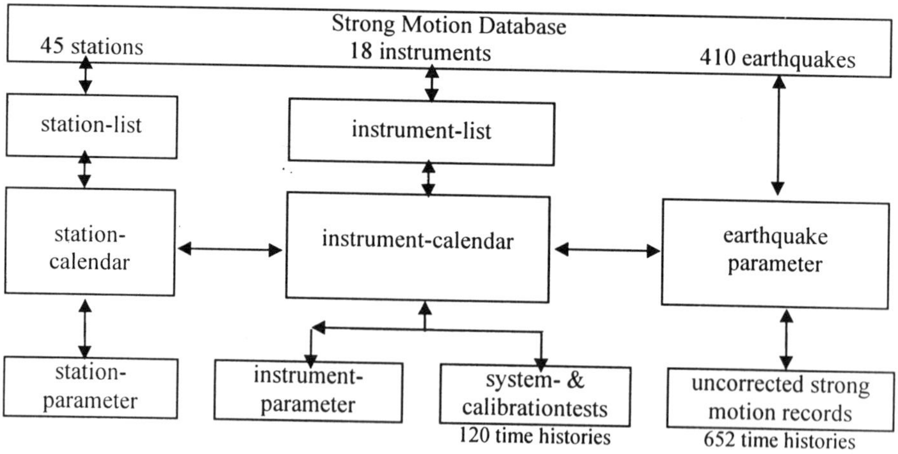

Figure 5. Flowchart of the strong-motion database SM1_BASE.

IMPLICATIONS OF THE DURATION OF STRONG GROUND MOTION: (OBSERVATIONS FROM THE U.S. LOMA PRIETA EARTHQUAKE OF OCTOBER 17, 1989)

M. TIV
International Institute of Earthquake Engineering and Seismology
P.O. Box 1939513913, Tehran, I.R. Iran

Abstract

A database consisting of 78 horizontal strong ground motion records from a single earthquake, the U.S. Loma Prieta earthquake of Oct. 17, 1989, has been considered to study the statistical relationship of duration of strong ground motion with several parameters associated with the characteristics of an earthquake record. These parameters are: 1) distance from the source of energy, 2) root-mean-square (rms) value of the acceleration record, 3) frequency content of the record, and 4) the mean response spectral values. Three different definitions were adopted for extracting the strong ground motion duration - the "*bracketed*" duration, the "*accelerogram intensity*", and the "*energy arrival rate*" definitions. Results are reported for two geological groups namely, rock and alluvium.

Introduction

Duration of strong ground motion has been used as a measure of estimating the intensity of a record as well as other applications having to do with stationary characteristics and the computation of power spectral density of a record (Moayyad and Mohraz, 1982). Different procedures have been proposed to extract the strong ground motion portion of a record. Bolt (1969) and Page et al. (1972) used a "*bracketed duration*" which is the time interval between the first and the last acceleration peaks exceeding a specified value, usually 0.05g. Page and his colleagues concluded that for a given earthquake magnitude, duration decreases in the distance from the source. Chang and Krinitzsky (1977) used the bracketed duration and proposed a set of plots for estimating the duration of strong motion for rock and alluvium as a, function of distance for different earthquake magnitudes. According to their study, for a given magnitude, the duration of strong ground motion for alluvium sites is about twice the duration for sites on rock. Trifunac and Brady (1975) defined the so called "*accelerogram intensity*" of strong ground motion as the time interval in which a significant contribution to the integral of the square of acceleration, $\int a^2 dt$, takes place. They selected the central ninety percent as the duration of strong ground motion contribution and concluded that the average duration of strong ground motion for alluvium is about 10-12 seconds longer than that for rock and it increases by approximately 1.0-1.5 seconds for every 10-km increase in distance. In a more recent work by Chang et al. (1996) duration of strong ground motion from the Northridge earthquake was studied using the same definition. They concluded,

S. Balassanian et al. (eds.), Earthquake Hazard and Seismic Risk Reduction, 119–125.

"there is a slight tendency for duration to increase with distance" for both rock and soil sites. Another procedure suggested by McCann and Shah (1979) is based on the average "*energy arrival rate*", i.e., the segment between two cut-off instances in time beyond which the rate of change of the cumulative root-mean-square acceleration becomes negative and remains so for the remainder of the record.

The magnitude 7.1 Loma Prieta earthquake (Plakfer and Galloway, 1989) provided a large number of recorded accelerogram from which it was possible to carry out the present statistical study on the duration of strong ground motion, The acceleration, integrated velocity and displacement time histories, and the response and Fourier amplitude spectra used in this study include records from one tape compiled by the United States Geological Survey (1989) and three tapes compiled by the California Strong Motion Instrumentation Program of the U.S. Department of Conservation (1989). The accelerogram included in this study were recorded on free field, and in the basement or on the first floor of small buildings to avoid soil-structure interaction effects. Those with a peak ground acceleration of less than 0.05g were excluded from the database. List of the records used in this study can be found in Mohraz and Tiv (1991).

DISTANCE vs. DURATION

Since the earthquake motion was recorded in a number of recording stations with a variety of source-to-site distances, it was possible to divide the data into three subgroups of near, mid, and far-field. Of the 78 horizontal records, (50 on alluvium and 28 on rock,) those with a source to site distance of up to 20 km were grouped as near-field, between 20 and 50 km were grouped as mid-field, and those with greater than 50 km were grouped as far-field. The average duration of strong motion for the three definitions of duration described earlier and the three distant subgroups were computed and the results are reported in Table 1. As may be seen from this table, the duration computed from the "bracketed" definition decreases with an increase in distance (with a high correlation coefficient.) On the other hand, both the "accelerogram intensity" and the "energy arrival rate" definitions give duration, which are the smallest for the near-field and the <u>largest</u> for the <u>mid-field</u> for both geological conditions. Figure 1 shows distribution of duration according to "accelerogram intensity" definition vs. distance.

TABLE 1 *Average duration of strong motion in seconds*

	Bracketed		*Accelerogram Intensity*		*Energy Arrival Rate*	
Distance	Rock	Alluvium	Rock	Alluvium	Rock	Alluvium
Near-field	11.6	14.4	8.5	10.5	10.0	11.6
Mid-field	6.2	12.5	15.6	18.6	15.9	19.8
Far-field	1.9	6.7	11.6	11.5	14.7	14.8

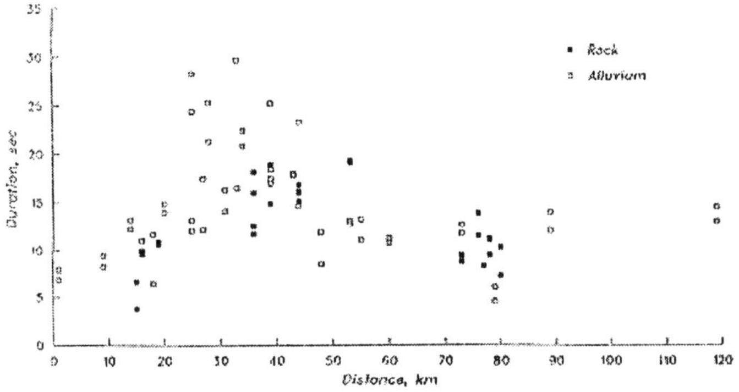

Figure 1. Accelerogram intensity definition of strong motion as a function of distance

DISTANCE vs. ROOT-MEAN-SQUARE

Similar to Table 1, rms values of the acceleration records were computed for the three definitions of duration and are reported in Table 2. As may be seen from this table, rms values for alluvium sites are somewhat larger for all distance categories; however, for all three definitions of duration have similar trend when plotted vs. distance. As an example, Fig.2 shows rms values vs. distance for the "accelerogram intensity" definition of duration. The trend in this figure, as well as other figures for other definitions, is exponentially decaying.

TABLE 2. *Average root-mean-square values in em/sec/sec*

	Bracketed		*Accelerogram Intensity*		*Energy Arrival Rate*	
Distance	Rock	Alluvium	Rock	Alluvium	Rock	Alluvium
Near-field	86.5	87.9	103.7	102.7	98.0	101.3
Mid-field	43.9	51.9	30.8	46.2	30.6	46.7
Far-field	35.3	52.5	19.8	44.4	18.6	40.2

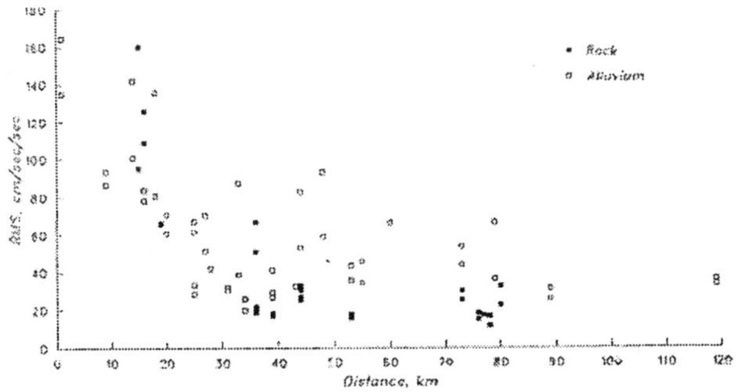

Figure 2. Root-mean-square acceleration for the duration computed according to the accelerogram intensity definition of strong motion vs. distance.

RELATIONSHIP WITH FREQUENCY CONTENT

Normalized Kanai-Tajimi power spectral density functions for the two geological groups and the three distant subgroups are shown in Fig's.3 and 4. As depicted from these figures, near- field records are predominated by high-frequency oscillations as opposed to the mid-field records, which are predominated by low-frequency motions. Considering once again Table 1 for the case of "accelerogram intensity" and "energy arrival rate" definitions, near-field records are associated with having short duration and mid-field records are associated with having long duration. Thus, one can infer that the longer the duration of strong motion, (according to these two definitions,) the lower is the frequency content of the corresponding record. On the contrary, in the case of "bracketed" duration, the longer the duration the higher is the frequency content.

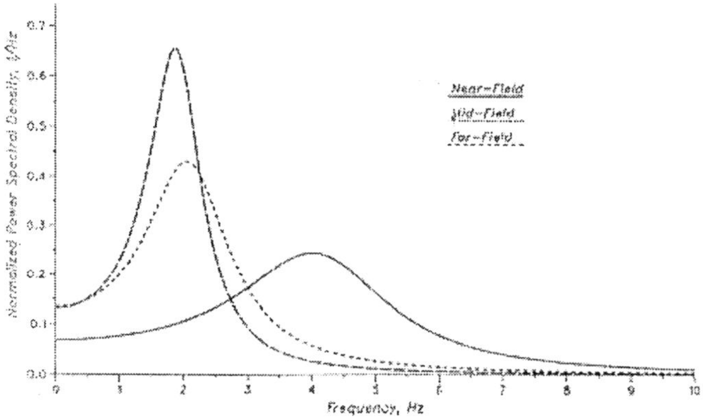

Figure 3. Normalized Kanai-Tajimi power spectral densities for rock

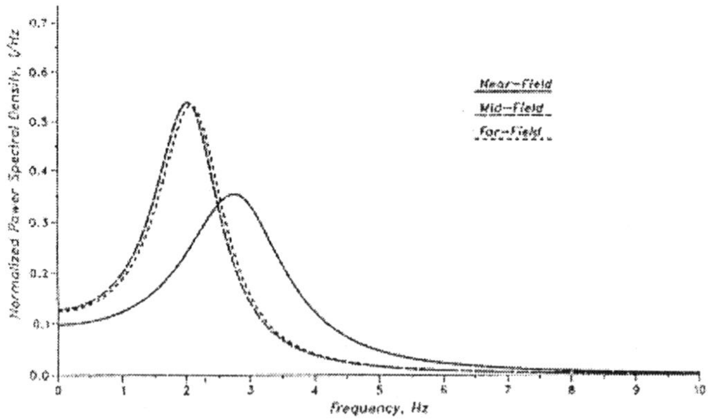

Figure 4. Normalized Kanai-Tajimi power spectral densities for alluvium

RELATIONSHIP WITH RESPONSE SPECTRA

It is by now obvious that what kind of relationship there may exist between the duration of strong motion and response spectra. In the case of the "bracketed" duration, Fig.5 shows the mean response spectra for alluvium sites for different duration. Records with longer duration in this figure have higher response values. However, this is quite the opposite for the "energy arrival rate" definition. Figure 6 shows response spectra for different duration according to the "energy arrival rate" definition and as can be seen, response of short duration records are higher for this definition. These two figures may not be compared together since different categories of duration were considered for each definition. This was done to allocate a reasonable number of records in each category. Results for the "accelerogram intensity" was very similar to the "energy arrival rate" definition and therefore is not presented here. Similar trend was observed for rock sites.

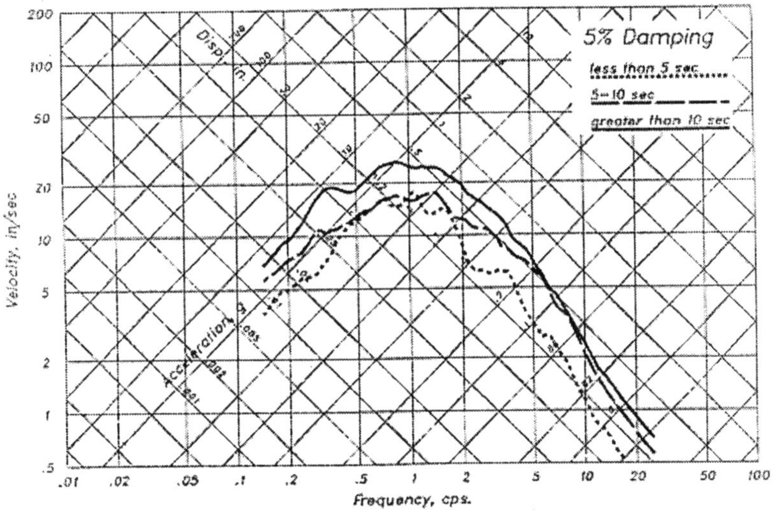

Figure 5. Mean response spectra of different groups of duration for the bracketed definition of strong motion.

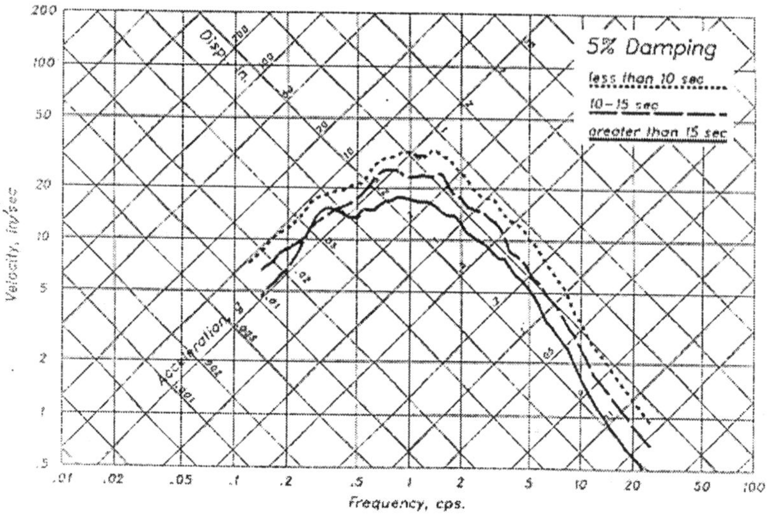

Figure 6. Mean response spectra of different groups of duration for the energy arrival rate definition of strong motion.

CONCLUSIONS

For any given source-to-site distance, regardless of the criteria considered for defining the duration, duration of strong motion on rock is in general shorter than those from alluvium sites. The "bracketed" duration correlates strongly with distance and decreases as distance increases for both geological conditions. In the case of "accelerogram intensity" and "energy arrival rate» definitions duration of strong motion increases with distance, reaches a maximum around 35-45 km from the source, and then decreases as distance is increased for both geological conditions. As expected, rms values, which represent the intensity of the records and correlates strongly with peak ground acceleration, decrease exponentially as distance is increased. The contradiction that exists between the trend observed for the «bracketed» definition vs. distance and that observed for the "accelerogram intensity" or the "energy arrival rate" definitions vs. distance may be argued by considering the concepts defining each definition. The bracketed duration is based on the acceleration level (0.05g in this study) and at longer distances the acceleration peaks become smaller and a shorter duration is therefore to be expected. The other two definitions are based on the relative rather than absolute measures – one on percentile contribution to the acceleration intensity at the site and the other on the relative contribution to the energy arrival rate at the site. According to those two definitions, in the near-field the motion is more intense (see Table 2) within a shorter period as compared to mid and far-fields where it is less intense but spans over a longer time.

In the case of "accelerogram intensity" and "energy arrival rate" definitions, the longer the duration of strong motion the lower is the frequency content of the corresponding record. On the contrary, in the case of "bracketed" duration, the longer the duration the higher is the frequency content. Response spectral values become higher as "bracketed" duration gets longer as opposed to the "energy arrival rate" definition where higher response is associated with short duration records.

REFERENCES

Bolt, B. A. (1969). Duration of strong motion, Proc. 4th World Conf. Earthquake Eng., Santiago, Chile, 1304-1315.

California Department of Conservation, Division of Mines and Geology (1989). CSMIP strong-motion records from the Santa Cruz mountains Loma Prieta, California earthquake of October 17 1989, *Report OSMS 89-06*, 195 pp.

Chang, F. K., and E. L. Krinitzsky (1977). Duration, spectral content, and predominant period of strong motion earthquake records from western United States, U.S. Army Engineer Waterways Experiment Station, miscellaneous paper S-73-1, Vicksburg, MS.

Chang, S. W., D. B. Jonathan, and R. B. Seed (1996). Engineering implications of ground motions from the Northridge earthquake, *Bull. Seism. Soc. Am.* **86,** no. lb, S270-S288.

McCann, W. B., and H. C. Shah (1979). Detertnining strong-motion duration of earthquakes, *Bull. Seism. Soc. Am.* **69**, no. 4, 1253-1265.

Moayyad, P. and B. Mohraz (1982). A study of power spectral density of earthquake accelerogram, NSF Report PFR 8004824, Civil and Mechanical Engineering Dept., Southern Methodist University, Dallas, Texas.

Mohraz, B. and M. Tiv (1991). Spectral shapes and amplifications for the Loma Prieta earthquake of October 17, 1989, in Proc. 3rd U.S. Conf. Zifeline *Earthquake Eng*., Los Angeles, California, 562-571.

Page, R. A., D. M. Boore, W. B. Joyner, and H. W. Caulter (1972). Ground motion values for use in the seismic design of the Trans-Alaska pipeline system, *USGS Circular 672*.

Plakfer, G. and J. P. Galloway, Editors (1989). Lessons learned from the Loma Prieta, California, earthquake of October 17, 1989, *USGS Circular 1045*.

Trifunac, M. D., and A. 6. Brady (1975). Astudy of the duration of strong earthquake ground motion, *Bull. Seism. Soc. Am.* **65**, 581-626.

United States Geological Survey (1989). U.S. geological survey strong-motion records from the Northern California Loma Prieta earthquake of October 17, 1989, *U.S. Geol. Surv. Open-File Rept. 89-568*, 85 pp.

SEISMICITY AND CULTURAL MONUMENTS OF GEORGIA

V. ZAALISHVILI *, O. SULABERIDZE *, T. CHELIDZE **,
O. VARAZANASHVILI**, Z. JAVAKHISHVILI**
**Zavriev Institute of Structural Mechanics and Earthquake Engineering,*
Georgian Academy of Sciences;
8,M.Aleksidze str., Tbilisi 380093, Georgia
***Institute of Geophysics, Georgian Academy of Sciences;*
1, M.Aleksidze str., Tbilisi 380093, Georgia

1. Abstract

There are two main connections between seismicity and the state of cultural monuments in earthquake prone countries.

The first one is connected with using both historical chronicles and results of revision of monuments for parameterization of ancient earthquakes, namely, for reconstruction of date, location of epicenter, definition of hypocenter depth, macroseismic effect and intensity. This approach has been extensively used by Georgian seismologists for quantification of ancient event; recently the effects of 1088 Tmogvi, 1263 Samtskhe, 1350 Lechkhumi-Svaneti earthquakes of magnitudes close or equal to 7 have been investigated.

The second approach consists in assessment of destructive effects of recent earthquakes on cultural monuments and elaboration of cost-effective methods of their reconstruction and conservation which could guarantee their resistance to possible future seismic events. This implies involvement of seismologists, civil engineers, geologists, archaeologists and material scientists.

Racha 1991 earthquake (M~7) damaged numerous architectural monuments of Middle Ages. The level of damage varied from zero to the full destruction (Mravaldzali). It is interesting to note that in some cases the monuments, founded on bedrock were more damaged than these founded on soft soils.

The monuments, damaged by Racha earthquake, are now partially or fully reconstructed. The reconstruction of the brilliant example of Georgian medieval architecture – XI-th century Nikortsminda cathedral – was performed in an original way: the shifted cupola of the church of 6 m diameter was lifted by a big inflating balloon, turned and fixed in the correct position. The materials, related to damage (III degree) and reconstruction of Oni synagogue (XIX-th century) with cupola-shaped roof are presented.

S. Balassanian et al. (eds.), Earthquake Hazard and Seismic Risk Reduction, 127–136.

2. Introduction

In many countries of the world the level of seismic hazard increases and accordingly the development of seismic investigations made it necessary to define strong earthquake parameters for time that considerably exceeds the instrumental period of observations. In the present work the main attention is paid to the connection of basic parameters of strong earthquakes in Trans-Caucasian seismoactive region by historical-archaeological materials.

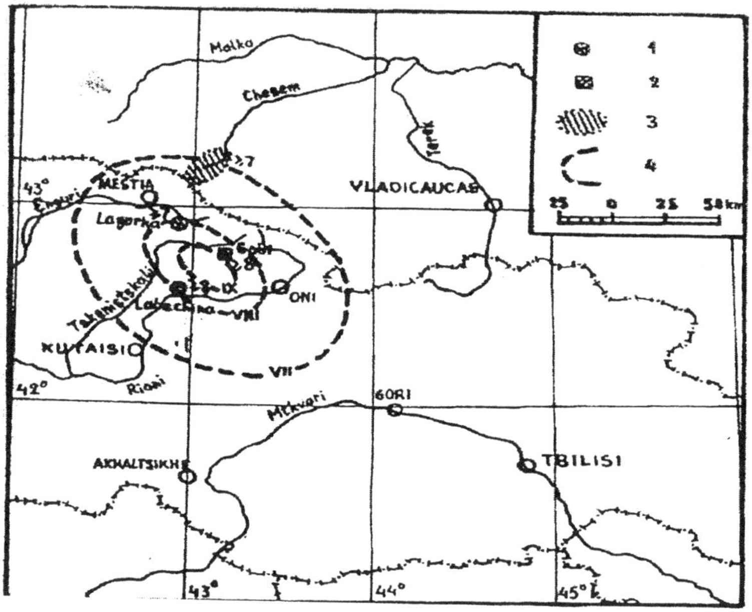

Figure 1. Scheme of macroseismic data and isoseismals of the 1350 earthquake.
1 - points with macroseismic characteristics, their intensity;
2 - paleoseismodislocations, their intensity;
3 - areas with intensity I ≥7;
4 - isoseismals.

At the same time it is very important to reveal and reconstruct parameters of earthquakes which caused maximal intensity of shaking on the territory of Georgia. The study will take into account also the knowledge on earthquakes, which happened in the vicinity of territory of Georgia. In particular on the basic of these data [1,2,3] we can conclude that the seismic event of the 14-th century had maximum macroseismic display in Georgia. More specifically, the earthquake epicenter probably was in the area of Letchkhumi ridge (Fig. 1). On the basis of chosing isoseismal models of various magnitude levels with regard for macroseismic data of this event, the main parameters of earthquake, which we define as Lechkhumi-Svanetia earthquake, can be determined as following; 1350 ($\Delta t=\pm 50$ years); $\varphi=42{,}7$, $\lambda=43{,}1$ ($\Delta E=\pm 0{,}5°$); h=15 km ($\Delta h=8$ - 30 km); M=7,0 ($\Delta M=\pm 0{,}5$); $I_0=9$-10 ($\Delta I=\pm 1$) [4]. Probably seismoactive structure or Gobi refers to this earthquake because both its place and the initial time practically coincides with considered seismic event [4]. In those centuries a shaking of similar intensity on the territory of Georgia might be caused by a well-known (from historical sources) strongest Samtskhe earthquake. However, its epicenter was at 100 km distance from indicated region and therefore the intensity could not be more than 6-7, the more so, that seismic vibrations spread across three tectonic zones in the direction on Great Caucasus.

On the basis of the analysis the following can be accepted as main parameters of Samtskhe (Fig. 2) earthquake: 1283 year ($\Delta t=\pm 10$ years); $\varphi=41,7$, $\lambda=43,2$ ($\Delta E=\pm 0,5°$); h=15 km ($\Delta h=8$ - 30 km); M=7,0 ($\Delta M=\pm 0,5$); $I_0=9$-10 ($\Delta I=\pm 1$) [4].

In the historical past similar seismic events occurred in Samtskhe-Djavakheti, for example Tmogvi earthquake, April 22,1088 years [1,2]. For some rceason this earthquake is attributed to moderate earthquakes in the catalogues. Detailed acquaintance with primary sources, describing the earthquake [4] revealed, that by its scale and intensity it is not much inferior to Samtskhe earthquake.

Tmogvi earthquake was of larger scale than it is reported in the catalogues (Fig.3). Its parameters could be: April 22, 1088 year ($\Delta t=\pm 1$ day); $\varphi=41,5$, $\lambda=43,3$ ($\Delta E=\pm 0,5°$); h=15 km ($\Delta h=8$ - 30 km); M=6,5 ($\Delta M=\pm 0,5$); $I_0=9$-10 ($\Delta I=\pm 1$) [4].

In conclusion, we shall consider in short the meaning of the results, received on the basis of reconstruction of the historical materials about the strongest earthquakes, for seismic hazard estimation of the territory of Georgia. Firstly, Lechkhumi-Svanetian earthquake has shown that within the boundaries of the south slope of the main Caucasian ridge, in the vast region, between the areas of maximum intensity of Ckhalta (1963) and Racha (1991) earthquakes, seismic events was fixed not only by materials of paleoseismo-geology, but also by historical data, and this confirms the existence of the entre zone with intensity of 9 along the northern boundary of Georgia. Secondly, Tmogvi earthquake and especially Samtskhe earthquake revealed that the extreme part of Samtskhe-Djavakheti had once been distinguished by very high intensity. Now this region is calm. Thirdly, according to available historical materials there are not enough arguments to assert that in historical past a series of strongest earthquakes occurred in the region of the Mtskheta, though according to instrumental data there is no doubt that moderate level seismicity existed in this region. The lack of information about destruction of Mtskheta requires further investigation in this direction.

The second direction of research concerts impact of recent seismic events on cultural monuments.

Racha (1991) earthquake intensity was the highest in Ambrolauri and Oni districts. Architectural and historical monuments of Ambrolauri district include 83 units. About 30 of them belong to the category of national importance. The earthquake damaged most of the monuments. The most well-known are: Nikortsminda cathedral (XI-th cent.), Barakoni cathedral (XVII-th cent.), Archangel church (X-XI cent.) near Zemo-Krikhi village, Mravaldzali church (XI-th cent.), etc. [1].

Archangel church is situated on a slope of a great mountain, 5 km from the town Ambrolauri. Originally it belonged to the so called hall type. In XIX-th century the church was enlarged. This worsened its firmness. In the sixties of our century the church was fortified and restored. A reinforced concrete belt was arranged along the main hall contour. In spite of this belt, after the earthquake the vault came down completely, northern and southern walls of the church were utterly destroyed. Only a western wall (restored in the sixties) was intact. Strength of concrete in the belt proved to be low.

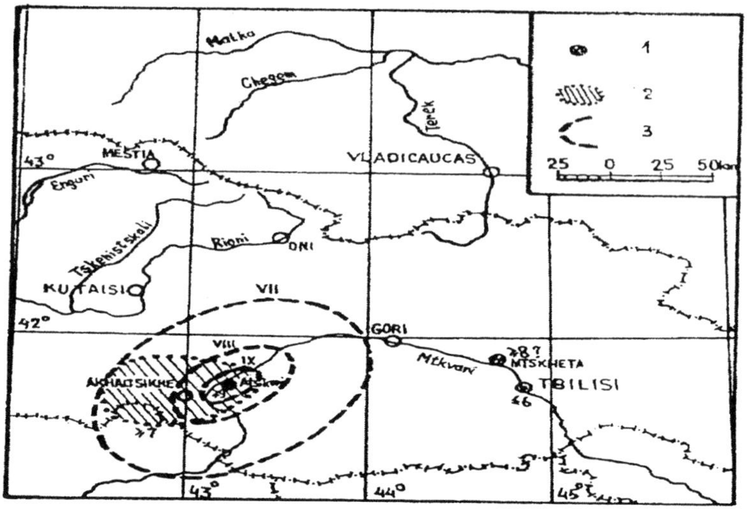

Figure 2. Scheme of macroseismic data and isoseismals of the 1283 earthquake.
1- points with macroseismic characteristics, their intensity;
2- areas with intensity I≥7;
3- isoseismals.

Barakoni cathedral has a cupola-cruciform construction. Neatly trimmed stable sand stones are used as building materials for walls. The cathedral is disposed on a low place. After the earthquake insignificant damages – falling out of separate decorative stones were observed. It was fully restored in September 1991.

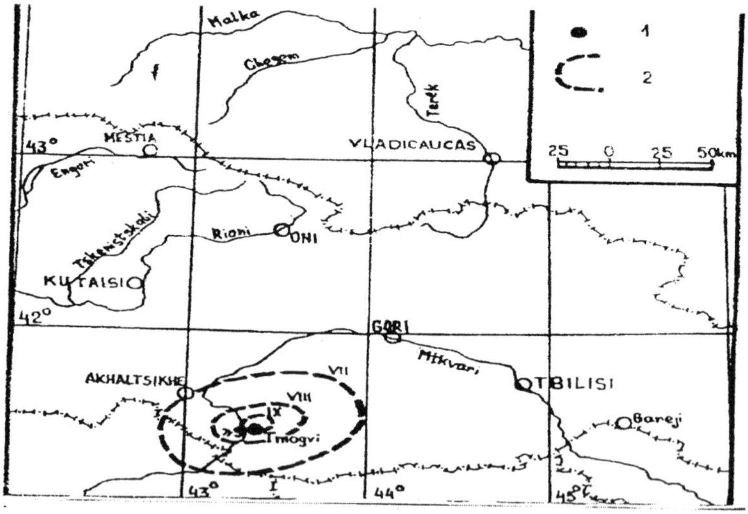

Figure 3. Scheme of macroseismic data and isoseismals of the April 22, 1088 earthquake.
1- points with macroseismic characteristics, their intensity;
2- isoseismals.

Mravaldzali church (hall type) is situated on a slope of a mountain. The church was transformed into cupola construction in the last century. A western part of it was disassembled and a cruciform annex was raised there, and the cupola rested up on it. In 1990 restorers had to take some measures. The construction of the cupola was reinforced, a new roofing was arranged. In spite of this, after the earthquake the church was razed to the ground (Fig.4a, b).

The church of the Virgin (XIV-XV cent.) in village Bugeuli is situated at the edge of a plateau, 6-7 km from Amblolauri. The church is small, 8x12 m, hall type. The walls are made of well-fitted basalt blocks. The so called "cyclopean" laying is used. In the sixties of this century restoration works were carried out. A reinforced concrete belt was arranged at a cornice level. But it was impossible to lock it because of the architectural detail in the form of a stone head of a ram. This determined lack of strength of the construction. After the earthquake the north-west corner of the building was split. The northern wall was destroyed. The lining of the western wall, where one of the entrances was placed, exfoliated and a part of big blocks came down.

In Oni district the number of monuments reaches 150.

Saint Giorgi church in Khirkholisi village. This village is situated on the north slope of Racha ridge, 5 km from town Oni (at a height of 1280 m). The earthquake intensity in this village was 7. The foundation of the church is of root rocks. Before the earthquake the overhead cover was absent. After the earthquake the walls came down on three sides of the church

Figure 4a. Mravaldzali church before the earthquake.

Saint Giorgi church in Parakheti village is situated in 12 km from town Oni. The village is divided in two parts, upper-on a slope, and lower-on a level surface. The earthquake intensity in the upper part of the village was 8 and in the lower part 7 [5]. The church is situated in the upper part of the village, on a hill, and a single-have building, 12×6 m, of lacerated stones. The foundation is of root rocks. The earthquake destroyed the church completely.

Figure 4b. Mravaldzali church after earthquake.

Mtavarangelozi church (XIX-th cent.) in Zemo Bari village is situated on a slope of Racha ridge, 20 km from town Oni (1200 m above sea-level). The earthquake intensity was 8. The church is a single – nave basilica c f lacerated stones, 18×8 m. The foundation of the church is of root rocks. After the earthquake fragments of longitudial walls fell out, the overhead cover came down.

The church (XIX-th cent.) near Gomi village is situated on a top of a mountain. The earthquake intensity was 8 in this village. The earthquake caused no damage to the church.

Oni synagogue (1895) is situated in the center of the town, on flat country. The building is right-angled with portico, 14,9×18,5 m [6]. Walls are made of a local stone with lime-cement solution of average firmness. Outside walls are reveted with trimmed stones. After the earthquake the building was seriously damaged: cracks appeared at angular and extreme arches, in vaults. A wall and some sculptural figures came down. At present Oni synagogue is restored (Fig.5).

The character and the degree of monument damage and destruction are often quite precise indicators for estimation of earthquake intensity; hence, it is an actual and important task to collect and analyze corresponding data. After Racha earthquake (1991) very significant number of architectural monuments [5] was damaged in west Georgia. At present some of them are restored and even former functional purposes are returned to them. As an example of a successful solution of this problem serves

Nikortsminda cathedral, one of the most prominent monument of Georgian architecture of XI century (Fig.6).

Figure 5. Synagogue in Oni after earthquake.

Figure 6. Nikortsminda temple.

The cathedral is of a cupola-cruciform construction, with some late annexes [7]. The facade laying of the walls is of a neatly trimmed travertin. Portals and the drum of the cupola are richly decorated with deep fretwork on a stone. Interior walls and the cupola of the cathedral are fresco painted. Thickness of walls at a drum level is 1,5 m and includes facade laying, interior revetment and plaster on lime solution. The exact date of building completion is 1014 year.

The cathedral is situated in Nikortsminda village at a height of 1100 m above sea-level. The very first earthquake tremors (April 29) caused impressive damages of the upper registers of the cathedral. Through vertical cracks appeared in the walls of the drum and then it was striped in two. Through cracks were seen in the cupola (diameter 6,4 m) too, its fragment (about 6 m^2) came down. A typical scheme of cupola destruction was realized- a number of radial and one annular crack, limiting the central rigid disc (about 3m in diameter). At the same time this disc was displaced downwards about 6-10 cm relative to the rest of the cupola. It became wedged and this prevented subsequent destruction of the cupola. Besides, after the earthquake a sizable part of the facade laying of the drum came down and separate stones were found at 20-25 m distance from their primary location.

Shortly after the earthquake the first measures were taken in order to avoid subsequent damages from possible aftershocks. At the top of the drum, where the cupola rests upon it, temporary belts of steel frame 10 mm were set over plank linings. Then inside the cathedral a power metal scaffolding was installed, proofing up the most damaged parts of the cupola. Above the cupola a temporary supplemental roofing was erected. In order to ensure general steadiness of the building during reconstruction, the sub-cupola drum was reinforced by spatial pivotal system. At drum base and where the cupola rests, steel belts (with 10x100 mm section), which were tightened with bolts, were set. After all spade-works were finished a final stage of cathedral reconstruction began: directly under the central disc, at a distance of 5-6 cm from it, a plank scaffold was erected, where about 20 big oxygen cushions (cylinders) were played (Fig. 7). The cylinders were connected with a compressor by a system of flexible hoses. The lifting of the wedged central disc (about 4t) up to the initial position was realized by filling the cylinders with air. After the annular crack was filled with modified lime solution, reinforcing and concreting of the shell upon the cupola was done. In such a way it was formed a double – layer construction in the form of a stone vault and a reinforced concrete envelope, working as a single whole.

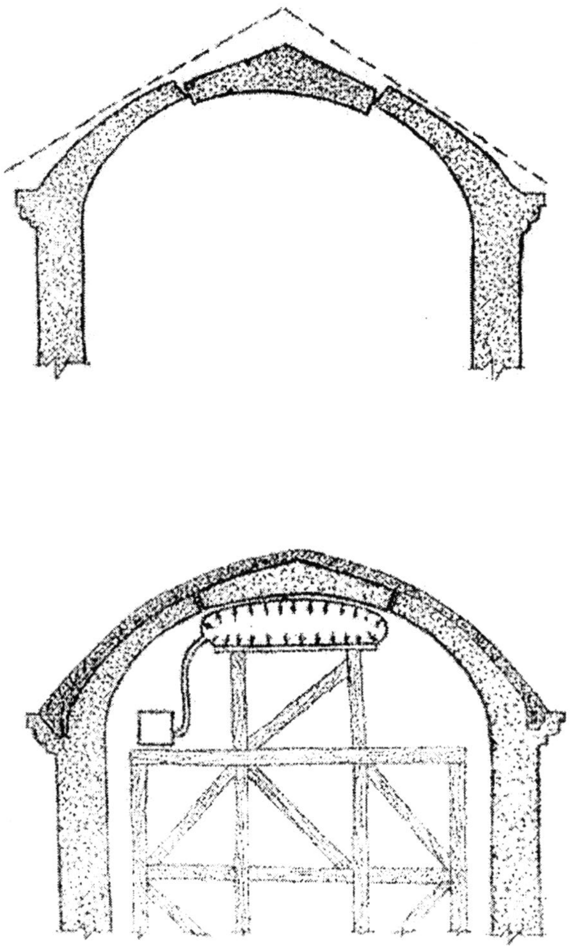

Figure 7. Initial (a) and final (b) stages of reconstruction of cupola of Nikortsminda temple.

The great advantage of the used method is that the lifting of a part of the cupola is performed by using soft cylinders, which do not damage a fresco.

3. Conclusions

Damages are approximately identical in both districts. There are historical monuments which practically did not suffer from the earthquake, though in respect of construction the buildings do not differ much.

The greatest differences are in soil conditions. There are first of all, geo-morphological conditions, it means, peculiarities of the relief. As a rule, uniform buildings were more seriously damaged, even destroyed, on slopes and peaks of a mountain. Great number of buildings, foundation of which were on the so called soft soils, practically did not suffer from the earthquake. The stated facts must be the result of the differences in seismic energy absorption, on the one hand, and of resonant phenomena in the system soil-building, on the other hand. Essential influence on construction vulnerability also had their constructive peculiarities and durability of building elements. Massive, badly clamped buildings were much more damaged than light ones. Wooden buildings practically were not damaged. For example, in the village Bokva, where the earthquake intensity was 9, Saint Giorgi church was razed to the ground. The roof of the church, as well as the roofs of standard one-or two- story country dwelling houses (made of stones), lay on destroyed walls. Wooden houses were moved, sunk to one side and nevertheless, stayed undestroyed.

4. References

1. New Catalog of Strong Earthquakes in the USSR (1977) Nauka Publ. House, Moscow (in Russian).
2. New Catalog of Strong Earthquakes in the USSR from Ancient Times though 1977. (1982) NOAA. USA.
3. Paleoseismogeology of the Great Caucasus (1979) Nauka Publ. House, Moscow (in Russian).
4. Varazanashvili O., Papalashvili V. Reconstruction of Strong Earthquake Parameters by historical Data of XI-XIV cc in Georgia, Tbilisi, Metsniereba, 1998 (in press).
5. Gogmachadze S., Zaalishvili V.. Kipiani D., Damages of Historical and Architectural Monuments. Engineer Analysis of Consequences of Racha Earthquake 1991 in Georgia, Tbilisi 1991, Metsniereba Publ, House, Tbilisi 1991, 91-108 pp.
6. Danielashvili M.,Oni sinagogue. Engineer Analysis of Consequence of Racha Earthquake 1991 in Georgia, Tbilisi 1991, Metsniereba Publ, House, Tbilisi 1991, 222-227 pp.
7. Sulaberidze O., Nikortsminda Cathedral. Engineer Analysis nf Consequences of Racha Earthquake 1991 in Georgia, Tbilisi 1991, Metsniereba Publ, House, Tbilisi 1991,217-1221 pp.

GEOMORPHOLOGICAL ASPECTS OF THE FORMING OF EARTHQUAKE INTENSITY EFFECT IN HIGHLANDS

V. ZAALISHVILI*, S. GOGMACHADZE**

* *Institute of Structural Mechanics and Engineering Seismology;*
8, M.Aleksidze str., Tbilisi 380093, Georgia.
Center of Applied Geophysics, Engineering Seismology and Seismic Protection of Structures;
1, M.Aleksidze str., Tbilisi 380093, Georgia.
Institute of Applied Mathematics and Informatics, Vladikavkaz, Russia;
10,Vatutin str., Vladikavkaz, 362040, (North Ossetia) Russia.
Joint Institute of Physics of the Earth;
10, B. Gruzinskaia str., Moscow, 123242, Russia.

** *Institute of Structural Mechanics and Engineering Seismology;*
8, M.Aleksidze str., Tbilisi 380093, Georgia.
Center of Applied Geophysics, Engineering Seismology and Seismic Protection of Structures;
1, M.Aleksidze str., Tbilisi 380093, Georgia.

1. Abstract

The analysis of the data of engineering macroseismic survey of the destructive earthquakes effect has shown that the degree of seismic effect is in close correlation with the features of geomorphological conditions. On the other hand, this relation is not simple. There can be vividly observed a multifactor phenomenon of correlation. The above-mentioned circumstance imposes certain (definite) restrictions on the direct use of the data of macroseismic survey, that is of superficial character to some extent. It occurs due to brief terms, given for direct study of various engineering-seismologic situations and reducing the analysis to the established or descriptive type. It leads to the urgency of sorting out the survey data on the basis of correcting information. Such approach allows to specify the reasons for changes of earthquake intensity in each particular case.

In the paper there are presented the main factors, that cause the damage of buildings and edifices. The roles of nonuniform deformations including strains have been shown in the formation of seismic effect.

There were given examples (Spitak earthquake, 1988; Racha earthquake,1991) of changes in the earthquake intensity effect, stipulated by the peculiarities of physics-mechanical characteristics of grounds-bases of buildings and edifices, by the type and structural features of the latter.

S. Balassanian et al. (eds.), Earthquake Hazard and Seismic Risk Reduction, 137–144.

In the paper it has been shown that intensity is defined not only by the value of the slope of the relief, but also by a particular location of site in a hierarchy of massif. There were given the results of the analysis of the earlier conducted studies, connected with the need of practice and special experimental investigation (full-scale and laboratory tests).

The effect was achieved in the form of explosions and monochromatic vibrations within ultrasonic range etc. Despite minor differences in conditions, the results of studies, are mainly characterized by a high degree of correspondence.

The conclusion is as follows: the account of geomorphological conditions of site should be carried out by using the so called "relief coefficient". It allows to consider besides the value of the slope of relief, particular location conditions of the site under study.

The latter circumstance increases the motivation for establishing the reasons of abnormal changes in the intensity effect of strong and destructive earthquakes.

2. Introduction

It is known that the major part of seismoactive zones is located in highlands (mountainous regions), that are characterised primarily by rugged country of the relief and various engineering – geologic conditions. The results of the engineering survey of strong and destructive earthquake effects show that the above-mentioned features predetermine the intensity of earthquake effect. The account of the effect of the relief and engineer – geologic site structure on the intensity of destructive earthquake effects is of great scientific and practical interest for conducting corresponding of antiseismic measures both on the territories of the regions that have already been studied and those – still under investigation. In table 1 of Building Codes and Regulations of former USSR and other Building Codes of Russia (1997), Armenia, Eurocode [1 - 4], soils are subdivided into three categories according to seismic features. However such subdivision is mainly connected with their physical- mechanical properties and it does' t reflect real seismic soil features under seismic effects. The effect of the relief on the intensity of earthquake in the indicated Codes hasn't been even considered, whereas in the remmendations on seismic microzonation it is advised to increase seismicity of sites investigation 1 degree greater with the of steepness of 15 and higher degrees. Thus the study of the effect of the relief and peculiarities of soil conditions are becoming more actual for obtaining sufficiently motivated results during the works on seismic microzonation (SMZ).

It should be also mentioned, that this problem hasn't been all over the thoroughly enough studied the world so far.

3. Forming of Intensity Under Earthquakes

Behavior of building and structures under earthquake is defined by a number of factors, that can be conventionally subdivided into three groups. The first group includes magnitude of earthquake (moment of magnitude), focal depth, epicentral distance, the angle of seismic waves passing. The second group includes the features of engineering – geologic site' s structure type of soils-layers, physical state, power of layers, the angle of stretching, underground bedrock, surface relief etc. The third group includes type of buildings located on a site, their structural features, type of the employed materials, technical state, building location according to the relief and orientation relative to the earthquake focus etc. Thus we can vividly see multifactor correlation of links. The mentioned circumstance imposes certain limits to use directly the data of macroseismic survey. It occurs to brief terms given for direct study of various engineering - geologic situations and reducing the analysis to established or descriptive type . It leads to the urgency of sorting out survey data on the basis of correcting information. Such approach to specify the reasons for changes of earthquake intensity in each particular case.

Seismic effect or earthquake intensity is formed by particular arrangement of the above mentioned factors. There are considered physical phenomena formed by concrete set of such factors; the so-called resonance phenomenon in a soil-structure system , nonuniform sagging of buildings and structures, strengthening (attenuation) of the falling signal by sail layer etc.

Following the Spitak earthquake (Armenia, 1988), large destruction in Gumri (Leninakan), was aggravated, in our opinion, by resonant phenomena. At the same time it is of interest to study the damages of a complex of 5-storey apartment houses during the Racha Earthguare (Georgia,1991) .The complex was consisted of 4 identical (similar) masonry 5 story buildings. The buildings in plane were arranged by ledges. The soils of buildings bases were macroporous, clayey soils, bedded by watery gravel s. Damages of buildings 1-4 were monotonously increased and changed from the first to the third degree relatively. Special investigation have shown that the power of clay is modified from 8 m (build. 1) to 11m (build. 4). In the borchole at the build. 4 the level of water table was found at the depth of 5.5m. At the build.1, the level of gravels (h=8.0m) Thus, monotonous increase of the water table level led to the monotonous aggravation of damages. Moreover the bore pit opened up directly under the foundation of build. 4, showed that the soils were jammed by water (H=3.0m) The indicated soils subject to water lose considerably their carrying capacity. This fact stipulated nonuniform sagging of soils before earthquake were as the earthquake only revealed the difference in stress - strain state of buildings [5].The examples of thin kind could be further continued.

Geomorphological conditions of the region influence considerably the intensity of earthquake effect. On the other hand, during the Spitak earthquake, buildings and structures of the hospital in town Kalinino, located on the top of the hill were slightly damaged (0-1 degree of damage),

However the buildings located at the bottom of the hill had 1-2 degree of damage.

It is accounted for the fact that there were different soils of bases: bedrock and clay soils relatively.

During the Racha earthquake many buildings of rural type located at the bottom of hills were slightly damaged (0-1 degree). While buildings of the same type located along the slope ($\alpha > 20^0$) got considerable damages (2-3 degree) etc.

4. Investigation of the Geomorphological Conditions of the Region.

Studies have been carried out for Sukhumi mountainous massif [6,7]. While creating a physical 3-D model we followed the corresponding terms of similarity (Fig.1). In particular, velocity of propagation and density of medium in the relative layers of nature and model were retained the same (identical).

The territory of city was in plane: 9x9 km. with 9000x9000 m. The scale of a model: M1:10.000. The most hazardous frequencies for the model 0.9x0.9 m. correspond to 10-100kHz. In the process of studies frequency of generator was 62 kHz. It corresponds to oscillation frequency of nature. It is equal to 6.2 Hz.

Bedrock of mountainous massif is represented by weathered from the surface, cracked, carsted limestone.

Gypsum has been employed as a model material. In the course of studies, there was simulated a wave field of Rayleigh waves, propagating from the earthquake focus in the shelf area in the vicinity of Sukhumi.

The tests have been conducted using a device, which consisted generator of ultrasonic oscillations with the period T=3-20 ms and receivers (piezoelectric transducer of acceleration GTC-19).

While using the impulse with T=15msec, when the velocity of Rayleigh wave propagation in gypsum was V_R=1.21 mm/s, then the length of oscillation wave for the model was λ=18mm.

The source of ultrasonic oscillations was located on the surface of model in the area of shelf at the point of control.

The surface of model was divided into 96 squares and wave field of Rayleigh waves was studied 67 modal points (Fig.2-Fig.3).

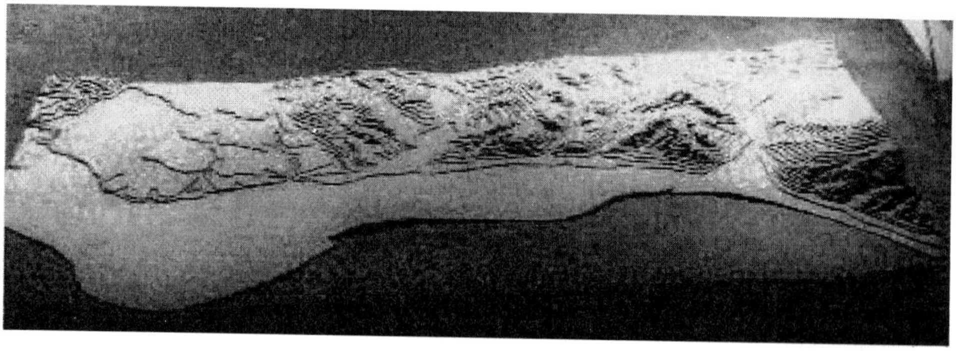

Figure 1. General view of a model.

At each station there were conducted 7 measurements and the obtained results were reduced to average. In the experiment there was considered possibility of reflection from the lateral face of the model.

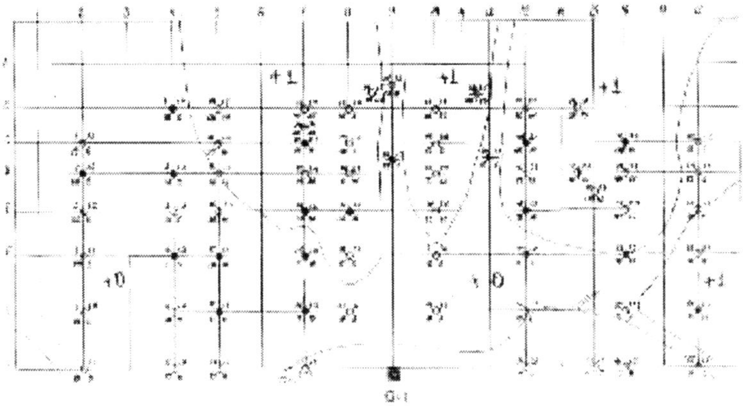

Figure 2. Division of the territory into individual sites (model).

Figure 3. Conducting of the experiment.

The effect of the relief features on Rayleigh waves (Fig.4) passing was assessed relative to its amplitude on a reference-circle i.e. relative to oscillation amplitude at corresponding distances on a similar model with homogeneous relief. Coefficients of amplitude level increase were specified by consecutive movement along the profile on the surface of model and compared with the amplitude at the point of control.

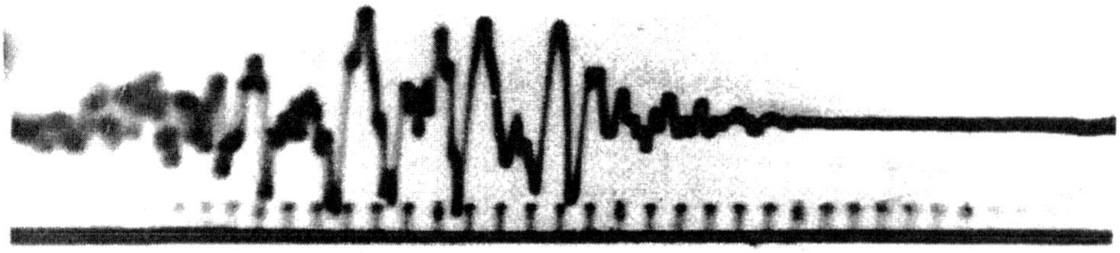

Figure 4. Characteristic accelerogramme at one of the stations.

The analysis of the data shows, that at an angle of slope d ≥20°, strengthening gains k=2.0, it corresponds to an increment in 1 degree, according to the conventional concepts. In a few points increment achieves 2 degrees. The indicated points correspond to separately standing peaks. In some cases at d<<20°, strengthening is however rather high and it also corresponds to 2 degrees. It can be explained by the arrangement (location) of the indicated points between dominant peaks. Thus we can draw a conclusion, that only the angle of relief, slope isn't sufficient to use for the relief characteristics. The height of survey doesn't always give adequate correspondence. In Fig.2 it have been given longitudinal (radially from the Sukhumi shelf) sections of the mountainous massif with the corresponding signal strengthening. It is clearly seen, that depending on the angle of slope, increase considerably varies (0÷3.0). On the other hand, transverse sections (almost parallel to coastline) show, that besides the angle of slope of survey, the height of survey influences substantially the effect of strengthening (Fig.2). Moreover, when local slope is not observed, nevertheless, strengthening is rather considerable. The gorges correlate well with smooth reduction of strengthening etc.

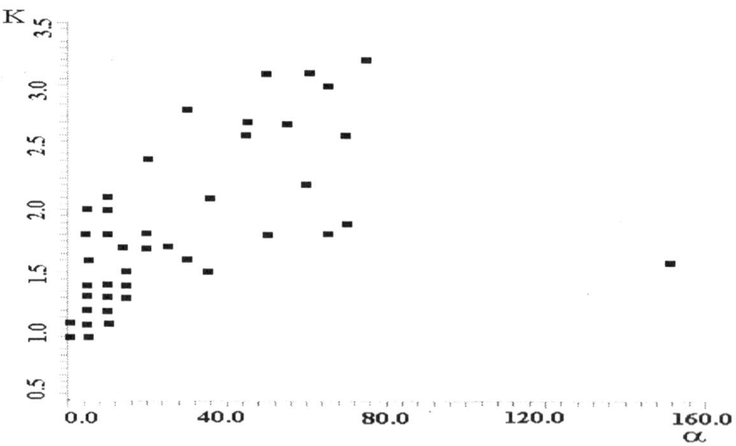

Figure 5. Dependence of increase on the angle of slope of the relief

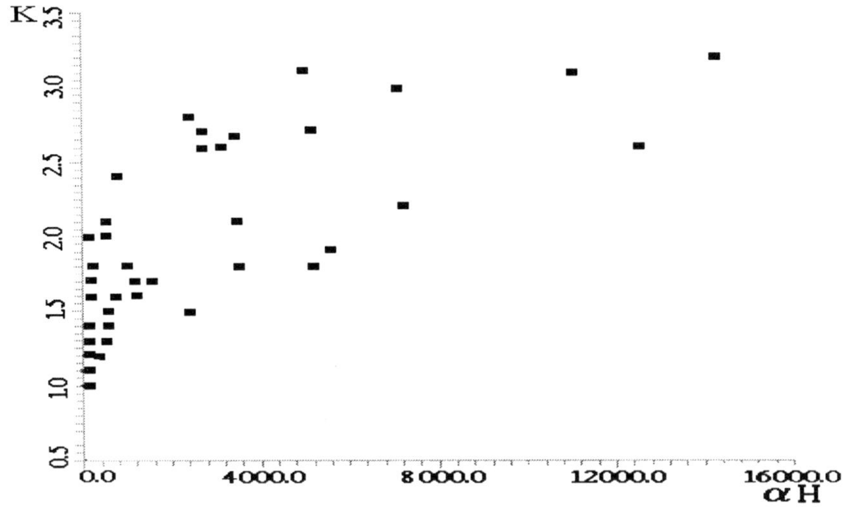

Figure 6. Dependence of increase on the height of survey

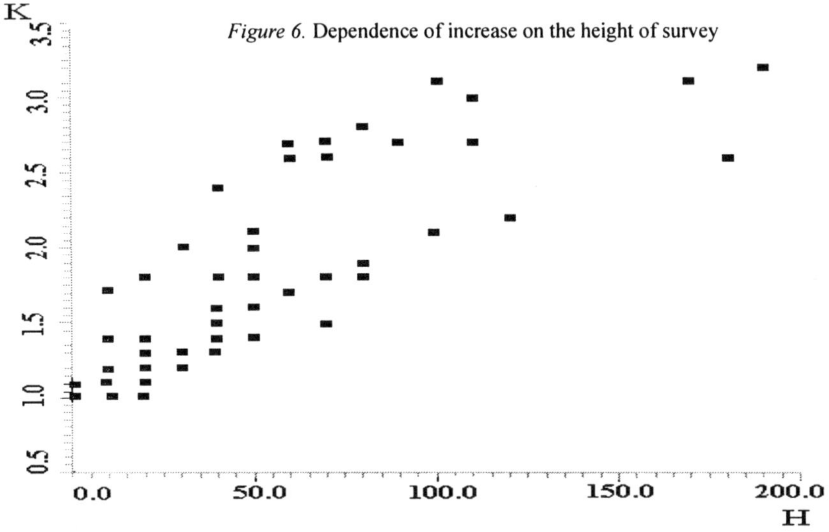

Figure 7. Dependence of increase on the "coefficient of relief"

The analysis of relations (Fig.5-Fig.6) reliability shows that in order to increase of results it is necessary to consider both factors. It led to the introduction of the so-called "coefficient of relief" R, which allows to take into account both factors. The corresponding dependence (relationship) shows (Fig.7) that reliability of the results considerably increases:

$$R = \alpha H \qquad (1)$$

Formally at $\alpha = 1^0$ (gradus) and height H=650 -700 m, i.e. at R=650-700 (gradus x m), increment of intensity would be one degree.

Thus, effect of relief can be rather safely considered by employing the coefficient of relief. Following special studies, the obtained results allow to introduce the mentioned parameter into Building Codes and Design Regulations of Georgia. The results of work can be also useful in methodic recommendations on seismic microzonation of territories of cities and large sites.

5. Conclusions

The analysis of the results of experimental survey shows that characteristics of relief effect on seismic earthquake effect are insufficiently such characteristics as the angle of slope or only the height. It was necessary to employ such a parameter, which could enable to consider both of the mentioned factors simultaneously.

From the physical point of view, the relief possesses a certain potential energy. For the assessment of the mentioned energy, the so-called "coefficient of relief" has been introduced.

The reliability of the account of relief considerably increases using coefficient of relief, that can be represented as a product of the angle of slope into corresponding height (proceeding from initial conditions).

The results of studies will be used in the developing Building Codes and Design Regulations and the recommendations on seismic microzonation of Georgia.

6. References

1. Building Codes of USSR. Moscow, 1982
2. Building Codes of Russia, (Second edition - project). Moscow, 1997
3. Building Codes of Armenia (project). Yerevan, 1997
4. Building Codes of Uzbekistan. Tashkent, 1997
5. Gogmachadze S.A., V.B. Zaalishvili, D.G.Kipiani, A.V. Odisharia. Examples of the soil conditions and topography features influence on the damage degree of buildings." Engineer Analysis 1991 year of the Racha earthquake consequences in Georgia ". 141-145 pp.
6. Zaalishvili V.B., S.A.Gogmachadze. Research wave field of impulse and vabration sources. Report of Institute of Structures Mechanics and Earthquake Engineering. Tbilisi, manuscript, 1989, 240 pp.
7. Gogmachadze S.A. The results of solution of spatial problems using design-theoretic methods of seismic microzonation. Seismic microzonation of territories of cities and large sites. Metsniereba, 1987.Tbilisi. 116-119pp

ON ESTIMATION OF DOMINANT VIBRATION PERIODS OF SITE

O. KURANOVA*, A. ODISHARIA, V. ZAALISHVILI****
**Institute of Applied Mathematics and Informatics,*
10, Vatutin str., Vladikavkaz 362040, (North Ossetia) Russia
***Institute of Structural Mechanics and Earthquake Engineering,*
Georgian Academy of Sciences
8, M.Aleksidze str., Tbilisi 380093, Georgia

1. Abstract

Analysis of data of macroseismic inspection of consequences of strong and destructive earthquake shows, that quite often the cause of destruction or of amplification of seismic effect is connected with resonant phenomena in the soil-construction system.

There is a lot of example of such behaviour: Mexico (1985), Spitak(1988), etc.

This makes definition of dominant frequencies of vibrations of soil layer on which the construction is or will be built very important for increasing object's safety.

In the paper some methodical aspects of estimation of dominant periods are analyzed. The techniques, suggested by K. Kanai and Y. Nakamura are compared. The last one, based on the comparasion of spectra of horizontal and vertical oscillations (H/V ratio) unexpectedly has found very wide application all over the world.

In both methods the study of microtremors in an essential point. It has been proved, nevertheless that their application is not always justified. The examples of determination of dominant vibration periods of soil layer by Nakamura approach and by modern powerful artificial vibration sources (pulse sources and vibrators).

It is shown that at intensive loading the oscillation periods variation on attain 20-30% or more of initial, value.

Instrumental recordings et of magnitudes M=1.8-8.2 earthquakes obtained in varies regions of the world with various soil conditions.

It is shown that it is possible to standardize physical state of soil layer or their seismic properties using values of dominant periods of vibration of soil, which is important for practical purposes.

2. Introduction

Macroseismic- engineering inspection of strong and destructive earthquake consequents shows that very often the cause of destruction and amplification of seismic effect is connected to resonant phenomena in the soil-construction system.

S. Balassanian et al. (eds.), Earthquake Hazard and Seismic Risk Reduction, 145–151.

There are several examples of such behavior: Mexico (1985), Spitak (1988) etc. That makes definition of dominant vibration frequencies of soil layer on which the construction is or will be built very important for increasing object's safety.

The paper discusses some methodical aspects of dominant vibration periods on a specific example.

The building ground of crucial projecting equipment is located on a left bank of the river Don.

According to the data of engineer-geological investigations with the account of genesis, lithological structure and the physic-mechanical quality of ground-soil, five engineer-geological elements are distinguished. This five will serve as a foundation of equipment.

1. Loamstone layers near the bank of the river are represented by yellow-brown loamstones. Their thickness equals 0,7-1,3 m. Because of low physic-mechanical features and poor capacity formation cannot be used as the foundation of equipment and thus needs to be eliminated.

2. Sandy sediment formation is represented. Thickness of sandy broomtail equals 6.5-8.0 and the depth of the loamstone layers shifts from 1-2 to 10 sm.

3. Formation of sands of different (middle and smaller) thicknesses is represented. Sandy clays have the depth 25-30 m.

4. Alluvial formation is represented by saturated sands. The top of the layer reaches the depth 37-39 m saturated clays.

5. The top of the layers reaches the depth 45-50 m.

Analysis above-exposed data of engineer-geological investigations enables us to distinguish twelve different sites of the investigated territory, with the typical engineer-geological structure of soil and the diverse quality of irrigation. It must be underlined that inspected soil conditions – mainly similar in their engineer-geological construction, pretty much differ from each other by the power of estimated soil vibrations and composition of the sands. According to laboratory investigations sands of different saturation do not differ from each other much by their physic-mechanical features. The main difference between them comes from different formation of soil layers. But on the other hand more considerable differences might occur in some situations.

It must be mentioned that the positive characteristic of the investigation itself is the informational saturate of the ground on the basis of geologic studies.

A big number of wells considerably increase the ability of distinguishing several typical sites. Each instrumental station, with the little exception, is limited to an appropriate welling.

Indicated choice of typical sites gave us opportunity to scope majority of features of engineer – geological structure of soil layers, and also hydrogeologic and geomorphologic conditions of districted territory.

While carrying out corresponding calculations we investigated the site composed of less-moist and unirrigated sands with the thickness of 12-14 m. Level of soil waters exceeds 8-10 m.

On the basis of analyzing engineer-geological data, twelve typical sites were distinguished on the investigated ground. These twelve sites have principle characteristics of soils engineer-geological structure and different levels of soil waters.

3. The experimental investigations

Instrumental investigations were carried out on the typical sites of the ground territory in order to register the microseism and soil-vibration caused by power artifical sources – ГСК-6М. Instrumental net included 60 investigating stations equipped by vertically oriented velocimeters WILLMORE and accelerometers A-1632 (A-1633) with three components. The registration of signals was carried out with the help of digital-seismic station. As a result 364 microseismic and 242 soil vibration records were obtained. Data-processing program GeMis_WIN was used.

On the basis of instrumental information and special investigations we designed the method of evaluating soil vibrations [1,3,5,6]. This method was used to carry out the real investigation.

Analysis of spectral curved soil-vibrations showed next:

In the formulation of peaks characteristics of soil engineer-geological construction plays a big role. Increasing length of inspected piece of the record increases the stability of displaying frequency peaks on spectrogram. Availability of having large number of engineer-geological investigations describing characteristics of soil's engineer-geological structure enables us to carry out very important investigations in order to develop the method of "direct-inverse" problem – solving of engineering seismology. Thus the investigation aimed at establishing corral factors between the quality of soil's engineer-geological construction and corresponding spectrograms.

As mentioned above, in pressing paper [8] we investigated how the continuance of analyzing particular piece of record influences the quality of spectrogram.

In order to identify the effect of analysis or the number of spectrogram quality calculations consisting of interrelated vertical and horizontal signals appropriate estimations was carried out. As identified, the main features of spectrogram and the location of concrete frequency peaks are characterized with high stability during the welling "break out". From 500 /c to 62-5/c. The same time different soil conditions of investigated stations are characterized with diversing spectrogram "figures".

It has to be mentioned that in some cases on spectrogram of interrelated horizontal and vertical signals single peaks might show up. This kind of peaks in typical for irrigated sands. The possibility of receiving reasonable and dependable spectrograms of soil vibration were investigated with the help of impulse source ГСК-6М engaged in work as the soil layer shows particular reaction to the signals as weak as microseisms. Analysis of corresponding spectrograms for soils of investigated station №3 shows that the spectrograms with the minimal permission 62.5/c are most adequate with the initial spectrogram. This must be caused by the fact that the influence of microseism continuance is much higher that the longitude of soil vibration caused by impulse source. Therefore, reducing the permission will lead to particular rapprochement of corresponding "figures" of peaks on spectrograms and as a result - to the increase of processing standards. It must be mentioned that the particular part of information gets lost. In order to register the signal of low frequency appropriate for the given depth amplification of record time is essential. Sometimes during the registration of microseisms LF (low frequencies) peaks might occur. That is occurred by imperfection of measuring equipment (Seismoreceiver).

Therefore, for the future we are planning to investigate opportunities of appropriate approach on the ground of using more powerful vibration source for stronger soil oscillation. It will enable us to obtain more qualitative data of influence-time, to raise the permission of spectrogram and finally - to perfect the final data.

Besides we are going to check the possibility of detailing the frequency evaluation by processing the records of weak and strong earthquakes.

On the basis of above-discussed and also planned investigations the very important principle was established- the frequency of soil oscillation does not depend on the course and type of vibration sources. Here we also must add that if we increase the magnitude of influence, the vibration period will approach the period typical for real seismic influences. Perfection and further development of elaborated method will enable us to distribute well-grounded distributions of soil-vibration occurring on a particular territory.

4. The Analyzing of Result

Analyzing result of application following inference can be made: The new method is much more informative than the traditional one when investigating directly the amplitude Furrier spectra. This is clearly shown on appropriate spectrograms. They include suitable amplitude Furrier spectra for accelerometer consisting of three component (X, Y, Z) – channel 2 (Fig.1, Fig.2) and the connection of engaged specters from horizontal to vertical for the channels №2 and №4.

Analysis shows that the characters of peaks are quite adequate for different engineer-geological conditions.

For example, soil irrigation characterized by quite typical peak transferring throughout the high frequency region of the spectra. It clearly distinguished peaks can not be seen an amplitude spectra (X, Y and Z are being investigated all together) it will be accurately displayed on the spectra -connections (X / Z and Y / Z).

So, for the sample site faceted on the territory of investigation dominant soil frequency for connection γ/Z is 3.7Hz. Evaluation of the given layer with the use of well-known formula will look like this [2]:

$$\mathbf{H = V_S / 4 f}$$

f - vibration frequency;
H - soil formation thickness;
V_S - velocity of shear wave in a soil.

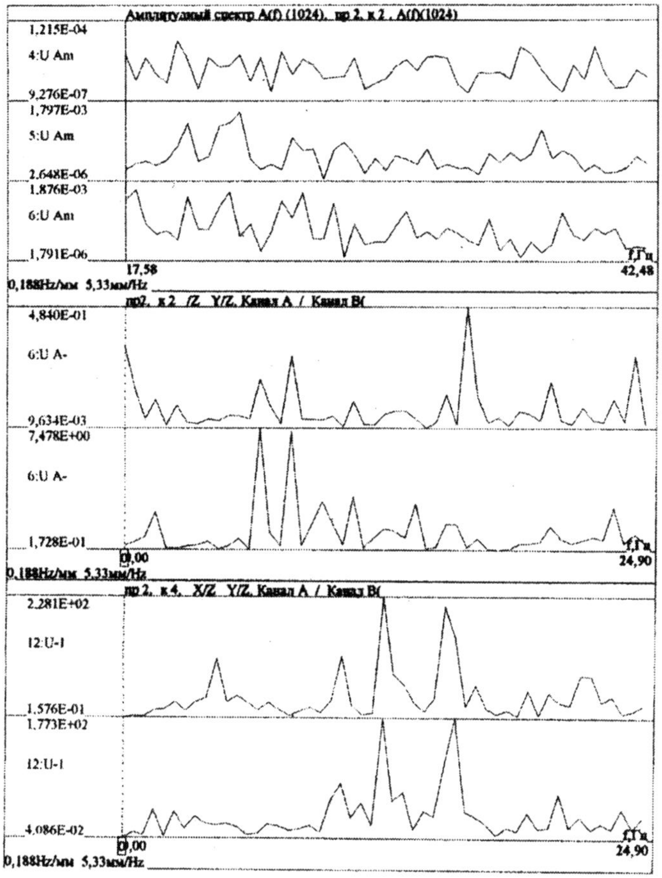

Figure 1. Amplitude Furrier spectra and Spectra Ratio X/Z, Y/Z

Calculations show, that when V_S = 240 m/c /23/, H = 15.8 m. For the connection X / Z when f = 9.5 Hz, H = 6.3m, etc.

Below we show the compassion of frequency measured for building 1 and building 2 ground soils.

Building 1 ground soil is meanly characterized by vibration LF - 0.8, 0.9, 1.5 (Hz) and building 2 ground soil most of the time experiences vibrations HF (high frequencies) - 0.6, 1.3, 4.9, 5.8 etc. Frequency for building 1 ground soil equals 1.0-1.5 Hz and for building 2 it reaches 3.0-5.0 Hz. By the results of previous investigations carried out on the basic of traditional seismic method dominant frequency for the building 1 shifted from 5.0 to 8.0 Hz and averaged 5.8 Hz. Thus, using above-discussed method increases dominant frequency.

On the basic of obtained material dominant vibration frequency distributions schemes were designed for two frequency ranges (LF and HF regions). They are quite adequate with real ground conditions. Thus the analysis of the scheme in LF range shows that if we increase the soil schock the dominant frequency will increase (regions of building 1). The same time soil irrigation is characterized by implication of dominant vibration frequency and the appearance of specific high frequency CHF "water" peak on the region 4.

Obtained of earthquake records using the mentioned method show that with the increase of motion magnitude in the boundaries m = 1.8 - 8.2 the period of own soil vibration increases by the level of seismic influence. Besides, this change has very accurate nonlinear character.

Conducted analysis show that if we perfect the developed method it will help us to receive more dependable evaluation of frequency features.

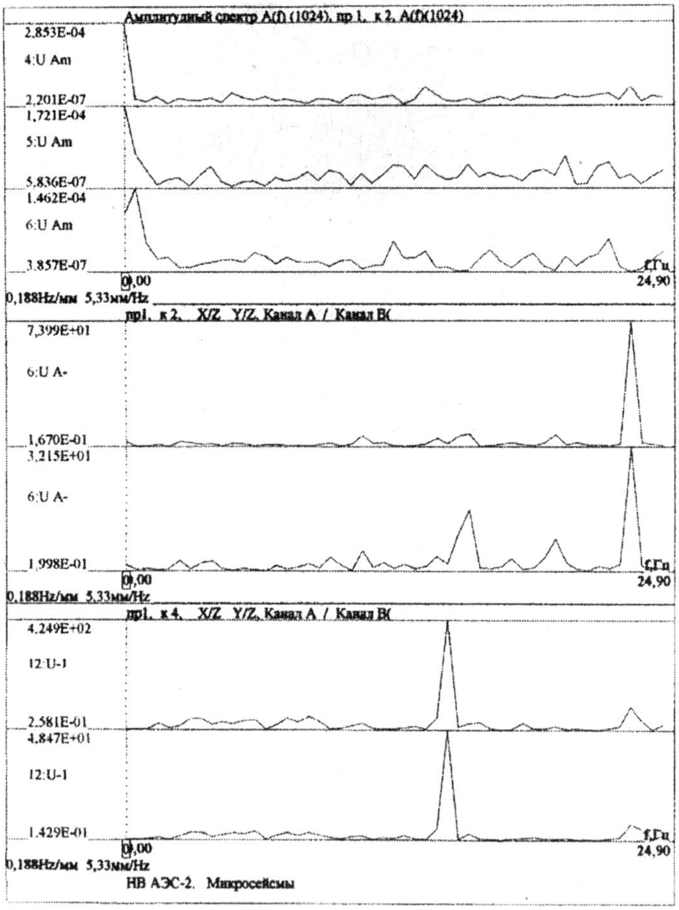

Figure 2. Amplitude Furrier spectra and Spectra Ratio X/Z, Y/Z

5. Conclusion

By using specially developed method dominant vibration frequency distribution schemes were structured (in the range of LF and HF). These schemes were in accordance with the engineer-geological composition or features of soil-layers and calculated data. According to the scheme designed for LF range ground soils of building 1 are characterized by 1.0-1.5 Hz frequency, and the soils of building 2-experience 3.0-5.0 Hz frequency. It should be underlined that the dominant frequencies shift from 5.0 to 8.0 Hz and average 5. 8 Hz as the results reveal. Thus, utilitarian of developed method increases ability of detailing soil vibration frequency distribution scheme structuring.

6. References

1. Hudson M.B., Skyers B.D., Lew M. // Vertikal Strong Ground Motion Characteristics of the Notridge Earthquake. In Proceedings of the 11th World Conference of Earthquake Engineering. Mexico., Acapulko, Abstr., 1996.
2. Kanai K. Relation Between the Nature of Surface Layer and the Amplitudes of Earthquake Motions. Bull. Earthquake Res. Inst., № 30, Tokyo Univ., 1952, pp. 31-37.
3. Kobayashi H., Vidal F., Feriche M.D., Evaluation of Dynamic Behaviour of Building Structures With Microtremors For Seismic Microzonation Mapping. //In Proceedings of the 11th World Conference of Earthquake Engineering. Mexico., Acapulko, Abstr., 1996.
4. Lermo J., Rodriquer M., Singh S.K. The Mexico Earthquake of September l9, 1985 - Natural Period of Sites in the Valley of Mexico from Microtremor Measurements and Strong Motion Data. Earthquake Spectra, N 4, 1988, pp.805-814.
5. Mohammadioun G., Mohammadioun B. Vertical strong ground motion in the near field and soil non-linearity. // In Proceedings of the. 11th World Conference of Earthquake Engineering. Mexico., Acapulko, Abstr., 1996.
6. Tokimatsu K., Arai H., Asaka Y. Three-dimensional Soil Profiling In Kobe Area Using Microtremors. //In Proceedings of the. 11th World Conference of Earthquake Engineering. Mexico., Acapulko, Abstr., 1996.
7. Zaalishvili V.B. «Seismic microzoning of the town territory». // In Proceedings The Second Russian-Chines Regional Seminar on Earthquake Engineering, May 1992, Mosc., Russia, pp. 58-64.
8. Zaalishvili V.B. «The influence of engineering and geologic features of soil lauer on the formation of wave field by impulse and vibration sources». // In Proceedings of the 9-th European Conference of Earthquake Engineering. Mosc., 1990, Vol. 4A, pp. 169-175.
9. Kramer S., L. Geotechnical Earthquake Engineering, Printice Hall, International Series in Civil Engineering and Engineering Mechanics, 1996, pp. 254-294.
10. Bonnet G, Heitz J. Non-linear seismic response of a soft layer. Proceedings of 10-th European conference on earthquake engineering, pp.361-364

A MAJOR NEOTECTONIC STRUCTURE WITHIN THE ARABIA – EURASIA COLLISIONAL OROGEN: THE ARARAT – ARAKS CONJUGATE FAULT SYSTEM AND ITS POSSIBLE IMPACT ON THE SAFETY OF THE METSAMOR NUCLEAR POWER PLANT

H. K. MKRTCHIAN
Black Mountain Oil and Gas Company; Ojai, California, 93023 USA;
American University of Armenia, Baghramian 40, 37021 ARMENIA

1. Introduction and Neotectonic Setting

Armenia is situated within the Alpine-Himalayan collisional belt, in southwestern Eurasia. It is an area of high seismic activity, as recently exemplified by the destructive Spitak earthquake of 7 December 1988 (M=7.0). The earthquake has dramatized the need for modern methods and techniques to be applied in geological and seismological studies.

Plate tectonic models (NUVEL 1 and 1A) [5,6] based on analysis of global oceanic spreading centers, transform fault systems and earthquake slip vectors indicate the Arabian plate is moving in a north-northwest direction (338°) relative to Eurasia at a rate of about 25-28mm/yr, averaged over 3m.y (fig. 1). These models also indicate 22mm/yr dextral-slip in 120° direction and 17mm/yr shortening in 030° direction [10]. Recent Global Positioning System (GPS) data suggest that in the fixed Eurasia reference frame, the Arabian platform moves in 322°±13° direction, at 20±3mm/yr rate [14,17]. Satellite Laser Ranging measurements are also in agreement with the GPS results [14,18]. Available scientific data suggest a convergence rate across the Armenian Highland of ±15mm/yr, absorbed along known and unknown active fault zones, in the form of dextral and sinistral slip on strike-slip faults and dip-slip on thrust and reverse faults. Presently there is not enough data available to allow more detailed quantitative estimates of aseismic and coseismic slip along the individual faults.

The Armenian Highland is the most contracted and deformed portion of this collisional belt, and has a mosaic structure, where the strain is released along internal fault zones. The important faults [2,3,9,11,15] are shown in figure 1.

This study documents the existence of the major dextral Ararat fault zone, striking along the Araks river, through the Ararat and Nakhijevan valleys. The fault is 220km in length, and extends from the area south of the Metsamor Nuclear Power Plant, southeastward, to the town of Ordoubad. The segmentation of the Ararat fault, historical seismicity, comparative analysis with other major faults in the same seismotectonic regime within 100km of it, and earthquake scaling parameters allow to conclude that the Ararat fault can generate M≥7 earthquakes. This in its turn raises

S. Balassanian et al. (eds.), Earthquake Hazard and Seismic Risk Reduction, 153–161.

serious questions regarding existing seismic hazard assesment of the Ararat valley in general, and the Metsamor Nuclear Power Plant in particular.

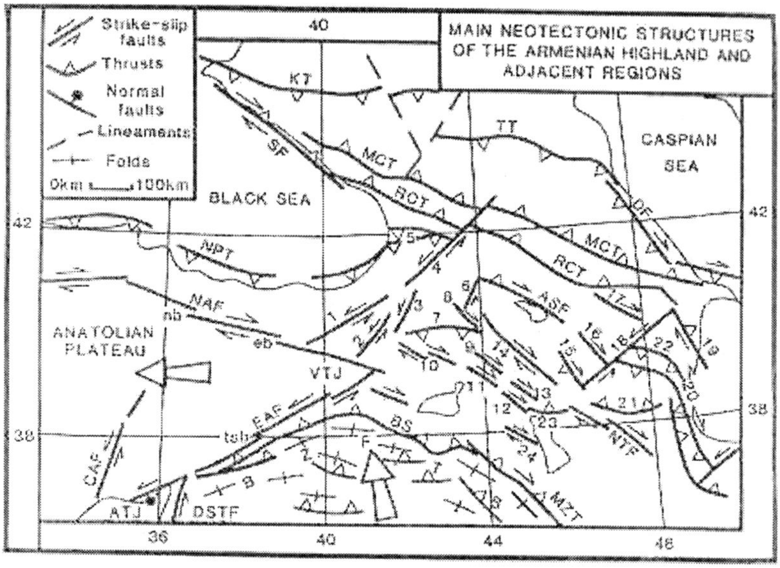

Figure 1. Main neotectonic structures of the Armenian Highland and adjacent regions. Abbreviations and numerical indexes: **VTJ**-Vard triple junction; **ATJ**-Adana triple junction; **BZFTB**-Bagesh-Zagros fold & thrust belt; **nb**-Niksar basin; **eb**-Erznka basin; **tsb**-Tsovk basin; Faults: **ASF**-Amasia-Sevan; **BS**-Bagesh suture; **CAF**-Central Anatolian fault; **DF**-Dagestan fault; **DSTF**-Dead Sea transform fault; **EAF**-East Anatolian fault; **KT**-Kouban thrust; **MCT**-Main Caucasus thrust; **MZT**-Main Zagros thrust; **NAF**-North Anatolian fault; **NPT**-North Pontides thrust; **NTF**-North Tavriz fault; **RCT**-Recent Caucasus thrust; **SF**-Sochi fault; **TT**-Terek thrust; **1**-Northeast Anatolian; **2**-Erzrum; **3**-Horasan-Narman; **4**-Borjomi-Kazbeg; **5**-Ajaro-Trialeti; **6**-Ackhurian; **7**-Kaghzvan; **8**-Ervandashat; **9**-Gailatu; **10**-Tutak-Karayazi; **11**-Chaldiran; **12**-Khoy; **13**-Maku; **14**-Ararat; **15**-Khustup-Giratagh; **16**-Akera; **17**-Kura; **18**-Araks; **19**-Lenkoran; **20**-Talesh; **21**-Sabalan; **22**-Dasht-Moughan; **23**-Marand; **24**-Salmas.

2. Kinematic Model and Geometry of the Ararat – Araks Conjugate Strike-Slip Fault System

The existence of the Ararat fault zone is presented for the first time - a major northwest striking (320°NW) dextral fault zone, identified by analysis of satellite imagery, geomorphology and historical seismicity. Nikonov [13] earlier suggested the presence of a fault striking along the river Araks in the Ararat valley.

The Ararat fault zone is defined as a 220km long lineament striking through the valley of the Araks river from the town of Ordoubad in the southeast to the town of Metsamor (site of the nuclear power plant) in the northwest. The most striking geomorphological indicator is a lineament striking through the middle part of the Ararat and Nakhijevan plains, along which the Araks river flows. The Araks fault is about 240km long, and strikes about 50°NE, along the Araks river valley, from the town of Megri in the southwest to the town of Sabirabad in the northeast. It has left-lateral slip and was delineated by Jackson and McKenzie [9], as a well defined lineament. In 1976, an M=5.6 earthquake was generated near Araks fault's northeastern end [9].

The dextral Ararat and sinistral Araks faults constitute a major conjugate fault system, in a pure-shear setting (fig. 1,2). Based on the recognition of this major conjugate fault system, a new neotectonic-kinematic model is suggested for the area located between the

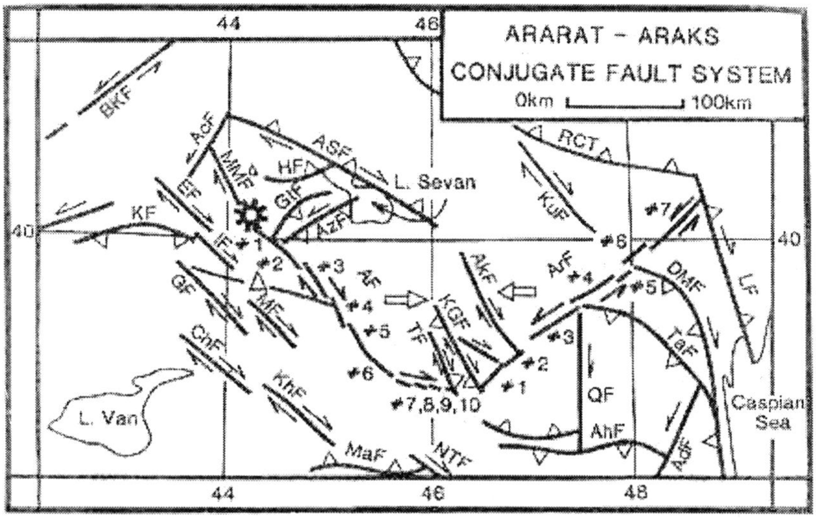

Figure 2. Ararat – Araks conjugate fault system and neighboring faults. Fault abbreviations: **AF**-Ararat; **AcF**-Ackhurian; **AdF**-Ardabil; **AhF**-Ahar; **ArF**-Araks; **AkF**-Akera; **ASF**-Amasia-Sevan; **AzF**-Azat; **BKF**-Borjomi-Kazbeg; **ChF**-Chaldiran; **DMF**-Dasht-Moughan; **EF**-Ervandashat; **GF**-Gailatu; **GtF**–Getar; **HF**-Hrazdan; **IF**-Igdir; **KF**-Kaghzvan; **KGF**-Khustoup-Giratagh; **KhF**-Khoy; **KuF**-Kura; **LF**-Lenkoran; **MF**-Maku; **MaF**-Marand; **MMF**-Mastara-Maralik; **NTF**-NorthTavriz; **QF**-Qareh; **RCT**-Recent Caucasus thrust; **TF**-Tashtoun; **TaF**-Talesh; **#1, #2, ..., #10**-segments of Ararat and Araks faults; starred circle-Metsamor Nuclear Power Plant; tall triangle-Mt. Ararat (5165m); small triangle-Mt. Aragats (4096m); open arrows-shortening between Ararat and Araks faults.

Ararat and Araks fault zones. The uplift of mountain ranges in southern Armenia can be explained as follows: due to the dextral slip on the Ararat fault and sinistral slip on the Araks fault, the region situated between these faults is being contracted, giving rise to submeridional mountain chains where highest peak is Mount Kapoutjough in Zangezour range – 3906m. Uplift of these mountain ranges is accommodated by the major Tashtoun, Khoustoup-Giratagh, Elpin, Sharour oblique-thrusts, and related faults, as evidenced by historical and modern earthquakes, geomorphology and geology.

Based on this kinematic model, long-term geological slip rates along the Ararat and Araks faults can be estimated. The rise of the mountain ranges along thrust faults to their present elevation of 4km has occurred since first continental collision between the Arabian and Eurasian plates, i. e. for at least the last 3m.y. 3km of the 4km elevation is considered to be a direct result of compression due to slip on the Ararat and Araks faults. 3km uplift in 3m.y. averages to a 1mm/yr uplift rate. In order to accommodate the suggested uplift, a minimum horizontal slip of 0.5mm/yr is attributed to both - Ararat and Araks fault zones.

2.1. THE GEOMETRY OF THE ARARAT AND ARAKS FAULT ZONES

The Ararat - Araks conjugate fault system represents a crucial neotectonic structure within the Armenian Highland. To the north the Ararat fault probably joins with the Mastara-Maralik dextral fault (fig. 2). To the southeast the termination of the Ararat fault occurs along north-northwest striking oblique-thrusts and reverse faults bordering the Zangezour mountain range on the southwest. The Araks fault terminates along the Khoustup-Giratagh oblique-thrust fault zone in the southwest and along the oblique-dextral Lenkoran fault in the northeast (fig. 2).

2.1.1. Segmentation Model for the Ararat and Araks Fault Zones

Based on the geometry and known discontinuities along the trace of the faults I propose a preliminary segmentation model for the Ararat and Araks faults. The Ararat fault is composed of 10 major segments (fig. 2). The Araks fault is composed of 7 major segments. Due to space constraints only brief descriptions of those segments are summarized in table 1 below.

TABLE 1. *Segmentation data for the Ararat and Araks fault zones (L = Length in kilometers; rs. – restraining; rl. - releasing)*

Segments of the Ararat fault zone				**Segments of the Araks fault zone**			
Seg. No.	**L**	**Strike**	**Type & size of the terminations**	**Seg. No.**	**L**	**Strike**	**Type & size of the terminations**
# 1	35	124° SE	Joins Maralik fault to the NW	**# 1**	25	50° NE	Terminates at Koustup-Giratagh thrust
# 2	18	145° SE	20° bend - # 1&2	**# 2**	18	45° NE	5km SE bend - # 1&2
# 3	30	152° SE	7km rs. step - # 2&3	**# 3**	35	50° NE	12km rs. step & 20km long uplift - # 2&3
# 4	22	162° SE	8km rs. step - # 3&4	**# 4**	50	60° NE	15° rs. bend - # 3&4
# 5	25	150° SE	4km rs. step - # 4&5	**# 5**	10	65° NE	10km rs. step - # 4&5
# 6	20	120° SE	2km rs. step & 30° bend - # 5&6	**# 6**	25	44° NE	11km rl. step - # 5&6
# 7	25	110° SE	2.5 km rs. step #6&7	**# 7**	65	46° NE	15 km rs. bend - # 6&7; Terminates at the Lenkoran fault
# 8	20	110° SE	1.5 km rs. step #7&8				
# 9	12	110° SE	1.5 km rs. step #8&9				
# 10	10	110° SE	1 km rs. step # 9&10				

3. Historical Seismicity of the Ararat Fault Zone

The city of Dvin (ancient capital of Armenia), situated 15km southeast from Yerevan and 40km east-southeast from Metsamor, was devastated during strong earthquakes in 854, 858, 863, 869 and 893 A.D [1,8,12,19]. The New Catalog [12] gives a low magnitude value of 5.2 ± 0.7 for the 854A.D. earthquake. The next strong earthquake in Dvin occurred in 869A.D. during which 120,000 people died. The earthquake of December 28, 893A.D. was the strongest one, after which the Armenian capital was relocated to Ani. The evidence from the Arabic author Ibn al Asir suggests that the total number of deaths during the 893A.D. Dvin earthquake reached 150,000 and only 100 houses remained intact [19]. Armenian historian Tovma Artsruni puts the death toll around 70,000. In any case, this was a catastrophic event, with social and economic consequences probably worse than those of the 1988 Spitak earthquake. In the New Catalog [12] the intensity of this earthquake is 8-9, magnitude 5.3 ± 0.7, which is a low value. The same low value of M=5.3 (8-9 intensity) was given by Gabrelian *et al.* [8]. Nikonov [13] suggested an intensity 10 (M≥7) for the 893A.D. earthquake, a value, which seems realistic. Other historic earthquakes which have struck the area are: 970A.D. Zvartnots (M≥6.5) [16], 1812 Jougha [1], 139A.D. Ararat valley [8] and 1937 Parakar (Yerevan) [8]. The absence of any other known seismogenic source in this area indicates that these earthquakes originated on the Ararat fault. These historical earthquakes clearly show that the Ararat fault is an active tectonic feature, and will continue to be so in the future.

4. Comparative Analysis of the Ararat Fault with the Adjacent Faults

A comparative analysis of the Ararat fault with other major strike-slip faults in the area within a 100km distance can be used to asses possible earthquake magnitudes that can be generated along the Ararat fault.

Due to space constraints the data is given in a brief format in table 2 below. The data show that faults of comparable geometry and kinematics located within 100km of the Ararat fault have generated numerous earthquakes with magnitudes greater than 7, and that Ararat fault zone cannot reasonably be deemed a lesser hazard.

TABLE 2. *Strong earthquake generated at faults located within 100 km distance from the Ararat fault zone (L=Length; M=Magnitude)*

Fault	L km & strike	Sense of slip	Major events M	Date	Important historic earthquakes	Ref.
Amasia-Sevan	200, NW	Dextral-thrust	7.0;	1988	1853, 1869, 1935, 1945,	[8]
Ervandashat	65, NW	Dextral-oblique	>6;	1605	1319, 1935, 1938	[8]
Igdir	35, NW	Dextral-oblique		1962	118	[8]
Gailatou	100, NW	Dextral-oblique	7.4;	1840	1834(M=6)	[1]
Chaldiran	55, NW	Dextral-oblique	7.4;	1976	1696 (M>7), 1988	[20]
Khoy	75, NW	Dextral-oblique	5.4;	1977	1843, 1900, 1970	[1]
Makou	40, NW	Dextral-oblique	6.0;	1968	1319	[1]
Salmas	40, NW	Dextral-oblique	7.2;	1930	1837	[1]
North Tavriz	280, NW	Dextral-thrust	7.7;	1780	858 (M=6), 1042 (M=7.6), 1304 (M=6.7), 1641 (M=6.8), 1717 (M=5.9), 1721 (M=7.7)	[1]
Marand	65, E-W	Thrust	6.3;	1786		[1]
Azat	45, NE	Thrust-sinistral	>6.5;	1679		[8]

5. Assessment of Expected Earthquake Risk for the Ararat Fault

In this section the possible earthquake size for the Ararat fault is assessed based on the following earthquake scaling parameters: **historical seismicity**, **fault-rupture-length** and **fault-rupture-area** [4,7].

The Dvin earthquake of 893, with M≥7 (intensity more than 10) is the largest known **historical** event within the Ararat valley, demonstrating that the risk of a future M≥7 earthquake is real.

The fault-rupture-length method involves measuring the potential length of an earthquake rupture, and the estimation of the potential earthquake size through empirical relations. The most direct determination of the fault length is by the segmentation technique. Bonilla *et al.* [4] developed the following magnitude-fault length relation:

$$M_s = 6.04 + 0.704(\log L) \qquad (1)$$

where **L** is the surface rupture length in km-s. The standard deviation for this relation is 0.3 magnitude units, and the relationship is valid over the range of M_s from 5.5 to 8.

Bonilla *et al.* [4] examined regressions of 21 events of M≥6 size, estimated fault widths from aftershock and microseismic activity, and suggested the following empirical relation between **earthquake size and fault area**:

$$M = 0.823\log A + 4.96 \qquad (2)$$

for M≥6, having a standard deviation of 0.34 magnitude units.

For the Ararat fault zone the fault-segment lengths are described in a previous section.

The depth of a seismogenic zone is about 15km, but given the presence of deeper earthquakes in the region, and indications of another decoupling zone at depth of 25km (S. Akopian, pers. comm.), another depth value of 25km is also used. Calculations are presented in the table 3. From the data provided in the table 3 it is obvious that the Ararat fault is capable of generating earthquakes of M≥7 magnitude (intensity ≥ 10 on MSK-64 scale) in the vicinity of the Metsamor power plant, raising the necessity of revaluation of seismic risk for the region.

TABLE 3. *Estimates of possible earthquake magnitudes along the Ararat fault zone and its segments* *Ms = 6.04 + 0.704 logL – fault-rupture-length method; M = 0.823logA + 4.96 – fault-rupture-area method*

FAULT	LENGTH (L)	Ms	M	M
	Km		15 km depth (D)	25 km depth (D)
Whole AFZ	220	7.67	7.85	8.03
1/2 AFZ length	110	7.48	7.61	7.8
1/5 AFZ length	44	7.2	7.28	7.46
Segment #1	35	7.12	7.2	7.38
Segment #2	18	6.92	6.96	7.14
Segment #3	30	7.08	7.14	7.32
Segment #4	22	6.98	7.03	7.21
Segment #5	25	7.02	7.08	7.26
Segment #6	20	6.95	7.0	7.18
Segment #7	25	7.02	7.08	7.26
Segment #8	20	6.95	7.0	7.18
Segment #9	12	6.8	6.81	7.0
Segment #10	10	6.74	6.75	6.93

6. A Few Considerations Regarding Seismic Risk Along the Ararat Fault Zone and Conclusions

The Ararat fault is located in the densely populated Ararat valley, where about 2/3 of Armenia's 3.5 million population reside. The capital of Armenia, Yerevan, with population over 1 million, is situated 15-25km east-northeast of the Ararat fault. Most of the city's buildings are designed for much smaller earthquakes than an earthquake of M≥7. The situation is further complicated by the Metsamor Nuclear Power Plant located immediately north of the northern portion of the Ararat fault and 25km west from Yerevan.

High-risk structures, such as a nuclear power plant, are commonly designed for a **maximum credible earthquake** - the largest or maximum earthquake that appears possible in an area or along a fault [4,7], which in the case of the Ararat fault zone is 7.85±0.34. The major hazards of active faulting in a continental environment are the destructive effects of earthquake shaking ("strong ground motion"), displacement or offset along the rupture (during surface faulting), tectonic deformations, landslides and rockfalls and soil-liquefaction. When the Metsamor Nuclear Power Plant was designed and built, the area was considered as a possible sight for earthquake magnitude ≤6 (7-8 intensities in MSK scale). The recognition herein of the active Ararat fault zone, capable of generating an M≥7 earthquake (intensity 10 or greater), with associated strong ground shaking and soil-liquefaction, raises serious concerns regarding the seismic safety of the power plant. Besides the power plant itself there are subsidiary structures of vital importance to the power plant, such as the pump-station which supplies cooling water to the reactor. The collapse of the pump-station can cut off the cooling water supply and result in complete or partial meltdown of the reactor. The main water supply station is near the Metsamor museum, 7km southeast from the power plant. This structure is built on a recent cinder cone, but the pipeline runs through areas of unconsolidated Quaternary lacustrine sediments, with a water table 0.5-1m below the surface. The height of the water table and widespread soil-liquefaction reported in the Ararat valley during the 1840 Gailatou earthquake [1] indicate that liquefaction of the unconsolidated deposits will occur during a strong earthquake. This in its turn may cause the collapse of the water pipeline at one or several locations, interrupting the flow of cooling water to the reactor. Tectonic deformation can cause leakage and other problems from the radioactive waste depository (being built next to the power plant), resulting in contamination of the Ararat valley's freshwater aquifers, with long lasting and irreversible consequences. The seismic hazards associated with a strong earthquake along the Ararat fault can have catastrophic consequences not only for the republic of Armenia, but also for the entire surrounding region. A multidisciplinary research study project of active faulting of this region, involving geology, seismology, geodesy, geophysics, earthquake engineering and other related spheres of science, should be immediately undertaken.

An in-depth understanding of the detailed structure of the fault zone will enable us to undertake appropriate steps toward earthquake hazard reduction in the whole region.

NOTE: At the request of the NSSP papers should not exceed 8 pages total, and this paper is a condensed version of the original 36 page paper presented and submitted at the Conference.

References

1. Ambraseys N. N. and C. P. Melville, 1982 - A history of Persian earthquakes; Cambridge University Press, 221p.
2. Barka A. A. and K. Kadinsky-Cade, 1988 - Strike-slip fault geometry in Turkey and its influence on earthquake activity; Tectonics, v. 7, No. 3, pp. 663-684.
3. Berberian M, 1981 - Active faulting and tectonics of Iran; AGU, Geodynamic Series, v. 3, pp. 33-69.
4. Bonilla M. G., R. K. Mark and J. J. Lienkaemper, 1984 - Statistical relationships among earthquake magnitude, surface rupture length and surface fault displacement; Bull. Seism. Soc. Am., v. 74, pp. 2379-2411.
5. De Mets C., R. G. Gordon, D. F. Argus and S. Stein, 1990 - Current plate motions; Geophys. J. Int., v. 101, pp. 425-478.
6. DeMets C., R. G. Gordon, D. F. Argus and S. Stein, 1994 – Effects of recent revisions to the geomagnetic reversal time scale on estimates of current plate motions; Geophys. Res. Letters, v. 21, pp. 2191-2194.
7. dePolo C. M. and Slemmons D. B., 1990 - Estimation of earthquake size for seismic hazards; Neotectonics in Earthquake Evaluation, GSA, Boulder, Colorado, pp. 1-28.
8. Gabrelian A. H., H. H. Sargisian and G. P. Simonian, 1981 - Seismotectonica Armyanskoy SSR (in Russian) (Seismotectonics of the Armenian SSR); Yerevan State University, 282 p.
9. Jackson J., and D. McKenzie, 1984 - Active tectonics of the Alpine-Himalayan belt between western Turkey and Pakistan; Geophys. J. Royal Astr. Soc., v. 64, pp. 561-586.
10. Jackson J., 1992 - Partitioning of strike-slip and convergent motion between Eurasia and Arabia in Eastern Turkey and the Caucasus; JGR, v. 97, no. B9, pp. 12,471-12,479.
11. Mkrtchian H., K, 1995 - Collisional tectonics of the Arabian and Eurasian plates over the Armenian Highland, Anatolian and Iranian Plateaus; Pan Armenian Science Review, v. 2, no. 3, pp. 16-23.
12. New Catalog of strong earthquakes in the USSR from ancient times through 1977; editors - N. V. Kondorskaya and N. V. Shebalin; Academy of Sciences of USSR, July 1982.
13. Nikonov A. A., 1989 - Zemlya Zemletryasenii (in Russian - The Soil of Earthquakes); Priroda, No. 12, pp. 39-46.
14. Noomen R., T. A. Springer, B. A. C. Ambrosius, K. Herzberger, D. C. Kuijper, G.-J. Mets, B. Overgaauw and K. F. Wakker, 1996 - Crustal deformations in the Mediterranean area computed from SLR and GPS observations; Journal of Geodynamics, v. 21, no. 1, pp. 73-96.

15. Philip H, A. Cisternas, A. Gvishiani and A. Gorshkov, 1989 - The Caucasus; an actual example of the initial stages of continental collision; Tectonophysics, 161, pp. 1-21.
16. Piruzian S. M., A. S. Matevosian and L. V. Shaksuvarian, 1984 - (in Russian); Istoriko-Phylological Journal, no. 3, pp. 189-199.
17. Reilinger R. E., S. C. McClusky, M. B. Oral, R. W. King, M. N. Toksoz, A, A. Barka, I. Kinik, O. Lenk and I. Sanli, 1997 (a) - Global Positioning System measurements of present-day crustal movements in the Arabia-Africa-Eurasia plate collision zone; JGR, v. 102, no. B5, pp. 9983-9999.
18. Smith D. E. and R. Kolenkiewicz, 1994 – Horizontal crustal motion in the central and eastern Mediterranean inferred from Satellite Laser Ranging measurements; Geophys. Res. Letters, v. 21, no. 18, pp. 1979-1982.
19. Stepanian V. A., 1942 - Kratkaya khronologiya naibolee znachitelnikh zemletryasenii v ictoricheskoi Armenii i v prelegayuschikh chastyakh (in Russian) (Short cronology of the important earthquakes of the historical Armenia and adjacent areas), USSR Academy of Sciences, Yerevan.
20. Toksoz M. N., E. Arpat and F. Saroglu, 1977 - East Anatolian eartquake of 24 November, 1976; Nature, v. 270, pp. 423-425.

ONE INEQUALITY OF ELASTOKINETICS AND ITS APPLICATION IN SEISMOLOGY

G. GABRICHIDZE
Institute of Srtuctural Mechanics and Earthquake Engineering, Georgian Academy of Sciences
8, M.Aleksidze str., Tbilisi 380093, Georgia

1. Abstract

There is obtained an inequality that can connect the parameters of the motion of the earth surface with the energy released at the earthquake focus. On the basis of obtained inequality, a possibility is being discussed concerning the application of special experimental-analytical technique.

2. Introduction

Let's consider an arbitrary elastic body which is pinched on the A_u part of the surface and free from loading on the surface A_σ . Let's consider two systems of mass forces that induce the motion of that body. The first system of causes and effects is denoted by the index (1), while the other one - by the index (2). Correspondingly, in the first state we have mass forces $X_i^{(1)}$ caused by them displacements of $u_i^{(1)}$ inside the body, related to them strains $\varepsilon_{ij}^{(1)}$, stresses $\sigma_{ij}^{(1)}$ and volume expantion $e^{(1)}$ In the second state we have the same parameters but with index (2). Besides the abovementioned, there are also given the values of ρ- density of material, λ, μ –Lame constants.

In elastokinetics it is known a reciprocity theorem that connects stress and strain of these two states.Using Laplace transformation

$$\left[\sigma_{ij}(x,t)\right] = \overline{\sigma}_{ij}(x,p) = \int_0^\infty \sigma_{ij}(x,t)e^{-pt}dt$$

reciprocity theorem is written as [1]:

$$\int_V \overline{\sigma}_{ij}^{(1)}\overline{\varepsilon}_{ij}^{(2)}dv = \int_V \overline{\sigma}_{ij}^{(2)}\overline{\varepsilon}_{ij}^{(1)}dv \qquad (1)$$

S. Balassanian et al. (eds.), Earthquake Hazard and Seismic Risk Reduction, 163–167.

By presenting the left-hand side of the equation (1) in the form

$$\int_V \bar{\sigma}_{ij}^{(1)}\bar{\varepsilon}_{ij}^{(2)} dv \equiv \int_V \left[(\bar{\sigma}_{ij}^{(1)}\bar{u}_i^{(2)})_{,j} - \bar{\sigma}_{ij,j}^{(1)}\bar{u}_i^{(2)}\right] dv$$

employing equation of equilibrium, general ratious connecting stresses and straines and considering the boundary conditions along with zero initial conditions, after using the Ostrogradskii-Gauss formula, a tensor equation (1) in the open form can be written as:

$$\bar{A}^{(1)(2)}(p) = \int_V \left[\bar{x}_1^{(1)}\bar{u}_1^{(2)} + \bar{x}_2^{(1)}\bar{u}_2^{(2)} + \bar{x}_3^{(1)}\bar{u}_3^{(2)}\right] dv =$$

$$= \int_V \Big\{ 2\mu\big[\bar{\varepsilon}_{11}^{(1)}\bar{\varepsilon}_{11}^{(2)} + \bar{\varepsilon}_{12}^{(1)}\bar{\varepsilon}_{12}^{(2)} + \bar{\varepsilon}_{13}^{(1)}\bar{\varepsilon}_{13}^{(2)} + \bar{\varepsilon}_{22}^{(1)}\bar{\varepsilon}_{22}^{(2)} + \bar{\varepsilon}_{23}^{(1)}\bar{\varepsilon}_{23}^{(2)} +$$

$$+ \bar{\varepsilon}_{33}^{(1)}\bar{\varepsilon}_{33}^{(2)}\big] + \lambda \bar{e}^{(1)}\bar{e}^{(2)} \Big\} dv + \int_V \rho p^2 \left[\bar{u}_1^{(1)}\bar{u}_1^{(2)} + \bar{u}_2^{(1)}\bar{u}_2^{(2)} + \bar{u}_3^{(1)}\bar{u}_3^{(2)}\right] dv \tag{2}$$

Formula (2) shows the work of the external forces of the first system on the displacements of the second system, thus it is designated as $\bar{A}^{(1)(2)}(p)$. On the basis of (2), one can also write an expression for the work $\bar{A}^{(1)(1)}(p)$ and $\bar{A}^{(2)(2)}(p)$

$$\bar{A}^{(1)(1)}(p) = \int_V \left[\bar{x}_1^{(1)}\bar{u}_1^{(1)} + \bar{x}_2^{(1)}\bar{u}_2^{(1)} + \bar{x}_3^{(1)}\bar{u}_3^{(1)}\right] dv =$$

$$= \int_V \Big\{ 2\mu\big[\bar{\varepsilon}_{11}^{(1)}\bar{\varepsilon}_{11}^{(1)} + \bar{\varepsilon}_{12}^{(1)}\bar{\varepsilon}_{12}^{(1)} + \bar{\varepsilon}_{13}^{(1)}\bar{\varepsilon}_{13}^{(1)} + \bar{\varepsilon}_{22}^{(1)}\bar{\varepsilon}_{22}^{(1)} + \bar{\varepsilon}_{23}^{(1)}\bar{\varepsilon}_{23}^{(1)} +$$

$$+ \bar{\varepsilon}_{33}^{(1)}\bar{\varepsilon}_{33}^{(1)}\big] + \lambda \bar{e}^{(1)}\bar{e}^{(1)} \Big\} dv + \int_V \rho p^2 \left[\bar{u}_1^{(1)}\bar{u}_1^{(1)} + \bar{u}_2^{(1)}\bar{u}_2^{(1)} + \bar{u}_3^{(1)}\bar{u}_3^{(1)}\right] dv \tag{3}$$

$$\bar{A}^{(2)(2)}(p) = \int_V \left[\bar{x}_1^{(2)}\bar{u}_1^{(2)} + \bar{x}_2^{(2)}\bar{u}_2^{(2)} + \bar{x}_3^{(2)}\bar{u}_3^{(2)}\right] dv =$$

$$= \int_V \Big\{ 2\mu \Big[\bar{\varepsilon}_{11}^{(2)} \bar{\varepsilon}_{11}^{(2)} + \bar{\varepsilon}_{12}^{(2)} \bar{\varepsilon}_{12}^{(2)} + \bar{\varepsilon}_{13}^{(2)} \bar{\varepsilon}_{13}^{(2)} + \bar{\varepsilon}_{22}^{(2)} \bar{\varepsilon}_{22}^{(2)} + \bar{\varepsilon}_{23}^{(2)} \bar{\varepsilon}_{23}^{(2)} +$$

$$+ \bar{\varepsilon}_{33}^{(2)} \bar{\varepsilon}_{33}^{(2)} \Big] + \lambda \bar{e}^{(2)} \bar{e}^{(2)} \Big\} dv + \int_V \rho p^2 \Big[\bar{u}_1^{(2)} \bar{u}_1^{(2)} + \bar{u}_2^{(2)} \bar{u}_2^{(2)} + \bar{u}_3^{(2)} \bar{u}_3^{(2)} \Big] dv \tag{4}$$

The right-hand parts of the expressions (2), (3) and (4) can be presented as local sums:

$$\bar{A}^{(1)(2)}(p) = \int_V \Big[\bar{x}_1^{(1)} \bar{u}_1^{(2)} + \bar{x}_2^{(1)} \bar{u}_2^{(2)} + \bar{x}_3^{(1)} \bar{u}_3^{(2)} \Big] dv = \int_V \Big[\sum_k a_k(p) b_k(p) \Big] dv \tag{5}$$

$$\bar{A}^{(1)(1)}(p) = \int_V \Big[\bar{x}_1^{(1)} \bar{u}_1^{(1)} + \bar{x}_2^{(1)} \bar{u}_2^{(1)} + \bar{x}_3^{(1)} \bar{u}_3^{(1)} \Big] dv = \int_V \Big[\sum_k a_k^2(p) \Big] dv \tag{6}$$

$$\bar{A}^{(2)(2)}(p) = \int_V \Big[\bar{x}_1^{(2)} \bar{u}_1^{(2)} + \bar{x}_2^{(2)} \bar{u}_2^{(2)} + \bar{x}_3^{(2)} \bar{u}_3^{(2)} \Big] dv = \int_V \Big[\sum_k b_k^2(p) \Big] dv \tag{7}$$

On the basis of the Hölder's inequality [2]

$$\left| \sum_k a_k b_k \right|^2 \le \left| \sum_k a_k^2 \right| \left| \sum_k b_k^2 \right| \tag{8}$$

one can write

$$\left| \int_V \Big[\bar{x}_1^{(1)} \bar{u}_1^{(2)} + \bar{x}_2^{(1)} \bar{u}_2^{(2)} + \bar{x}_3^{(1)} \bar{u}_3^{(2)} \Big] dv \right|^2 \le$$

$$\le \left| \int_V \Big[\bar{x}_1^{(1)} \bar{u}_1^{(1)} + \bar{x}_2^{(1)} \bar{u}_2^{(1)} + \bar{x}_3^{(1)} \bar{u}_3^{(1)} \Big] dv \right| \left| \int_V \Big[\bar{x}_1^{(2)} \bar{u}_1^{(2)} + \bar{x}_2^{(2)} \bar{u}_2^{(2)} + \bar{x}_3^{(2)} \bar{u}_3^{(2)} \Big] dv \right| \tag{9}$$

or transferring to the originals, and coming back to tensore form,

$$\left|\int_0^t d\tau \int_V x_i^{(1)}(x,\tau)u_i^{(2)}(x,t-\tau)dv\right|^2 = \left|\int_0^t d\tau \int_V x_i^{(2)}(x,\tau)u_i^{(1)}(x,t-\tau)dv\right|^2 \le$$

$$\le \left|\int_0^t d\tau \int_V x_i^{(1)}(x,\tau)u_i^{(1)}(x,t-\tau)dv\right| \cdot \left|\int_0^t d\tau \int_V x_i^{(2)}(x,\tau)u_i^{(2)}(x,t-\tau)dv\right| \qquad (10)$$

or

$$\left|A^{(1)(2)}(t)\right|^2 = \left|A^{(2)(1)}(t)\right|^2 \le \left|A^{(1)(1)}(t)\right|\left|A^{(2)(2)}(t)\right| \qquad (11)$$

It should be mentioned that the content and structure of the obtained expression won't be altered providing not only mass forces are considered, but also the forces that can be applied to outer and inner boundary surfaces of the body. Taking into account the above-mentioned, inequality (11) can be formulated as:

The work of a first system of dynamic forces on the displacements, caused by a second system of dynamic forces at a time is lower than the mean proportional of works done by each system of forces on self-induced displacements.

Let's return to the problem under investigation. One system of causes and effects can be referred to the case of earthquake occurrence and designated by the index (1). Then $A^{(1)(1)}(t)$ will express the work of forces in the focal zone on self-induced displacements, i.e. energy that is released in the focal zone.

The other system of causes and effects that is designated by index (2) can be imulated by carrying out physical experiment. On the surface of the earth at the x_k point an arbitrary dynamic force of $X_1^{(2)}(x_k,t)$ can be applied in the given direction and the law of displacement $u_1^{(2)}(x_k,t)$ of point x_k is recorded.

For the above mentioned two systems of causes and effects, inequalities (10) and (11) can be written as

$$\left|\int_0^t x_1^{(2)}(x_k,\tau)u_1^{(1)}(x_k,t-\tau)d\tau\right|^2 \le \left|\int_0^t x_1^{(2)}(x_k,\tau)u_1^{(2)}(x_k,t-\tau)d\tau\right| \cdot \left|A^{(1)(1)}(t)\right| \qquad (12)$$

In these expressions $x_1^{(2)}(x_k,\tau$ and ${}_1^{(2)}(x_k,t-\tau)$ are known functions as a result of experiment. Thus, inequality (12) connects directly the law of displacement of point

x_k on the earth surface with the energy release in the focal zone during the earthquake. Inequality (12) may be applied for the investigation of direct and indirect problem - if the law of displacement of the point x_k on the earth surface during the earthquake is known, the energy $A^{(1)(1)}(t)$ that is released in the focal zone may be estimated, and vice versa, if the law of energy release at the earthquake focus zone is assumed, the law of x_k point displacement on the earth surface can be estimated.

5. References

1. W.Novacki. Theory of Elasticity. Moscow, 1975 (Russian).
2. G.Corn, T.Corn. Mathematical Handbook for Scientists and Engineers. Moscow, 1970 (Russian).

EARTHQUAKE PREDICTION RESEARCH FOR CURRENT SEISMIC HAZARD ASSESSMENT

S. BALASSANIAN
Armenian National Survey for Seismic Protection (NSSP)

1. Introduction

Earthquake prediction problem includes two aspects: scientific and social.

Unsettledness of the scientific aspect of that most complicated problem in its complete extent, i.e. prediction of the place, time and magnitude of an earthquake, is automatically excluding its social aspects' solution possibility, i.e. warning of the population, according to the experts majority opinion.

Such situation turned out today particularly in developing countries experiencing economical difficulties, when seismoresistance of the existing buildings and constructions is significantly lower than seismic hazard level. As a result of that, any strong earthquake in the territory of those countries is being accompanied by numerous casualties and high social-economical losses.

Taking that situation into consideration it seems expedient to us to make an attempt to use the intermediate scientific results in the field of earthquake prediction research for Current Seismic Hazard Assessment (CSHA).

It is shown in this paper what and how intermediate scientific results in the field of earthquake prediction research could be used for CSHA aimed to take preventive measures for protection of the population against strong earthquakes.

2. Earthquake Prediction Research in Armenia

Armenia is situated in the collision zone between Arabian and Eurasian plates (Fig.1). All the area between the northern boundary of the Arabian plate and the Greater Caucasian thrust represents a zone of diffusive distribution of earthquakes epicenters with different magnitudes - from numerous microquakes M=1~2 up to unique strong events M=7-7.4 [1]. The seismically active zone is caricaturized by moderate mean depth (h) of earthquake hypocenters (h up to 15 km). The geological structure of microplates formed between the Arabian and Eurasian plates is more similar to that of the Eurasian plate. From a tectonic standpoint this zone can be presented as a higher rank microplates formation area - the Anatolian microplate (A), the Black Sea microplate (B), the Caucasian (C), South Caspian (SC) and the Iranian (I) microplates.

The Arabian plate is moving at a rate of 10 mm/year towards the Eurasian plate, according to the modern GPS observations [2]. The Greater Caucasian trust is formed

S. Balassanian et al. (eds.), Earthquake Hazard and Seismic Risk Reduction, 169–209.

toward the motion of two deep-see basins - the Black Sea and the South-Caspian one, with dip to the North, i.e. towards the Eurasian plate.

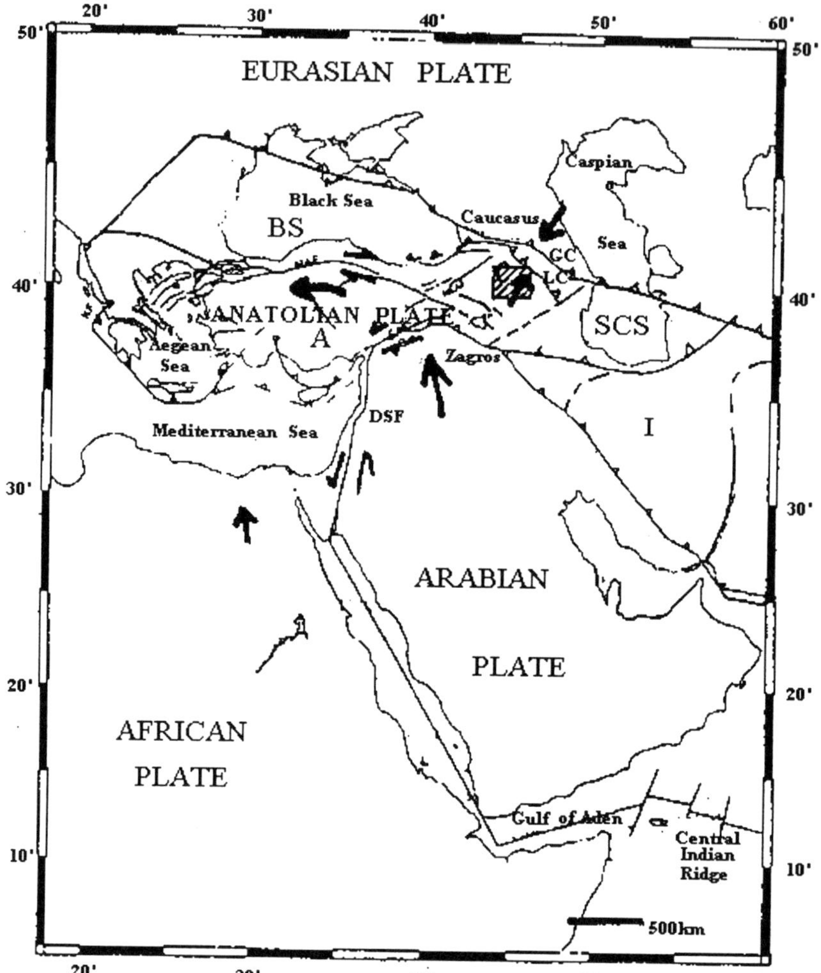

Figure 1. Collision zone of the Arabian and Eurasian plates (the data of McKenzie, 1972; Relinger et al, 1995)

Microplates: **A** - Anatolian, **BS** - Black Sea, **C** - Caucasian, **SCR** - South-Caspian, **I** - Iranian.

Thrust-faults: **GC**- the Greater Caucasus, **LC** - the Lesser Caucasus.

→ - directions and rates of movements of plates and microplates in collision zone (relatively to the Eurasian plate)

⇌ - strike-slip faults with arrows showing the sense of motion

Faults: △△△ - thrust faults with triangles on overriding block

—— - active faults

▨ - territory of Armenia

Earthquake prediction research in Armenia has began approximately 30 years ago [3]. In spite of that, it has had the same non-systematic nature as it is having today in many countries. The measurements were carried out irregularly in some casually selected observation sites by means of one or several not sufficiently grounded chosen methods with the retrospective data analysis by various research institutions where along with the basic activities also additionally the activities on the earthquake prediction research were carried out. After the devastating 1988 Spitak earthquake and the establishment of the National Survey for Seismic Protection (NSSP) of Armenia in 1991 there were large-scale studies started in the earthquake prediction research [4] including:

- permanent multiparameter (40 characteristics) real time monitoring of the lithosphere, (its solid, liquid, gaseous phases) atmosphere and biosphere by means of 150 stations installed in the observation sites selected by the special methodology;
- data recording, processing and multiparameter analysis in the Armenian NSSP's Unified Center for Seismic Hazard Assessment (UCSHA);
- design of the National Data Bank including all the seismological, geological, geophysical, geochemical and other kinds of information characterizing state of the lithosphere, atmosphere and biosphere in Armenia and adjoining areas.

There were the following main problems defined on the first-stages of those studies:

- how much the physico-geological states of Armenia are favorable for pre-, co, post- seismic processes reflection in the various characteristics of the lithosphere, atmosphere, biosphere;
- what are the peculiarities of the seismic process' various stages occurrence in the lithosphere, atmosphere and biosphere;
- what characteristics of the lithosphere, atmosphere and biosphere are the most informative for pre-, co-, post- seismic stages determination.

2.1. NSSP's NATIONAL OBSERVATION NETWORK

Multiparameter NSSP's National Observation Network including all the observation sites in Armenia is illustrated in Fig.2 [5]. The following basic principles of the National Network are in the basis of its design:

- monitoring of all three phases of the lithosphere (solid, liquid, gaseous), atmosphere and biosphere;
- multiparameter monitoring based upon defined characteristics, showing sufficient informativity during the single observations carried out earlier or known as an informative parameter from the international experience;
- hierarchical structure of the network including the local, regional and global observation sites;
- selection of the observation sites in the manner that would allow to control all the active faults in Armenia in any case along with participation in the local, regional or global monitoring (Fig.2);
- using of high sensitive energy-active points for selection of the observation sites [6];

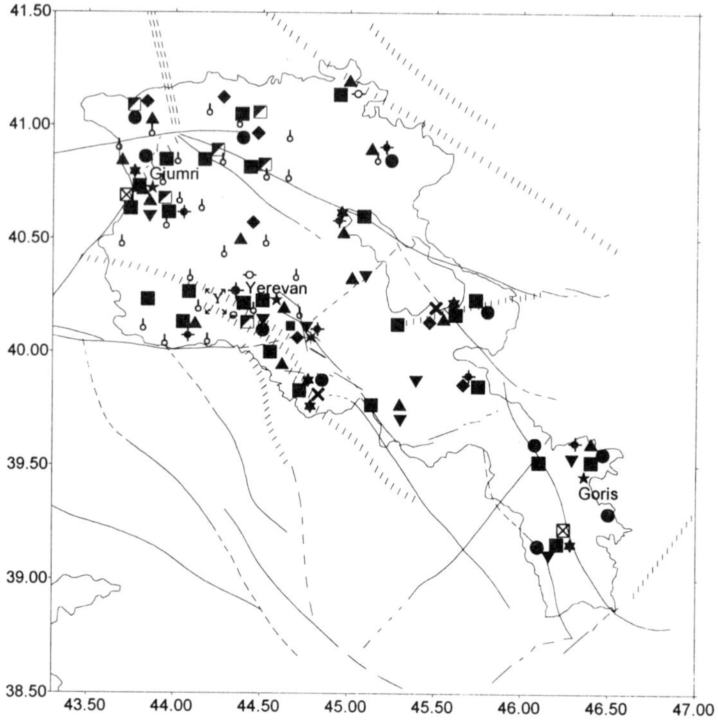

Figure 2. National Multiparameter Observation Network of the Armenian NSSP

Seismic stations:
- ✱ Global Seismographic Network (IRIS)
- ● Regional
- Local telemetry
- Strong ground motion stations (SMACH)

Seismogeodetic stations (GPS):
- Global
- Regional

Hydrogeodynamic stations
- ▲ Regional

Seismogeochemical stations:
- Regional
- ✕ Local - Le Centrale CR2M
- ⊠ Inter - READINESS

Geophysical stations:
- ■ Radon
- ◆ Geomagnetic
- ▼ Electromagnetic
- ↖ ↗ Y ↙ ↘ VLF radio wave receiving system (RNS "Omega")
- Atmospheric pressure

Seismobiological stations
- Regional

- data transmission in the real or quasi-real time from the ordinary observation sites to the NSSP's Regional Centers and then from the Regional ones to the NSSP's UCSHA; data transmission from the NSSP's UCSHA to the other Global and Regional International Centers (Fig.3);
- using of all the possible types of data transmission (telephone, radio, satellite) from observation sites to UCSHA; using of the satellite communication and Internet amongst NSSP's UCSHA and other International Centers.

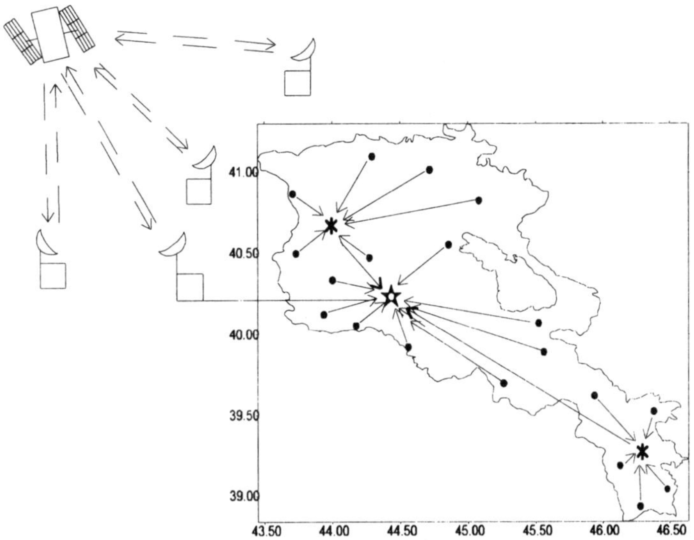

Figure 3. Scheme of internal and external information flows in NSSP's National Observation Network

● Ordinary station

✱ Regional centers

☆ Unified National center (NSSP Armenia)

 International and National Seismic Centers

2.2. MULTIPARAMETER MONITORING

Fig.4 demonstrates the results of 14-year observations of Helium contents (He) dissolved in groundwaters at the Kajaran Station (Armenia). From the given data it's well seen that all the strong seismic events in the region - the Narman earthquake of 1983 (Turkey, M=6.8), the Spitak earthquake of 1988 (Armenia, M=7.1), the Roudbar earthquake of 1990 (Iran, M=7.7), the Racha earthquake of 1991 (Georgia, M=7.1) have been accompanied by co-seismic anomalous changes in contents of the dissolved in water He that are restored during approximately 0.5 year. It is necessary to mention that in the more detailed plots of time series of such parameters as Cl, HCO_3 and others there are clearly pronounced also pre- seismic changes (Fig.4b,c). Pre- seismic stage is characterized by changes in frequency of chemical components' variations – from anomalous increase to anomalous decrease ("quiescence") (Fig.4b,c). Thus, Fig.4 is clearly indicating that all the strong earthquakes ($M \geq 5.5$) in the region are being accompanied by pre-, co-, post- seismic changes of chemical and therefor physical parameters of groundwaters in the territory of Armenia.

Picualiarities of seismic regime in the territory of Armenia before, during and after the 1988 Spitak Earthquake (M=7.1) are shown in the Fig.5.

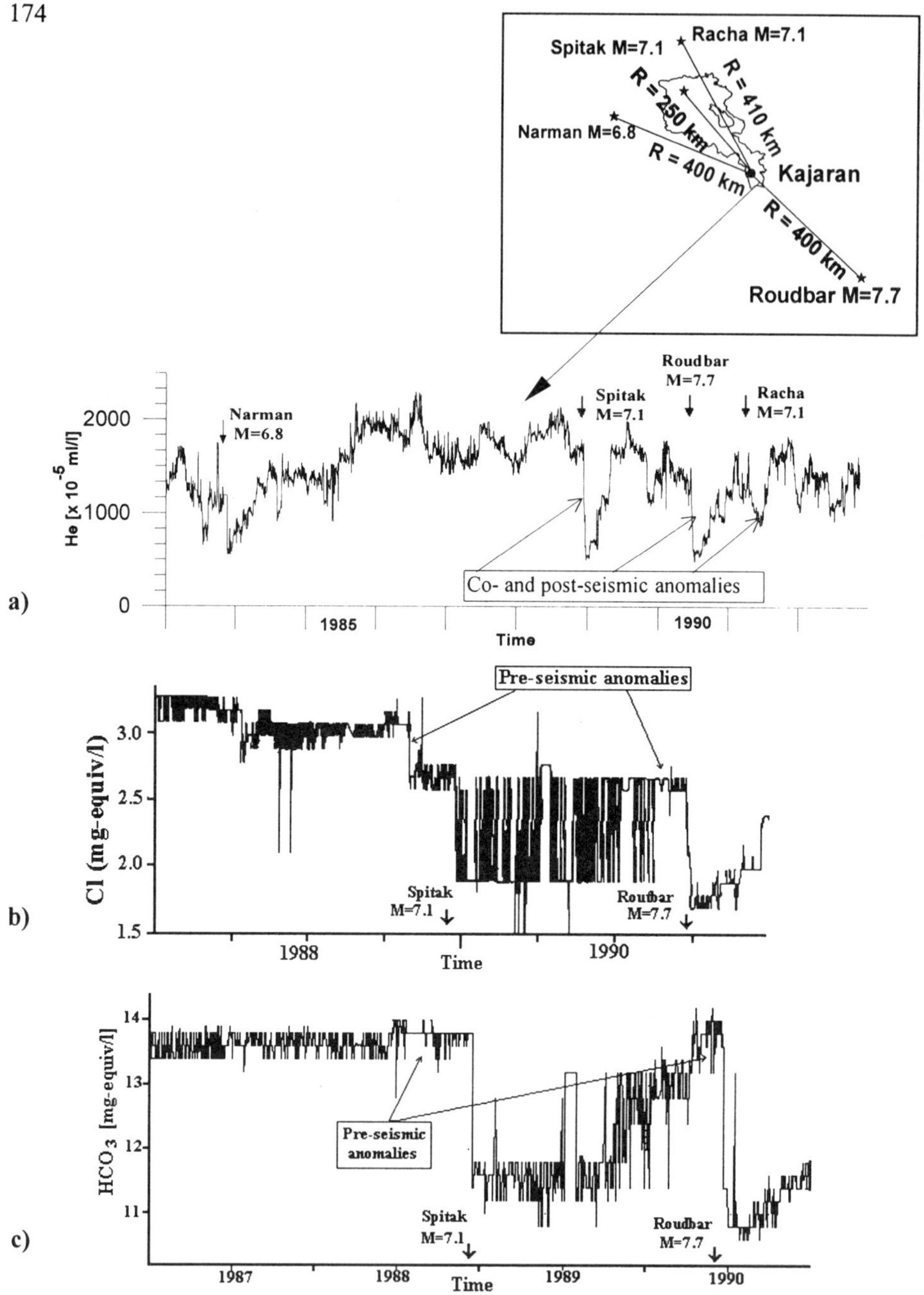

Figure 4. Pre-, co-, post-seismic changes in contents of different chemical components in ground waters at the "Kajaran" station:
a) helium (He) dissolved in ground waters;
b) chlorine (Cl) ion variations;
c) hydrocarbonate (HCO_3) ion variations.

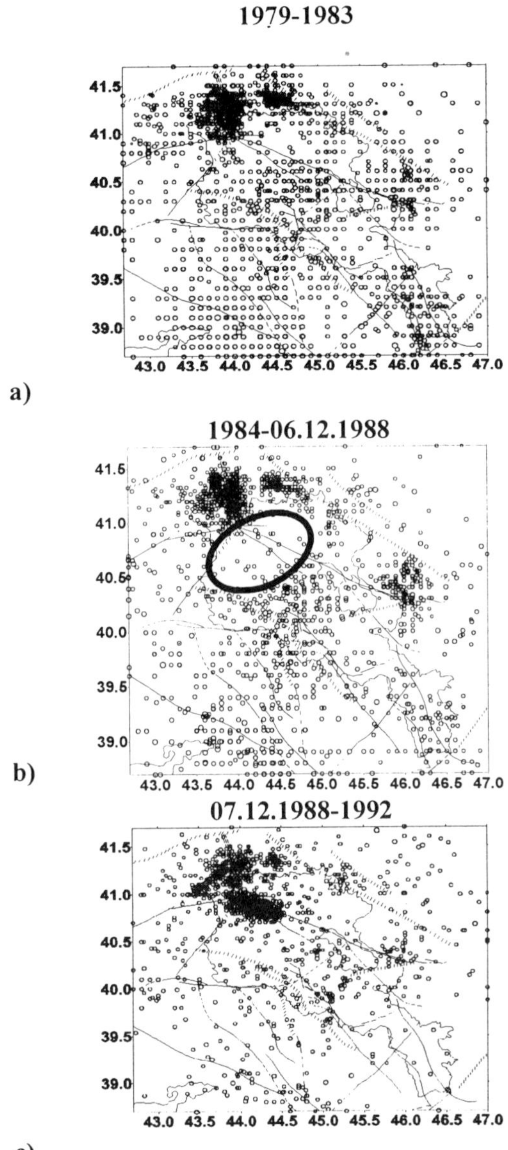

Figure 5. Maps of earthquake epicenters for the time period prior to (a), during (b) and after (c) the Spitak event: M = 7.1, 1988, Armenia

O - seismic quiescence in the Spitak earthquake source zone.

It is clear from the Fig.5 that the Spitak earthquake was preceded by four years (from 1984 to 06.12.1988, Fig.5.b) seismic quiescence within its future source zone.

Fig.6 shows anomalous variations of geomagnetic field at the "Giulagharak" Station (Northern Armenia) with respect to the "Jermouk" Station (Central Armenia) during preparation of the Narman (M=6.8, 1983, Turkey) and the Spitak (M=7.1, 1988, Armenia) earthquakes. It is clear from the presented data that increase of geomagnetic field (ΔT) on 30nT was have being observed during 3,5 years in the Northern part of Armenia in epicentral zone of Spitak earthquake (Giulagharak Station) in relation to the central part of the Country (Jermouk Station) right before the 1988 Spitak earthquake (M=7.1). In spite of that, ΔT had reached to its maximum values 1,5 years before the seismic event. Then, the annual period of relative minimum of ΔT was observed. Sharp changes of the geomagnetic field ΔT were observed in the Giulagharak Station also a 0.5 years before the 1983 Narman Earthquake (Turkey, M=6.8).

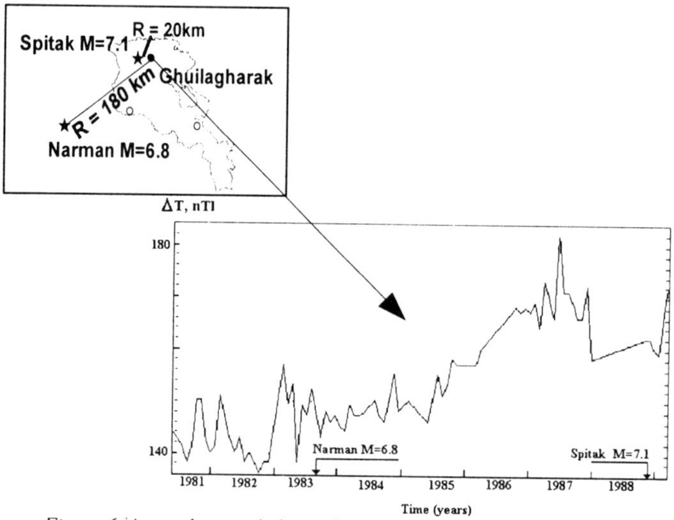

Figure 6. Anomalous variations of geomagnetic field at the Ghiulagharak station with respect to "Jermouk" station during preparation of the Norman (M=6.8, 1983, Turkey) and the Spitak (M = 7.1, 1988, Armenia) earthquakes.

● "Ghiulagharak" station
○ "Jermouk" and "Janfida" stations
★ earthquake epicenter

Differences of mediums' geomagnetic reaction upon both of those strong seismic events is the fact that the stronger geomagnetic anomaly being observed during the longer period of time before seismic event is corresponded to the stronger Spitak earthquake with the source located closer to the observation site. Attention also, along with that conclusion, should be paid to the sircumstance that the observation sites in other locations have not reacted to above mentioned strong seismic events. It means that in the territory of Armenia there is a heterogeneous geomagnetic reaction of medium upon the same seismic events being observed .

Fig.7 demonstrates the variations of geoelectric parameters at the observation sites in Armenia during preparation of the Parakar earthquake (M=3.7, 1997, Armenia). The observed parameter of geoelectric field is the equivalent of integral energetic changes $E^{+,-}$ (S. Balassanian, 1990, [6]; 1997, [7]). This parameter is characterizing the irreversibility of non-stationary geoelectric processes which is expressed in the difference of the shape of the curve of formation of non-stationary processes from the shape of the curve of its decay. Non-stationary processes at the observation stations of the NSSP are stimulated in the medium by succession of unipolar pulses of the stabilized current. In so doing, for registration of electric component of the field of non-stationary processes is connected with sum of non-linear effects of different physical nature stimulated in deformed medium by "quick" pulses of current. Value of non-linear effects directly depends on change in degree of non-linearity of deformed geological medium. The sum of non-linear effects besides other factors, is connected with intensity of natural electromagnetic radiation caused by deformed rocks.

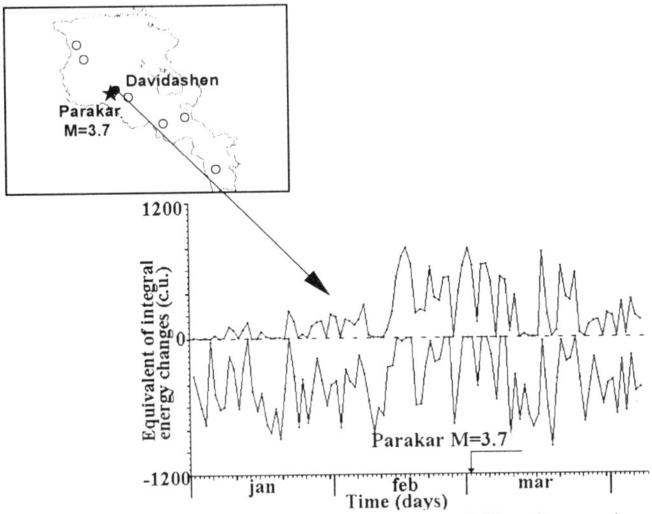

Figure 7. Anomalies of the parameter of the irreversibility of non-stationary processes $E^{+(-)}$ in the territory of Armenia during Parakar earthquake (M_s=3.7, 1997, Armenia) preparation and occurrence.

● Station with anomalous respond to seismic event;

○ Stations without respond to seismic event

The described method of irreversibility of non-stationary processes has been elaborated for studying the rock deformations at the level of deformation of the zones of separated charges (ZSC), formed at the interface of solid and liquid phases in a real geological medium (S. Balassanian, 1990, [6,7,8]).

It is clear from the data presented in the Fig.7 that there was and is a strong anomaly E^+ being observed 15 days before, in time and during 25 days after the Parakar earthquake. From Fig.7 it is also seen that some geoelectric stations did not respond to the Parakar earthquake. This fact is well explained by two factors – non-homogenety of geological medium and its non-homogeneous deformability under the impact of the regional elastic strain forces.

Fig. 8 shows alterations of the groundwaters table in Armenia during preparation and occurence of the Barisakho earthquake (M=6.4, 1992, Georgia). It is clear from the Fig.8 that the Barisakho earthquake was have being accompanied by pre-, co-, post-seismic changes of the groundwaters table in Armenia. In spite of that, the pre-seismic changes began 20 days before the earthquake. Then the co- seismic stage has clearly occured and during the following 7 days there was post- seismic re-establishment of groundwater table observed.

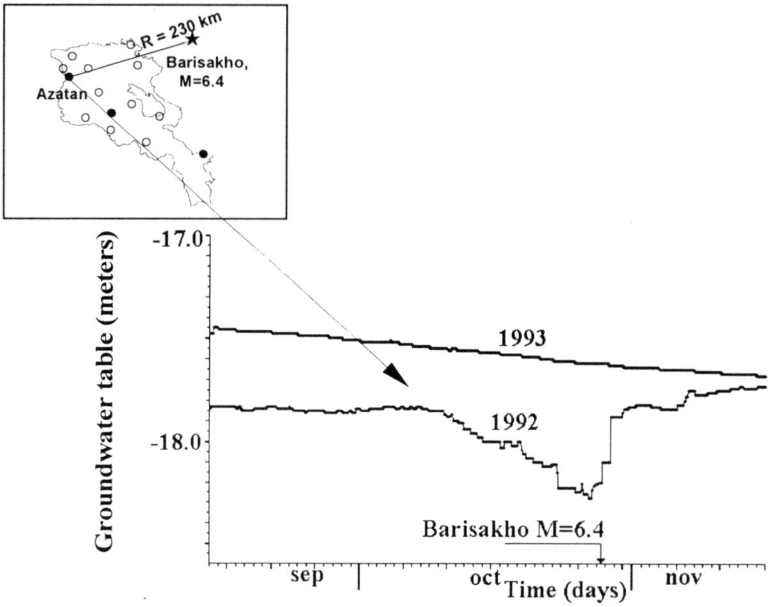

Figure 8. Anomalous groundwaters table alterations during Barisakho earthquake (M_s=6.4, 1992, Georgia) preparation and occurrence .

At the same time, like in case of the geoelectric parameter (E), a dissimilar response of hydrogeodynamic stations of the National Observation Network to the preparation and occurrence of the Barisakho earthquake is observed: from the anomalous response to the normal one. Moreover, this response does not depend on distance between observation sites and earthquake epicenter (Fig.8). The physical explanation of the fact observed is the same as in case of geoelectric parameter (E)- non-homogenety of the geological medium and its non-homogeneous deformability in the field of regional forces of elastic strains.

From the standpoint of studying the relation between preparation of strong earthquake and its various geophysical, geochemical precursors, the cardinal question, in opinion of many experts, is discussed: at what distance from the earthquake epicenter the precursors are observed? The logic of arguments is usually as follows: (1) if the precursors are observed at a remote distance (more than 100 km), then they are not precursors but casual coincidence; (2) if the precursors are observed at a short distance (10 km), in this case it might be related to the earthquake preparation if the anomalous effect is observed also at the other observation sites in the vicinity of the epicentral zone.

In connection with the above-mentioned the data listed in the Fig.9 is of the interest. This data is showing that the process of a strong earthquake preparation causes big changes in the chemical contents of the groundwaters at the distances of hundreds and thousands km from the seismic event epicentre. This phenomenon which is evident also in the other examples listed by us (Fig. 4-8) can be explained only by the following way.

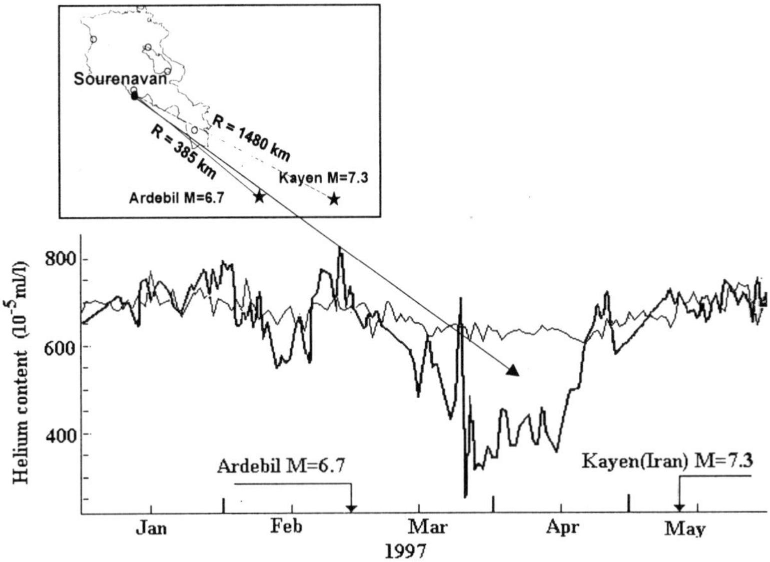

Figure 9. Changes in the helium (He) content in the ground waters of the central part of Armenia at the "Sourenavan" station during preparation of the Ardebil (M_s = 6.7, 1997, Iran) and the Kayen (M_S = 7.3, 1997, Iran) earthquakes.

——— Helium (He) mean value for 1986-1996;

——— Helium (He) content for 1997;

Phenomena accompanying the seismic event preparation are not always related to the earthquakes. But these events (the earthquakes and the phenomena accompanying its preparation) are always relating to the regional deformation of the medium under the impact of the regional forces of elastic strains. In this case the anomalous phenomenon and earthquake can be at any distance from each other in the field of external regional forces.The deformed medium will be ruptured in the most weakened place, while the anomalous phenomenon in the general case will be observed where the deformation forces are able to produce a temporary sourse of anomaly. In so doing, there might be cases when anomalous phenomena would coincide in space with the future epicentral zone. The same arguments are applicable for explaination of the fact that why the anomalous phenomena are observed in one site, while in the other are not. Fig.10 confirms the above-mentioned conclusions.

Thus, summerizing the data given in the Fig.4-10, it may be really stated that during the preparation of strong earthquakes in the collision zone between Arabian and Eurasian plates in the territory of Armenia in all three phases (solid, liquid and gaseous) of the deformed medium, the strong changes of their physical, physico-chemical and chemical parameters can be observed. The anomalous phenomena can therewith be observed at any distances from the earthquake epicenter and (likewise) may not be observed in its vicinity. This regularity is being explained by dependance of the anomalous phenomena and earthquakes in general case not the one (the phenomenon) on the other (the earthquake) but both (the phenomenon and the earthquake) on the regional elastic strain forces, the non-homogenety of geological medium and its non-homogeneous deformability.

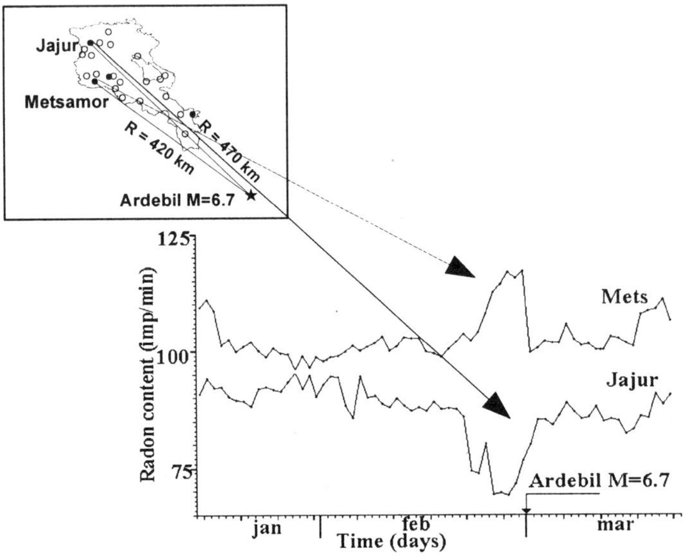

Figure 10. Changes in the contents of the undersoil gas Radon (Rn) in Armenia during preparation of the Ardebil (M_S = 6.7, 1997, Iran) earthquake

It is interesting to mention that the anomalous phenomena coinciding by its time with the earthquake preparation period in the same seismoactive zone can be observed also in the atmosphere and biosphere.

Fig.11 illustrates the monitoring data of very law frequency (VLF) radiowave propagation from the "Omega" radionavigation system (RNS) along with the radioroads controlled by the NSSP: Reunion Island (France) – Yerevan (NSSP, Armenia), *f*=12.3 kHz; Aldra (Norway) – Yerevan (NSSP, Armenia), *f*=12.1 kHz. It is seen from the data given that during the preparation of the Ardebil Earthquake (M=6.7, 1997, Iran), along the radioroad (Reunion Island /France/ - Yerevan, /NSSP, Armenia/) crossing that earthquake preparation zone the typical "seismogenic" anomalies of VLF radiowave propagation i.e. Phase Cycle Prolonged Losses (PCPL) effect □□ was observed. (S.Balassanian et al, 1997, [5]). The PCPL effect is giving an evidence about disturbances in ionosphere above the strong earthquake preparation zone. This disturbances disrupt normal propagation of the VLF waves emitted by transmitters of the RNS OMEGA and, correspondingly, deteriorate their receiption by receiving □7-38 stations set at the NSSP. The □7-38 receiver measures phase shift (□□) between reflected from ionosphere VLF waves (radiated by the transmitter of the RNS OMEGA) and standard signal formed by the C□B-74 atomic frequency standard. In the paper of S. Balassanian et al, 1997, [7] it is substantiated that ionosphere excitation over the strong seismic event preparation zone takes place of the influence intensely oscillating low-frequency electromagnetic waves falling on ionosphere from seismogenic source formed in strong earthquake preparation zone. This sourse represents an area of uniformly oriented ZSC, under the influence of growing elastic strain forces.

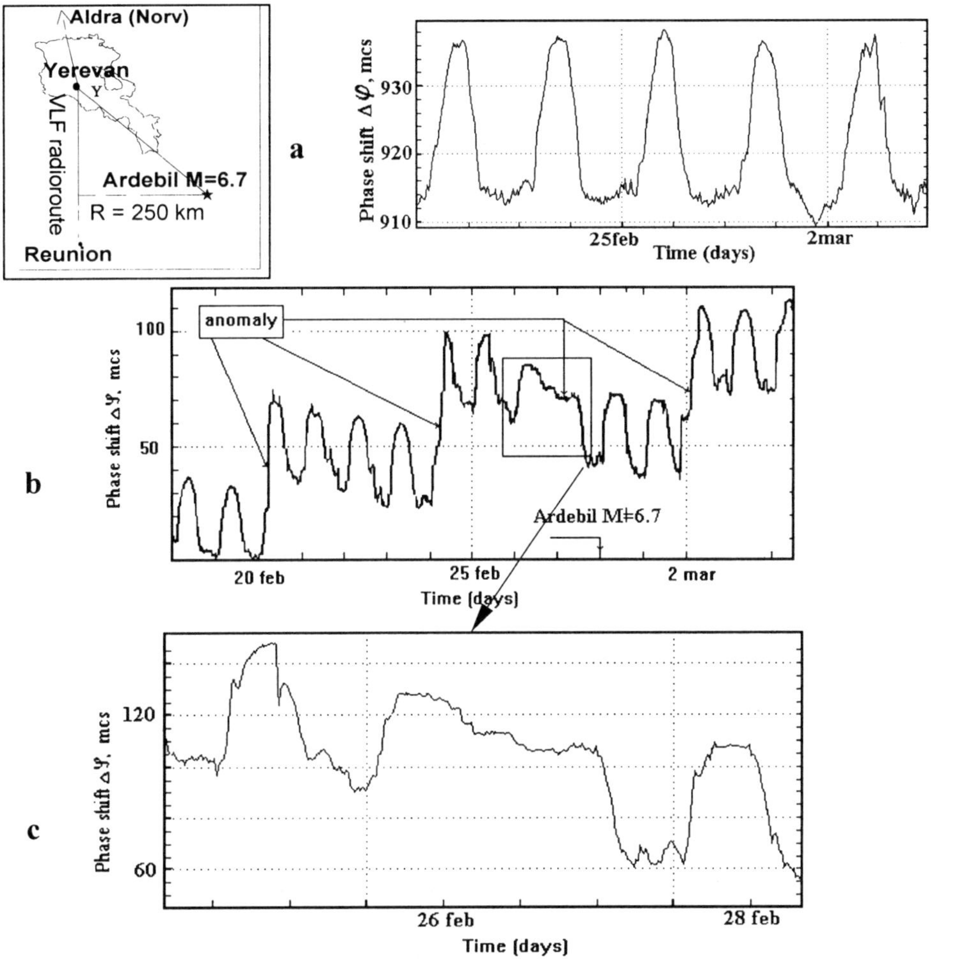

Figure 11. VLF radiowave propagation during preparation of the Ardebil (M_S = 6.7, February 28, 1997, Iran) earthquake along the VLF radioroades of RNS "Omega":
a) Aldra (Norway) - Yerevan (NSSP, Armenia): -Normal Δφ;
b) Reunion Island (France) - Yerevan (NSSP, Armenia):-Anomalous Δφ;
c) Zoom of the fragment of anomaly from fig. 11b

Most probably, the activity of aquarium fish during earthquake preparation (Fig.12) is also explained by the presence of "seismogenic" electromagnetic radiation in a broad frequency range cost, to our opinion (S. Balassanian et al, 1997, [7]), by the deformed ZSC of the rocks. It is seen from the Fig.12 that the anomalous activity of the aquarium fish during preparation of the Noyemberyan earthquakes (1997, M=4.4,

Armenia) was provoced not by the external factors, in particular by atmospheric pressure.

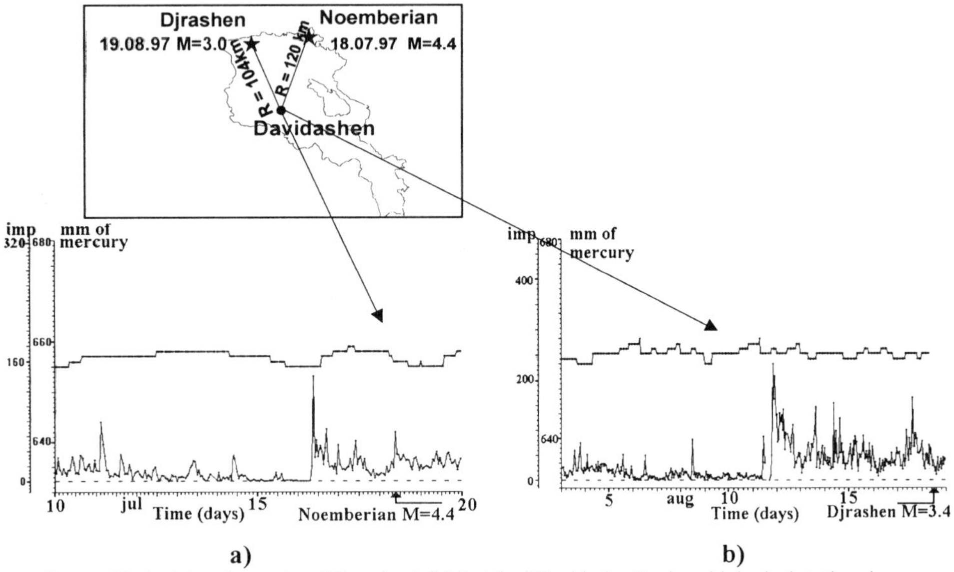

Figure 12 Activity of aquarium "Don sheat-fish" at the "Davidashen" seismobiological station in comparison with atmospheric pressure during preparation of the Noemberian and Djrashen earthquakes in Armenia: a) M=4.4, July 18, 1997; b) M=3.4 August 17, 1997.

Fig.13 shows a sufficiently good correlation between the variation of the parameter of irreversibility if geoelectric fields, behaviour of aquarium fish (the "Don sheat-fish") and seismic regime. The behaviour of the aquarium fish i.e. the number of comings of the "Don sheat-fish" from the bottom of the aquarium to its surface was registered by a special recorder of. In both cases the activity of the fish (1) and the irreversibility of non-stationary electromagnetic processes (2) (S. Balassanian, 1990, 1997, [6,7]), may be influenced by the electromagnetic radiation of the seismogenic nature.

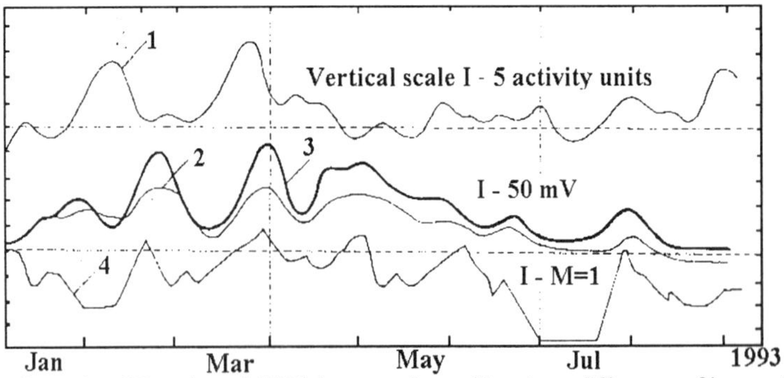

Figure 13 Activity of aquarium "Don sheat-fish"(1) in comparison with voltage difference of irreversibility of non-stationary electromagnetic processes at their formation (2) and decay (3) (at the "Davidashen" observation station), as well as in comparison with the seismic regime of the region (4)

Examples given in the Fig.12 and 13 also well confirm the above-drawn conclusion concerning the connection of the anomalous phenomena and earthquakes with the same source, i.e. the non-homogeneously deformed geological medium in the impact zone of the regional forces of elastic strains.

The following main conclusions could be made from the above-presented data:

- pre-, co-, post- seismic changes of physical, physico-chemical parameters of medium could be observed in the territory of Armenia in all the three phases of the lithosphere: solid (Fig. 5, 6, 7), liquid (Fig.8), gaseous (Fig.4, 9, 10);
- pre-, co-, post- seismic changes of the various parameters of the lithosphere could take place in any long distance from earthquake source within the seismoactive zone located in the fields of influence of one and the same regional forces of elastic deformations, for example, in the collision zone of the Arabian and Eurasian plates;
- pre-, co-, post- seismic changes of various parameters of the lithosphere in the view of its heterogeneity are taking place by various modes in various observation sites from the complete absence of any reaction to the clearly displayed anomalies;
- pre-, co-, post- seismic changes of various parameters of the lithosphere are mainly depend on local geology of the observation site;
- the most sensitives to the external physical impacts generally and to the elastic deformations accumulation in particular are so called energy-active points [6, 8];
- according to the author who has discovered that phenomenon [6,8], energy-active points are typical for multi-phase, poli-disperce, multi-component, heterogeneous medium defined by high concentration of ZSC [6,7,8];
- any physical, physico-chemical, chemical parameters of energy-active points could be informative in the territory of Armenia.

The physical nature of pre-, co-, post- seismic changes can be determined by the following manner:

- remarkable changes in physical, physico-chemical and chemical parameters of the deformed medium must be accompanied by formation of temporary sources of different physical fields that may be identified in all three phases (solid, liquid, gaseous) of the lithosphere;
- temporary sources of different physical fields are formed at macro- and micro-hierarchic levels of the geological medium during accumulation of elastic deformations and disintegrated during its discharge;
- temporary sources are formed in deformed lithosphere (initial seismogenic sources), inducing secondary sources in atmosphere and biosphere at stages that are close to a seismic event;
- temporary sources represent a system with changing properties creating different dynamic physical fields surely registered on observation surfaces.

Energy-active points situated in the most permeable and weakened zones of the lithosphere are having the maximum changeability of the physical properties due to the specific structure [6,8].

At the end, it is necessary to point out that pre-, co-, post- seismic changes of the medium out of energy-active points are being displayed significantly worse were physical properties of the medium, under the external physical impacts, are being changed significantly slower. As a result of that, in general case, it is difficult to accumulate satisfying statistics of seismogenic fenomena reccurance in connection to the various earthquakes which are differing from each other by magnitude, source mechanism, azimuth and distance from the source to the observation site, without differentiation of the geological medium into local energy-active points and wide spread energy-passive massivs of rocks.

In all the cases, our experience is showing that in general case the number of anomalies in sensitive observation sites is exceeding as a rule the number of seismic events. It could be related with the creep events and uncertainty of seismogenic anomaly "image".

More clear picture in any cases can be obtained by the special studies of energy-active sites (points) in seismoactive zones.

3. Energy-Active Sites (Points) in Seismoactive Zones

Energy-active points' investigation in Armenia presented in this paper by the Realtime Data Information Network in Earth Sciences (READINESS) Project. READINESS has been carried out by the NSSP (Armenia) jointly with GFZ, FU BERLIN, SAW (Germany) and ERD, KO, TUBITAK, EU (Turkey) by Prof. J. Zschau's initiative and financial assistance of the GFZ (Potsdam, Germany). Multiparameter groundwater monitoring with continuous recording of: pH, water temperature (t^0_W), Radon (Rn), specific electrical conductivity (C), water quantity (Q) and some other parameters at some locations additionally has been carried out in the frame of the READINESS Project (Fig.14) [9].

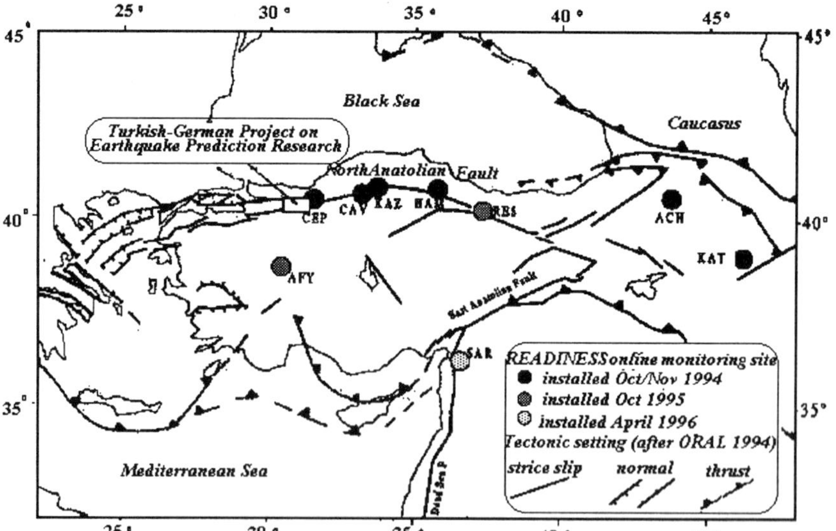

Figure 14. Location of the READINESS stations in the Arabian and Eurasian plates collision zone [9].

Data recording and transmission is being carried out by means of the special standardized equipment with every hour data transmission from observation site to the center of data processing, particularly UCSHA NSSP, through the METEOSATE with sampling interval of 10 minutes.

One of the main ideas of the READINESS Project is a multiparameter monitoring of the Arabian and Eurasian plates' collusion zone - from the Western Turkey to the Eastern Caucasus aimed to pre-, co-, post- seismic processes study in that seismoactive zone [9].

READINESS Stations in Armenia have been installed in two sites. According to the NSSP's experts proposal the first site is the Kajaran station located in the energy-active point according to data presented in the Fig.4, second site is the Akhurik station located in the energy-passive medium. Position of two stations "Kajaran" (later on K) and "Akhurik" (later on A), installed, respectively, on the Southern and Northern parts of Armenia by "READINESS" project with plotted on epicenters of the seismic events in the region for the period of time between September, 1998 and February, 1999 is shown in Fig. 15. The period of time between September 1988 and February 1999 is choosen since the well data quality in that period and strong seismic event absence in the region which are usually being accompanied by strong seismotectonic disturbance and reconstruction of the medium.

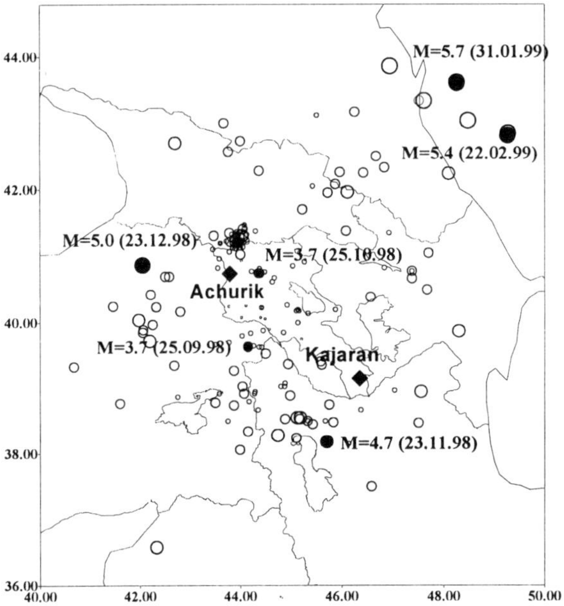

Figure 15. Location of Kajaran and Akhurik READINESS stations in the territory of Armenia with earthquake epicenters for the period of time between September, 1998 and February, 1999.

The (K) station is hydrogeochemical well, sinking into methomorphized volcanic-sedimentary rocks, depth of borehole is 170 m, depth of aquifer is 90 to 140 m (fault zone). The selfpouring of borehole is: water discharge ~ 0,7 l/sec, gas discharge ~ 0,3 ml/sec, water mineralization ~ 1,2 g/l, composition – HCO_3, Na – Ca, water temperature

~ 15,7°C. The main parameters of measurement are: pH, water temperature, Rn concentration (after degassing the system) conductivity, particular discharge. The other parameters: temperatures, humidity, and in site (daily): pH, HCO_3, Cl, So_4, (Na+K), Ca, Mg, He.

The (A) station, as well as the (K) station, is hydrogeochemical well sinking into lake sedimentary rocks. Depth of borehole is 1750 m, depth of aquifer is 140 to 170 m. The selfpouring of borehole is: water discharge – 10-15 l/sec, gas discharge ~ small. Mineralization of water – 1,8 g/l, composition – Cl-HCO_3, Ca-Na-Mg, water temperature - 24°C. The main parameters of measurement are the same as for the (K) station except the daily in site He.

3.1. DATA PROCESSING

So called (Balassanian, 1998, [8,10]) "DF" (dynamic fields) special technology used for the READINESS data processing and analysis.

Through comparison of the data obtained from all the observation sites, the "DF" allows to determine the followings for the given time interval: the maximum dynamics δP of the parameter under study (P) at each observation sites is:

$$\delta P = P_{\max} - P_{\min}, \qquad (1)$$

where $P_{\max}$ and $P_{\min}$ are the maximum and minimum values of P for the given time interval, in our case for a day (24 hours), as the "ADF" effect is under study;
the local relative dynamics at each observation sites (x) is:

$$P_L(t,x) = P(t,x) - \overline{P}(x),$$

where $P(t,x)$ is time dependence of parameters being measured during a day,

$$\overline{P}(x) = \sum_t P(t,x)/n,$$

n is the number of measured $P(t,x)$ at the observation site during a day and $P_L(t,x)$ are anomalous values, i.e. "ADF" effect

$$\Delta P_L(t,x) = P_L(t,x) - P_T(t), \qquad (3)$$

$P_T(t)$ is relative daily dynamics during the whole time interval of observations equal to

$$\sum_T P_L(t,x) / N,$$

N is the number of days; the function of mutual anomaly correlation between the daily and relative daily dynamics for the whole time interval of observations is:

$$C_{\Delta L,R}(\tau,x)=\sum \Delta P_L(t,x)P_T(t+\tau),\qquad (4)$$

where a phase shift is $\tau \in \left\{\frac{-t_{\max}}{2}, \frac{t_{\max}}{2}\right\}$;

the spectrum of the daily relative dynamics anomaly obtained by means of spectral transformation (the Fourier discrete transformation)

$$\Delta P_L(t,x)\rightarrow \left|\Delta S_L(\nu,x)\right|^2,\qquad (5)$$

where $\nu \in \left[0,\ 1/2\Delta t\right]$, Δt is a discretisation interval equal to time span of observations,

$$\Delta S_L(\nu_k,x)=\sum_{i=1}^{n}\Delta P_L(t_i,x)e^{-j2\pi(i-1)\Delta t\nu_k},\qquad (6)$$

$$k=0,1,\ldots,[n/2];$$
$$j=\sqrt{-1},\ t_i=(i-1)\Delta t+t_1;\ i=1,\ldots,n;$$
$$\nu_k=k\Delta\nu;\ \Delta\nu=1/\left(2\Delta t[n/2]\right);$$

the index of energy emission (or absorption) is:

$$\Delta E(x)=\sum_{\nu} S_{\Delta L}(\nu,x)\qquad (7)$$

where $S_{\Delta L}$ is a spectrum difference between the daily and relative daily dynamics for the whole time interval of observations time histories

$$S_{\Delta L}(\nu,x)=\left|S_L(\nu_k,x)\right|^2-S_T(\nu_k)\qquad (8)$$

where $S_T(\nu_k)$ is a spectrum of the relative daily dynamics for the whole time interval of observations is:

$$S_T(\nu_k)=\frac{\sum_T \left|S_L(\nu_k,x)\right|^2}{N},\qquad (9)$$

$S_L(\nu_k,x)$ is a spectrum of daily relative dynamics; the index of energy redistribution is:

$$movE(x) = \sum_{\nu} \left| S_{\Delta L}(\nu, x) \right| - \left| \sum_{\nu} S_{\Delta L}(\nu, x) \right| \qquad (10)$$

All the above mentioned spectral characteristics are transformed into real power spectra W(W/m^3) related to medium volume unit by means of multiplying by the index corresponding to the given physical field.

Besides the characteristics mentioned above, the "DF" allows to determine various indexes of dynamics of being-measured characteristics, the medium activity index, etc. Thus, the "DF" software package allows to find out energy-active sites (or points), where intensive nonlinear dynamic fields (ADF) effect may occur at a certain time before, during or after the seismic event due to the impact of increasing deformation force of elastic stresses.

3.2. DATA ANALYSIS

Amongst the groundwater parameters being recorded in the READINESS stations, pH and t^0_w are presented in this paper as the most stable parameters in a sense of its measurement simplicity and reliability.

The results of pH and ground water temperature t^0_w of (K) and (A) stations in compare with seismic regime of the region and also with strongest seismic event that takes place during each month for the same period of time are shown in Fig. 16-19.

One month interval of time based upon the NSSP many years experience shows that one month is the time for quasi-equiponderous condition rehabilitation of the various mediumal characteristics in the territory of Armenia after seismic event in the magnitude range of $6.0 \geq M \geq 2.0$ ($M=2.0$ is a lower representative magnitude in the National earthquakes catalogue of the NSSP for Armenia and adjoining areas).

Initial variations of pH and t^0_w under the (a) index, its average daily values (b), maximum daily dynamics (c), variance (d), index of daily emission (absorption) of energy (e), index of energy redistribution (f) made by formulas 1-10 are shown in Fig.16-19.

The data presented in Fig.16-19 shows that:

1. The characters of pH containment alteration, its dynamic and spectral characteristics of (K) and (A) stations are significantly differ from each other .

Behaviour of pH dynamic and spectral characteristics on the (K) station distinctively repeats the behaviour of the region's seismic regime. In spite of this, pH characteristic alterations are significantly preceding the seismic regime alterations of the region. The other peculiarity of pH dynamic characteristics is the fact, that strong nonlinear effect of pH anomalous daily dynamics (so called "ADF" effect Balassanian, [6,8]), and afterwards 10 days long "quiescence" of pH variations (Fig. 6d) precedes the strongest earthquake in the region with M=5.7 occurred in the Northern Caucasus at 30.01.99 (Fig. 16). Also earthquake swarms that took place in November in the territory a)

b)

c)

d)

e)

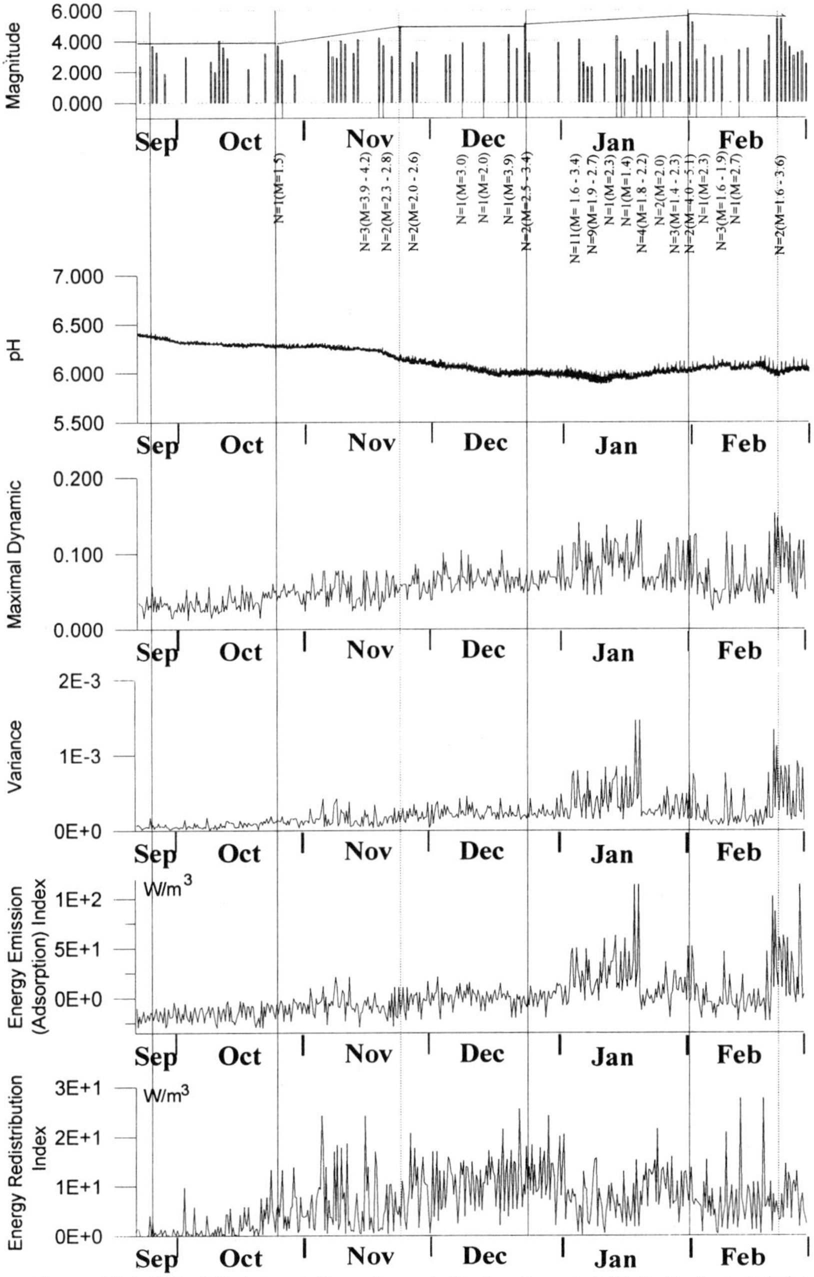

Figure 16. Initial daily (a) variations, dynamic (b,c) and spectral (d,e) pH characteristics of Kajaran station's groundwater in compare with seismic regime of the region (in magnitudes):

/‾ - the rounding of seismic regime of region

| - stronger seismic event for each month

N =5 (M=1.6-3.6) - earthquake's number with certain magnitude

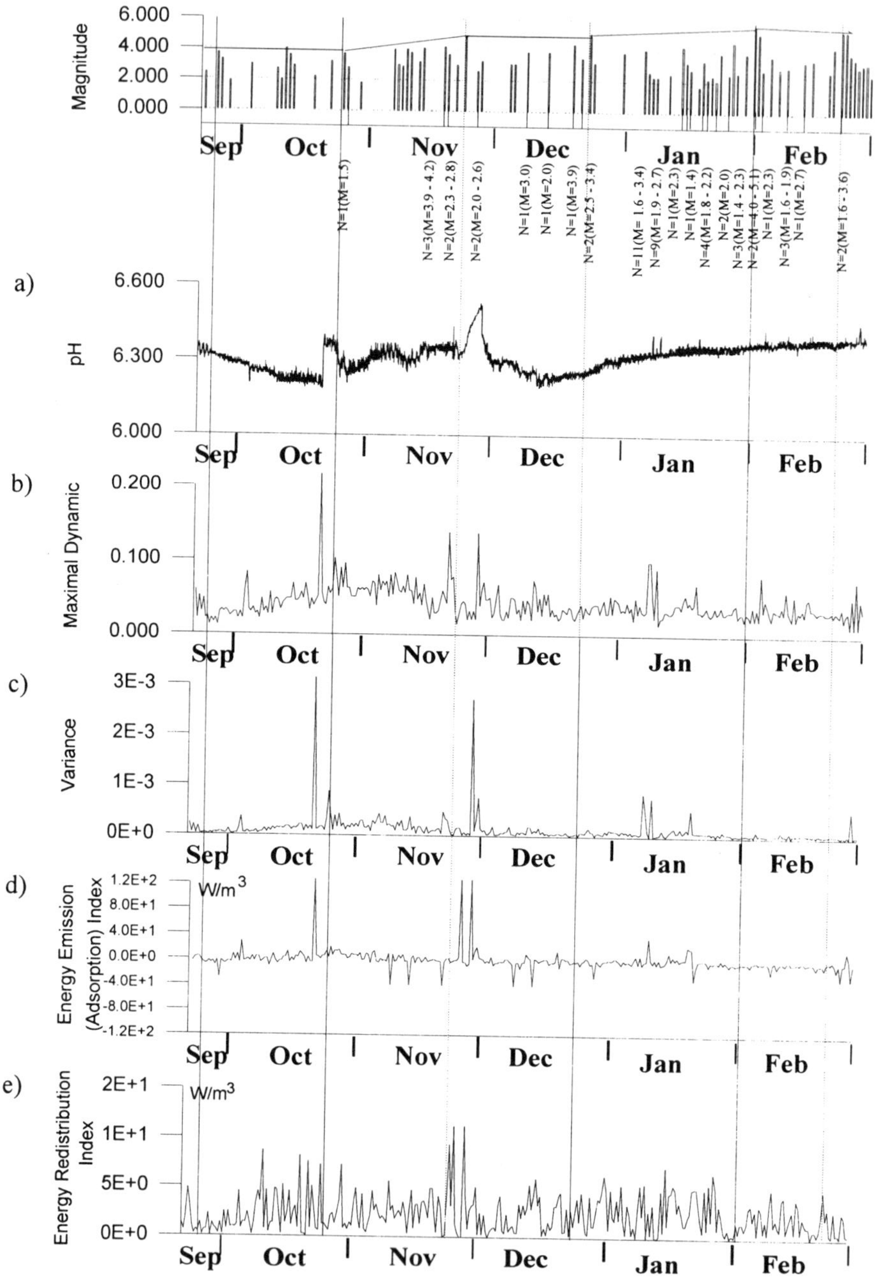

Figure 17. Initial daily (a) variations, dynamic (b,c) and spectral (d,e) pH characteristics of Akhurik station's groundwater in compare with seismic regime of the region (in magnitudes).

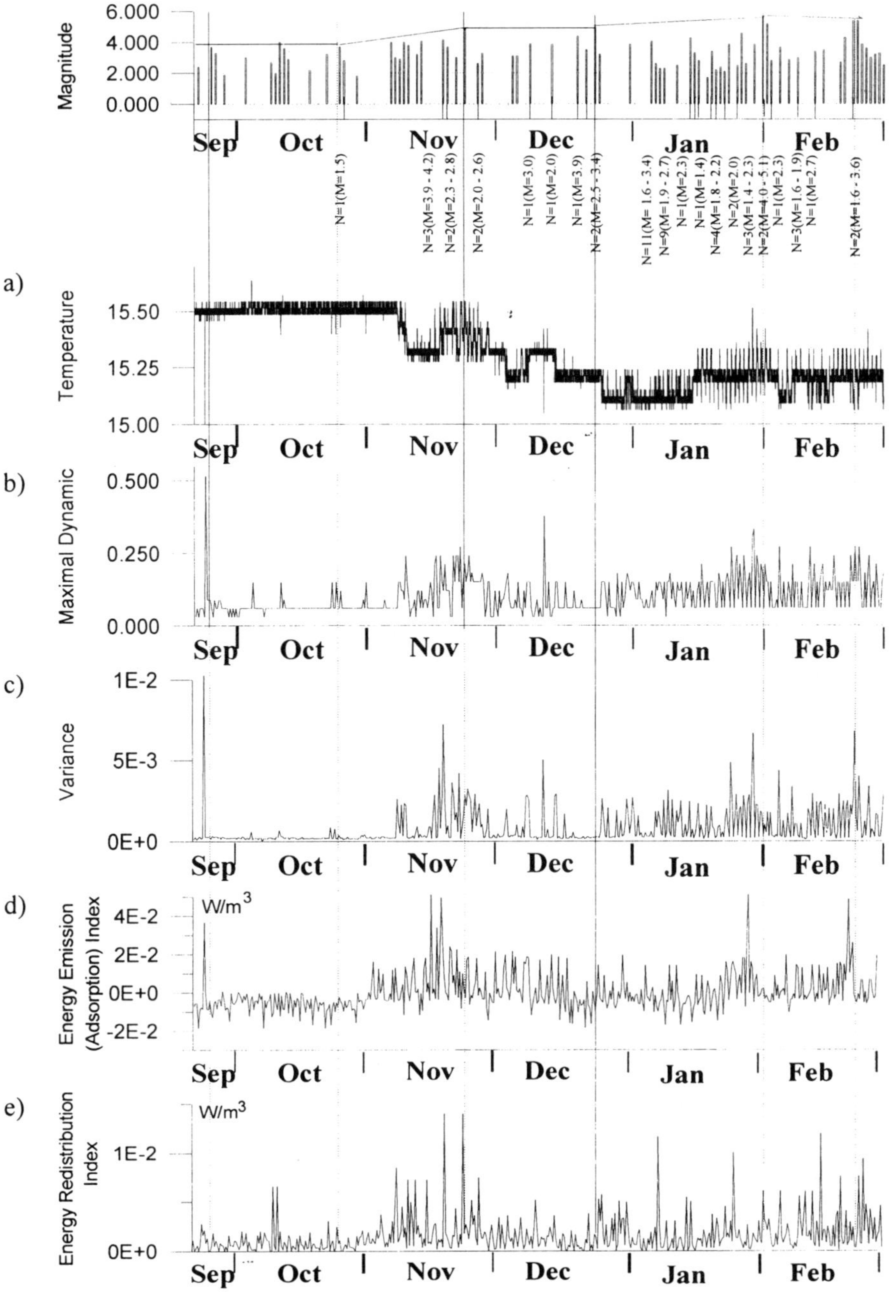

Figure 18. Initial daily (a) variations, dynamic (b,c) and spectral (d,e) temperature characteristics of Kajaran station's groundwater in compare with seismic regime of the region (in magnitudes).

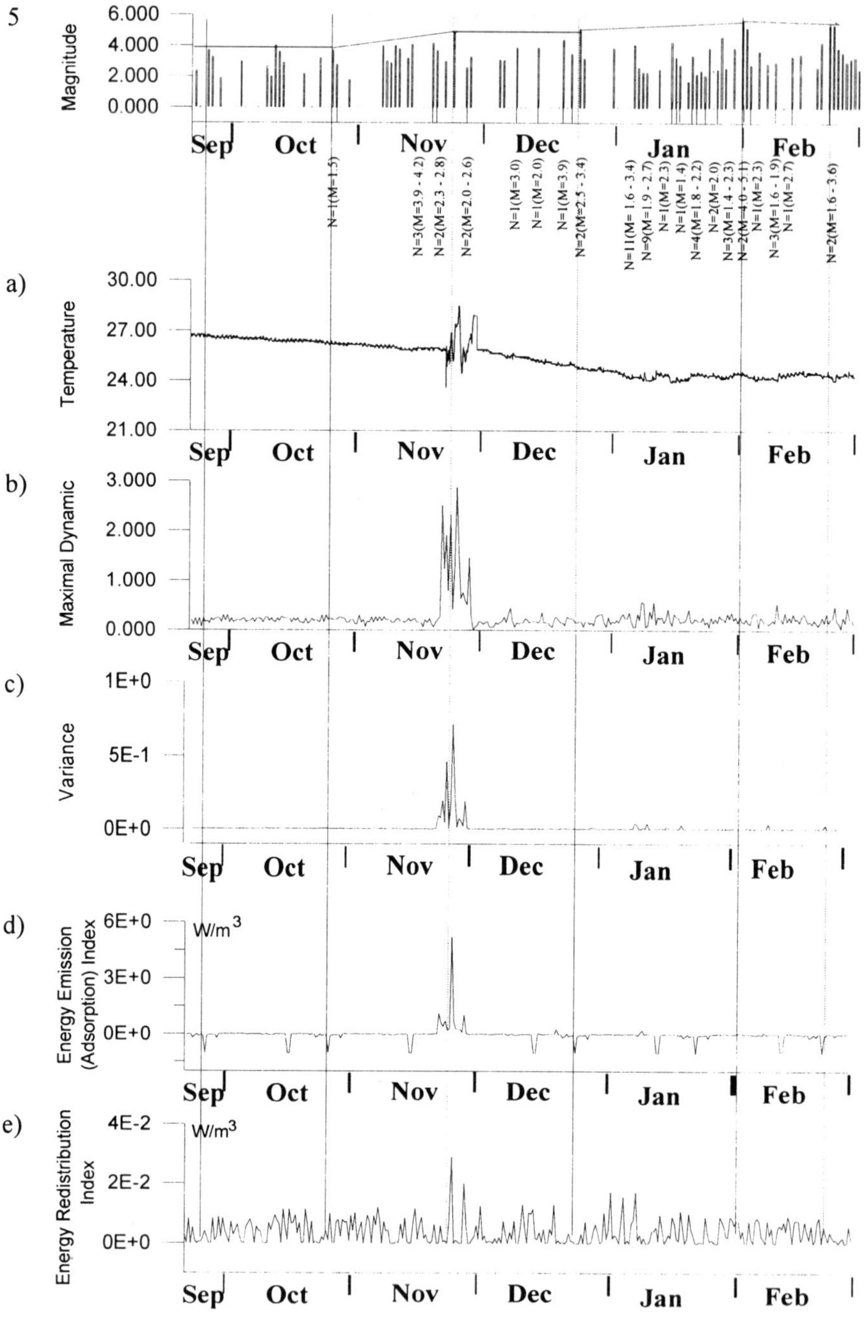

Figure 19. Initial daily (a) variations, dynamic (b,c) and spectral (d,e) temperature characteristics of Akhurik station's groundwater in compare with seismic regime of the region (in magnitudes).

of North-Western part of Iran, in January and February on the Northern border of Armenia with Georgia (Fig. 16 swarm indicated by N≥3 on the seismic regime magnitude diagram, where N is a number of events) being distinctively displayed in pH dynamic characteristics.

Alterations of pH dynamic and spectral characteristics of (A) station have a local eruptive character that does not reflect the general behaviour of region's seismic regime. Impulses of pH dynamic characteristics have in one cases co- in another – pre- seismic character. The only more or less similar pH dynamic parameter of the both (A) and (K) stations is the energy redistribution between various frequencies of pH variations (Fig.17f and Fig. 16f) which looks like a monthly "packages".

2. The character of t^0_w alterations and it's dynamic parameters in the both (K) and (A) stations is more similar. In spite of that, the first and rather typical peculiarity is the fact, that majority of strongest seismic events for each month are being distinguished (pre-, or co-, or post - seismically) by a negative values of Energy Emission Index (EEI) (Fig.18d, Fig. 19d). Anomalies of EEI negative values means spectrum suppression of t^0_w daily dynamics by some dominating force. Coming from well coincidence of EEI anomalies with overwhelming majority of strongest earthquakes for each month, it might be supposed, that the dominant has a seismogenic nature, i.e. it is bonded with critical deformations of rocks under the impact of regional forces of elastic stresses. Convergence of EEI in the (K) station with strongest earthquake for the period of February is some exception to the mentioned regularities. But, the EEI, at the same time, well coincides with the earthquake (M=5,0), occurred at December, 23 in the Northern part of Armenia.

Another, for feature both stations is strong anomalous dynamics of t^0_w from the end of November 1998. It is probably does not connect directly to specific earthquake, but certainly has a regional nature, that should be clarified.

One of the distinctive peculiarities of both stations is the fact, that the background dynamics of t^0_w of the (K) station is raising from October to February. As in the case with pH it is also getting well correlated with the seismic regime of the region. Well ability of energy redistribution index (ERI) t^0_w to be correlated pre-seismically (which is very significant) with majority of the strongest seismic events for one month period is another peculiarity of the (K) station. It means, that a redistribution of frequencies within energetic variations spectrum is taking place in the (K) station before strongest seismic event.

So, by at list two measurable parameters of groundwater – pH and t^0_w the (K) station has a unique sensitivity to seismic regime of the region within a minimum 500 km radius from observation site.

Concerning the other parameters, so the data obtained by our colleagues from the GFZ [9] shows that the Kajaran energy-active point is sensitive even to the Earth's tides (Fig.20).

It is clear from the Fig.20 that the characteristics of the Earth tides are clearly pronounced in the spectrum of daily cycles of radon, water temperature, specific electrical conductivity. It means that the Kajaran station's strain sensitivity edge to deformations is lower than 10^{-8} [9].

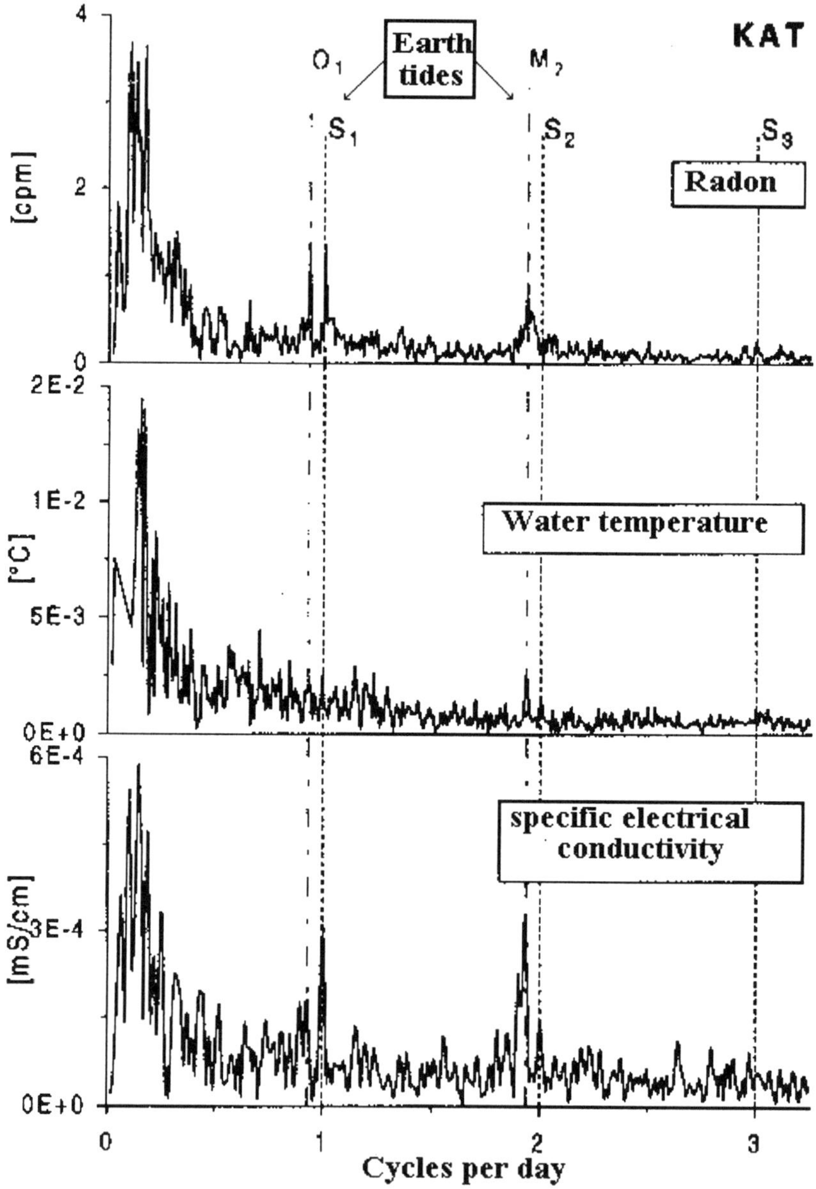

Figure 20. Earth tides pronounced in different groundwater parameters at the "Kajaran" energy-active site

Besides of energy-active points investigation within the relatively quiet period of seismic regime there is "READINESS" data recorded at the Kajaran station in the period of the strongest seismic event occurred in the Western part of Turkey (Izmit earthquake, M=7.4, August 17, 1999) presented below.

Curves presented in the Fig. 21 are giving an evidence that the Kajaran station is indeed having a unique sensitivity to pre-, co-, post- seismic processes taking place all over the Arabian and Eurasian plates' collusion, i.e. from the Western part of Turkey to the Eastern part of Iran. Analyzing a data presented in the Fig. 21 it could be the same conclusion made as we have made in case of the other NSSP's observation stations' data analysis (Fig. 4-13).

"Kajaran" READINESS station located in the distance of 1300 km from the Izmit earthquake source, i.e. much further than the other READINESS stations situated in the territory of Turkey and Northern Armenia (Akhurik), has responded on the pre-, co-, post- seismic stages of Izmit earthquake better than the others. It confirms our conclusion that the sensitivity of a station to the earthquake is being determined not by distance from source to the observation site but by the local geological conditions in which the observation station is situated, i.e. site sensitivity to the regional deformations (energy-active point phenomena).

Another important conclusion made is the fact that not all the parameters of the observation site are responding similarly on one and the same seismic event. Thus, it is clear from the Fig. 21 that conductivity (C) and water quantity (Q) are responding on the Izmit earthquake better than the other parameters of the Kajaran station. All the 3 stages (pre-, co-, post-) of seismic event were clearly reflected in the C and Q variations. In spite of that, decrease of C and increase of Q were foregoing the Izmit earthquake, started simultaneously and leap-type from the middle of March 1999, i.e. 5 months before the seismic event (first anomaly). Then, sharper decrease of C and increase of Q started 2,5 months before the seismic event (second anomaly). Concerning the first negative anomaly of C being observed at the end of November 1998 then it does probably not have a seismogenic nature since is not being confirmed by the increase of Q.

Decrease of C and the simultaneous increase of Q can be explained by the strong compression of the inter-pore and inter-crack space in the Kajaran energy-active area of high sensitivity, under the impact of the increasing elastic strain, foregoing the Izmit earthquake. Such a strong compressions of the medium could resulted to the injection of the near-surface fresh water into the mineralized fissure waters. As a result of that the conductivity C will decrease and the water quantity Q increase.

Besides above listed pre-seismic variations there is a co- seismic effect of sharp decrease of C and the same sharpness increase of Q, clearly occurred in data of C and Q presented in the Fig. 21. After that, the post-seismic stage of 3 months reinstatement of the normal water conductivity and its table is seen.

Daily time interval in which co-seismic effect of C and Q being observed is a matter of a big interest for deeper understanding of the seismic processes correlation with the medium parameters' variations. It is seen from the data presented in Fig. 22 that after the seismic event at t_0=00.01:38 (origin time) there are three impulsive variations of C and Q occurred: with relative lagging of C upon Q for 50±10 min (where 10 min is a measurements' time interval).

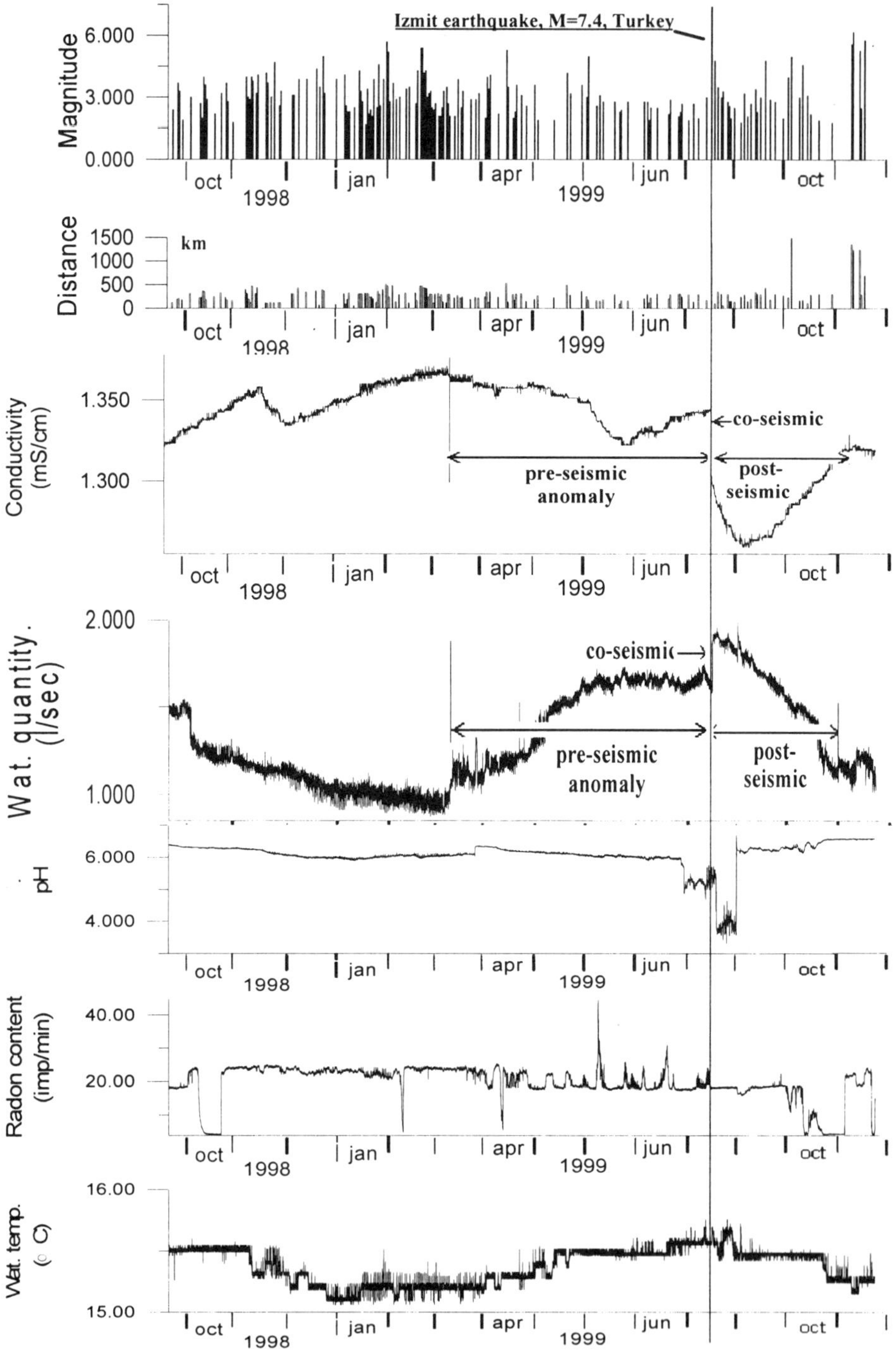

Figure 21. Curves of conductivity, water quantity, pH, radon content, temperature during the Izmit earthquake.

It is clear from Fig. 22 that indicated impulsive variations of C and Q are having the following peculiarities:

- all the three impulses occurred after the roughly equal time intervals, i.e. 1 hour ±10 min.;
- all the three C impulses are lagging upon Q for 50-60 minutes;
- there is an increase (decrease) of C being accompanied by decrease (increase) of Q for all the three cases;
- the first and the third impulses of C and Q are reversible but the second one is irreversible;
- the first and second impulses are relating with decrease of C and increase of Q, but the third impulse is having a reverse direction, i.e. increase of C and decrease of Q - against irreversible C and Q common level variations' background after the second impulse.

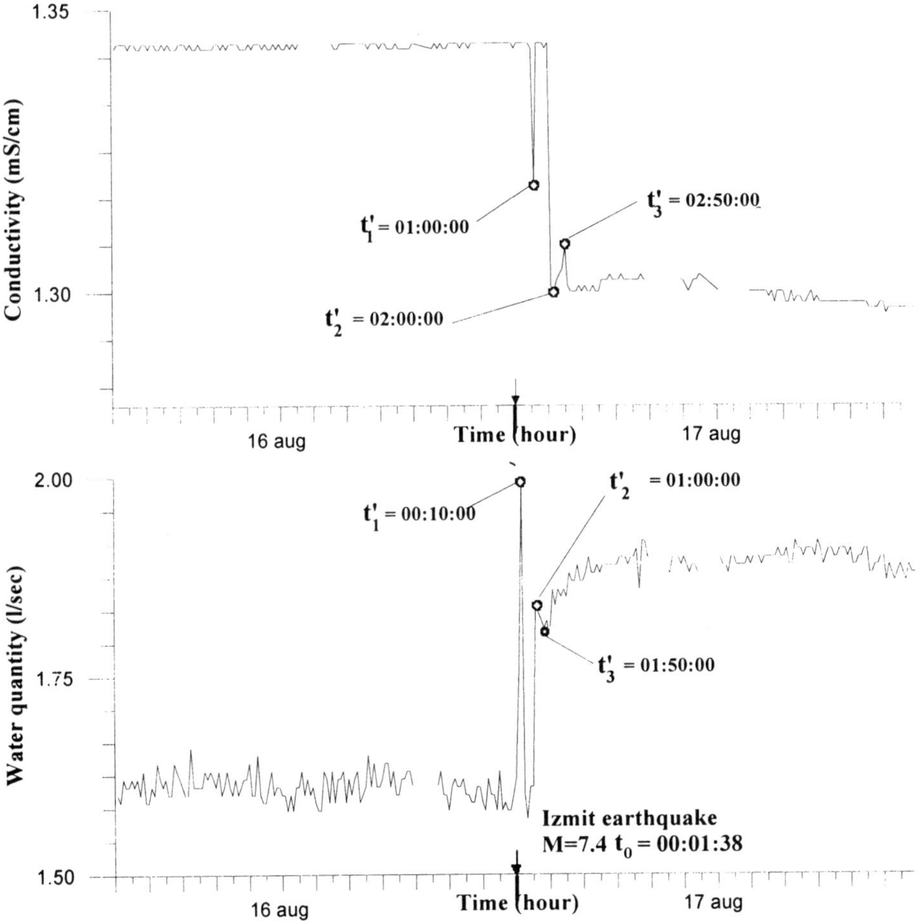

Figure 22. Co-seismic effect during the Izmit earthquake.

Analyzing the co- seismic effects of the Kajaran station it could be assumed that only the first one is correlated with dynamic wave propagating from the earthquake source to the Kajaran observation station. It is being justified to the fact that during $\Delta t_0=t_1-t_0=8$ min 22 sec (where $t_0=00:01:38$ - origin time, $t_1=00:10:00$ - first Q impulse time in the observation site) the supposed wave can cover 1300 km distance with approximate velocity of 2590 km/sec which is corresponding to the wave velocity in the elastic medium. Elastic wave of the observed case has resulted to the reversible compression of the observation site which leads to the sharp Q increase and to the sharp C decrease 50 minutes later. Time of 50±10 min could be a time of fresh surface waters' mixing with the mineralized fissure waters within the energy-active point. Concerning another two impulses, one of which is irreversible one, then those are originated approximately 1 hour ($t_2=01:00:00$) and 1 hour 50 min ($t_3=01:50:00$) after the Izmit earthquake. It means, that this impulses can not be related to the dynamic wave propagation from the earthquake source but are having some another nature. Thus, we are dealing with the new effect relating with peculiarities of the energy-active points' phenomenon in seismoactive zones.

Concerning the other parameters of groundwater of the Kajaran area (Fig.21) i.e. pH, Rn and t_w^0 then it is clearly seen here the relation of variance increase of the indicated parameters, i.e. amplification of the "ADF" effect along with numbers and magnitude increase of the seismic events. It is well conforming with the "ADF" effect's peculiarities defined during the analysis of relatively quiet seismic regime of the region (Fig.16-19). There is a strong anomaly of pH in the period of the Izmit earthquake being stood out in data presented in Fig. 21. It is hard to be sure in its seismogenic nature since 20 days after the Izmit earthquake there was determined by means of field controlling measurements that the pH data unit was technically faulty at the moment of controlling measurements.

3.3. DISCUSSION

So, the results presented above are confirming the fact as defined by us earlier [6] about the existence of the local energy-active geological sites in the lithosphere upon which the "ADF" effect is being observed under the external physical impacts [6,8].

Such sensitivity of the energy-active sites could be explained by the anomalous changeability of its physical parameters to the external physical impact. This phenomenon is typical for the most weakened and since that anomalously permeable zones of the lithosphere.

The complex geological-geophysical data analysis carried out by us is showing that the Kajaran station is located within one of the typical energy-active sites of the lithosphere which is in the state of seismic "quiescence " during recent several years (Fig.23)

The area of Kajaran is located in one of the most complexly structured areas in Armenia on the junction of three active faults. Typical geological features are widespread magmatic intrusions of granotoidal, alkaline and subalkaline composition. Kajaran is situated within the North Siunik block (an anticlinal uplift) limited to the East by the Khoustum-Giratakh fault, and to the West by the Tashtoun fault.

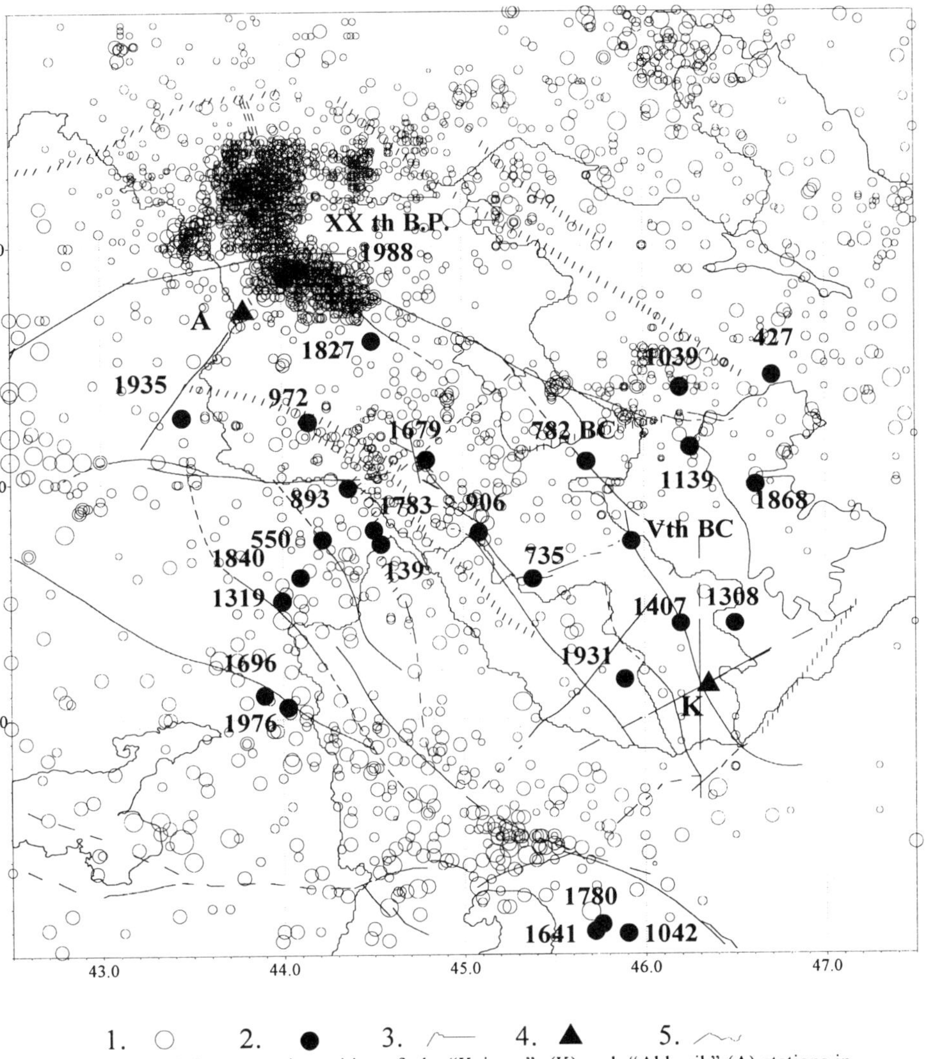

Figure 23. Seismotectonic position of the "Kajaran" (K) and "Akhurik" (A) stations in compare with seismic regime after the Spitak earthquake (M=7.1, 1988).

1. Epicenters of earthquakes after the Spitak earthquake (1988)
2. Epicenters of historical earthquakes (M ≥6.0)
3. Active faults
4. Positions of the "Akhurik" (A) and "Kajaran " (K) stations
5. National borders outline

The followings are giving an evidence about the anomalously high weakness and permeability of the Kajaran area besides of developed tectonic structures (from the stabilized to the active faults) :

- sources of historical and modern strong earthquakes (M_{max}=6.5) (Fig. 23)
- existence of endogenous main deposites, giving an evidence about the injection of deep magmatic fluids (Fig.24.a)

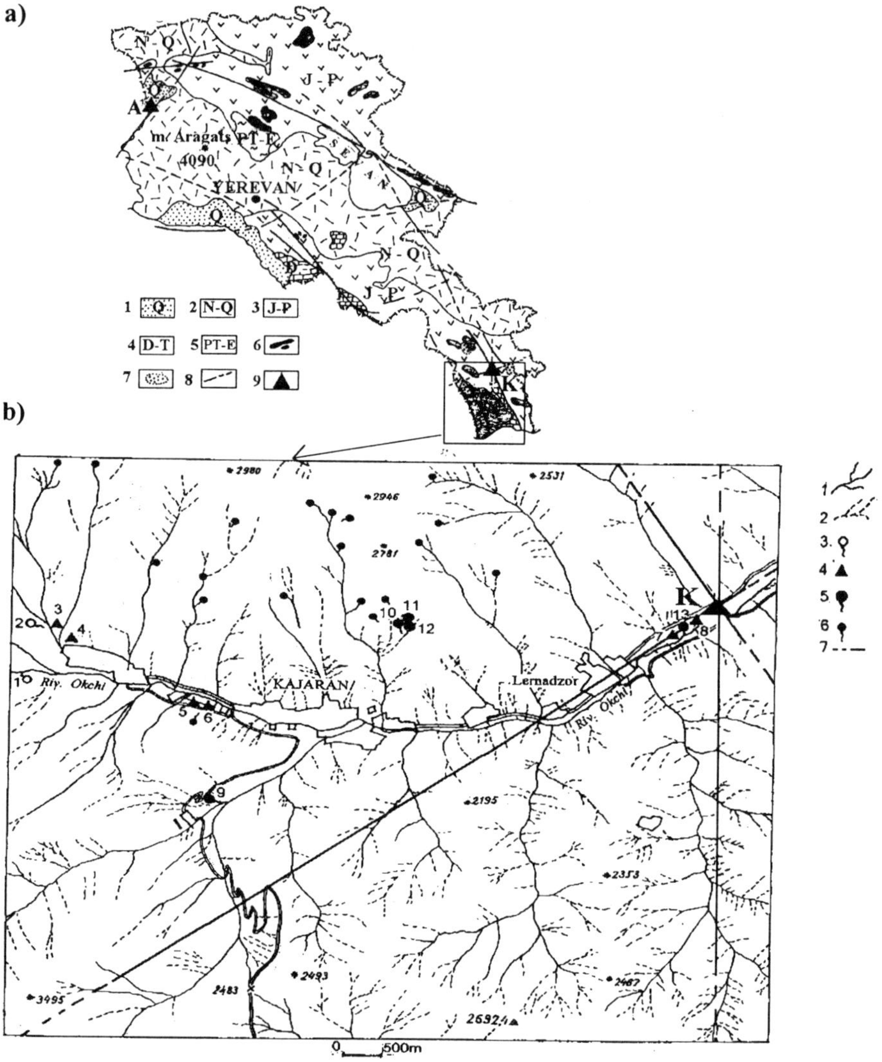

Figure 24.

a) Schematic geological map of Armenia

1. Recent alluvial, deluvial, lake deposits
2. Neogene-Quartemary (not segmented) lavas, tuffs, molasse deposits, clays, sands
3. Jurassic-Paleogene porphyrites, tuffs, tuffs sandstones, limestones, marls
4. Devonian-Trassic limestones quartz sandstones, clays, schists, coal limestones.
5. Upper Proterozoic Lower Cambrian gneisses, phillits,marbles,metamorphic schists
6. Hyperbasites, gabbro
7. Contour of mines
8. Active faults
9. Station location

b) Zoom of Kajaran area with location of mineral springs

1 - waterflows;
2 - creeks;
3 - natural mineral springs;
4 - borehole with mineral waters;
5 - main natural waters;
6 - spring with drinking waters;
7 - active faults.

- wide spreadeness of fissure waters and huge amount of springs with various mineralization and temperatures (Fig.24.b)
- presence of 3He/4He=3.9x10-6% in the isotope composition of He, confirming the fact about significant deepness of propagation of the high permeability zones.

Above-stated factors are giving the explanation about the anomalous heterogenic, multi-phase, multi-component origin of the Kajaran energy-active area.

Such kind of consistence is resulting to the anomalously high changeability of the energy-active point's physical and other parameters caused by the high concentration of easily deformable ZSC (at the microlevel) and zones of high jointing (at the macrolevel) under the impact of the regional elastic stress forces.

Anomalous changeability of physical parameters is turning the energy-active point into the highly sensitive body reacting to the pre-, co-, post-seismic processes in the form of "ADF" effect.

It would be possible to explain the "ADF" effect's peculiarities pronounced in the various parameters of groundwaters in Kajaran area in the period of the Izmit earthquake after discussion of the Earth's crust deformations' peculiarities in the collision zone of the Arabian and Eurasian plates. The picture of the Earth's crust movements directions' and velocities' distribution in the collision zone of the Arabian and Eurasian plates according to 1991-1998 GPS observations is shown in Fig.25 [2].

It is seen from Fig.25 that the Anatolian plate is being squeezed out from the collision zone in the North - West with the average rate of 24±2 mm/year and Caucasus plate being compressed with the rate of 10±2 mm/year under the impact of the Arabian plate's movement towards Eurasian one.

Cross-section of the possible seismotectonic scheme of the Arabian plate's movement in the Caucasus region is presented in Fig. 26, [1]. It is seen from the presented scheme that the wedge splitting of the Eurasian plate by the Arabian plate should resulted to the exclusive role of splitting processes in the forming of the Caucasus seismotectonic patterns:

1. Rock deformations and ruptures all over the splitting zone, including whole its surface, should make it up a major source zone of possible earthquakes.

   That conclusion is being confirmed by the diffusive distribution of the seismicity in Caucasus (Fig.15, 23) [1].
2. High temperatures and pressures being formed in the splitting zone should result to strong variations of its physical parameters and rocks' plastic state development.

   It is being confirmed by the existence of certain limit 20-25 km deeper of which the earthquake sources in the region of Caucasus collision zone never being observed [1].
3. Arabian wedge movement along the splitting zone should result to the various velocities of formed and forming active blocks (Fig.26) having different size, height, petrographic composition and physical properties.

   It is being confirmed by the GPS observations' data (Fig.25) [2]
4. Different velocity of active blocks which are differing by its weight and physical parameters will result to forming of compression, extension, rotation zones with forthcoming formation of trust and reverse faults, strike-slip faults and other seismotectonic structures due to the crust's rupture in these zones.

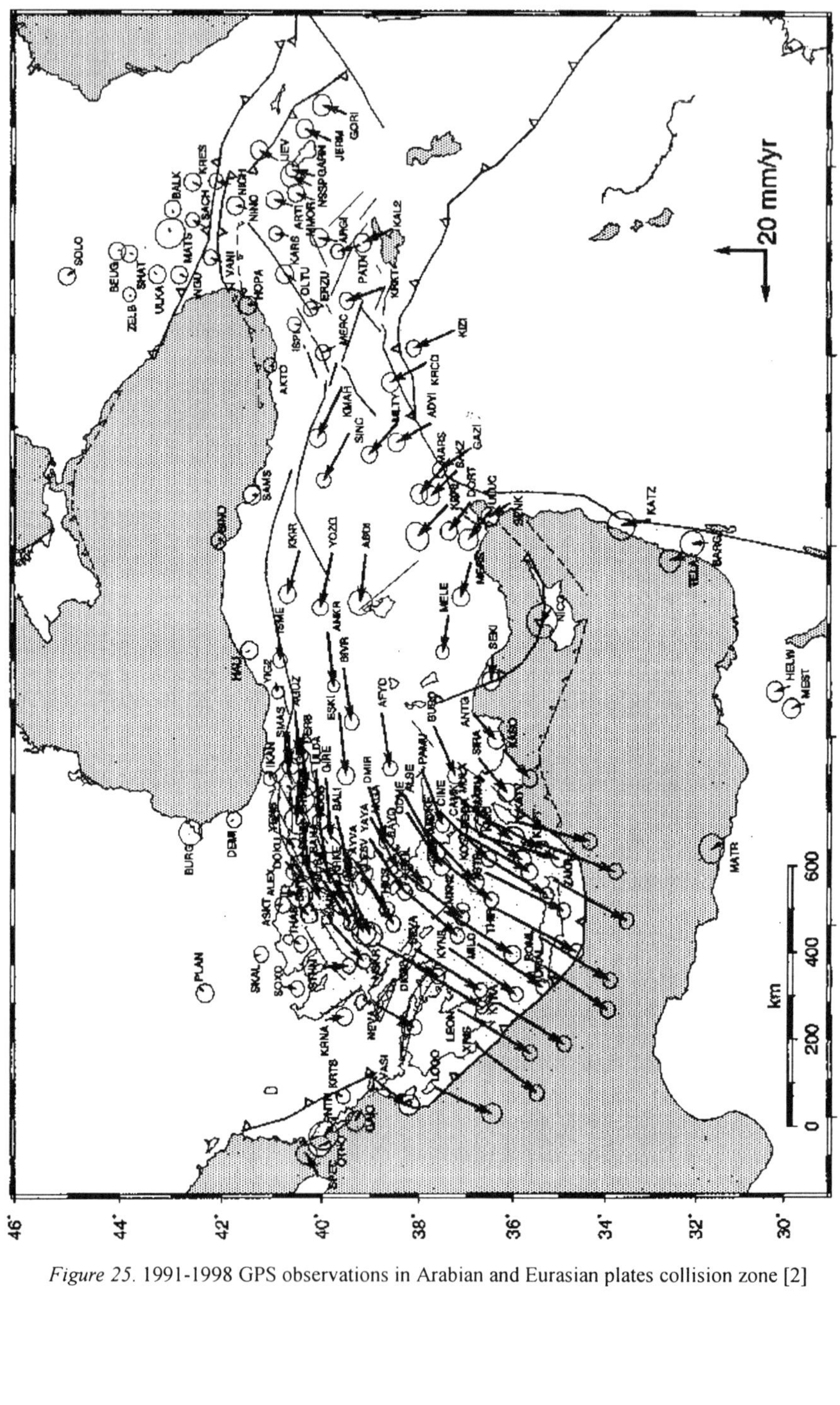

Figure 25. 1991-1998 GPS observations in Arabian and Eurasian plates collision zone [2]

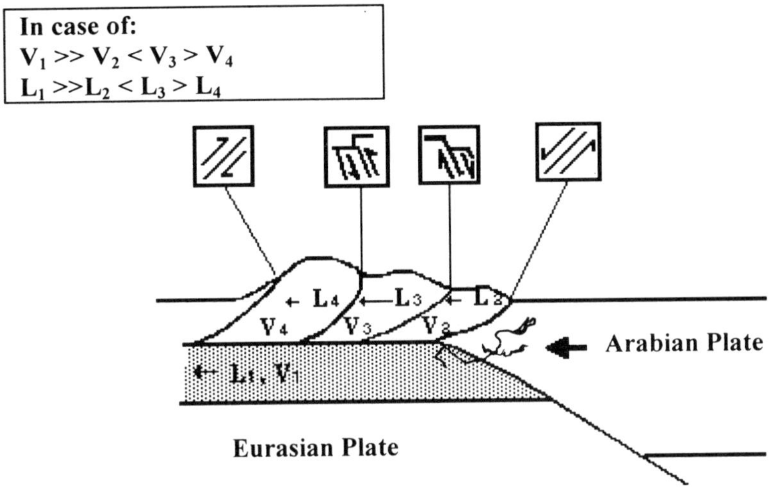

Figure 26. Cross-section of the possible seismotectonic scheme of the Arabian and Eurasian plate's collision zone in the Caucasus region.

- active blocks (strong , solid part of the Earth's Crust)

- splitting zone (plastic zone in the Earth's Crust)

- types of fault motion

V_2... - active blocks movement velocity

←L_2... - active blocks displacement

- Arabian plate's movement

5. Forming of the new rupture surfaces after each seismic events will lead to the stress redistribution in the collision zone and, accordingly, the probability of transformation of the local compression zone into the extension or strike-slip zone and vice versa. That fact is being confirmed by temporal-dimensional analysis of earthquake focal mechanisms in Caucasus (Fig. 27)
6. Arabian plate's wedged movement will deform the active blocks' boundaries, braking, in particular, its angles adjoining to the splitting surface along the wedge's movement, since velocity of the movement in the lower part of the block adjoining to the splitting surface will always be higher than the block's upper part's velocity which is adjacent with the open surface of the Earth.
   Such bow-shaped form of surfaces restricting the active blocks (Fig.26) is being well confirmed by seismic prospecting data carried out by means of Reflection survey [11].
7. Variety of the velocities of various blocks' movement both horizontally and vertically will create anomalously high stresses along the active faults' surfaces restricting active blocks.

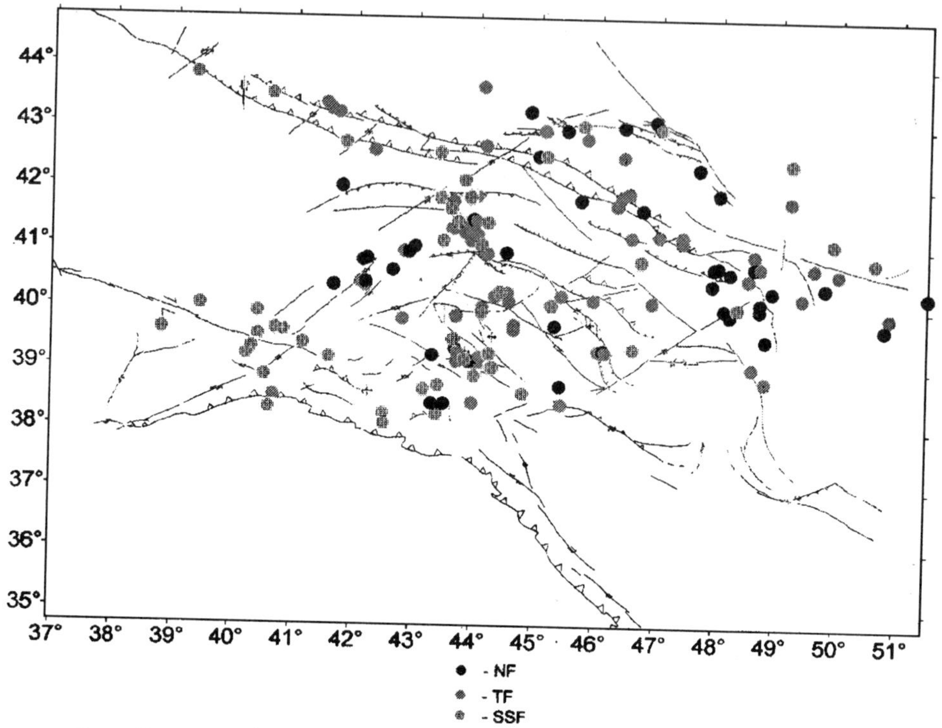

Figure 27. Selected earthquakes' focal mechanisms with M≥ 4 in the time period since 1962 in Caucasus region (focal mechanisms determination by A. Tovmasyan, NSSP).

In connection with that, the highest deformations under impact of Arabian plate's movement should be expected on the boundaries (in ties) of various active blocks' junctions and other most weakened and permeable zones of the lithosphere being deformed.

Coming from the above-stated it is not difficult to suppose that the Kajaran station, exactly situated in such zone of active blocks junction with high permeability of medium, should respond with the exclusive sensitivity to the Arabian plate's movement.

Under that, both impulses of Q at t_2 and t_3 time moments were being accompanied by C impulses at t_2' and t_3' time moments with lagging for the one and the same time $t_2'-t_2=1$ hour and $t_3'-t_3=1$ hour (fig. 22).

Thus, one of this two probable explanations of the "ADF" co-seismic impulsive responses of the "Kajaran" energy-active site to the Izmit seismic event could appear by the following manner. The first Q and C co-seismic impulses are relating to the reversible compression of the Kajaran energy-active site under the impact of dynamic seismic wave being propagated from the Izmit earthquake's source to the observation site in the distance of 1300 km. The second co-seismic Q and then C impulses are relating to the irreversible compression of the Kajaran observation site under the impact of the Arabian plate's impulsive crush movement towards the Eurasian plate after the Earth's crust's rupture in the Izmit area. The third co-seismic Q and then C impulses are relating to the reversible extention of the energy-active site, stipulated possibly, by reverse rebound of the Arabian plate or reversible reverse reconstruction of the energy-active site.

Another less probable explanation of the co-seismic Q and C impulses are relating to the triple deformation (compression-compression-extension type) of the Kajaran energy-active site with the relaxation period τ=50 min ± 10 min under the impact of the seismic wave propagated from the Izmit earthquake's source.

4. Current Seismic Hazard Assessment

So, approaching to the Current Seismic Hazard Assessment lets enumerate those intermediate scientific results in the sphere of earthquake prediction research which could be used for the purposes of preventive measures taken for the protection of the population against strong earthquakes.

It is first of all an availability of:

- the energy-active points within the seismoactive zones that highly sensitive to pre-, co-, post-seismic processes;
- ADF effect being recorded on the surface above the energy-active sites, in which the pre-, co-, post-seismic phenomena is being exposed ;

It is necessary to overcome the main peculiarity of the seismic event's occurrence, i.e. the uncertainty principle of the seismic events' coordinate and time for the current seismic hazard's parameters assessment, its probable place, time and magnitude.

Fig. 28 illustrates the time axis (T) and three increasing in time areas (S_1, S_2, S_{cr}) of seismic event preparation till the moment of T_{cr} when an earthquake occurs somewhere on the area S_{cr}.

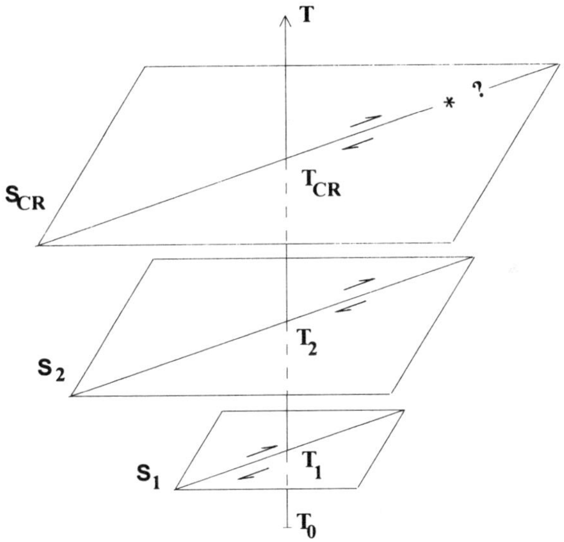

Figure 28. "Uncertainty principle".

From the pictures in Fig. 28, which well conforms to the seismological experience, the following is evident. The smaller is the deformation area S_1, the more certain are the coordinates of the possible seismic event, and the more uncertain is its time (T_{cr}). And vice versa, the greater is the deformation area S_1, the closer we are (i.e. the more certain is) to T_{cr} and the further (i.e. the more uncertain) from the possible seismic event coordinates.

Overcoming the uncertainty principal, it is possible by means of detection and use of more than three energy-active points with pronounced and studied "ADF" effect.

Probable place and time of the current seismic hazard occurrence are being defined by the "ADF" effect, observed upon the several energy-active points by means of the "Expert" software (Fig. 29). The "Expert" software package has been developed in the NSSP (Balassanian, [8]) for CSHA main characteristic's determination.

The idea of "Expert" includes the following:

1. Definition of the permanent examining empirical relation between the seismic events and different measured parameters variations including "ADF" effect of geological medium on the base of retrospective analysis of the many-years geological medium's multiparameter monitoring in Armenia;

2. Calculation and mapping of the critical level of current seismic hazard (in specific hazard units) for an earthquake with $M \geq 4.5$ (dangerous for Armenia level when depth of hypocenter is $h \leq 10$km in the densely populated area), on the base of defined empirical relation by way pointed above (point 1);

3. Testing of current variations of different measuring parameters of the medium on the matter of "seismogenity", occurence of "seismogenic" anomalies by established for Armenia criterions, definition of probable time, place and magnitude of a current seismic hazard by retrospectively defined empirical relation between the CSHA parameters and seismic events that take place;

4. Comparison of current seismic hazard defined value at any moment of time, in any place with its known critical level in the same place for definition of current seismic hazard degree.

The magnitude of probable seismic event could be assessed approximately by three characteristics:

- the lower level of the magnitude could be defined, as

$$M_{cur} \geq M_{crit};$$

  where M_{cur} is a current seismic hazard (in specific hazard units); M_{crit} is the critical level of seismic hazard, which is corresponded to the earthquake certain magnitude alredy occurred in some place where current hazard is being observed.

- the upper level of the magnitude could be defined by the ionospheric precursors (Fig.11) being observed during the earthquake preparation with $M \geq 5.5$ and also by Benioff graphs, as

$$M_{crit} \leq M_{cur} \leq M_{Benioff} \gtrless M_{ion}$$

where $M_{Benioff}$ is the magnitude determined from Benioff plot, M_{ion} is the magnitude defined from ionospheric precursors.

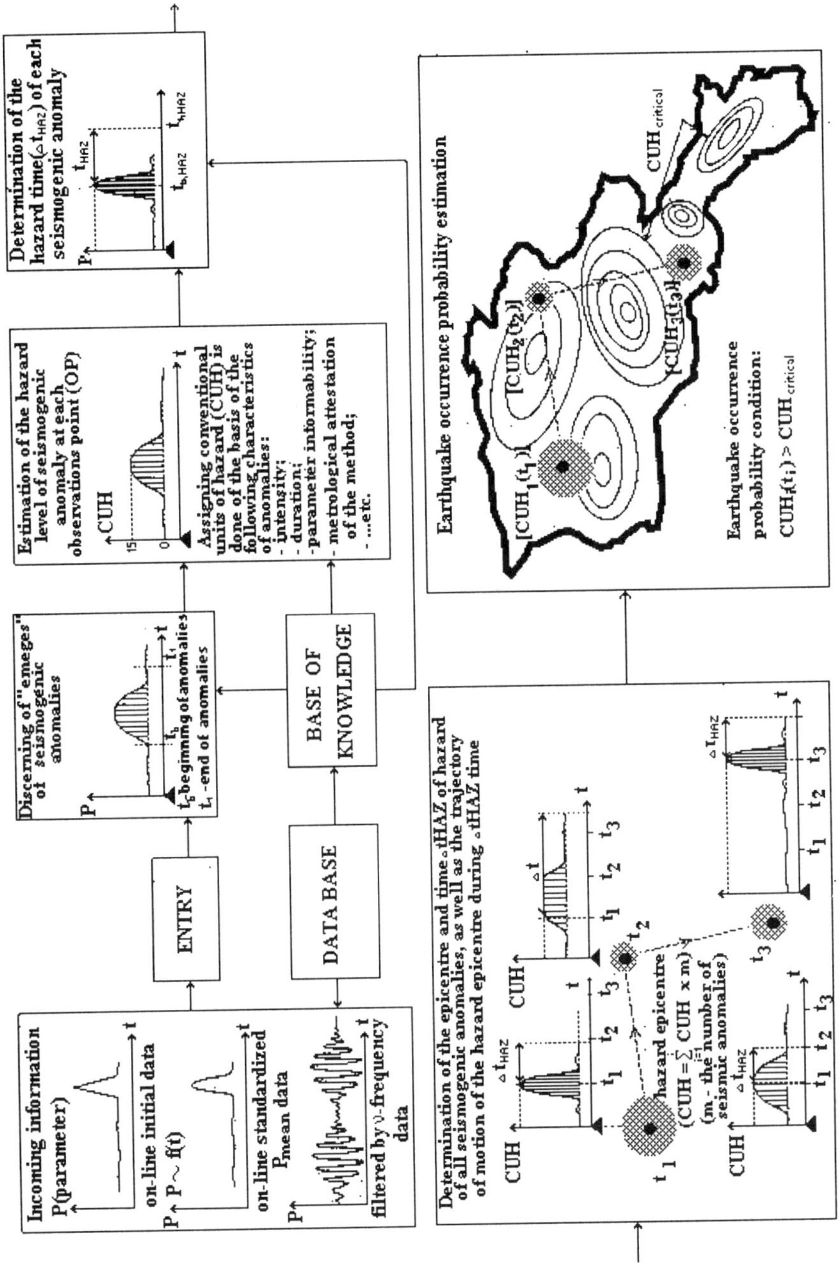

Figure 29. "Expert" software scheme.

Assessing the main parameters of the CSHA by the way shown above it could adequate preventive measures be taken by decision makers for the protection of the population as it is shown in our paper "Seismic Risk Reduction Strategy in the XXI Century", published in this Proceedings.

Conclusion

Experience of the Armenian NSSP shows the following:

1. CSHA is a constructive approach towards step by step solution of most complex problem of earthquake prediction.
2. Disclosure of high sensitive energy-active observation sites to the seismic regime of region, on the background of wide-propagation energy-passive massifs of rocks is the main requisite for the successful CSHA.
3. Disclosure of high sensitive energy-active sites (point) within being researched medium may be used for the current seismic hazard's main parameters (place, time and magnitude) assessment.
4. Assessment of CSHA parameters should be based on many-years retrospective analysis of multiparameter monitoring of various geological medium parameters in compare with the seismic regime of a region.

Today, before the fundamental solution of earthquake prediction problem, particularly in the developing countries, it is necessary to use all the intermediate solutions of that most complex problem, one of which could become CSHA.

Acknowledgements

I am grateful to Prof. J. Zschau, Dr. H. Woith and Dr. C. Milkerait of GFZ (Germany) for "READINESS" initiative, technical and financial assistance of that excellent research Programme.

I am also grateful to Ashot Avanessian, Avetis Arakelian (NSSP, Armenia), Ashkhen Tovmasyan for placing of geological and seismological data and Zaruhi Baghdassaryan, Emma Gevorkian, Armen Wardanyan (NSSP, Armenia) for paper design support.

References

1. Balassanian S. Yu. Seismogenesic and Distructive Earthquakes in the Caucasus; *Historical and Prehistorical Earthquakes in the Caucasus. Kluwer Academic Publishers. Series 2*: Environment, vol. 28, pp.**1-37**.
2. Reilinger R. et al. (1997) Preliminary Estimates of Plate Convergence in the Caucasus Collision Zone from Global Positioning System Measurements. *Geophys. Research Letters*, vol. 24, N14, pp. **1815-1818**.
3. Balassanian S. Yu. et al (1995) Retrospective Analysis of the Spitak Earthquake. *Annali di Geofisica*, Italy,vol.38.N3-4, Sept.-Oct., pp.**345-372**.

4. Balassanian S. Yu. (1994) Earthquake Prediction Problem and the Ways of its Solution. *XXIV General Assembly ESC*, September 19-24 Athens, Greece, p. **1029-1038**.
5. Balassanian S. Yu. et al (1995) Seismological and earthquake prediction observation network of the NSSP. *Proceeding of scientific meeting on the seismic protection*, Venice, Italy pp.**77-83**.
6. Balassanian, S. Yu. (1990) Dynamic Geoelectricity, Novosibirsk, *Nauka*, Siberian Department, p. **232**.
7. Balassanian S. Yu. et al (1997) The Investigation of Electromagnetic Precursors to Earthquakes in Armenia. *Annali di Geofisica*, XL,N2, Mardi,1117, pp.**200-225**.
8. Balassanian, S. Yu. (1999) The Anomalous Daily Dynamics of Local Geophysical and Geochemical Fields (ADF) Effect study in the Connection with Earthquake Preparation and Occurrence, "*The Physics and Chemistry of the Earth*", vol. 24, N8, pp. **741-752**.
9. Zschau J., Woith H., Ergunay, Balassanian S. (1998) READINESS: Realtime Data Information Network in Earth Sciences- *2-nd International Conference on Earthquake Hazard and Seismic Risk Reduction*, abstract volume p.**210**, Yerevan, Armenia, September 15-21,1998.
10. Balassanian S. (1997) Earthquake Prediction Research in Armenia. *The 29th General Assembly of IASPEI, Abstract Volume*, Greece, p. **138**.
11. Schoukin Yu. K. (1990) Geological-geophysical conditions in the source zone of Spitak earthquake 1988, Report, Armenia Geological funds, Moskow, Yerevan.

THE STAGES OF EARTHQUAKE PREPARATION: LABORATORY EXPERIMENT AND FIELD STUDY

G.A.SOBOLEV, A.V.PONOMAREV, Yu.S.TYUPKIN
United Institute of Physics of the Earth, Russian Academy of Sciences
UIPE RAS, Bol.Gruzinskaya 10, Moscow, 123810, Russia.

1. Introduction

The complex of laboratory and field researches of physical substantiation of new methods of seismic hazard evaluation and prognosis of earthquakes was carried out.

It was established as result of laboratory experiments, that the model of a seismic source passes some main stages during its development: accumulation of small-sized ruptures up to a critical level; association of cracks with formation larger and localization of this process in a zone of an unstable strain. These stages are accompanied by prognostic effects of clusterization of acoustic emission, quiescence and forshocks activation.

On a basis of the physical modelling regularities the algorithms were offered and the programs for revealing places of the developing sources, time and magnitude of appropriate earthquakes were compiled. The medium term forecasts in real time of strong earthquakes on Kamchatka are carried out. The information basis of work was the regional catalogue of earthquakes (more than 60 thousands of events) and laboratory catalogue of acoustic emission (more than 100 thousands of events). The suggested approach is shown to be effective for using in real time warning systems.

2. Laboratory experiment

The study of acoustic emission in the course of deformation of rock samples in the laboratory is extensively applied as one of the approaches to the establishment of regularities in seismicity. The acoustic emission is akin to seismicity in nature by a common physical basis, since both these events are caused by dynamic fracturing in the solid matter, though the sizes of these fractures differ by many orders of magnitude. In the laboratory experiment, the study of the development of macrofailure by applying the usual hydraulic presses is hindered by the fact that the time interval from the incipient focus to complete destruction of the sample is measured in seconds. In order to prolong this interval, an original loading scheme was applied [1–3]. The loading rate decreased with the growth of acoustic activity as a result of the feedback between the acoustic sensor and the directing facility of the press with servocontrol. In this paper, we used the

S. Balassanian et al. (eds.), Earthquake Hazard and Seismic Risk Reduction, 211–223.

data of acoustic emission registration obtained by these experiments. The results are demonstrated by the example of AE42 experiment.

A cylindrical sample of Westerly finegrain granite 76.2 mm in diameter and 190.5 mm long was subjected to deformation by an axial load under confining pressure of 50 MPa. Six piezoelectric sensors for acoustic reception were mounted in a "screw" pattern on the surface of the sample. The system of the acoustic events collection provided digital record of the first arrivals and of amplitudes of the first maximums of signals for each of the six piezosensors. The error in co-ordinates was 1–3 mm depending on the amplitude of the acoustic event. A detailed account of the methods of the experiment is given in [3].

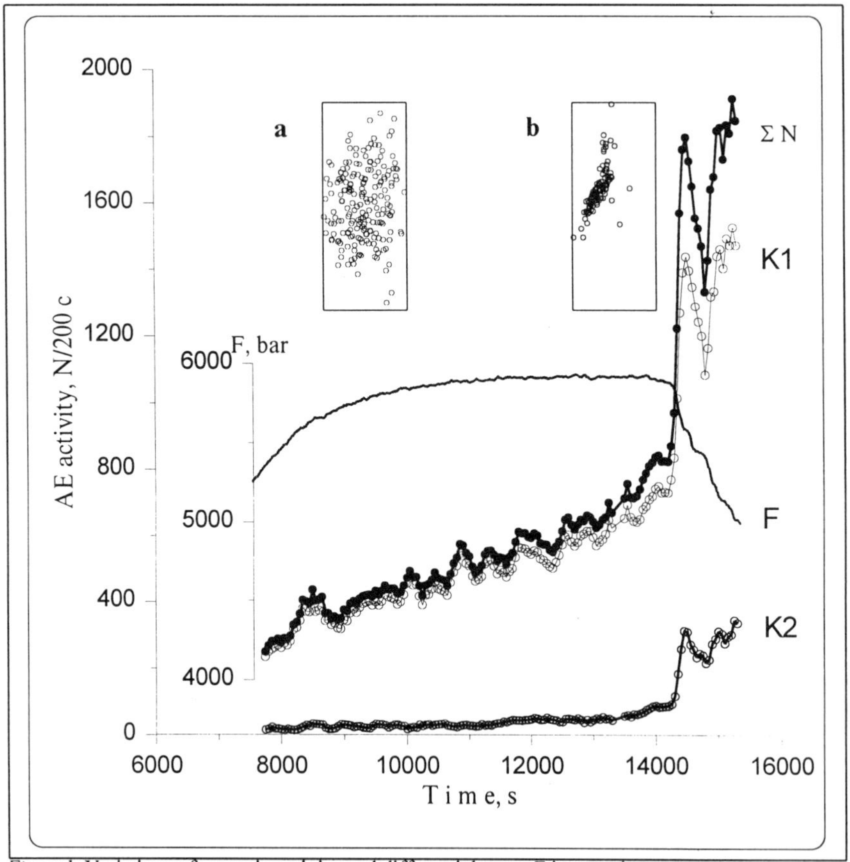

Figure1. Variations of acoustic activity and differential stress F in experiment. K1 – activity of weak events, K2 – activity of larger events, $\sum$ N – total acoustic activity. The top of the figure presents schemes of macrofractures based on the location of acoustic emission (AE) – random AE distribution for 7800-13900 s time interval (a) and localization of failure for 14000-15300 s time interval (b).

The main purpose of this paper is to reveal the stages of development of a macrodestruction source. Therefore, we used the acoustic signals that appeared in the sample during the period, when the axial load *F* approached the ultimate value and also at the post-peak stage of loading up to the moment of artificial unloading of the sample. This loading history is shown in Fig. 1, and it takes from 7553 to 15349 seconds. During that time interval, 26504 acoustic signals were recorded, and we succeeded to determine their hypocenters with a better accuracy than 3 mm. Their amplitude A changed from 5 to 146 mV, which makes K vary from 1.4 to 4.33 after conversion to conventional energy characteristic by formula $K = lgE = 2 \cdot lgA$. The $K = 1.4$ value is a representative energy estimation obtained from an analysis of Gutenberg-Richter law for the employed catalogue of acoustic events. The acoustic signals, recorded during the analyzed experiment, were divided into three conventional energy classes $K1$ = 1.42–2.43, $K2$ = 2.44–3.43, and $K3 \geq$ 3.44, which approximately corresponds to the size duplication of fissures.

Besides the F load curve, Fig. 1 shows the plots of changes in the number of acoustic events within the entire sample in the 200s window with a 50s shift. The plots are constructed for the events of all energy classes and for the classes of $K1$ and $K2$. The statistics of events of $K3$ class is too low to conveniently draw it on a plot in comparison with weaker signals. Fig. 1 implies, that the experiment can be divided, in the first approximation, into two intervals. The first interval is characterized by the slowly changing load F and the growth of the quasiconstant rate of acoustic activity. This interval takes from 7753s to 14200s. At this stage, the spatial distribution of the appearing microcracks is random. Then the acoustic activity suddenly grows and is accompanied by reduction of load. The moment of time 14200s was noted in paper [3] as the beginning of formation of a macrosource of destruction, when most of the hypocenters of acoustic signals are concentrated in the layer up to 5mm thick, which obliquely crosses the sample and runs from its surface to the center. It should be emphasized that, on the whole, in all energy classes of events, the acoustic activity grew in the course of the experiment, though there was a period of relative quiescence in the 14553–14803s time interval.

The behaviour of the relative frequency of events of various energy classes is quite different. In Fig. 2, plots K1, K2 and K3 represent the portion of events of the corresponding classes in the given window, which is similar to the relative probability of their appearances. The window of 500 acoustic signals long and with shift of 50 events was applied. Starting at 1700s time, i.e., long before a clearly expressed appearance of the macrodestruction source, a gradual attenuation is observed in the relative frequency of events of the weak classes with the simultaneous growth of more strong events.

The formation of the focus apparently takes place at the loading stage, when the material in the area of the future macrorupture reaches the metastable state. Under these conditions, the appearance of another fissure causes, with a greater probability, the formation of new cracks, or lengthening of already existing ones, particularly in the adjacent areas. As a result, the probability of the appearance of clusters of acoustic signals grows.

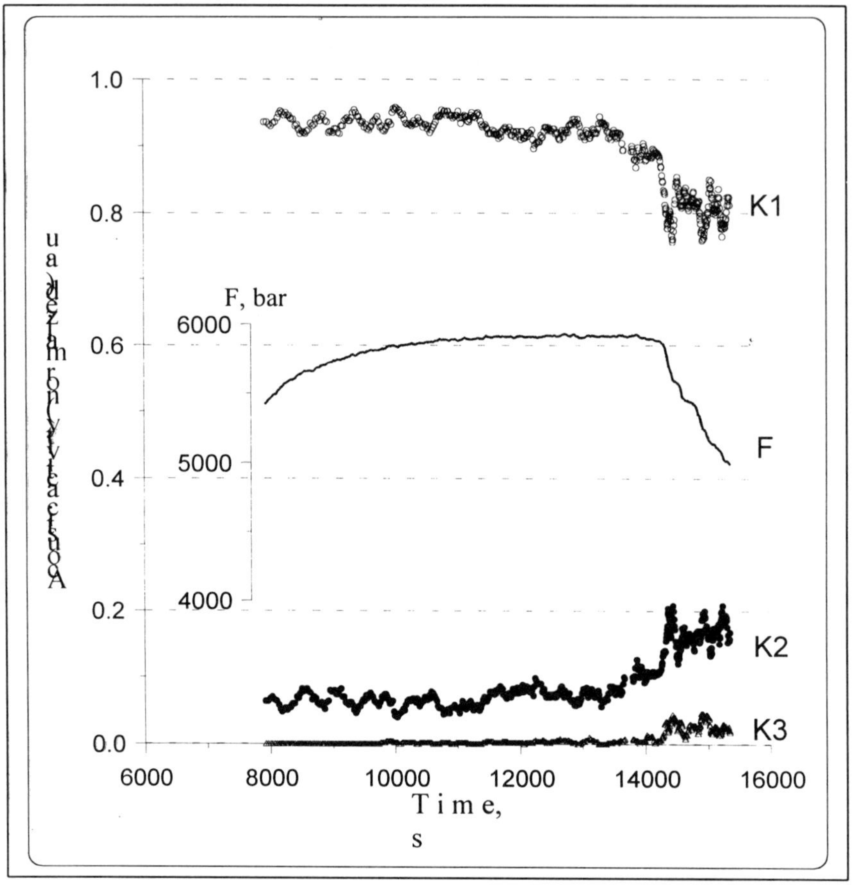

Figure 2. Relative activity of different acoustic energy events (K1, K2, K3) during macrofailure source formation.

We studied this effect by the following procedure. It was assumed that a cluster is formed by the appearance of two or more acoustic signals, if the location of their hypocenters, the difference in the time of appearance, and the energy satisfy the following conditions. The distance between the hypocenters should be less than critical, calculated by formula $R_{cr} = 3 \cdot R + B$, where R is the size of the fissure calculated from formula $lgR = D \cdot lgE + C$. Coefficient 3 corresponds to the concentration criterion of destruction [5], which indicates that before macrodestruction of a solid body, a critical density of cracks is accumulated in its volume with the fissures distributed at distances three times as long as their sizes. Coefficient D, equal to 0.244 [6], determines the relation between the size of a fissure and the energy, and was chosen by analogy with seismological data, whereas coefficient C, equal to 1.3416, was selected experimentally. Coefficient B, equal to 1.5mm, was introduced for registration of errors in determination of hypocenters. The R_{cr} value varied from 1.8mm to 3mm for the recorded acoustic events of all energies. The time between the two events should be less than critical, which is calculated by the formula $T_{cr} = a \cdot 10^{b \cdot K} - t_0$. Coefficient *b* was chosen equal to 0.5, which corresponds to the relative reduction of the number of signals with higher energy, in accordance with the reiteration law. Coefficient *a* was selected empirically and was equal to 10, while parameter to allowed a reduction of T_{cr} for events of the low energy class. A similar approach was applied by us earlier to distinguish the groups of earthquake [4]. Besides observing the indicated conditions by P_{cr} and T_{cr}, in the present paper we referred only those groups to clusters, in which the previous acoustic signal was less than the next. Thus, the aftershocks were eliminated. As a result, 4313 clusters were found, which makes 16.3% of the total number of recorded acoustic signals.

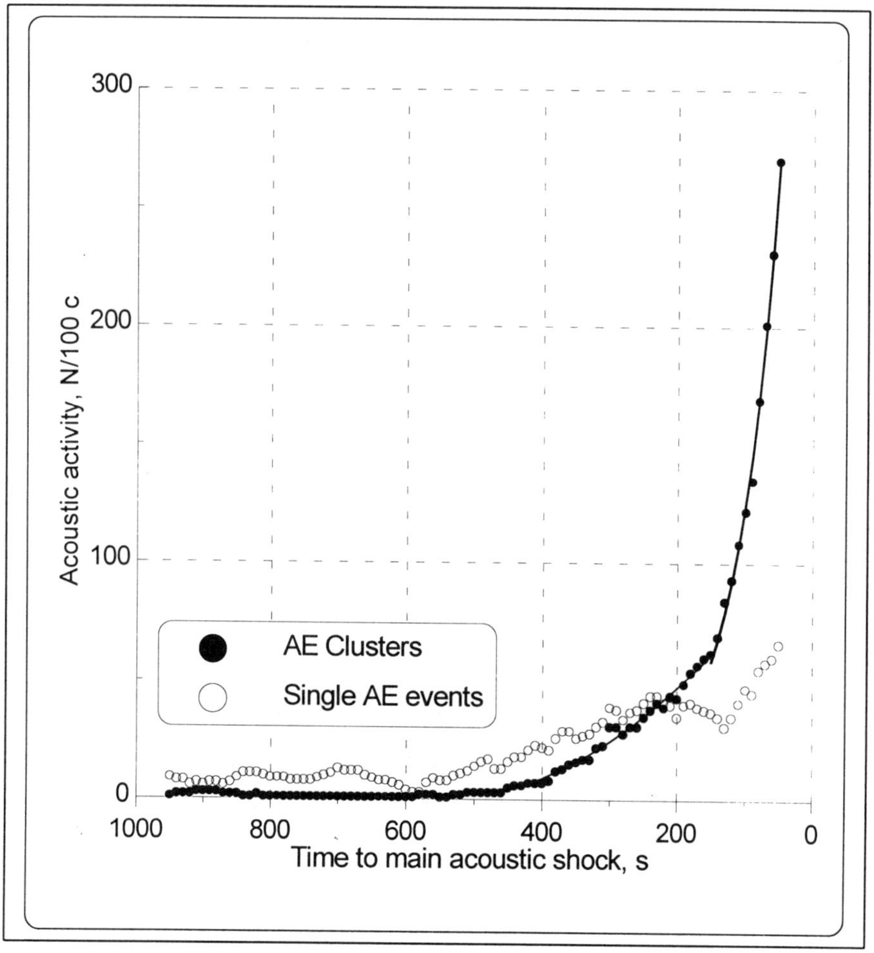

Figure3. Activity of single acoustic events and events that initialised clusters of K1 and K2 before the large events.

An analysis was carried out of the variations of acoustic activity and of the number of clusters before strong impulses of K3 class. The following procedure was used for this purpose. From all the impulses of K3 class those impulses were selected, the areas around the hypocenters of which were not overlapped by the R_{cr} dimension. In cases, when at one and the same place consecutively appeared several events of K3 class, then the first in time was selected. After that, all single impulses and clusters of K1 and K2 classes were recorded, which appeared in the corresponding spheres up to the selected event of K3 class. Fig. 3 shows the results of an analysis of distribution in time of single events and events that initialised clusters of K1 and K2 before the events of K3 class. The plots were obtained by the method of the epochs superposition and with the times of appearance of all events of K3 class assumed to be 0. The axis of ordinates shows the number of events in the 100s window with 10s spacing, and the points on the plots are referred to the half of the interval of the window. The results are given for the last 1000s, because the statistics for more remote times is so scarce that both curves lie practically on zero level of acoustic values. Fig.3 implies a considerable difference in the trend of the plots for single events and for clusters, particularly with approach to the moment of a strong event. Similar results were likewise obtained in the AE-36 experiment.

3. Field observation

The results of laboratory research show that the formation of the macrodestruction focus in rocks is reflected in the acoustic regime anomalies, such as a temporary quiescence, localization of the hypocenter of acoustic events, foreshock activation, the appearance of clusters in the area of the future macrorupture [13]. Let us examine how clearly the regularities detected in the laboratory experiments are manifested in the course of strong earthquakes preparation by the example of Kamchatka. An analysis of seismicity was carried out on the basis of the regional catalogue which is now being compiled by the Experimental Seismic Methods Group (ESMG) of the Geophysical Service (GS), Russian Academy of Sciences (RAS). The aftershocks were eliminated from the catalogue according to the program kindly offered by V.B.Smirnov. In the central part of the Pacific seismoactive zone adjacent to the coasts of Kamchatka, the existing seismological network provides the representativity of the catalogue at the level of energy class $K \geq 8.5$ [14]. In the 90's, 3 strong earthquakes occurred in this zone: on March 2, 1992 (52.92° N, 159.83° E), M=7.1; on November 13, 1993 (51.79° N, 158.83° E), M = 7.1; and on December 5, 1997 (54.95° N, 163.23° E), M = 7.7.

The quest for the stages of seismic quiescence and the following foreshock activation was based, by analogy with [11,12], on the RTL algorithm, which is the product of three functions: RTL=R×T×L, where

$$R(x,y,z,t)=\left[\sum_i \exp(-\frac{r_i}{r_o})\right]-R_s$$

$$T(x,y,z,t)=\left[\sum_i \exp(-\frac{t_i}{t_o})\right]-T_s \qquad (1)$$

$$L(x,y,z,t)=\left[\sum_i \exp(-\frac{l_i}{r_i})^p\right]-L_s$$

In these formula the x, y, z, t are the co-ordinates and the time of the anlyzed strong earthquakes; r_i, t_i are the epicentral distances and the times of the current earthquakes of energy classes $8.5 \leq K < 14$ in relation to the epicenter and the time of a strong earthquake; l_i is the size of the focus of the occurring earthquake calculated by the formula $lgl_i = 0.244lgK_i - 2.266$ [6]; $r_o = 50$ km is the coefficient that characterizes the function of the effect of the occurred current earthquake depending on the extent of its removal from the epicenter of a strong earthquake; $t_o = 1$ year is the coefficient characterizing the function of the effect of the occurred current earthquake depending on remoteness in time from the moment of a strong earthquake; $p = 1$. Calculations were carried out in the 1-year time window at 10 days shift. The R_s, T_s, L_s corrections eliminated the linear trends of the corresponding functions. Owing to the differences indicated earlier in the behaviour of the shallow and deep seismicity of Kamchatka and taking into account the results of an earlier analysis [11], only those current earthquakes were used for calculations the hypocentral depth of which was more than 20 km.

One of the advantages of the RTL algorithm in the search for the anomalies of seismic quiescence and foreshock activation is its high sensitivity with respect to the background variations of many years. The reduction of values from the zero level of perennial background corresponds to the stage of seismic quiescence, while the subsequent re-establishment to zero level characterizes the stage of foreshock activation.

Fig 4 shows the plots of the RTL parameter in the units of the standard deviation σ before the three strong Kamchatka earthquakes mentioned above. For convenience of comparison, all the plots are shown in the 7 years interval, though the σ value for every plot was calculated in the perennial interval from January 1, 1962 up to the moment of the corresponding strong earthquake. The localities of the epicenters of these earthquakes are shown at the upper end of Fig. 4 on the inset map. Their numbers 1,2,3 correspond to the numbers of the RTL plots.

Fig. 4 implies that before all the analyzed earthquakes there were stages of quiescence followed by activation. Before the earthquake of March 2, 1992 (№ 1) the quiescence started 2.2 years before the event and was 1.7 years long. Before earthquake of November 13, 1993 (№ 2) this stage started 3 years prior to the event and lasted 1.2 years. Before the earthquake of December 5, 1997 (№ 3) the beginning of quiescence was recorded 3.7 years prior to the event, and it lasted 1.7 years. The corresponding stages of foreshock activation for the three indicated earthquakes were 0.5, 1.8 and 1.7 years. In all cases the values of deviation from the background exceeded several times the level of perennial σ values. It should be emphasized that the RTL plots in Fig. 4 relate to the localities of epicenters of the corresponding strong earthquakes.

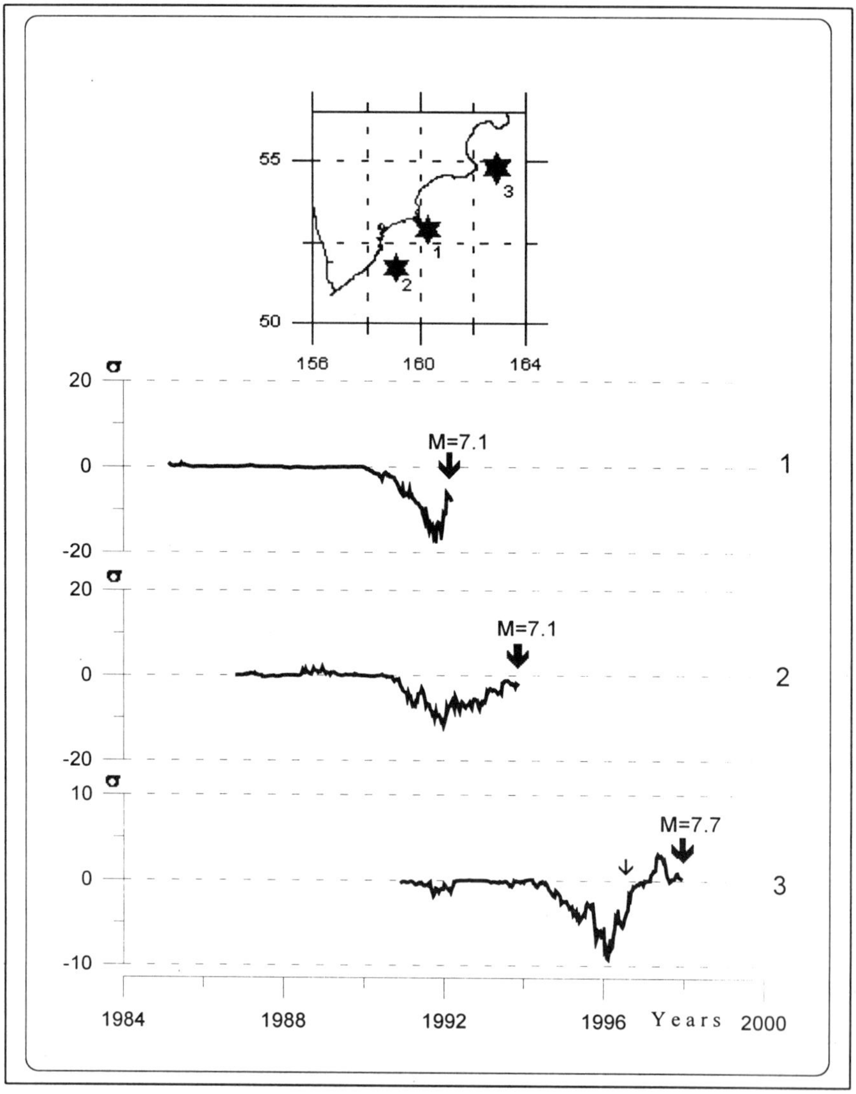

Figure 4. Locations of the three strong Kamchatka earthquakes of March 2, 1992, November 13, 1993, and December 5, 1997 (at the top) and the plots under the corresponding numbers 1, 2, 3 of the RTL parameter in the units of standard deviation from the perennial background at the place of their epicenters. Thin arrow in plot 3 shows the time of issue of prognostic information to the Expert Council on Earthquake Prediction at the Ministry of Extreme Situations of Russia.

The authors in papers [11,12] demonstrate the areal distribution of anomalies of seismic quiescence and foreshock activation before the earthquakes of March 2, 1992 and November 13, 1993.

It was recorded that the linear dimensions of the quiescence anomalies are about 100 km, whereas the epicenters of strong earthquakes are located in the marginal zones of the anomalies. The anomalies of the foreshock activation were located near the epicenters of strong earthquakes, and their linear dimensions for the indicated two seismic events make about 50 km.

Let us show these properties for the earthquake of December 5, 1997 with magnitude 7.7 taking note that the mid-term prediction of this earthquake was obtained 16 months prior to the event (white arrow in Fig. 4).

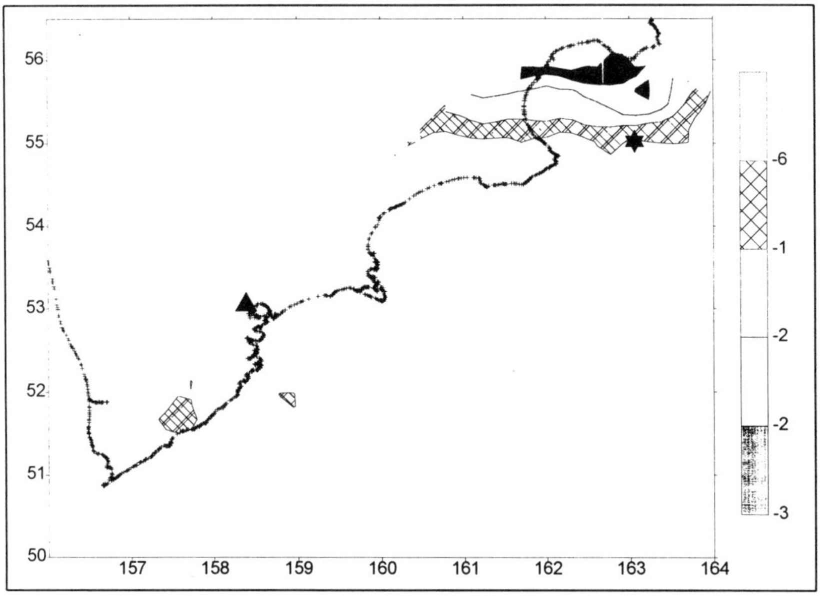

Figure 5. Location of the seismic quiescence anomaly before the earthquake of December 5, 1997. The scale on the right corresponds to the isoline value in the fractions of the standard deviation from the perennial background. An asterisk shows the position of the instrumental epicenter, and the triangle indicates the location of Petropavlovsk-Kamchatski.

Fig. 5 shows the areal location of the seismic quiescence anomaly before the earthquake of December 5, 1997 according to the calculations of the RTL parameter in the cells of the geographic net with 15 km spacing by latitude and longitude. For every cell the minimal value was sought at the interval of 1 year up to the transit of plot 3 in Fig. 4 across the minimum; i.e., from February 1995 till February 1996. The seismic quiescence anomaly is located in the north-east, in the Kamchatka Bay region. It is difficult to asses the size of the anomaly because it stretches beyond the representative catalogue K > 8.5. The earthquake of December 5, 1997 occurred at the distance of 110 km from its center.

The development of the foreshock activation map involved the application of the following algorithm. The difference ΔS was calculated between the accumulated area of seismogenic ruptures during the last S_s year before the strong earthquake and the average yearly perennial S_{av} value for the 0.5° × 0.5° areas of the geographic net. The rupture areas were estimated in both cases by the formula $S = \Sigma\, 10^{2/3(Ki - Ko)}$, where K_i is the energy class of the current events and $K_o = 8.5$. Fig. 6 shows the map of parameter ΔS before the earthquake of December 5, 1997. The size of the anomalous zone is about 50 - 100 km, while the epicenter of the strong earthquake practically coincides with the locality of the maximum. It is interesting to note that one of the significant anomalies in Fig.6 includes the strongest aftershock of 6.6 magnitude of the December 5, 1997 earthquake.

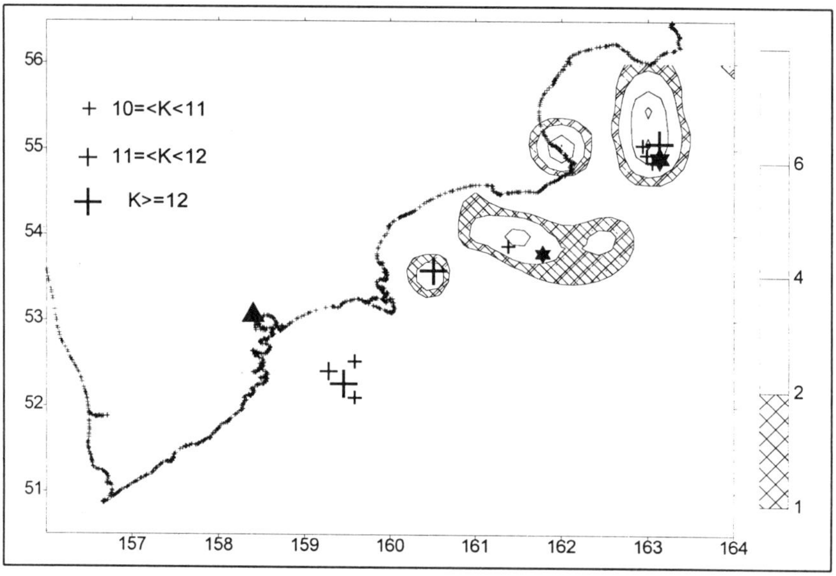

Figure 6. Distribution of anomalies of accumulated areas of seismogenic ruptures one year before the earthquake of December 5, 1997. The scale to the right corresponds to the value of the isoline in the fractions of the standard deviation from the perennial background. The asterisk shows the location of the instrumental epicenter, and the triangle indicates Petropavlovsk-Kamchatski. The smaller asterisk shows the place of the strongest aftershock on December 5, 1997. The crosses indicate the epicenters of earthquakes of different energy classes K, which initiated the clusters and which appeared during the year before the strong earthquake.

The laboratory experiments testify that before the appearance of the macrorupture there is a sudden increase of clusters of acoustic events in its vicinity [13]. In order to verify this circumstance the same algorithm was applied to the analysis of Kamchatka seismicity as that used in the processing of laboratory data. It was assumed that a cluster is the occurrence of two or more earthquakes, if the localities of their hypocenters, the difference in the time of appearance, and the energy satisfy the following conditions. The distance between the hypocenters should be less than critical calculated by formula $R_{cr} = 3l_i + \varepsilon$. In this case l_i is the size of the seismogenic rupture, calculated by the formula $lgl_i = 0.244lgK_i - 2.266$, and ε which is the correction for the inaccuracy of determination of the hypocenters assumed at 10 km. Coefficient 3 corresponds to the concentration criterion of destruction [5] that was applied in the analysis of seismicity [8, 9]. The R_{cr} value in the earthquake range of energy classes $8.5 \leq K_i < 14$ changed from 12 to 52 km. The time between the two events should be less than critical calculated by formula $T_{cr} = a \cdot 10^{b \cdot Ki}$. Coefficient b was chosen equal to 0.5, which corresponded with the relative decrease in the number of earthquakes of higher energies in accordance with the Gutenberg-Richter law, but it was chosen empirically equal to 0.01. Apart from conforming to the mentioned conditions set by R_{cr} and T_{cr}, only those groups were referred to clusters in which the previous earthquake was less than the next. In this way, the re-treatment of the aftershocks was carried out. As a result of the analysis of the catalogue under the given conditions it was found that the earthquakes, which initiated one or more shocks, compose 7% of their total number; i.e., the selection of clusters was sufficiently rigorous.

Let us further discuss how far the detecting of clusters promotes the establishment of anomalies in the seismic regime. The crosses of different sizes in Fig. 6 indicate the epicenters of earthquakes of the energy classes K more than 10, which initiated the clusters and which appeared in the interval of 1 year before the corresponding strong earthquake. Before the earthquake of December 5, 1997, in the vicinity of its epicenter, 3 clusters with K > 10 and 1 cluster with K > 12 were recorded. Let us note that the group of clusters of 1997 appeared in the center of the Avachinskii Bay, where as yet no strong earthquake was recorded.

4. Discussion of results and the inference

In the course of the analysis of the seismic regime of Kamchatka with the purpose of finding the different stages in the preparation of a strong earthquake focus, we arrived at the conclusion that the determination of the stable regularities in this area is a much more complicated task than that which could be accomplished from the data of study of acoustic emission by the deformed rock sample [13]. In particular, we could not obtain an unambiguous indication of a regularity in the attraction of weak seismicity towards the geocenter of the future strong earthquake, or establish the law of accretion of the number of clusters with the approach of its moment. The following circumstances can be the cause of this.

In spite of the self-similitude of certain elements of structure and processes in rocks of different geometric scale, such as the fractal structure, the hierarchy of blocks, the laws of iteration of events of different energy, etc. [7], an object of large volume has certain new and complicated properties affecting the behaviour of such objects, and likewise during the process of development of instability. For example, it is not clear how the behaviour of rocks in the depths of the Earth is affected by the remnant stresses, gas saturation, brines, the geochemical and thermodynamical processes.

As to the Kamchatka seismoactive zone proper, it is pertinent to note that it already exists during a geological period of time, whereas in the laboratory experiment the zone of macrostructure is formed in the course of the experiment. Furthermore, in the structure of the Kamchatka zone the regimes of low seismicity down to the depth of the order of 20-40 km considerably differ from those at a much greater depths [11], while in the sample the background acoustic activity is distributed in the volume more regularly. The statistics of the number of earthquakes of the representative classes is not as ample as the catalogues of acoustic activity and, therefore, it is hardly probable to distinguish significant anomalies in seismic activity during the relatively short intervals of averaging. The last factor possibly explains the absence of obvious anomalies of localization of seismicity and clusterization in to focal area of the earthquake of June 8, 1993 with magnitude 7.4 in the south of Kamchatka, though the stages of seismic quiescence and the subsequent activation were revealed there by the RTL algorithm [11].

5. Conclusions

The formation of an earthquake source occurs by gradual enlargement of ruptures and their concurrent localization at the place of the future macrorupture.

Antiphase variations were revealed in the relative frequency of appearance of acoustic events with high and low energy.

A difference is observed in the regularities of changes in the acoustic activity of single events and of clusters of low energy classes in the focal zone of the event of a higher energy class before its appearance. The factor allows to distinguish consecutive stages in the appearance and formation of the source, including the quiescence and foreshock activation.

These peculiarities of the acoustic regime are confirmed by the data obtained during the earthquake preparation period.

The results obtained in the present research have also acquired a prognostic aspect by strengthening the physical basis for detection and interpretation of seismic precursors.

6. Acknowledgements

The work has been executed under financial support of RFBR, projects 96-05-65439 and 97-05-65906.

7. References

1. Lockner D.A., Moore D.E., Reches Z. (1992) Microcrack interaction leading to shear fracture. In: Tillerson J.R., Wawersik W.R. (Editors), Rock Mechanics Proc. 33rd US Sympos., 807-816.
2. Lockner D.A., Stanchits S.A., Kuksenko V.S., Byerlee. J.D. (1995) Interaction of Microcrack Damage Inferred from Acoustic Emissions. EOS (AGU Transactions), Fall 1995 AGU Meeting, San Francisco 1995, Abstracts, F566.

3. Ponomarev A.V., Zavyalov A.D., Smirnov V.B., Lockner D.A. (1997) Physical modelling of the formation and evolution of seismically active fault zones. Tectonophysics, 277, 57-81.
4. Sobolev G.A., Vasiliev V.Yu. (1991) On the patterns of clustering low magnitude events before strong earthquakes on Caucasus. Izv. AN SSSR, Fizika Zemli, 4, 24-36 (in Russian).
5. Kuksenko V.S. (1986) Transition Model from micro- to macro- fracture of solid bodies. In: Physics of strength and plasticity. Leningrad, 36-41 (in Russian).
6. Riznichenko Yu. V. (1976) Dimensions of the crustal earthquake focus and the seismic moment. Researches in Earthquake Physics. M., Nauka, 9-27 (in Russian).
7. Sadovskii M.A., Bolkhovitinov L.G., Pisarenko V.F. (1982) On the discrete properties of rocks. Izv. AN SSSR, Fizika Zemli, 12, 3-18 (in Russian).
8. Smirnov V.B., Zavyalov A.D. (1997) Incorporating the fractal distribution of faults as a measure of failure concentration. Volcanology and Seismology, 18, 447-452.
9. Sobolev G.A., Zav'yalov A.D. (1980) On the concentration criterion of seismogenic ruptures. Dokl. AN SSSR, 252, 1, 69-71 (in Russian).
10. Sobolev G.A. (1993) The principal bases for earthquake prediction. M., Nauka (in Russian).
11. Sobolev G.A., Tyupkin Yu.S. (1996) Anomalies in low magnitude seismicity preceeding large Kamchatka earthquakes. Vulkanologiya i Seismologiya, 4, 64-74 (in Russian).
12. Sobolev G.A., Tyupkin Yu. S. (1998) The stages of preparation, seismological precursors and prediction of earthquakes of Kamchatka. Vulkanologiya i Seismologiya, 6, 17-26 (in Russian).
13. Sobolev G.A., Ponomarev A.V. (1999) The analysis of stages of acoustic activity in the laboratory experiment. Vulkanologiya i Seismologiya, 4-5, (in press), (in Russian).
14. Fedotov S.A., Shumilina L.S., Chernyshova G.V. (1987) The detail study of seismicity of Kamchatka and Komandor islands. Vulkanologiya i Seismologiya, 6, 29-60, (in Russian).

EARTHQUAKE PREDICTION: PRO AND CONTRA

T. CHELIDZE
Institute of Geophysics, 1 Alexidze str., 380093, Tbilisi, Georgia

This report is provoked by the discussion on the fundamental problem of predictability of earthquakes, which last years became very intensive. The main arguments of unpredictability proponents are published in the special section: "Assessment of schemes of earthquake prediction" in Geophysical Journal International (v.131, N 3, 1997), namely in papers of R. Evans, Y. Kagan, P. Leary, P. Stark and others, presented on the meeting of the Royal Astronomical Society and Joint Association of Geophysics, held in London on November 1996. They claim that the prediction of the time of strong earthquake is principally impossible, though the validity of seismic hazard assessment procedures is not doubted.

The main arguments against possibility of earthquake forecast are:

i. The Earth's crust is in the state of self-organized criticality (SOC) (Bak, Tang and Wiesenfeld, 1988; Sholtz, 1990). In the state of SOC due to the strong nonlinearity of system, the smallest change in initial conditions may cause totally different response. This leads to conclusion on "the inherent implausibility of reliable earthquake prediction, because many small earthquakes occur through any active seismic zone, demonstrating that the critical conditions for earthquake nucleation are satisfied almost everywhere. Apparently, any small shock can grow into a large event. Thus it is likely that the earthquake has no preparatory stage" (Kagan, 1997).

ii. The second group of arguments is related to the statistical aspects of forecast. The existing techniques are mainly aimed to obtaining optimal retrospective prediction and prescribing to forecasts in future the same reliability. After-the-fact search is not of much predictive value, as far as for small statistics of strong earthquakes it is possible to optimize quite satisfactorily the prediction scheme for existing data. At the same time it can turn out that this optimized approach is not valid for prediction of following events or for other regions. This can be called hunchback jacket effect; you can work out the jacket, ideally fitting the given hunch, but it can be utterly useless for another one, as each hunch is individual. There are also other doubts, related, for example, to the problem of correct formulation of null hypothesis (Stark, 1997).

iii. The third group of arguments considers experimental prediction data. "The 100-years history of attempts shows little or no progress in prediction of time, magnitude and location of impending earthquake in reasonably tight limits on the basis of experimental observations". According to R. Geller (1997) "extensive searches have failed to find reliable precursors".

The arguments are serious, but it seems that they are not so firmly grounded and decisive as the proponents of unpredictability claim in their reports in very radical

S. Balassanian et al. (eds.), Earthquake Hazard and Seismic Risk Reduction, 225–229.

forms. There were already some critical responses to the above position in EOS (Gusev, 1998; Lomnitz, 1998).

Here I would like to present my personal point of view on the problem. First of all let us consider the consequences of nonlinearity.

i. The statement on absolute unpredictability on nonlinear processes is not correct. (Lomnitz, 1998). The behaviour of nonlinear system is predictable on the limited time interval T. The limit of predictability time is given by the Kolmogorov-Sinai (KS) entropy $h(X)$ where X is a set of states. The system became unpredictable after the passage of time $T \sim 1/h(X)$ (Abarbanel et al, 1993) when the X became equal to the number of states available for the orbits of the system (this time for seismicity can be from several hundreds to several thousands of years). Alternatively, as KS-entropy equals to the sum of the positive Lyapunov exponents, "the limited predictability of the chaos is quantified by the set of Lyapunov exponents which can be determined from the measurements itself" (Abarbanel et al, 1993).

The examples of relatively successful limited predictability are short-term weather forecasts and volcano eruption forecasts. It has been shown that (Shaw, Chouet, 1991) that volcano tremors can be described mathematically by chaotic models which means that volcano areas are close to the critical state. Nevertheless, volcano eruptions are successively predicted on the basis of geochemical, deformation and seismoacoustic observations. As has been mentioned the weather is also reliably predicted for short time intervals (3-4 days) though the atmosphere is a typical nonlinear system. Taking into account significant difference in the rates of atmospheric and tectonic processes due to different viscosity of media, the 3-4-days predictability period for atmosphere should correspond to months and years in lithosphere.

About SOC. The fact of close resemblance of seismic catalogues and SOC calculations is not a definite proof that the lithosphere is just in the state of SOC. For example, mathematical analogues of Gutenberg-Richter relationship can be obtained by different approaches: from percolation model of fracture (Chelidze, 1987, 1993), from kinetic model of strength (Petrov, 1984), etc.

In reality, the lithosphere can be not just in the critical state, but close to it (Kagan, 1997). This is enough to make things quite different. For example, in systems, which are close to the critical state, there is still a characteristic size, namely, the largest finite cluster size L_c. The size of finite clusters grows as the system approaches the critical state and this allows prediction of the distance to it Δp, because the properties of system (so called characteristic functions) depend on the size of clusters. The time, needed to cover the distance Δp can be large enough even for small Δp, especially if the process follows logarithmic kinetics (Petrov, 1984)

If there are no predictors of the strong event, then there should not be also after-event effects. The most obvious and indisputable examples of after-effects are aftershock activity, which lasts months and years, and water level relaxation in deep wells of the similar duration. Than, if there is clear evidence of long strain relaxation period after strong event why should be rejected the possibility of strain accumulation period before it? If the relaxation of stress takes several years, why the accumulation process does not need any time? If any small shock can grow into strong event, why strong events are so rear?

i. Even proponents of unpredictability agree that the seismic hazard assessment problem is solvable. But if the strong earthquake can be triggered by any small event, what sense is in doing seismic zoning? Why in some places the maximal expected earthquake magnitude will not exceed 6 if any small shock can trigger event of any magnitude? The matter is that the unlimited heterogeneity, that is self- similarity to infinity, is realized only in mathematical models. Real physical and geological systems are finite. This introduces the another characteristic size – the size of system, which sets limit to earthquake size and makes the hypothesis of unbounded heterogeneity (Leary, 1997) unrealistic.

The last but not least in the nonlinear approach is the problem of connectivity.

The fractal models and SOC theory do not consider the problem. At the same time the analysis of connectivity may be very important if, say, pore pressure is decisive in nucleation of earthquakes (Sibson, 1994, Muir Wood, 1994). If the connectivity is essential, percolation theory seems to be the best tool for assessment of closeness to the critical state, as it is focused on the analysis of connectivity of elementary objects, such as pores, fractures or just overstressed volumes (Chelidze, 1987, 1993).

ii. As a rule, all precursors are considered as nonreliable by unpredictability proponents due to the small statistics of such events as the sequence predictor-earthquake and possibility on non seismic origin of anomalies. At the same time they do not consider such situations, as appearance of three, four or more predictive anomalies before a given large earthquake. Ignoring consistent appearance of several predictors reminds of one funny story: An undisciplined driver crosses the crossroads on the red light. When the policemen stopped him, he denied his fault, saying that there was not the red light on. So, you think, sir – said the policemen - that all these cars are out of petrol simultaneously?

Indeed, when there is the great chance to misidentify some event due to the high level of noise, physicist uses the coincidence technique. In the cosmic ray physics, for example, the particle of cosmic origin is identified, when several layers of sensors register the particle simultaneously. Thus the statistical validity of several consistent anomalies in various fields should be given significant weight in order to make correct statistical assessments.

Of course the problem of "hunchback's jacket" is a difficult one, but shuffling and bootstrap procedures may help to avoid biased assessments.

iii. In laboratory tests of delayed fracture of rocks, both intact and containing artificial fracture, as well as in stick- slip experiments a plethora of precursors of main rupture has been found on the basis of simultaneous monitoring of acoustic, electric, local strain fields, gas emission and other phenomena (Sobolev, 1996). What is so specific in the seismic process that leads to taboo on prediction?

I conclude that the statement of unpredictability of strong earthquakes follows only from abstract mathematical models, which do not adequately take into account real physical processes, having place in real geological systems. It seems that contradiction of competing paradigms of unpredictable nonlinearity and predictable strain or damage accumulation models can be resolved by decoupling of these two approaches. That means that both of them are valid, but they have different limits of validity.

Such models as SOC and cellular automata are mainly focused on simulation and prediction of fundamental features of seismic regime, that is for modeling seismic catalogues and are less helpful for prediction of the next strong event. These models do not allow understanding the nature of strain-related anomalies in various geophysical fields. But the preparation of the strong earthquake is not a purely seismic process. Even during the earthquake only small part of released energy is transformed into seismic waves. During preparation process the seismic component plays even less role: the main feature here is mostly aseismic deformation which may cause, nevertheless, strong anomalies in strain-sensitive geophysical, geochemical and geodynamical fields, due, say, to evolution of fracture network before strong event and redistribution of pore fluid. Percolation fracture model (Chelidze, 1987 1993; Herrmann and Roux, 1990) and kinetic theory of strength (Petrov, 1984) allow to understand the process of nucleation, coagulation and growth of clusters of microfractures during delayed fracture and predict evolution of physical properties of system during destruction. Theoretical predictions agree quite well with experimental data on delayed fracture of rocks (Chelidze, 1987). The most important point is that in percolation and kinetic models closeness to the critical state can be assessed by analysis of behavior of characteristic functions, which depend on the size of finite clusters of defects. It should be stressed also that percolation model naturally explains the spatio-temporal heterogeneity of response of geophysical fields to the strain variation in terms of anomalous strain-sensitivity of objects that are close to percolation state (Chelidze, 1987, 1993).

We can illustrate the above statement on decoupling of competing paradigms by the following example: in microscopic studies the analysis of overall pattern of rock (its stricture, etc) is made at small magnification (analogue of SOC model), but if the individual grain is to be studied, quite different magnification should be used (percolation model).

It seems that it is impossible to elaborate the unique theory, which explains equally well the overall seismic regime and the process of preparation of the individual strong event.

References

1. Abarbanel, H., Brown, R., Sidorovich, J. and Tsimring, L. (1993) The analysis of observed chaotic de in physical systems. *Rev. of Mod. Phys.*, **65**, 1331-1340.
2. Bak,P., Tang, C. and Wiesenfeld, K.(1988) Self-organized criticality. *Phys.Rev. A.*, **38**, 364-374.
3. Chelidze, T. (1987) Percolation Theory in the Mechanics of Geomaterials, Nauka, Moscow, (in Russian).
4. Chelidze, T. (1993) Fractal damage mechanics of geomaterials. *Terra Nova*, **5**, 421-437.
5. Geller, R. (1997) Earthquake prediction: a critical review. *Geoph. Jour. Intern.* 131, 425-451.
6. Gusev, A. (1998) Earthquake precursors: Banished forever? *EOS, Trans., AGU*, **78**, 71-72.
7. Evans,R. (1997) Assessment of schemes of earthquake prediction. *Geoph. Jour. Intern.* **65**, 413-421.

8. Herrmann, H. and Roux, S. (Eds), (1990) Statistical Models for the Fracture of Disordered Media. North-Holland, Amsterdam.
9. Kagan, Y. (1997) Are earthquakes predictable? *Geophys. Jour. Intern.* **131**, 505-526.
10. Leary, P. (1997) Rock as a critical-point system and the inherent implausibility of reliable earthquake prediction. *Geophys. Jour. Intern.* **131**, 451-467.
11. Lomnitz, C. (1998) Comment: Unpredictability of earthquakes – Truth or Fiction. *EOS, Trans. AGU*, August 4.
12. Muir Wood, R. (1994) Earthquakes, strain-cycling and the mobilization of fluids, in J. Parnell (Ed), *Geofluids: Origin, Migration and Evolution of Fluids in Sedimentary Basins.* The Geological Society, London. pp.85-98.
13. Petrov, V. (1984) Fundamentals of kinetic theory of fracture and its prognosis. *Prognosis of Earthquakes*, **N 5**, Dushanbe-Moscow, pp. 30-45 (in Russian).
14. Scholtz, C. (1990) The mechanics of Earthquakes and Faulting. Cambridge Univ. Press.
15. H.Shaw and B.Chouet, (1991) Fractal hierarchies of magma transportation and critical self-organization of tremors. *Jour. Geoph. Res.* **B6**, 10191-10207.
16. Sibson, R. (1994) Crustal stress, faulting and fluid flow, in Parnell J. (Ed), Geofluids: *Origin, Migration and Evolution of Fluids in sedimentary Basins.* The Geological Society, London, pp.69-84
17. Sobolev, 6. (1993) Fundamentals of Earthquake Prediction. Nauka, Moscow (in Russian).
18. Stark, P. (1997) Earthquake prediction: the null hypothesis. *Geophys. Jour. Intern.* **131**, 495-500.

METHOD OF ANALYSIS OF GEOCHEMICAL OSCILLATIONS TO EARTHQUAKE PREDICTION

A.A. BELYAEV[1], V.A.IGUMNOV[2]
[1]*Vernadsky Institute of Geochemistry and Analytical Chemistry EBS, Moscow.*
[2]*National Survey for Seismic Protection of the Republic of Armenia, Yerevan.*

The research of geochemical prognostic criterion of earthquakes is carried out in Vernadsky Institute of Geochemistry since 1979. The long-term geochemical observations in various seismically active areas of Tadjikistan, Uzbekistan and Armenia, and also computational experiments and methods of mathematical modeling lie in the basis of presented results. The developed method uses an approach based on a set of the fundamental preconditions.

– The research develops as applied aspect of theoretical study of a kinetics of opened non-equilibrium geochemical systems. Processes in both systems: as tectonic well as geochemical system, are studied from positions of this theory.

– Tectonic (as well as seismic) phenomena represent self-oscillatory processes with variable frequency of oscillations. Oscillatory disturbances at preparation phase of earthquake have the lowest frequency of the order of 10^{-5} Hz (corresponds to a few days period approximately). Such extremely low frequency oscillations can not be sensed by physical resonant systems.

– These oscillations can be detected in dynamics of opened non-equilibrium geochemical systems. The self-oscillations in such systems proceed without participation of Newtonian acceleration and, depending on quality of a particular observation system, can have demanded resonant frequency.

Until recent time it was almost wholly concentrated on study of an only natural parameter - the Helium contents in underground waters. Nevertheless, during research a number of general laws connected to occurrence of prognostic flags in geochemical system of flowing water wells is established.

The conclusions concern three basic concepts determined during research: *a tracer* - observable natural parameter; *a prognostic characteristic*, - mathematically selected the characteristic of changes of tracer, which abnormal meaning can be used for detection of process of an earthquake preparation; *prognostic flag* - anomaly in prognostic characteristic, **only if it admits extrapolation** of a system state (prediction). For example, contents of anyone ions or gases in underground waters can be used as a observed natural tracer. Even from positions of elementary statistic it is obvious that the single sample of any natural parameter cannot carry any prognostic information, and only some statistical ensemble of them can be essential. The specific changes in set of

S. Balassanian et al. (eds.), Earthquake Hazard and Seismic Risk Reduction, 231–238.

geochemical components traditionally considered as precursors of earthquakes are expressed in sharp positive or negative anomalies of the contents of various components in ground waters. In this case as *a formal prognostic characteristic* the sliding average of the component could be used, because such anomalies assume significant difference of mathematical expectations of a signal and noise. Such anomaly, in certain conditions, could act as *a prognostic flag.*

It is necessary to note that here and further we essentially distinguish two concepts: «precursors» – phenomena statistically **accompanying** preparation of earthquake, and «prognostic flags» - changes of tracer **physically linked** with preparation and **admitting to extrapolation** of a dynamic state of tectonic system. For example, sporadic disturbances of structure of underground waters in extensive territory around the preparing seismic center can be precursors. Let's consider, for example, traditional Radon precursor previous to strong earthquake outlying from observation point on hundreds of kilometers, To consider this precursor as prognostic flag of earthquake it is necessary to accept a hypothesis that the flow of the gas is formed in the center of preparation and is transported therefrom to the observation area. Otherwise it will be obvious that neither time nor place of forthcoming earthquake on such anomaly are possible to extrapolate. However, even elementary calculations show, that physically allowable gradients of geochemical components, the speeds of geochemical flows do not allow accept this hypothesis. Regarding to Radon, in comparison with other geochemical components, the illegitimacy of a hypothesis is aggravated by a short period of half-disintegration of this gas.

The scientific reasonable prediction corresponds only to theoretically certain extrapolation of a current state of an analyzed dynamic system. Just from one this example there can be made the following conclusion a priory: either the geochemical forecast, in the sense, as it was determined above, is impossible basically, or it is necessary to search prognostic characteristic in other, in comparison with traditional, features of dynamics of geochemical system.

Concept «precursor» assumes, that the seismic danger becomes urgent from the moment of occurrence of anomaly. Therefore search of new precursors is directed on detection of anomalies closest in time to forthcoming seismic events (Fig.1). Main criterion of an estimation of a precursor is its long term. «Prognostic flag», noticed in the moment t, assumes that the seismic danger will exist in a period

$$T_A = t + t_l \pm \Delta t, \qquad (1)$$

where t_l and Δt – parameters of forecast: prediction time (extrapolation period) – t_l, and precision of its estimate – Δt. The search of prognostic flags is not limited to the analysis of the phenomena directly previous to earthquakes. Opposite, the phenomena more and more remote from them in time are analyzed. So, at present, prognostic flags foregoing to strong earthquakes for more than two years are revealed [1]. In many respects this parameter depends on quality of an observant system. The saved experience specifies that the range of prediction period mainly depends on depth of an observant well.

Second parameter of the forecast - Δt in process of development of research consistently decreases by perfection of the specialized techniques of monitoring, by escalating of mathematical means of allocation of prognostic flags and by development of theoretical model of extrapolation, Nowadays range of this parameter, depending on quality of the saved time series, varies from ±3 days up to ±1 day [1]. Thus, revealed prognostic flags estimated by criterion of traditional precursors on parameter prediction time t_l are to be in the range from long-term to intermediate-term and are to be up to most short-term on parameter of estimate precision Δt.

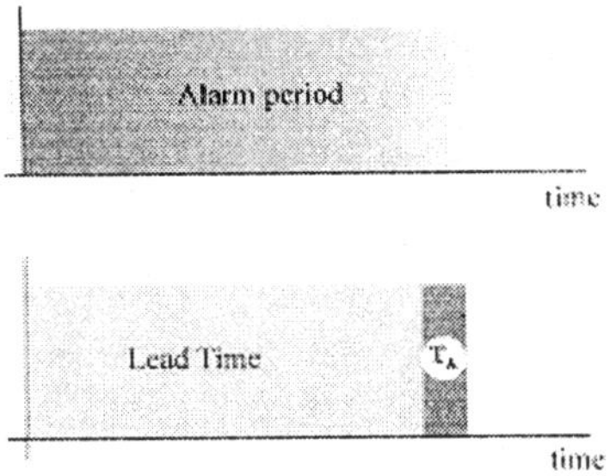

Figure 1. Distribution of alarm time of precursors (a) and prognostic flags (b). T_A- alarm period (1}.

The developed approach has a number of basic differences from traditional methods of the forecast. In general, it concerns of definition of prognostic anomaly. Traditionally, significant difference of an average of an observable parameter from a background level is considered as anomaly (Fig.2a). A formalism of this approach (the discriminant analysis) univocal assumes that variance of time series and its correlation structure in anomaly and background have no essential distinctions. In the present research the working model proceeds from the alternative assumption: prognostic anomaly and noise differ with temporary correlation structure, the differences of their averages are insignificant for the forecast (Fig. 2b, 2c). The formalism of such approach is based on application of analysis of variance or correlation (spectral) analysis of observation time series. As well as in previous case, this formalism assumes restriction: it is accepted, that anomaly and background have identical mathematical expectation.

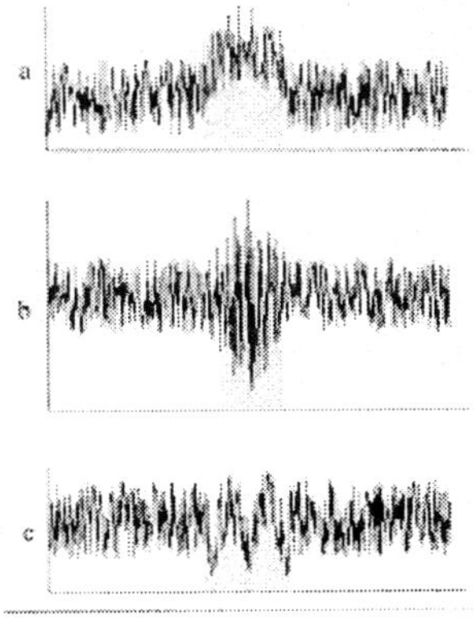

Figure 2. Three kinds of anomalies: a – level anomalies, b – variance anomalies, c – spectrum anomalies.

The statistical analysis of geochemical time series during preparation phase of earthquakes and subsequent to earthquakes intervals of time has shown [2] that the variations of sliding mean of a geochemical set reflect mainly postseismic reaction of the system. The prognostic importance of this characteristic is minimal, because its changes in the preseismic period have no significant difference from statistical noise. Moreover, it is known that there are strong anomalies in structure of underground waters generated after strong earthquakes. Thus, each occurred strong earthquake creates significant noise for the forecast of the subsequent seismic activity (Fig.3). On the basis of these results the developing approach assumes application of a mathematical procedure of a preliminary filtration sliding average to observable time series (Fig.4).

The research has shown [3] that the strong seismic events are preceded by a long period, during which the essential variations of correlation structure of geochemical time series are observed, In other words, the power spectrum of fluctuations of a geochemical series is shifted to a rather narrow band of frequencies. The research also has shown, that for each observant system one spectral frequency can be chosen, on which these variations are displayed brightly. As a rule, this frequency is the steady characteristic of the given observant well and changes only with essential reorganization of a hydro-geological mode. Though the described effect qualitatively was found out in dynamics of various gases dissolved in underground waters, the greatest degree of this effect was observed during the analysis of fluctuations of the Helium contents.

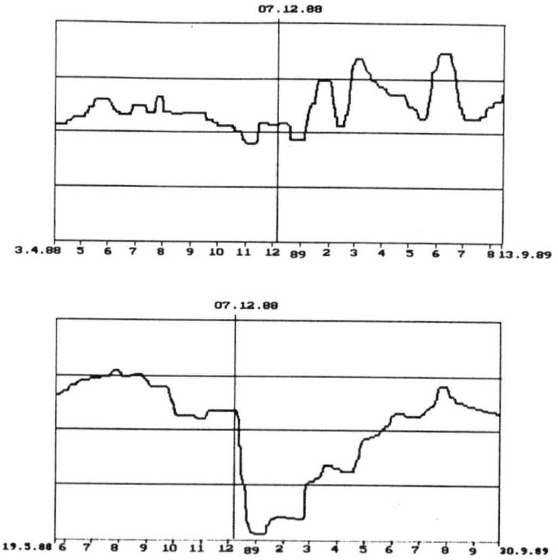

Figure 3. Running mean of daily time series of He – content (window 17 days) before and after Spitak earthquake 7.12.1988: a – Ararat site, b – Kajaran site.

Research established also that prognostic flags are linked with dynamics of the natural factors influencing on solubility of a gas. The measurements of one gas phase at the fixed level of a flow with the strictly fixed way of the sampling can be used as observable tracer.

The important aspect of this research is that it was possible to reveal not only individual prognostic flags, but also their linked series previous to each strong earthquake. Criterion of extrapolation is the regular reduction of intervals between consecutive signals. Theoretical model based on the results of these computations assumes, that the preparation of earthquake is linked with processes, which cause oscillatory variations of pressure in extensive area of preseismic activity, covering the seismic source. The fluctuations have nonlinear character [4]. The source of earthquake preparation is considered as the generator of oscillations of increasing frequency.

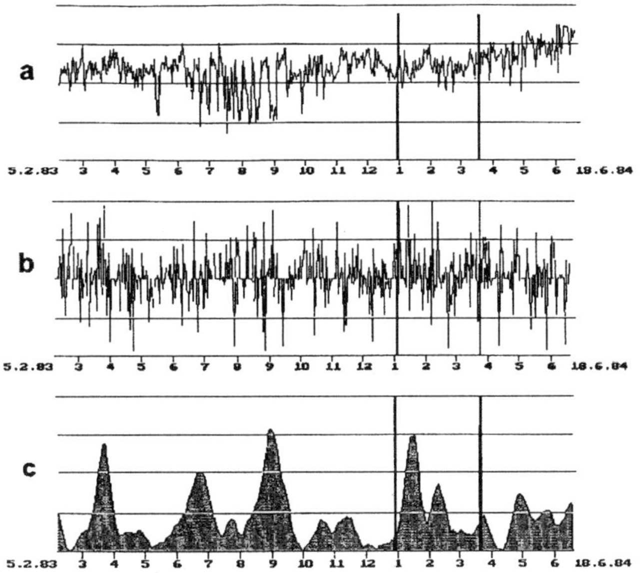

Figure 4. Procedure of noise filtration and detection of prognostic flags. Observation data: Helium content of ground water at Faisabad site (Tajikistan); a – initial time series, b – time series after noise filtration, c – detected prognostic flags: amplitude of oscillations of period T=3.25 days. Strong earthquakes: 30.12.1983, Hindu Kush, M=7.0, 19.03.1984, Kyzyl Kum, M=7.2.

The oscillations can be described by equation:

$$F = A\cos\beta\sqrt{t^* - t}, \qquad t < t^*$$

where t is time, t^* is the event moment, A and β – constants. Pendulum with linearly decreasing weight or volume can be as a mechanical analogue of such system. Oscillation frequency

$$\omega = {A\beta\pi}\Big/{\sqrt{t^* - t}}$$

aims at infinity with $t \to t^*$.

The observation system represents the resonator with steady resonant frequency ω_R. Resonance arises when frequency of disturbing oscillations is multiple to this frequency. It occurs in the moments, when the value $\theta = t^* - t$ will form decreasing sequence of squares of integers.

For the forecast of seismic danger at the moment t* the time series of the previous observations (display characteristic of the observations) is graphed on a non- linear countdown time scale on such graph the prognostic signals related to this date form

$$\tau = a\sqrt{t^* - t}$$

equidistant peaks (Fig5).

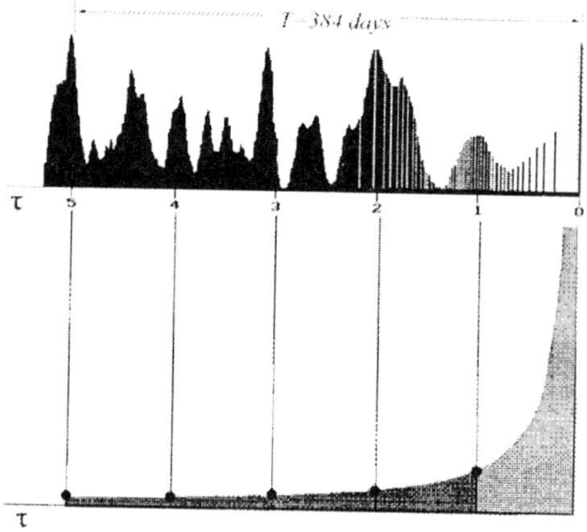

Figure 5. The sequence of prognostic flags before Spitak earthquake (Kajaran site) at nonlinear time scale τ=a(t_0-t)$^{1/2}$. Extrapolation equation: $F = 1/\tau$,

The disturbance frequency is also a function of epicentral distance. This enables the opportunity to obtain the algorithm of their spatial and temporary extrapolation [5]. These fluctuations have rather low amplitude and can be found out with the help of resonant observant system. In this respect the reaction of geochemical system is qualitatively similar to the usual seismological equipment constructed on principles of resonant amplification of oscillations. But geochemical fluctuations have much smaller resonant frequency in comparison with any physical oscillatory system.

The estimation of long term of the forecast is determined by time of last flag, on which the extrapolation curve is calculated. On various observation sites this time interval is estimated from 2-3 weeks (depth of wells up to 100m) up to about 4-5 months (depth more then 1500m). Experimental calculations revealed [6] that the found out laws are caused by a resonance effect between oscillatory reaction of degassing in the solution flow and n long developing oscillatory disturbance in the preparation area of earthquake, The resonant frequency is the steady characteristic of the observant well and the amplitude of fluctuations on this frequency can serve as prognostic characteristic. As a prognostic flag can be used an increase of amplitude of fluctuations (relative or absolute) on this frequency.

The algorithm of the forecast uses vector of 700 – 800 consecutive measurements. On the one hand, it gives necessary representative sample of data and improves reliability of calculated extrapolation function. On the other hand, it creates certain difficulties in actuation of new observation wells. Not less than two years of "empty" daily observations are required for accumulation necessary vector of the data. Observations at operational mode can be began only after such long period. Besides, such approach determines the rather rigid requirements to a technique of observations, sampling and subsequent analysis. Statistical comparability of the data in long time series is a necessary condition. Even short periods of disturbance or change of the techniques will negatively affect on all calculations carried out over a long period.

These results have defined the basic preconditions of the further development of a method. At present, methods, which have been developed during the analysis of a single tracer, are extended on the analysis of multicomponent geochemical systems. The basic difference of the developed approach from traditional is that, instead of sets of parallel measured contents of chemical components, the correlation matrix of their interrelations in the certain interval of time is analyzed. Computations show, that the paired correlation of a set of geochemical components give the same prognostic effects as found earlier (Fig.6). The factor analysis of such systems shows that the ionic composition of underground waters is divided to some subsystems. Dynamics of these subsystems reveals the same prognostic phenomena (Fig.7).

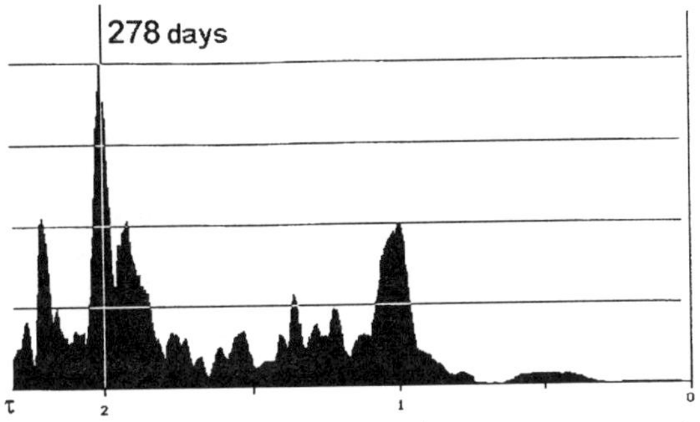

Figure 6. Power cross-spectral density ($f=1/3.5$ days^{-1}) for [Mg]-[HCO_3]. Kajaran site

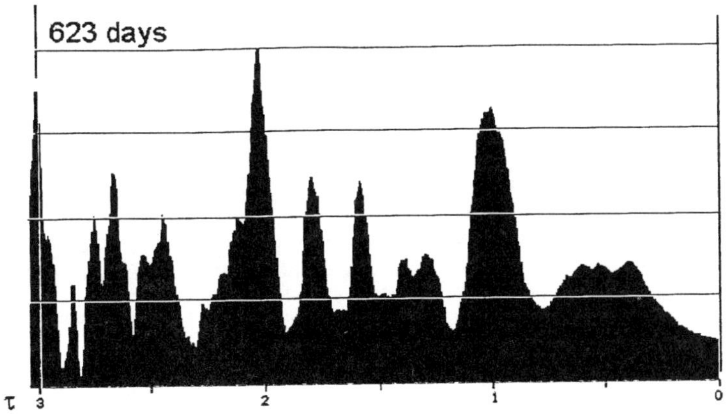

Figure 7. Power spectral density ($f=1/3.5$ days^{-1}) for [Ca-Mg-HCO_3 factor]. Kajaran site

Conclusion

The various geochemical systems, which dynamics depends on the natural factors affecting on pressure variations, were investigated. The obtained results show that the complex multicomponent systems as well as monocomponent ones submit to one common law: regularly repeated changes of a temporal correlation structure of these time series during long period of earthquake preparation. A correlation matrix {factors model) of multicomponent system can be used as prognostic characteristic.

References

1. Sultanhodjaev, A.N.,Belyaev, A.A.Ibragimova, T.L. (1988) Singularities by dynes of the helium dynamics during seismic activity. *Geochemistry* **11**,1642-1651. (in Russian)
2. Barsukov, V.L., Belyaev, A.A., Serebreanikov, V.S. at all (1982) Possibility af use of statistical estimates of of helium content variability in thermal waters for the earthquakes forecast. *Geochemistry* **11**, 1614-1620. (in Russian)
3. Barsukov, V.L., Belyaev, A.A., Serebrennikov, Y.S. (198') The analysis of a sequence of geochemical earthquake precursors. *Geochemistry* **12**, 1667-1673. (in Russian)
4.Belyaev, A.A., Romanov, A.V.(1987) Model of the oscillation system with alternative frequency of the oscillatins. *Geochemistry* **11**, 1607-1614. (in Russian)
5. Belyaev, A.A. (1991) Connection of geochemical prognostic signals with a place and time of forthcoming seismic event. *Geochemistry* **7**, 1061-1068. (in Russian)
6. Barsukov, V.L., Belyaev, A.A., Serebrennikov, V.S., Chervonenkis, A.Ya. (1984) Analisis of dinamics of seismic effect on the observed hydrothermal system.. *Geochemistry* **8**, 1147 – 1154. (in Russian)

EARTHQUAKE PREDICTION RESEARCH IN CHINA: ZHANGBEI M_S=6.2 EARTHQUAKE ON JANUARY 10, 1998

DING JIANHAI, CHEN XUEZHONG, LIU JIE, HUANG XUEXIANG, SUN ZHANAO, YU SURONG AND LI KAIWU
Center for Analysis and Prediction, State Seismological Bureau, Beijing 100035, China

Abstract

A national conference on seismic trend in China (NCSTC) is held at the beginning of every year by State Seismological Bureau (SSB), which distributes mid – term earthquake prediction, especially for the seismic trend in the next year and the most possible risk regions, which is called special risk regions for earthquake (SRREs). This lays a good base for the scientists in SSB to screen the future earthquakes and to do short-term earthquake prediction.

An earthquake (M_S6.2) took place between Zhangbei and Shangyi counties, Hebei province on Jan. 10, 1998, which caused serious damage and heavy casualties in the two counties and was also broadly felt in Beijing. The epicenter is in (41°06'N, 114°18'E) with depth of 10 Km. Seismic anomalies appeared 2 years earlier in the relative region, for example, a significant seismic belt emerged with M>4 earthquakes, regional shock activity increased and b value turned to be lower than before. Several days to several months before the main shock, the precursors were also found, such ground tilt in Chicheng and Zhangjiakou, geoelectric resistivity in Yangyuan and Baochang, water level in Yanqing, geomagnetic field and low – point displacement in Huhehaote and Beijing.

The boundary region of Shanxi Province, Hebei Province and Inner Mongolia, including Zhangbei county, has been taken as a SRRE of M5~6 in NCSTC in 1997 and 1998. In spite of the clear mid – term prediction, short – term to imminent earthquake prediction failed.

1. Introduction

Earthquake prediction is one of great scientific problems in the world. Destructive earthquakes have taken place constantly in recent years, which have caused great losses and heavy casualties, and most of all were not predicted. Therefore governments of a lot of countries have paid close attention to doing research and practice in earthquake prediction. It has taken more than 30 years for Chinese seismologists to work on earthquake prediction and gain some progress and experiences, but the difficult problem of earthquake prediction is still unsolved. The aim of earthquake prediction in China is to predict some kinds of earthquakes to some extent, especially the quakes in the regions with enough precursory monitoring networks. Scientists in China have been working hard to reduce casualties and losses by earthquakes.

From 1970s, a national conference on seismic trend in China (NCSTC) is held at the begin ning of every year by State Seismological Bureau (SSB), which distributes mid-term earthquake prediction, especially for the seismic trend in the next year and the

S. Balassanian et al. (eds.), Earthquake Hazard and Seismic Risk Reduction, 239–247.

most possible risk regions, which is called special risky regions for earthquake (SRREs). Short – term earthquake prediction is done on the basis of the results from NCSTC and has got some success. For instants, the prediction of Menglian earthquake (M_S7.3), Yunnan province, on July 12, 1995 and Jiashi earthquakes (M_S6.3 and M_S6.4), Xinjiang Uygur Autonomous Region, on April 6, 1997 were done successfully according to mid-term prediction of relative years (Chen Lide, 1997).

An earthquake (M_S6.2) took place between Zhangbei county and Shangyi county, Hebci province on Jan. 10, 1998, which caused serious damage and heavy casualties in the two counties and was also broadly felt in Beijing. Several anomalies emerged before the earthquake, such as seismic activity, ground deformation, geoelectric resistivity, geomagnetic field, ground water and etc. The boundary region of Hebei province, Shanxi province and Inner Mongolia to northwest of Beijing, including Zhangbei county, has been taken as a SRRE of M5 ~ 6. In spite of the clear mid-term prediction, short – term to imminent earthquake prediction failed. Analyzing the seismic activity and precursory anomalies before the Zhangbei earthquake will surely benefit future work on earthquake prediction research.

2. Ideology of Earthquake Prediction and Working Procedure

2.1. IDEOLOGY OF EARTHQUAKE PREDICTION IN CHINA

In China, earthquake prediction is divided into 4 stages, i.e. long – term (in ten years), medium-term (in one year), short – term (two to three months) and imminent (several days) earthquake prediction. Screening the evolution of seismic sources and monitoring the change of regional stress field are combined to judge the stage of seismogenic process and to do prediction for relative stages (Zhang Guomin, 1997).

Mid-term earthquake prediction is done to get source features from seismic field on the basis of the research of seismic sources and field. The preparation, development and occurrence of earthquakes all occur in the crust, and the seismic process of each shock is influenced by regional tectonic stress field. Earthquakes in China always take place in groups and are interfered each other, which may relate to the entire motion and dynamic evolution of the crust. Even for a single large earthquake, its preparation and the occurrence are not isolated in its source region, but closely depend on the regional stress field and the environment around the source. Therefore the mid-term earthquake prediction which will result in the SRREs in the next year should be based on the research of regional stress in source area and the environment around the source. The main point of research should be laid on:

(1) Group and statistical features of strong earthquakes. They could provide the normal distribution and background for earthquake activity in China mainland.

(2) Index system of medium to long term earthquake prediction. They can provide a synthetic index system for anomalies on the basis of the multi – subject research in temporal and spatial development of precursory anomalies (Fig. 1).

(3) Analysis and screening significant events and obvious changes in the main seismic belts and regions of China and of the world.

(4) Analysis and checking of seismic and precursory anomalies, the relationship among group features, evolution of anomalies and seismogenic process and the monitoring of SRREs.

(5) Synthetic methods for earthquake prediction, which can provide several standard and stylized prediction methods.

2.2. WORKING PROCEDURE FOR THE JUDGEMENT OF YEARLY SEISMIC TREND

Based on the results from the seismological bureau in each province and each branch institute of SSB, the Center for Analysis and Prediction (CAP) of SSB analyzes the seismic features of the past year and summarizes the reports from all the branches, and then presents a summary to be discussed in NCSTC. The final amended paper is sent to Chinese government.

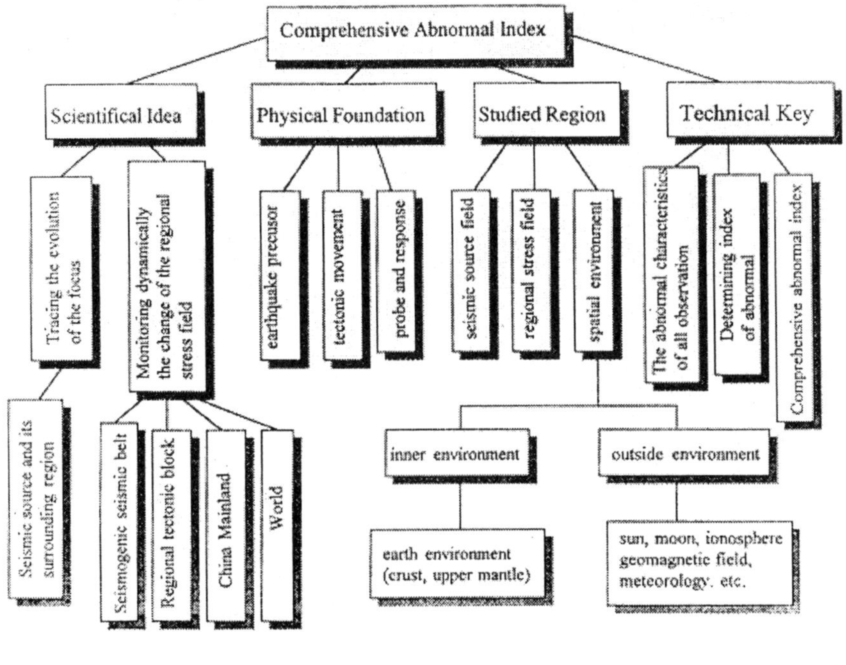

Figure 1. Comprehensive abnormal index system for mid-term earthquake prediction.

In the SSB's report to the State Council of China in 1998, there were two SRREs of MS-6 in eastern China, one of which laid in the boundary region of Shanxi province, Hebei province and Inner Mongolia to northwest Beijing, in which there is Zhangbei M_S6.2 earthquake on Jan. 10, 1998.

3. The Precursory Anomalies before Zhangbei M_S6.2 earthquake

The boundary region of Shanxi province, Hebei province and Inner Mongolia to northwest Beijing is located in the western segment of the Yanshan mountain seismic belt and the intersection region of Yinshan mountain seismic belt and the Shanxi seismic belt. There occurred several earthquakes of M_S>5.0 historically in the region. But, no earthquake of M_S>5.0 occurred in the range with distances less than 50 km from the epicenter of the Zhangbei M_S6.2 earthquake, with very low seismicity. In the range around the epicenter (with distance less than 200 km from the epicenter), there are several remarkable anomalies of seismicity, crustal deformation, geoelectric resistivity, geomagnetism and ground water level etc. The ratio of abnormal precursory items to the total precursory items is about 25%. From the comprehensive analysis, it was considered with a confidence level of 70% that an earthquake of M_S5~6 would occur in this area in 1998.

3.1. THE MID-TERM ANOMALIES O F SEISMICITY

More than two years before the Zhangbei earthquake, a seismic belt of M_L>4.2 is formed. About one year prior to the mainshock, regional seismicity increased obviously and the anomaly of lower b value emerged. A notable earthquake of M_S4.2 occurred near the epicenter seven months prior to the mainshock (Fig. 2).

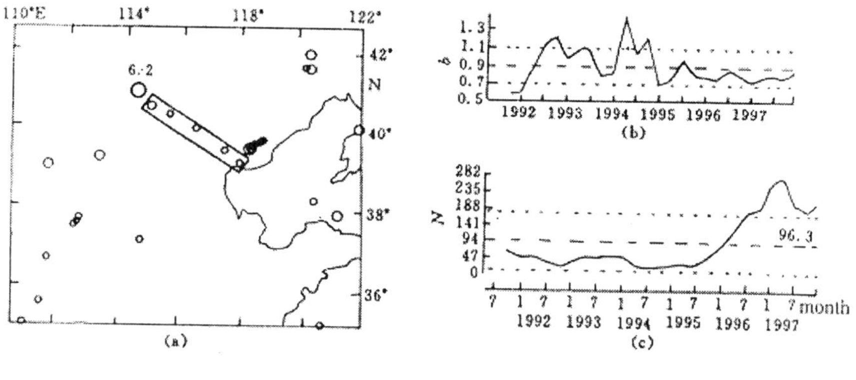

Figure 2. The mid-term anomalies of seismicity.
(a) The seismic band of M_L4.2 formed before the Zhangbei earthquake of M_S6.2; (b) The change of *b* value with time in the seismicitymnhanced region; (c) The frequency anomalies of small earthquakes.

3.1.1. A SEISMIC BELT OF M_L>4.2

After the Datong M_S5.8 earthquake on Mar. 26, 1991, earthquakes of M_S>3.7 (M_L>4.2) occurred along a northwestern belt in the range of 39°~ 41°N and 114° ~ 118°E from Apr. 1991 to Dec. 1997. An earthquake of M_S4.2 occurred at Zhangjiakou, Hebei province in the northwestern end of the belt on May 25, 1997. The Zhangbei M_S6.2 earthquake occurred also in this end of the belt, about 100 km away from the epicenter of the Zhangjiakou M_S4.2 earthquake.

3.1.2. THE ENHANCEMENT OF THE REGIONAL SEISMICITY

Despite the very low seismicity in the epicenter of Zhangbei earthquake, the seismicity in the range around the epicenter (39° ~ 42°N, 111° ~ 114°E) increased distinctly before the mainshock. The frequency of small earthquakes ascended gradually from 1996, and reached at the maximum in Mar. 1997. When it is close to the occurrence of the mainshock, the frequency dropped some what and remained at a relatively high level. The strain release began to accelerate from the end of 1996. In the seismicity – enhanced region, *b* value was at a lower level 2~3 years before the main shock.

3.2. THE IMMINENT ANOMALIES O F GROUND TILT

The ground tilt station at Chicheng, Hebei province is located 130 km away from the epicenter. The EW-direction ground tilt shows very obviously regular annual variation, but, it vanished from Nov. 1997 to Jan. 1998 before the mainshock. The ground tilted to the east rapidly on 16 Nov. 1997, and to the west steeply on Dec.17, 1997, and to the east again on Jan. 9, 1998. On 10 Jan. 1998, the Zhangbei M_S6.2 earthquake occurred (Fig. 3a).

The observation station of ground tilt at Zhangjiakou is located 68 km away from the epicenter, the nearest one. The EW – direction ground tilt changes smoothly. But, it tilted to the east by a big margin on Jan. 2, 1998, and began to return on Feb. 5, 1998 (Because the water in the instrument is changed, this anomaly is only as a reference) (Fig. 3b).

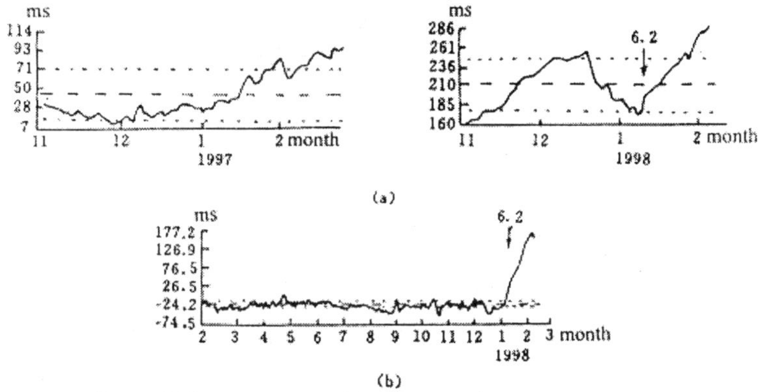

Figure 3. The short-term and imminent anomalies of ground tilt.
(a) The ground tile anomalies at Chicheng station; (b) The ground tilt anomalies at Zhangjiakou station.

3.3. THE MID – AND SHORT-TERM GEOELECTRIC RESISTIVITY ANOMALIES

The geoelectric station at Yangyuan, Shanxi province is 108 km away from the epicenter. The diurnal.average of geoelectric resistivity at this station descended steeply with the disappearance of its annual variation from June 1997, and dropped to the lowest value in Sep. 1997. While it is re turning to normal, the Zhangbei earthquake occurred (Fig. 4a).

The geoelectric observation station at Baochang is 119 km away from the epicenter, where the diurnal average of geoelectric resistivity dissolved its annual variations and was at lower level from Oct. 1997 (Fig. 4b).

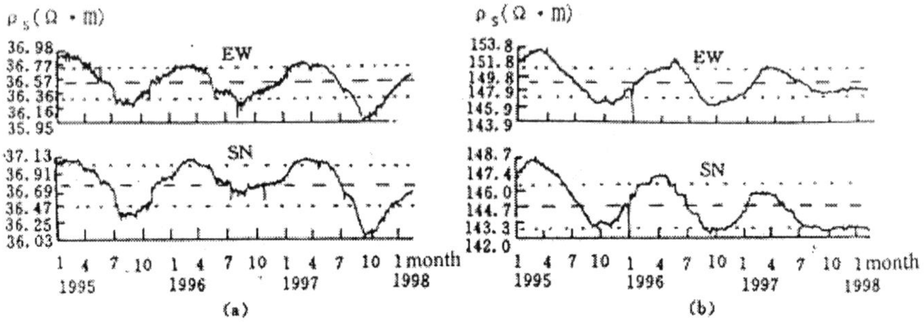

Figure 4. The mid-term and short-term anomalies of geoelectric resistivity. (a) The anomalies of geoelectric resistivity at Yangyuan station; (b) We anomalies of geoelectric resistivity at Baochang station.

3.4. THE MID-TERM, SHORT – TERM AND IMMINENT ANOMALIES OF GROUND WATER LEVEL

The observation station of ground water level at Wuliying (Yanqing county, Beijing) is. 160 km away from the epicenter. The ground water level at this station shows remarkable mid – term, short – term and imminent anomalies prior to the mainshock. These anomalies return to normal after the mainshock (Fig. 5).

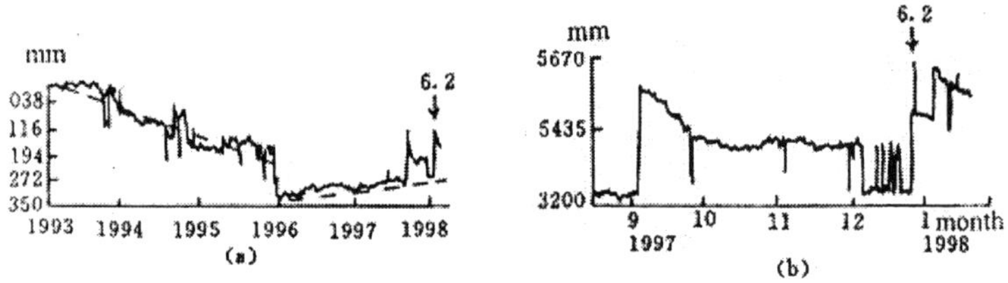

Figure 5. The mid – term, short-term and imminent anomalies of ground water level. (a) The tendentious anomalies of ground water level at Wuliying, Yanqing; (b) The impending anomalies of ground water level at Wuliying, Yanqing.

3.4.1. *Trend anomalies*

The ground water level descended tendentiously from 1994 to 1995, and dropped steeply in the end of 1995, and ascended slowly from Jan. 1996 to Sep. 1997. It took a jump on Sep. 18, 1997, and showed the short term variation.

3.4.2. *Short-term and imminent anomalies*

After the jump on 18 Sep. 1997" the water level dropped slowly. There was a pulse-shaped variation with a big margin one month before the mainshock.

The water level fluctuated strongly in the imminent phase. While the Zhangbei earthquake is occurring, the water level ascends by a big margin with a very obvious coseismic effect. It returns to normal to some extent after the mainshock.

3.5 THE SHORT-TERM AND IMMINENT GEOMAGNETIC ANOMALIES

The geomagnetic observation station at Huhehaote, Inner Mongolia is 210 km away from the epicenter. The differences of geomagnetic component Z between two observation stations, which are located at Huhehaote, Inner Mongolia and Changli, Hebei respectively, ascended abnormally from Nov. 1997 with an increment of 10 nT, and descended somewhat in the beginning of Feb. 1998.

3.5.1. *Low-point displacement anomalies*

The geomagnetic field shows the abnormal. lines with outstanding variation before strong earthquakes in China mainland. Using diurnal variation data of geomagnetic vertical component of 50 stations in China mainland, the temporal distribution of geomagnetic minimum displacement point is drawn out. The boundary line across which the time of minimum geomagnetic displacement each day will change remarkably could be found before the occurrence of most strong earthquakes. According to the results of earthquake cases, 27±4 days or 41±4 days after the boundary line emerges, an earthquake with M_S>6.0 could occur near the boundary (Ding Jianhai, 1994).

Three boundary lines emerged from 17 to 19 Dec. 1997. One on Dec. 18, 1997 is shown in Fig. 6. According to these figures, it was proposed before the mainshock that a moderate earthquake would occur near the boundary from 10 to 18 Jan. 1998 or from 24 Jan. 1998 to 1 Feb. 1998. As expected, the Zhangbei M_S6.2 earthquake occurred near the boundary on Jan. 10, 1998 (Fig. 6).

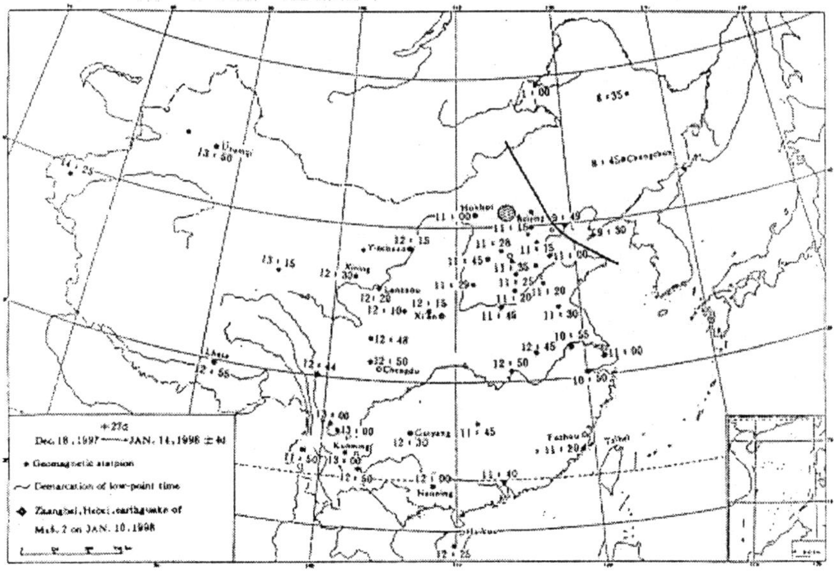

Figure 6. The distribution of low-point displacement on Dec. 18, 1997 before Zhangbei M_S6.2 earthquake.

3.5.2. *The anomalies of diurnal variation form of geomagnetic component Z*
The anomalies of the diurnal variation form include the remarkable changes on the amplitude and phase. The abnormal of the diurnal variation form of geomagnetic component Z at Beijing and Huhehaote are outstanding after Dec. 17, 1997. From one month before mainshock, there are larger changes at Beijing and Huhehaote, comparing with the other stations with longer distance from the epicenter.

3.6. THE THERMAL INFRARED ANOMALIES

Making use of observations observed by NOAA satellite of USA, the geothermal distribution in North China is analyzed. It is found that the temperature increased remarkably (2~6 k) along the Zhangjiakou-Bohai sea seismic belt from Dec. 28, 1997 to Jan. 10, 1998, and abates days after the mainshock. The distribution of temperature increasing coincides with the seismic belt and the demarcation line of abnormal time of low – point displacement, and could be related to the increase of the regional stress field. The results of rock tests show also that when rock is compressed, the thermal infrared radiation will enhance obviously (Cui Chengyu, 1993).

4. Discussion

One to two years before the Zhangbei $M_S6.2$ earthquake, the seismicity anomalies are relatively remarkable. The anomalies of ground tilt, geoelectric resistivity, geomagnetism and ground water level as earthquake precursors are observed several days to several months prior to' the mainshock. But most of the observation stations of earthquake precursors are located within the range with distances of 100 to 200 km away from the epicenter, and only the Zhangjiakou station is less than 100 km from the epicenter. In the whole view, the mid-short term precursory anomalies are abundant, the imminent anomalies are rare and not remarkable enough. So it is difficult to carry out the accurate imminent prediction of three parameters of an earthquake when no foreshocks occur.

The solution of problems on earthquake prediction awaits the scientific revelation of the rules of seismogenesis and occurrence of earthquakes, and needs long – term and unremitting exploration and scientific accumulations. If there is enough high monitoring ability of earthquake precursors in the regions, only the specified types of earthquakes, such as earthquakes with many remarkable precursory anomalies and foreshocks or the follow up strong earthquakes of swarms, could bc predicted probably to some extent under the existing conditions of science and technology.

It reflects the present level of earthquake prediction that the annual medium – term prediction is proposed and the imminent one is not given before the Zhangbei $M_S6.2$ earthquake.

References

1. Chen Lide, Luo Ping, Cai Jingguan et al., 1997. The mid-term, short-term and imminent predictions and features of precursory anomalies of the Menglian earthquake of M_S7.3, Yunnan province on July 12, 1995, earthquake, Vol. 17, No. 1, pp. 1-13.
2. Zhang Guomin, Li Li and Jiao Mingruo, 1997. Development in earthquake prediction research during the last decade in China and its prospects, Acta Geophysica Sinica, Vol. 40, Supp., pp. 396-410.
3. Ding Jianhai, Lu Zhenye and Huang Xuexiang, 1994. Seismogeomagnetics, Seismological Press, Beijing, pp. 253-261.
4. Cui Chengyu, Deng Mingde and Geng Naiguang, 1993. Research on the radiant characteristics of spectrum of rock under the different pressures, Scientific Bulletin, Vol. 38, No. 6, pp. 538-541.

INTENSE CRUSTAL DEFORMATION RATES ON NISSYROS ISLAND (GREECE), DEDUCED FROM GPS STUDIES, MAY FORESHADOW A FORTHCOMING VOLCANIC EVENT

E. LAGIOS

Department of Geophysics & Geothermics, National & Kapodistrian University of Athens, Panepistimiopolis, Ilissia, Athens 157 84, Hellas

Abstract

Nissyros Volcano lies at the SE part of the Hellenic Volcanic Arc (HVA) and, together with the Thera (Santorini) Volcano comprise the most active volcanoes in the Aegean Sea. The last major episode including freatic explosions and gas emanation tool place during the last century. Since then there was not any kind of similar major activity, until early 1997, when an intense local seismic activity broke out with a simultaneous reactivation of an older fault passing through Mandrakion at the NW part of the island. However, the observed seismic activity is not closely associated with the former fault, but seems to be spreading out to the north, encompassing the whole marine area between Nissyros and Western Kos islands.

A GPS network was established on Nissyros to monitor the crustal deformation of the area in June 1997. This network, as part of the implemented methods for the surveillance of Nissyros Volcano, was then remeasured three times, the last one in May 1998. It was found that significant displacements were observed in the coordinates of the stations, in both vertical and horizontal, with respect to a GPS station in eastern Kos. The overall observed displacements for the period June 1997 – May 1998 show intense ground deformation rates until six months after its establishment. A general uplift was measured at all stations of the network ranging from 14(±5)mm to 45(±10)mm (Sept. 1997), which in the subsequent remeasurements showed a slight decline, maintaining though the uplifted character of the deformation. The amplitude of the horizontal deformation was ranging from 13mm to 37mm (average error estimate about ±5mm) for the period of only three months (June-Sept. 1997), which increased for the next eight months, but not analogously, up to 53mm. The directions of the horizontal deformation observed at the GPS stations (June'97-May'98) indicate that the island is "opening up" along two major faulting zones, both commencing from the centre of the island, with one extending to the NW and the other one to the SW.

It is believed that the observed seismicity on Nissyros and the broader area to the north, a great part of which is of volcanic character (Stavrakakis, personal communication), may be attributed to gradually upwelling magma, which due to the change of the regional regime, produces:

S. Balassanian et al. (eds.), Earthquake Hazard and Seismic Risk Reduction, 249–259.

- The reactivation of older faults, which subsequently are responsible for the observed seismicity of tectonic character
- The defined relatively large uplift rates at almost all stations on Nissyros
- The fault-controlled "opening" of Nissyros along the two major faulting zones, introducing again relatively large rates of horizontal deformation.

A reactivation of an older volcanic centre (150,000yr) could possibly be taking place, located in the middle between Nissyros and West Kos, consistent to the local picture of the seismicity for the last two years.

Keywords: Nissyros Volcano, GPS studies, Hellenic Volcanic Arc.

1. Introduction

Nissyros is a small volcanic island at the SE part of the Hellenic Volcanic Arc (HVA). Together with the Thera (Santorini) Volcano, they constitute the most active volcanoes in the Aegean Sea. The last major episode including freatic explosions and gas emanation took place in Nissyros during the last century. Since then there was not any kind of similar activity, until about two years ago, when a local seismic activity broke out with a simultaneous reactivation of an older fault passing through Mandrakion (Fig. 1), at the NW part of the island.

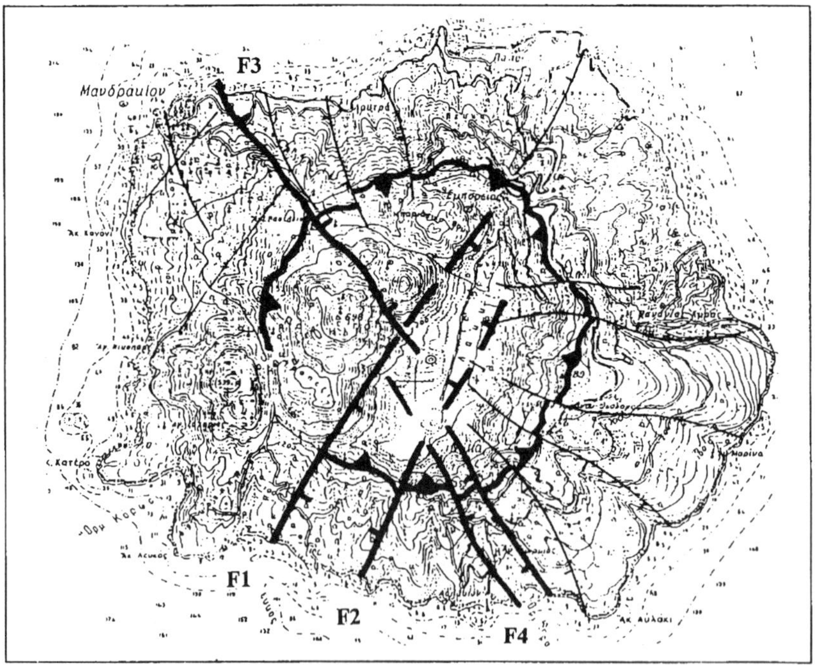

Figure 1. Nissyros Topographic Map showing the main faulting zones (after Papanikolaou & Lekkas, 1991) and the caldera rim.

However, as it will be shown, the observed seismic activity is not closely associated with the former fault, but seems to be spreading out to the north, encompassing the whole marine area between Nissyros and (western) Kos islands. It is reported (G. Stavrakakis, personal communication) that a significant number of the recently recorded earthquakes is of volcanic character, something which is very important with respect to Nissyros Volcano.

In the present paper, the so far obtained results of the Nissyros GPS Network will be presented relating to the measured crustal deformation on Nissyros. A reference to the likely causes producing the former deformation will be made, presenting a possible interpretation of the current tectonic activity in the broader area, taking into account the present picture of the on-going seismicity.

2. A Brief Geotectonic Account

An account of the earlier and recent geological and volcanological work on Nissyros has already given elsewhere (Hardiman, 1996; Lagios et al. 1998). However, the main faulting zones observed on Nissyros (Papanikolaou & Lekkas, 1991) are the following ones (Fig. 1):

- The faulting zone F1, with a downthrow to the SE and direction NE-SW.
- The faulting zone F2, commencing from the centre of the southern part of the island with a direction to the NE, terminating in the caldera area.
- The faulting zone F3, which starts from the NW part of the island and extends to the SE, truncating F1 at the centre of the island.
- The faulting zone F4 comprising a narrow small graben, commencing from the SE part of the island, having a direction to the NW at the southern part of the caldera area.

During the recent period of the geological evolution of the island, micromovements along all the aforementioned faulting zones took place, as well as the creation of minor faults with smaller downthrow. Finally, in the caldera area and along the major faults, intense superficial hydrothermal alterations and fumaroles take place.

3. The Gps Network

Strong crustal deformation usually takes place, with horizontal and vertical movements, before volcanic eruptions. The latter can even exceed 1 m (Sigmudsson, 1996) in the vertical direction. These deformations of the Earth's crust can be the result of: (i) The continuously increasing pressure in the magma chambers. (ii) The magmatic intrusion into superficial strata. (iii) The existence of large magnitude earthquakes.

Consequently, the geodetic methods for the surveillance of volcanoes are very important, concerning their contribution to the study and evaluation of the volcanic risk, as a means for early warning, as well as the study of the overall dynamics of volcanoes. The recognition of the need for the use of geodetic and relevant networks (microgravimetric), resulted in the establishment of such networks in Santorini (Lagios, 1995; Giannopoulos et al., 1996; Chailas & Lagios, 1996).

The Global Positioning System (GPS) (Hofmann-Wellenhoff et al., 1992) has also been applied on Nissyros for the surveillance of the local volcano. Detailed accounts have already been given elsewhere relating to the method and the system, as well as to the errors involved (Lagios et al., 1998). Since Nissyros is a small island, an accuracy of ±5 mm for the horizontal, and ±10 mm for the vertical direction is basically expected (Bilham, 1991).

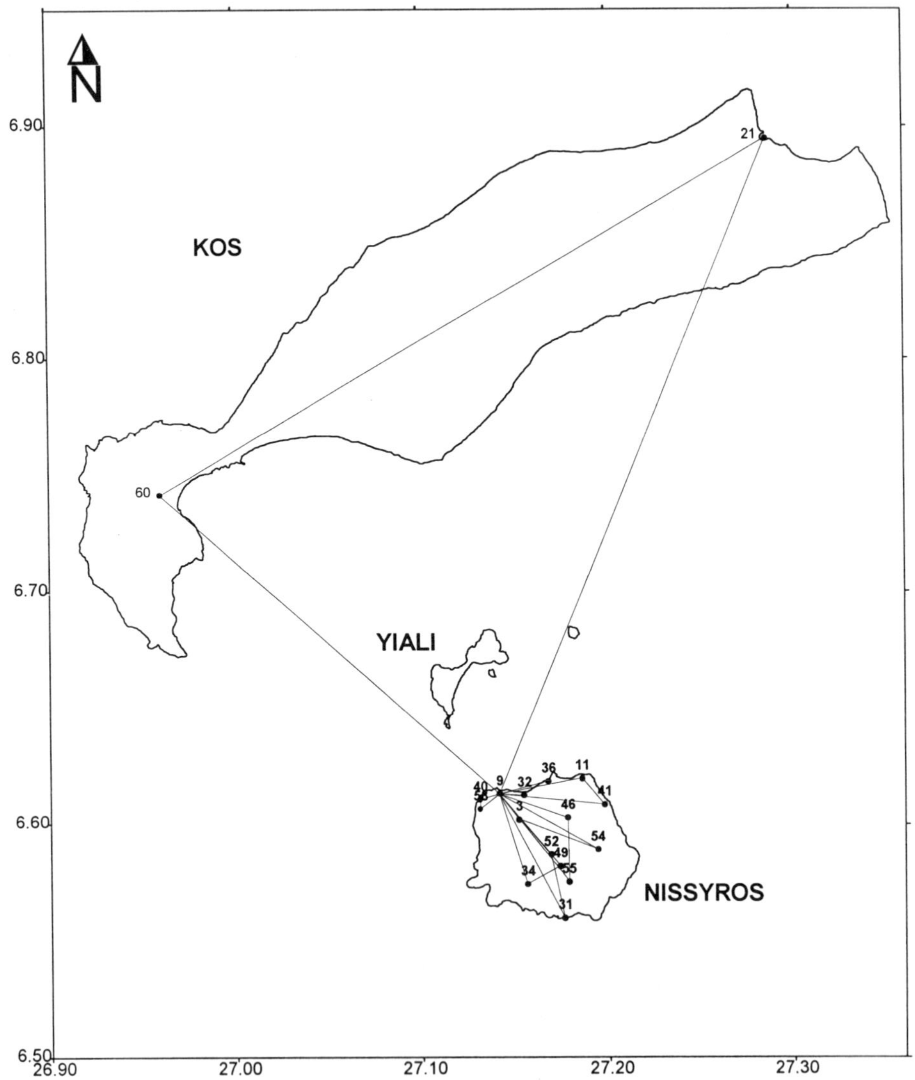

Figure 2. Baselines of GPS Network for the Surveillance of Nissyros Volcano

A geodetic GPS network consisting of 17 stations was established in the broader area of Nissyros in June 1997 (Fig. 2). Fifteen stations were established on Nissyros and two in Kos. The settlement of the GPS stations in Nissyros was mainly based on the need for a homogeneous coverage. The mean distance between two stations is about 2 km. It is thus expected that the ground deformation of the area will be studied in a satisfactory detail. The stations established on Kos serve two purposes: First, they contribute to the study of the ground deformation in a regional sense. Secondly, one of them (No 21), at the eastern part, was used as the reference base for the rest of the stations.

Three remeasurements of the network were performed, in September and December 1997, and May 1998. During the first reoccupation of the network, an additional station (No 60) was established in western Kos. During the second reoccupation of the network, station No 58 (Fig. 2) was substituted by No 28 since it was destroyed, and an additional station (No 38) was established in the centre of the western part of Nissyros.

The GPS observations were made by the use of geodetic GPS receivers WILD SR299 with capability in dual frequency data collection (L1 and L2). The recording period for each baseline on Nissyros was of the order of 3 hours. This is considered as adequate, given the small extent of the GPS network (diameter of Nissyros Island smaller than 7 km). However, for larger baselines, as those between No 9 (Nissyros) and No21 or No 60 (Kos) (Fig. 2), the recording period was about 24 hours.

3.1. THE DATA PROCESSING

The analysis and adjustment of GPS data were made using the GPS post-processing software LEICA Static Kinematic Software (S.KI. Version 2.2) 1997. The calculation of the baselines was made on the ellipsoid of the World Geodetic System 1984 (WGS' 84). The station No 21 (eastern Kos) was chosen as reference station (with fixed coordinates) of the network. The adjusted values for the three observational periods were considered for the detection of any crustal deformation.

3.2. GPS RESULTS

The coordinates of the GPS stations are presented in the form of geographic longitude, geographic latitude and elliptic height. The differences of these coordinates, for the periods June-September and September-December 1997 are presented in figure 3, and June'97–May'98 in figure 4, together with the representation of the main faulting zones and the rim of the caldera. These differences present the calculated horizontal displacements of the GPS stations with respect to the station No 21 (Fig. 2).

It was found, after the first remeasurement, that the observed change of the horizontal co-ordinates ranges between 13-37 mm, with a mean approximate error of about ±5mm (in extreme cases ranging between ±(7-11)mm). The direction of the change (Fig. 3) was mainly to the SE, while, for a small group of stations at the NW part of the island, the change was pointing to the SW.

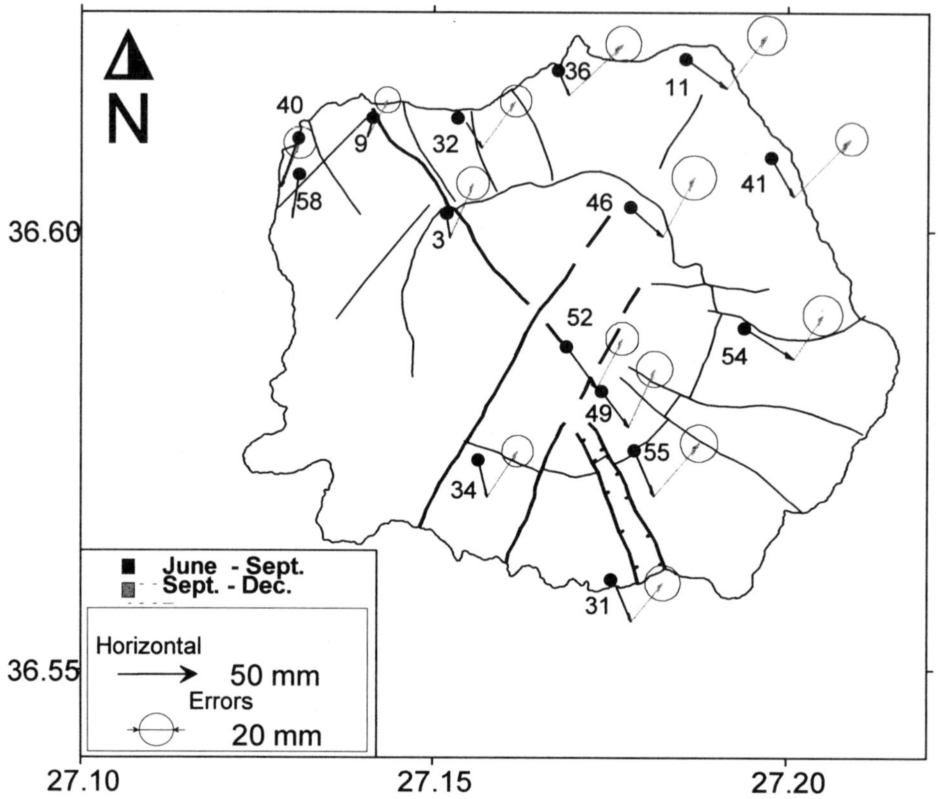

Figure 3. Horizontal displacements for the periods June-Sept. and Sept-Dec. (1997).

At the second reoccupation of the network, it is observed (Fig. 3) that most of the stations were horizontally displaced towards NE, ranging between 27±8 to 52±12 mm. Generally, the total horizontal displacements for most of the GPS stations in 1997 (June-December) point to the East, with magnitudes ranging between (3±10) mm to (53±12) mm, with the exception of some stations at the NW part of Nissyros, which seem to attain their original position.

The third remeasurement of the network resulted in the final deformation observed between June (1997) and May (1998) – (Fig. 4). Three groups of motion referring to the GPS stations are identified, which seem to be controlled by the two major faulting zones F1 and F3 (Fig. 1): One group of stations lying at the half northern part of the island and east of F3, having a motion to the East or ESE; another group of stations at the southern part of the island and east of F1, showing a motion to the SE, and, finally, a third group at the NW part of the island and west of F1 and F3, indicating a change to the SW or WSW. The amplitude of the observed change was maintained at approximately the same levels as found for the previous remeasurement. However, the direction of the motion has slightly changed more to the SE (for the majority of the stations), except stations at the NW part that showed a change to the SW. This annual resulted change in the horizontal coordinates of the network indicate a kind of "opening up" of the island along the two major faulting zones F1 and F3.

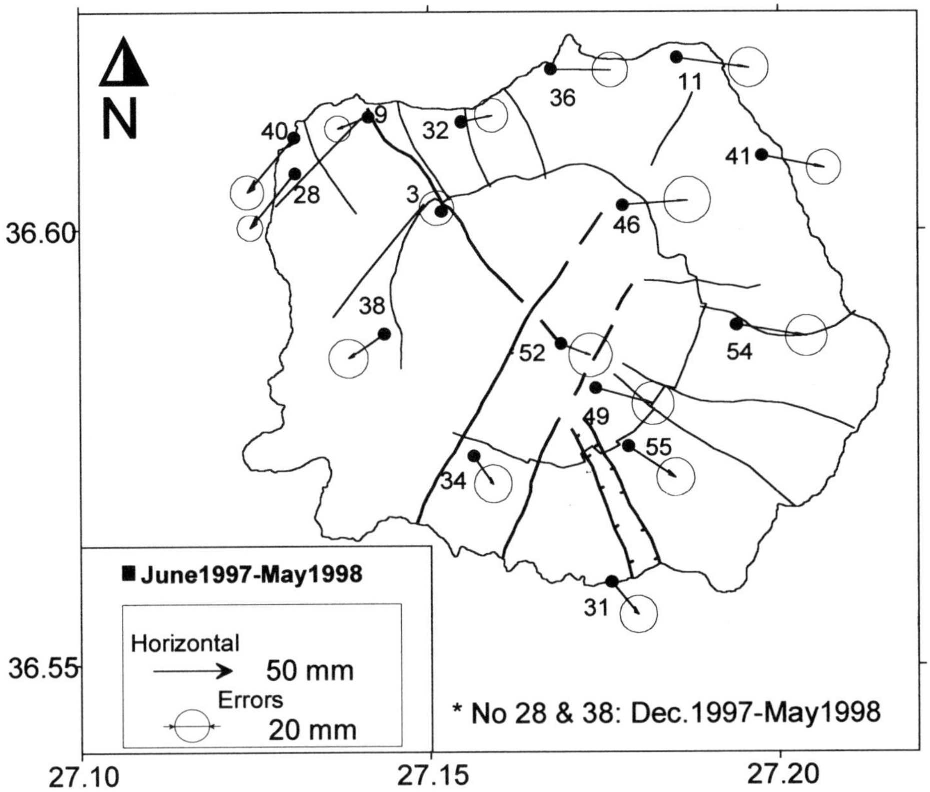

Figure 4. Horizontal displacements of the GPS stations for the period June'97-May'98

There were also changes observed in the vertical direction, indicating generally uplift. The vertical displacements were found between 14-45 mm (elliptic heights) during the first remeasurement (Lagios et al., 1998), only three months after, when the seismic activity was still in progress. During the second and third remeasurement, there was a negligible change, probably because the seismic activity was substantially decreased; however, the uplift of the ground was maintained. Figure 5 shows the final elevation change between June (1997) and May (1998), where uplift is observed at almost all GPS stations on the island and in some cases exceeding the value of 30 mm. It is to be mentioned here that the calculated error ranges between ±(5-10) mm.

4. Discussion - Conclusions

From the present GPS results and the increased Radon (Rn) emission change (Lagios et al., 1998), which is apparently related to the observed seismicity, it becomes clear that the broader area of Nissyros Island, as well as the marine area north of it, presents an intense tectonic activity. It appears that the former area is subjected to an intense degree of deformation. The former is generally supported by the following factors:

- An intense degree of horizontal deformation (resulting from GPS observations) that seems to be controlled by specific major faults, suggesting an "opening" of the island along major faults (Fig. 3 & 4).
- A considerable, rather general, uplift of Nissyros as inferred by the GPS measurements (Fig. 5).
- A continuous seismicity that presents periods of increased rates in magnitude and number of events. A great number of the recorded tremors was found to be of volcanic character.

Analytically, the GPS network was established in a period (June 1997) when the seismic activity had not ceased, but it was also not at an intense level. During the first two remeasurements of the GPS network, the most intense degree of deformation, both vertical and horizontal was noticed. During the third remeasurement of the network, it was found that the magnitude of deformation was maintained at approximately the same levels, only the direction of it changed. The intense degree of deformation observed especially after the first two remeasurements (period of deformation June-Dec.1997–(Fig. 3)) coincides with the relatively large magnitude earthquakes taking place in that period. Indeed, two earthquakes of magnitude M=5.1 and M=5.2 took place in August 1997, in the marine area between Nissyros and western Kos.

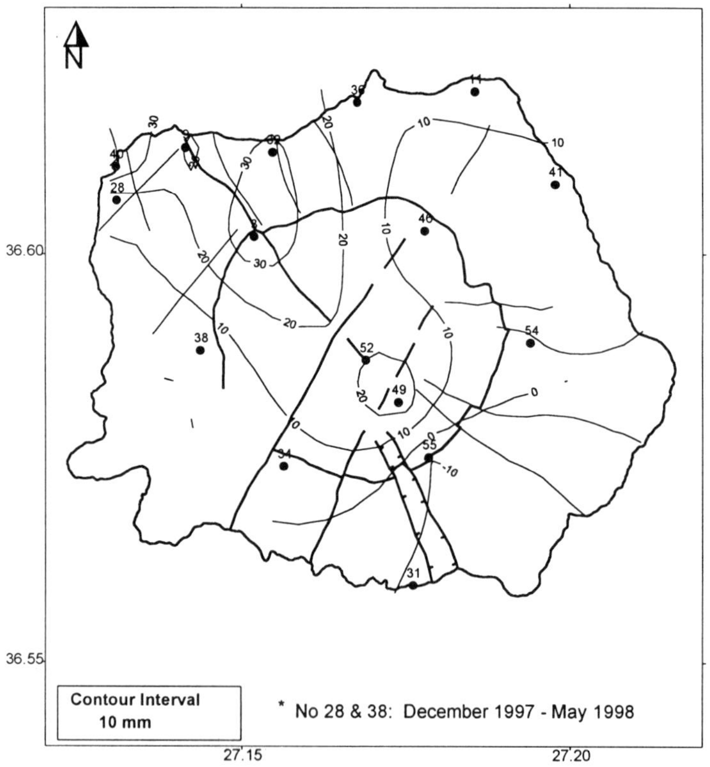

Figure 5. Vertical displacements of GPS network for the period June'97-May'98.

The intense seismic activity broke out in the broader marine area between western Kos and Nissyros in early 1997, or even earlier. This activity was characterised by fluctuations in the seismic energy release and the number of events per time period. During that time a reactivation of an older minor fault, passing through Mandrakion (Fig. 1), caused damages in some of the houses. These events led to the establishment of a local seismological network in the area by the Geodynamic Institute of the National Observatory of Athens to closely study the seismicity in this region. This local network was in operation periodically and for a limited period of time. Preliminary analysis of these seismic data has shown (Stavrakakis, personal communication) a systematic concentration of the epicentres in the marine area between Kos and Nissyros, and in the area of the northern part of Nissyros. Figure 6 shows a typical picture of the distribution of these tremors for the period March 1-17 (1997). This period is characterised as a quiet one, because in times of intense activity, the local network per week recorded more than 400 earthquakes. The analysis has also shown a very interesting conclusion that a great part of the observed number of earthquakes was of volcanic character (G. Stavrakakis, personal communication). Generally, it appears that the deeper (30-35 km depth) earthquakes tend to concentrate in the marine area between Yiali and Kos, while the most superficial ones at the upper crust tend to accumulate in the region between the northern part of Nissyros and Yiali.

The above significant outcomes in relation to the observed high rate of deformation (systematic uplift and "opening" of Nissyros), within less than a year, may indicate that probably upwelling magma from greater depths exerts great pressure in the regional area, resulting in the activation of older faults, as the one which passes through Mandrakion (Fig. 1). The tensional regime is probably deactivated through the major fault zones (e.g. F1 or F3 (Fig. 1)) - as the differentiation of motion of the GPS stations on both sides of the F1 and F3 faulting zones is quite clear. The intermediate depth earthquakes that usually take place at this part of the HVA could be responsible for a probable change of the deeper dynamic equilibrium of the area, and a likely initiation of upward magma motion. If this is the case, then it is probable that a reactivation of an older volcanic centre is possibly taking place. Such is the broader marine area between Nissyros and western Kos, constituting a large caldera, which was active 150,000 years ago (Vougioukalakis, 1994; Keller et al., 1989; 1991). The observed deformation in the area resulting from the GPS observations is an order of magnitude larger than the one expected from the seismic energy release in the same area (Papazachos & Kiratzi, 1996). Moreover, the observed directions of displacement are consistent to the general direction of the prevailing tensional regime in the area.

All the aforementioned factors compose an interesting dynamic picture of the Nissyros Volcano. It should be faced with consideration and caution, so that probable upward magma motions may be detected in time, a future volcanic event should be in preparation and eventually take place. Local permanent seismic network should be running continuously, not periodically, so that an objective picture of the seismicity is monitored and controlled in the broader area of Nissyros, which constitutes an older volcanic centre. Furthermore, together with the geodetic monitoring of the area, systematic monitoring of (real-time) gas sampling is proposed on a permanent basis.

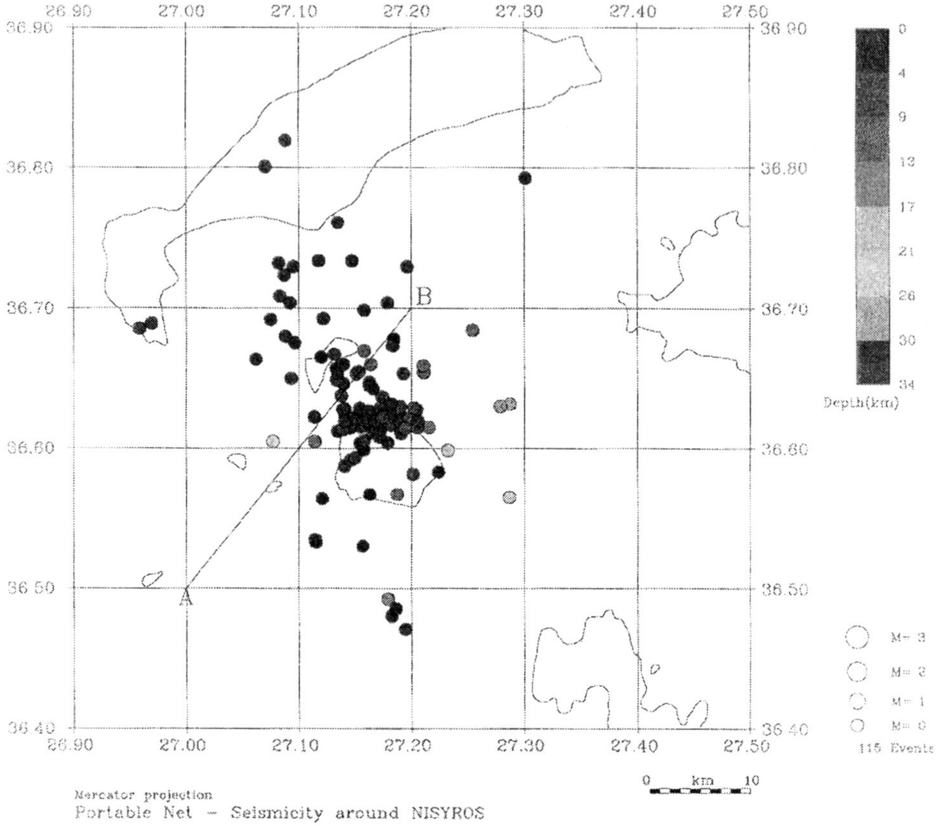

Figure 6. Distribution of earthquakes in the broader area of Nissyros as recorded by a portable seismic network for the period March 1-17, 1997 (G. Stavrakakis, Geodynamic Institute, National Observatory of Athens, Greece).

5. Acknowledgements

This research was financed by the European Centre on Prevention and Forecasting of Earthquakes through OASP (Athens, Hellas).

6. References

Bilham, R. (1991) Earthquakes and sea level: Space and terrestrial metrology on a changing planet. ***Rev. of Geoph.*** 29, **1-29.**

Chailas, S. & Lagios, E. (1996) Monitoring of the local gravity in Santorini Volcano, 1984-1995. *Proc. 2nd Workshop on European Laboratory Volcanoes*, May 2-4, Santorini, Greece. *European Commission D.G. SR&D.,* 297-309.

Giannopoulos, J., Lagios, E. & Sigmudsson, F. (1996) Global Positioning System (GPS) monitoring of Santorini Volcano. *Proc. 2nd Workshop on European Laboratory Volcanoes*, May 2-4, Santorini, Greece. *European Commission D.G. SR&D.*, 271-279.

Hardiman, J.C. (1996) Volcanology and dating of the caldera phase eruptions of Nissyros Volcano, Greece. *Proc. Course: The mitigation of volcanic hazards. European Commission D.G. SR&D*, 519-522.

Hofmann-Wellenhoff, B., Lichtenegger, H. & Collins, J. (1992) *Global Positioning System, Theory and Practice*. Spinger Verlag, Vienna, 326p.

Keller, J., Gillot, P.Y., Renen, Th. & Stadlbauer, E. (1989) Chronostratigraphic data for the volcanism in the Eastern Hellenic Arc: Nissyros & Kos. Abstract OS06-26, *Terra Abstracts*, 354.

Keller, J., Renen, Th. & Stadlbauer, E. (1991) Explosive volcanism in the Hellenic Arc: A summary and review. In: *Thera and the Aegean World III*, **2,** 13-26.

Lagios, E. (1995) Microgravimetric Monitoring of the Santorini Volcano, Greece. *Proc. Intern. Workshop "New Challenges for Geodesy in Volcanoes Monitoring"*, June 14-16, 1993, Walferdange, Luxembourg, Tire a part Des Cahiers du Centre Europeen de Geodynamique et de Seismologie, **8,** 293-305.

Lagios, E., Chailas, S., Giannopoulos, J. & Sotiropoulos, P. (1998) Surveillance of Nissyros Volcano: Establishment and remeasurement of GPS and Radon Networks. *Bull. Geol. Soc. Greece,* **32/4**, 215-227.

Papanikolaou, D. & Lekkas, E. (1991) Geological structure and evolution of the Nissyros Volcano. *Bull. Geol. Soc. Greece*, **25/1,** 405-419.

Papazachos, C.B. & Kiratzi, A.A. (1996) A detailed study of the active deformation in the Aegean and surrounding area. *Tectonophysics,* **253,** 129-153.

Sigmudsson, F. (1996) Crustal deformation at volcanoes. *Proc. Course: The mitigation of volcanic hazards. European Commission D.G. SR&D.* 237-257.

Vougioukalakis, G. (1994). Volcanic stratigraphy and evolution of Nissyros Island. *Bull. Geol. Soc. Greece,* **28/2**, 239-258.

TECTONIC EARLY WARNING SYSTEM THROUGH REAL-TIME RADON (Rn) MONITORING: PRELIMINARY RESULTS OF A GEOPHYSICAL METHOD FOR FORECASTING EARTHQUAKES

E. LAGIOS[1], G. SIDERIS[2], F. ZERVOS[2], P. TSOURLOS[2], R.A. NICHOLSON[3], AL. PONOMAREV[4], B. SALOV[4], S. BALASSANIAN[5], H. PETROSYAN[5], S. BUSHATI[6] & O. LIKA[6]

[1]Department of Geophysics & Geothermics, National & Kapodistrian University of Athens, Panepistimioupolis, Ilissia, Athens 157 84, Greece.
[2]GeoMentor, a European Economic Interest Grouping (E.E.I.G.), Sarantaporou 8 & Str. Tompra, Agia Paraskevi, Athens 153 42, Greece.
[3]British Geological Survey (BGS), Keyworth, Nottinghamshire, NG125GG, United Kingdom.
[4]Institute of Seismology of the United Institute of Physics of the Earth, RAS, B. Gruzinskaya 10, Moscow 123 810, Russia.
[5]National Survey for Seismic Protection, Davidashen-Massiv 4, Yerevan 375 054, Armenia.
[6]Center of Geophysical and Geochemical Exploration, L.9, Blloku Vasil Shanto, Tirana, Albania.

Abstract

Earthquakes are a major natural hazard in numerous parts of the world, and research into precursor signals of seismic activity has mobilised the scientific community for many years. Since earthquakes are a physical phenomenon, attempts for predicting these events in terms of date, magnitude and epicenter have traditionally called upon geophysical methods (seismology, deformation of the ground, electrical methods, etc.). In order to understand more the underlying processes involved, it is necessary to undertake extended periods of monitoring of individual gases at one or more locations, and statistically analyze the compositional data in relation to the observed seismic activity and the effects of extraneous variables. These geophysical anomalies indicate that the faults are zones of weakness acting as channels for deep degassing processes. Numerous successes in forecasting were subsequently recorded through the study of radon emanation in the soils and ground waters of seismic zones. Research into earthquake precursor signals requires continuous instrument surveillance of the selected geophysical parameters.

The final objective of this research work is to be able to estimate and prognoses a seismic event. Throughout a one-year period of our research work, we have collected sufficient soil-gas data over long periods at one or more recognized seismically active localities in Greece, Russia, Armenia, and Albania as a preliminary work to establishing

S. Balassanian et al. (eds.), Earthquake Hazard and Seismic Risk Reduction, 261–270.

a capability for earthquake prediction based on variations in soil gas composition. With the already studied results we are able to demonstrate the viability of the technique, followed by establishment of permanent monitoring stations. The information obtained from these surveys will be compared to data from on-going parallel studies in Greece where all data will be collected via satellite in real-time. Emphasis will be placed on the development of methodology and software suitable for the extended monitoring of gas composition on-site, supported by laboratory determinations to validate the field data where necessary. A number of data loggers (i.e. transducers/detectors) have already been installed at specific selected locations (base stations). Each base station includes a data-logger with which a number of geological parameters are detected. The data are transmitted to a modem, which is directly attached to the INMARSAT global satellite communication system. With the above mentioned multi-disciplinary approach we intend to establish a permanent central data acquisition-broadcasting system with a continuous monitoring of radon, water table level, groundwater temperature, atmospheric temperature and pressure, carbon dioxide, helium, nitrogen, etc.

1. Introduction

Research into earthquake precursors as a means for predicting earthquakes is of considerable current interest. All international organizations have recognized the importance of earthquake precursors as a potential tool for earthquake prediction; however, it is accepted that the manner and the way to use the precursors for prediction are not known, yet. Progress can be made by long-term high quality field observations, use of rigorous statistical techniques and quantitative approach to data analysis.

One of the widely used techniques for earthquake prediction research is measurement of radon concentration within seismo-active areas (King, 1981; Wyss, 1989). The behavior of soil volatile gases has been widely studied. The processes and mechanisms controlling the composition of a gas mixture can be explained in terms of thermodynamics. Traditionally, radon concentration measurements have been performed without continuous monitoring. To overcome the inconsistency of the data induced by this practice several instruments for continuous radon monitoring have been developed (Chambaudet et al., 1991; Khan, 1991; Kleis et al., 1991; Theodorson, 1996; Lenzen and Neugebauer, 1997).

Radon data are subject to spurious fluctuations due to meteorological parameters, temperature and pressure. Ways for correlating the radon data and the meteorological parameters to discriminate the signal from the noise have been suggested (Virk, 1993; Finkenstein et al., 1998). Several radon monitoring networks, either local or regional have been established all around the world in areas of active faults. Such networks have been reported among others in China (King, 1985), the North Anatolian Fault (Turkey) (Friedmann et al., 1988), Algeria (Cherouati et al., 1991), Pakistan (Qureshi et al., 1991), the French Alps (Borchielinni et al., 1991), India (Virk and Singh, 1992), Japan (Fujinawa, 1992; Wakita, 1996; Yasuoka and Shinogi, 1997; Nishizawa et al., 1998).

Geological research institutes and commercial companies in Greece, and the UK are collaborating with similar organisations from Russia, Armenia and Albania, under the European Union funded Inco-Copernicus programme, to develop new techniques for the automatic acquisition of gas geochemical data from remote sites using gas-sensing probes and telemetry. The data are integrated with seismic information acquired in parallel and used to determine whether the technique is a suitable method for monitoring ground disturbance associated with tectonic events. The project commenced in September 1997 and has been partially dependent on the development of the Barasol MC radon probe incorporating sensors for barometric pressure temperature and socks. These parameters combined with rainfall measurements are important for determining discrete alterations in gas flux, associated with changing meteorological conditions, thereby ensuring that variations in gas concentration, potentially correlated with deep geological movements, and are fully recognised. In particular the project's objectives are:

- Record the Tensional Properties of the Earth's Crust.
- Collect Soil-Gas Data over Long Periods in Seismically-Active Areas (in the Partner Countries).
- Develop State-of-the-Art Autonomous Equipment for Continuous Radon Monitoring.
- Monitor the Local Climate Conditions and Seismic Activity.
- Transfer the Data to the Base Stations in Real-Time via Satellite.
- Develop Methodologies (and Software) for real-time Data Correction, Interpretation and Presentation.
- Develop Methodologies for Data Inter-Comparison and Determination of Alert Criteria.

The main however objective is to increase the knowledge of how to use radon monitoring to estimate and predict a seismic event. To understand more the underlying process it is necessary to:

1. Select carefully the sites where the monitoring systems will be installed.
2. Undertake extended periods of monitoring of individual gases at one or more locations.
3. Statistically analyse the compositional data in view of
    - The local and/or regional seismic activity.
    - The effects of extraneous variables.

2. Technology

Within the framework of the project, the companies ALGADE and GeoMentor e.e.i.g developed a special site station under the directions given by the partners. The goal is to provide a "real-time" connection to a network based on the new Barasol MC probe. The station communicates using either the INMARSAT satellite network or a direct phone-line modem. The Barasol MC probe, developed by ALGADE, measures the radon-222 concentrations using a light-tight implanted silicon detector topped by an ogive marking the limits of the detection volume (Figure 1).

The Barasol MC probe (Figure 2) also measures and stores the atmospheric temperature and pressure. Finally, it contains a shock detector. It has the possibility to be connected with two radon-222 mobile heads and a raingauge (Figure 3).

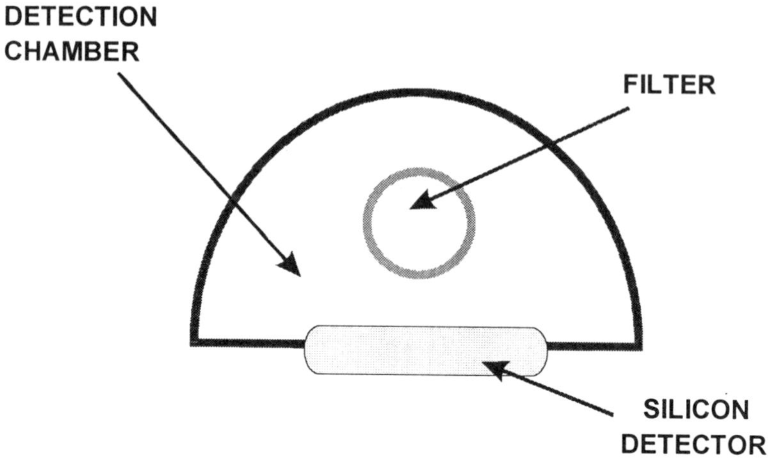

Figure 1. The principle of the apparatus for Radon-222 measurements.

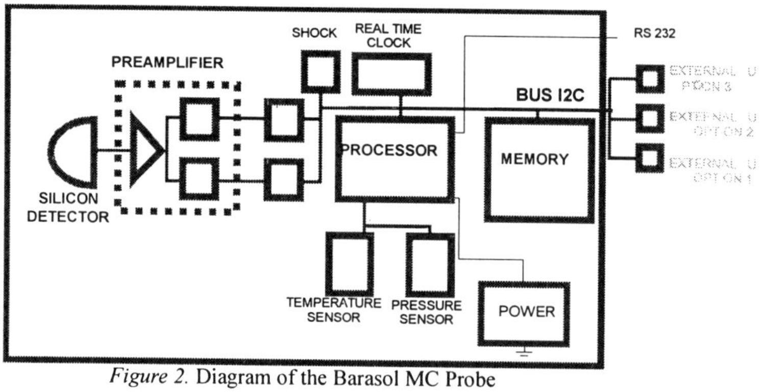

Figure 2. Diagram of the Barasol MC Probe

The probe is autonomous in view of power and data storage for one year and it records and stores all measured parameters every 15 minutes. The site station operates with a 220V supply and has a 12V battery-supply, which supports the system for 12 hours in case of a main power failure. The site station acts as an intermediate between the Barasol MC probe and the base station. It downloads the measured parameters from the Barasol MC probe, stores them and via a modem connection (via wired or wireless communication) sends the data to the base station when asked. The set-up in the field of the entire station is presented in figure 4.

3. Methodology

The efficiency of radonometric-observation stations depends on proper location. For that reason the selection of the sites was made after careful study of several parameters. In particular the sites were selected based on extensive studies of the local seismotectonic conditions. The selected sites should also be covered by a properly designed seismic observation network, in order to establish a seismological database to investigate the seismic conditions of the region and to allow comparisons with geophysical/geochemical data. Based on these principles the following sites were selected:

- The area of Comrie in Tayside, Scotland (U.K.)
- The island of Nissyros in Greece (SE Aegean Sea)
- The Northern Caucasus area in Russia
- The area of Yerevan (Armenia)
- The area of Tepeleni in Albania

To allow a better insight of the radonometric sensitivity of the selected areas individual radon measuring probes were installed for a short-term period into the partner countries prior to the installation of the permanent site stations. The preliminary Radon observations for a 2-month period from the area of Elbasan (Albania) are shown in figure 5.

4. Software

Signal processing software is under development. The signals related to gas exhalation will automatically be compared with the expected signals computed by linear prediction methods, and when the difference is significant, i.e. higher than a predetermined threshold, information will be given for an alert. The software involves:

- Development of the algorithm for automatic classifications of anomalies of Rn concentration time-series of seismogenic nature by means of special statistical processing and data analysis.
- Determination of the type of hazard for each type of anomaly depending on its amplitude, frequency, spectral content.
- Determination of a threshold above which the seismogenic anomaly is related to a potentially hazardous earthquake.
- Development of the algorithm for determination of a generalised threshold for a high seismic hazard zone as a result of observations from several observational stations in the area.

5. Preliminary Observations

The results from the preliminary observations are summarised as in the following:

1. The efficiency of radonometric observation sites strongly depends on observational site.

2. The stations, which are located directly close to deep faults with high permeability (for gases), are most sensitive to the earthquake preparation process.
3. The sensitivity of radonometric stations to the earthquake preparation process sharply reduces over small-depth (superficial) faults or away from them.
4. The preparation of a seismic event is most frequently mentioned in the plots of radonometric observations as an anomalous decreased values of the Rn content, having a duration of up to 1 month (and more) before, 1-60 days prior to the earthquake.

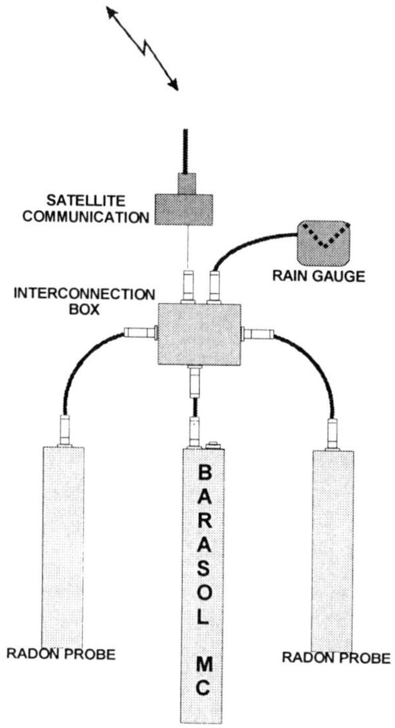

FEATURES:

WEIGHT: 2 Kgr

LENGTH: 500 mm
DIAMETER: 61 mm
MAX DEPTH: 10 m
POWER: 2 X 1,5V Bat.
OPERATION: 1 year

Rn DETECTION: 1pulse/h > 50 Bq.m^{-3}
Temp. Accuracy : 0.025°C
Pressure Accuracy: 1 hPa

Figure 3. A sketch-diagram of the Barasol MC Radon measuring probe

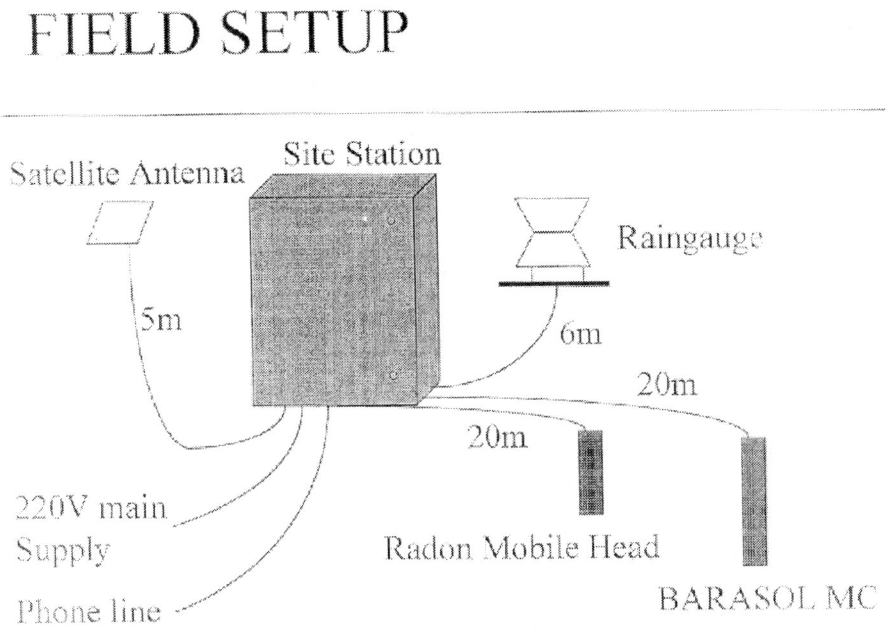

Figure 4. A sketch-diagram of the set up of the Radon station.

6. Conclusions

Throughout the one-year period of our research work, we have developed an efficient "real-time" satellite radonometric station and collected soil-gas data at one or more recognized seismically-active localities in Greece, Russia, Armenia, Albania and U.K., as a preliminary work, to establish the capability for earthquake prediction based on variations in soil gas composition.

Emphasis will be placed on the development of methodology and software suitable for the extended monitoring of gas composition in situ, supported by laboratory determinations to validate the field data where necessary.

A number of data loggers (i.e. transducers/detectors) have already been installed at specific locations (base stations) selected by the Project scientists. Each base station includes a data logger with which a number of geological parameters are being detected. The data are transmitted to a modem, which is directly attached to the INMARSAT global satellite communication system.

With the above mentioned multi-disciplinary approach we intend to establish a permanent central data acquisition-broadcasting system, of a continuous monitoring of radon, water-table level, groundwater temperature, atmospheric pressure and temperature, carbon dioxide, helium, nitrogen, etc.

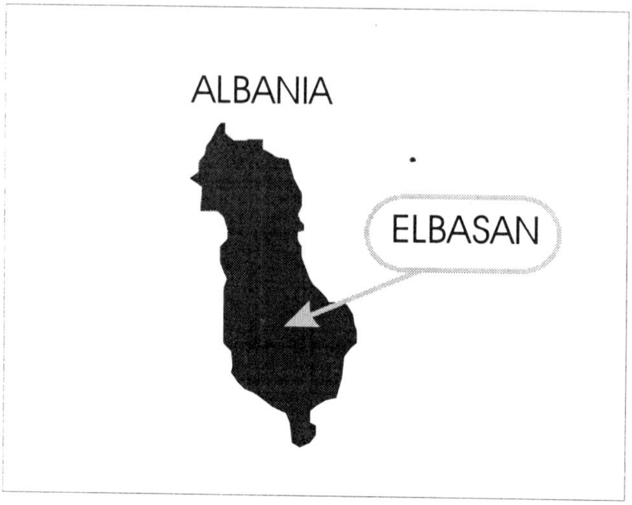

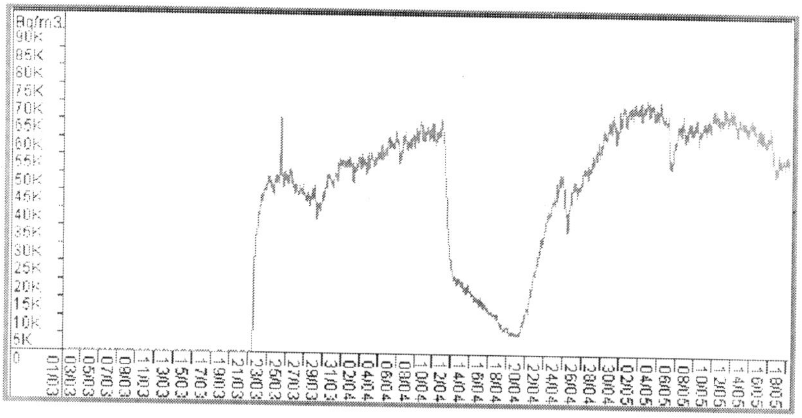

Figure 5. Preliminary Radon observations for a 2-month period from the area of Elbasan (Albania).

By establishing the stations, we achieve continuous and real-time monitoring, storage, and transmission via satellite of the field data to scientific centers for subsequent automatic processing and indication of alert thresholds.

7. Acknowledgements

This research was financed by the European Commission (Contract No IC15 CT96-0213, DG XII) within the framework of the INCO-COPERNICUS Program.

8. References

Borchielinni, S., Bernat, M., Campredon, R. (1991) Factors Controlling Radon Emissions from Sources in Regions of Accentuated Relief - The Influence of Seismicity Maritime Alps, France. *Earth & Planetary Science Letters*, **107/1**, 217-229.

Chambaudet, A., Cieur, M., Hussonois, M., Klein, D. (1991) A Portable System for the Continuous Measurement of Rn-222 in Hostile Geophysical Environments. *Nuclear Instruments & Methods* In *Physics Research Section B-Beam Interactions With Materials And Atoms*, **61,/2,** 244-250.

Cherouati, De, Chalal, Mt., Djeffal, S. (1991) Investigation of Radon Emanation along a Seismic Fault. *Nuclear Tracks and Radiation Measurements*, **19/1-4**, 299-303.

Finkelstein, M., Brenner, S., Eppelbaum, L. & Neeman, E. (1998) Identification of Anomalous Radon Concentrations due to Geodynamic Processes by Elimination of Rn Variations caused by other Factors. *Geophysical Journal International,* **133/2**, 407-412.

Friedmann, H., Aric, K., Gutdeutsch, R., King, C., Altay, C. & Sav, H. (1988) Radon Measurements for Earthquake Prediction along the North Anatolian Fault Zone - A Progress Report. *Tectonophysics*, **152/3-4,** 209-214.

Fujinawa, Y. (1992) A Note on a Relation Between the Synthetic and Individual Probabilities in the Quantitative Earthquake Prediction. *Journal of Physics of the Earth*, **40/3**, 479-486.

Khan, H. (1991) Radon - A Friend or a Foe. *Nuclear Tracks and Radiation Measurements*, **19, No.1-4**, 353-362.

King, C. (1981) Do Radon Anomalies Predict Earthquakes? *Nature,* **293, No 5830,** 262.

King, C. (1985) Radon Monitoring For Earthquake Prediction in China. *Earthquake Prediction Research*, **3, No.1**, 47-68.

Kleis, T., Enge, W., Woith, H. (1991) Plastic Detectors for Radon Monitoring in Earthquake Prediction. *Nuclear Tracks and Radiation Measurements,* **19, No.1-4,** 365-366.

Kleis, T., Enge, W., Woith, H. (1991) Radon Monitoring with Plastic Detectors in Earthquake Prediction. *Nuclear Tracks and Radiation Measurements*, **20, No.3,** 495-503.

Lenzen, M., Neugebauer, H. (1997) An Automatic Radon Sensor for Borehole Measurements. *Review of Scientific Instruments*, **68, No.7,** 2898-2903.

Nishizawa, S., Igarashi, G., Sano, Y., Shoto, E., Tasaka, S., Sasaki, Y. (1998) Radon, Cl- and SO4 Anomalies in Hot Spring Water Associated with the (1995) Earthquake Swarm off the East Coast of The Izu Peninsula, Central Japan. *Applied Geochemistry*, **13, No.1**, 89-94.

Qureshi, A., Khan, H., Jafri, E., Tufail, M. & Matiullah S. (1991) Radon Signals for Geological Explorations. *Nuclear Tracks and Radiation Measurements*, **19, No.1-4,** 383-384.

Theodorsson, P. (1996) A New Method for Automatic-Measurement of low-level Radon in Water. *Applied Radiation and Isotopes,* **47, No.9-10**, 855-859.

Virk, H. (1993) Radon and Earthquake Prediction in India - Present Status. *Nuclear Tracks and Radiation Measurements,* **22, No.1-4,** 483-494.

Virk, H. & Singh, B. (1994) Correlation of Radon Anomalies with Earthquakes in the Kangra Valley. *Nuclear Geophysics*, **6, No.2**, 293-300.

Wakita, H. (1996) Geochemical Challenge to Earthquake Prediction. *Proceedings of the National Academy of Sciences of the United States of America*, **93, No.9**, 3781-3786.

Wyss, M. (1989) Second Round of Evaluations of Proposed Earthquake Precursors. *Pure and Applied Geophysics,* **149, No.1,** 3-16.

Yasuoka, Y., Shinogi, M. (1997) Anomaly in Atmospheric Radon Concentration: A Possible Precursor of the (1995) Kobe, Japan, Earthquake. *Health Physics*, **72, No.5**, 759-761.

THE RESULTS OF THEORETICAL RESEARCHES OF THE GEODESIC-DEFORMED HERALD OF EARTHQUAKES

R.A. MOVSESSYAN
Professor
Yerevan Institute of Architecture and Construction

1. Introduction

As is generally known, the forecast of the earthquakes with probability, enough to warn the population don't realize regularly to date. It does not manage to know exactly time, location of the neared seismic shock (tremor) and to determine its magnitude. The mechanisms of appearing of the heralds and their connection with parameters of the future earthquakes are poorly studied. The physical models, describing well-known characteristics of heralds does not work out [1].

There are not mathematical correlation that permit to determine when in one or another volume of the entrails of the earth slowly proceeded processes will break with spasmodic change [2]. At the same time, in spite of difficulties mentioned above, acknowledged methodology of the forecasting works, based on complex use of set of heralds in combination with operative estimation of seismic conditions, permit to implement successful forecasts. The document [1] mention, that such kind of strategy provided 7 short-term forecasts (according to the data of UNESCO) and is realizing now in Japan, USA and CPR. The same strategy is realizing in the National Survey for Seismic Protection under the Government of RA (NSSP). The fact mentioned above, undoubtedly, raise hops of real possibilities of the decision of problem of forecast of heavy earthquakes by peeper studying of the heralds.

The number of heralds can use with a view of forecast, reach 20. The principal kinds are grouped in following succession: seismological, geodetic (deformed), hydrodynamic, electromagnetic, geochemical, biological. The heralds that become apparent by surface of the Earth outside the very seismic focus (center) reflect the process of accumulation and redistribution of deformations in various physical fields [1].

It is naturally to suppose, that preliminary forecasting of zones of origin of such deformations of various heralds more effective it determine the moment of seismic shock. Such approach is stated, in particular, in works [3, 4]. The expediency of the approach in stages to the decision of the most complicated problem of forecast is confirmed by following considerations. There is hypothesis of appearing of the heralds of earthquakes owing to griping - gradually increased motion of sides of active fault. The hypothesis is widely developing in different country [5]. Accosting to this hypothesis, during the tectonic cropping as well as during the preparation of earthquake,

S. Balassanian et al. (eds.), Earthquake Hazard and Seismic Risk Reduction, 271–279.

the level of subsoil wafers increase the gas radon is exude, various geophysical anomalies is appeared. High-exact observations record inclines of Earth surface and its general rising in zone of fault. It follows that before the complex observations of heralds it is expedient to differentiate along the active fount the sections where cropping is observed from sectors there resilient deformations of earth-crust take place, i.e. the preparation of earthquake takes place. This problem is purely geodetic, and possible ways of its solution are outlined, in particular, in works [6, 7]. The present article is dedicated to studying of mechanism of appearance of geodetic deformed herald, exposing the connections between its characteristics and parameters of future earthquake, working out the theory of herald.

2. Geodetic Precursor of Earthquakes

The geodetic herald, that become apparent as a resilient deformations of rocks in places of the preparation of earthquake is different from all others, at last, by three important features. It appears earlier than other heralds, that as mentioned, reflect in various fields accumulation and redistribution of deformations.

Geodesic herald reacts on cropping and on elastic deformations in different ways. At last, as well be shown below, for geodetic herald can be find mathematical dependence between deformed curve of herald and parameters of shock preparation of future earthquake. Taking into account the features of geodetic herald we worked out method of forecasting the zones of appearing of shocks of heavy crusting earthquakes [6, 7]. It based on hypothesis about heralds, appeared with cropping, theory of elastic rebound of Raid [8] and signs of forecast following from it, suggested by A.K.Pevnev [4], using the local geodetic buildings for determining right-angled co-ordinates of points of deformed curves.

Figure 1a shows the stripe of active fault on different sides of which are situated lithosphere blocks I and II. In order to differentiate zones of crop dislocations from zones of appearance of possible shocks of earthquakes along the stripe of fault one can use comparatively simple geodetic buildings in from of quadrangle with sides no more than 1 km. Quadrangles N1 and N2 are situated in zone of crop dislocations of seismogenic blocks I and II, moving in opposite directions at a speed of U_0 cm.year $^{-1}$. This speed is determine by means of repeated measurements of lengths of quadrangles.

Quadrangles N3, N4, N5 are situated in zone L of blocks engagement on surface S of fault /Figure 2/. The lengths of diagonals and sides of quadrangles, that situated near from surface of blocks engagement in due course will change negligibly. They can be considered invariable within the limits of exactness of linear changes. This exactness when using the precise the light range finder at distances, that don't exceed /1-1,5/ km, may be estimate by relative average quadratic error, close to $1 \cdot 10^{-6}$. It is possible that same exactness may be reached by means of using methods of satellite geodesy /GPS/ on the base of navigate systems NAVSTAR and GLONASS. Absolute errors of linear measurements by reaching pointed exactness in monitoring process must not exacted /1-2/mm, provided compulsory centralizing of devices above the tops of quadrangles.

On the sections EF, GH out of length α of zone of engagement of the blocks I and II on surface S, and also on the sections MN and M′N′, moved away from fault at the distance, exceeded some value B, take place creeping displacement of blocks. Moreover on the section EF and GH points 0,1,2,3... fastened on different blocks, dislocate during the time Δt relatively on other on the value $U_o\Delta t=2A_o\Delta t$, staying on each of the blocks in one vertical plane. The same character will have dislocations on the sections MN and M′N′, moved away from fault more than value B.

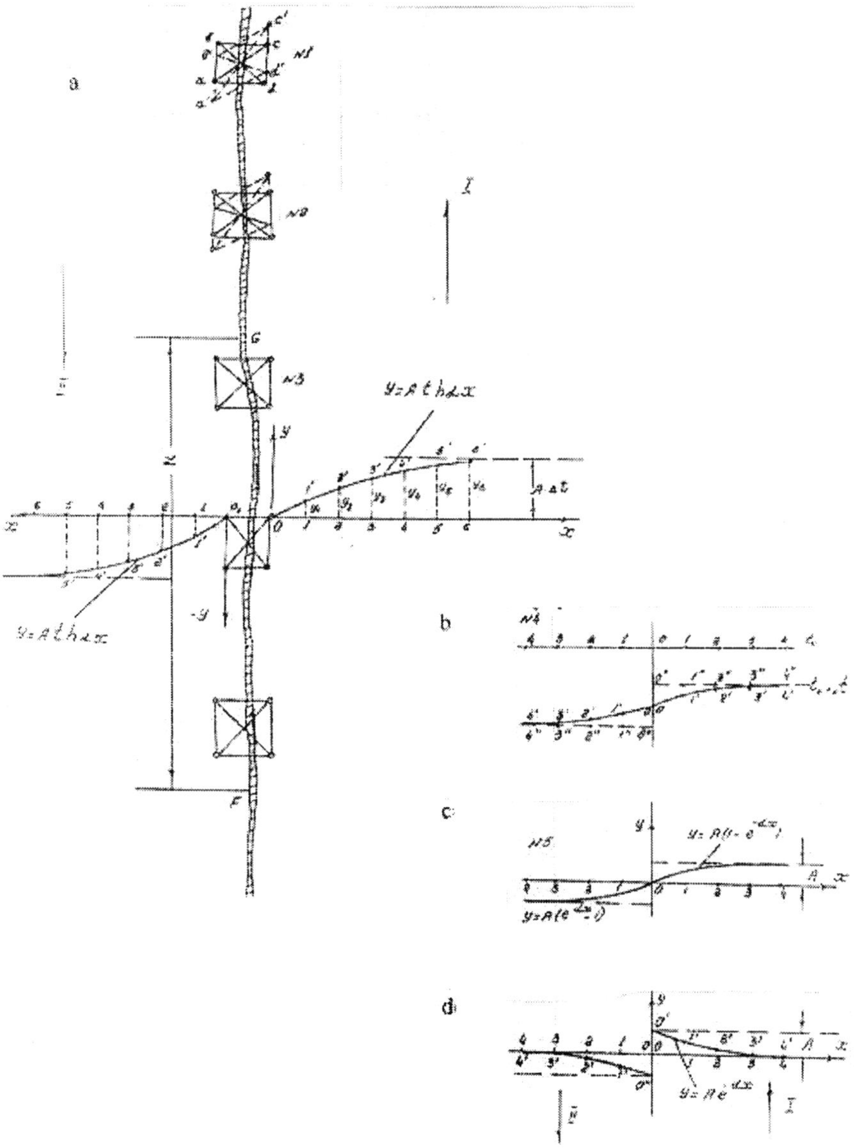

Figure 1.

On the section of engagement of the blocks FG a year later from fastening of the points 1, 2, 3... they will take the position 1′, 2′, 3′... on deformed curve. Two years later they will be on the another curve 1″, 2″, 3″..., being at the distance from break-up, exacted B, where take place creeping dislocations, points 4, 5, 6 will remove during two years on value $2A_0$. So will continue before maximum ordinate of the deformed curve will reach value $A_{max}=B4.7 \cdot 10^{-5}$, or comparative deformation of the curve will be equal $\frac{A_a}{B} \approx 4.7 \cdot 10^{-5}$. This value has got Japanese scientist Rikitake [9] at the result of the treatment of the data of many earthquakes. This average value may fluctuate in limits of $\pm(1.9 \cdot 10^{-5})$, and in individual cases may reach 10^{-4}.

When deformations will reach their limit values, strength of friction between the surfaces of engagement of the blocks already can't resist to the resilient stains, and blocks I and II spasmodically will dislocate at the direction O′, point O on block P – at the direction O″. "Resilient efficiency" and seismic shock will take place. All points on each of the blocks will turn out in one vertical plane. It will happen in that case, if points 1, 2, 3... were fastened in one range before the beginning of preparation to the earthquake, when earth-crust still doesn't strain. This case is shown on Fig.1b fore the moments of time t_o and $t_o+\Delta t$ [4, 10]. Now let's imagine, that points 1, 2, 3... fastened in earth-crust in one range in moment of time $t_o+\Delta t$, when earth-crust is already deformed and first fastened in moment t_o on one straight line the points appeared on deformed curve. If at this moment (Fig.1c) will take place earthquake, points 0, 1, 2, 3... on blocks I and II will appear on different curves-0′, 1′, 2′... on block 1 and 0″, 1′, 2′... on block P (Fig.1d). The distance O′O″ will be equal to $2A_{max}$. The curve 0′, 1′, 2′... on block 1, on which will appear geodesic point after earthquake, is near to exponential curve.

$$y = A \cdot e^{-\alpha x} \tag{1}$$

This conclusion is drawn on the base of data of many earthquakes that took place in places, where geodetic measurements was accomplished before and after the earthquake [10]. If after earthquake the position of geodetic points 0′, 1′, 2′... approximated by curve 1, then before the earthquake the equation of the first branch of deformed curve had the view

$$y = A\,(1 - e^{-\alpha x}) \tag{2}$$

and the left branch

$$y = A\,(e^{\alpha x} - 1) \tag{3}$$

according to the adopted system of coordinates.

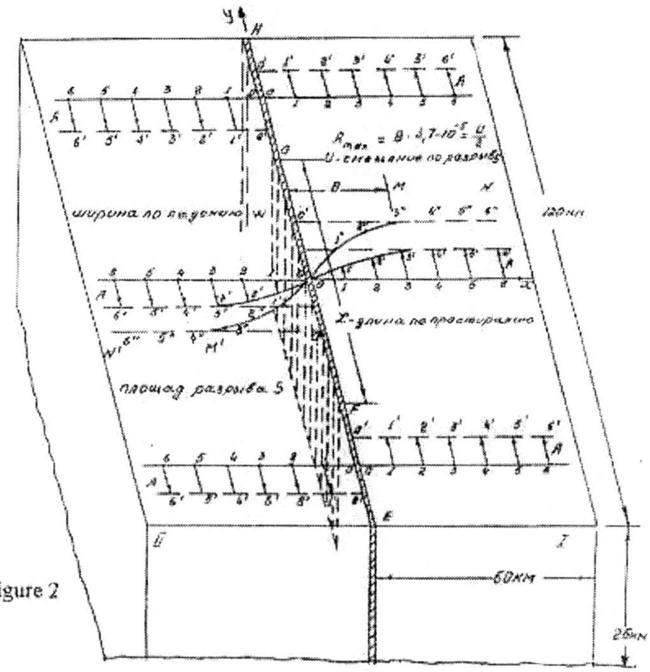

Figure 2

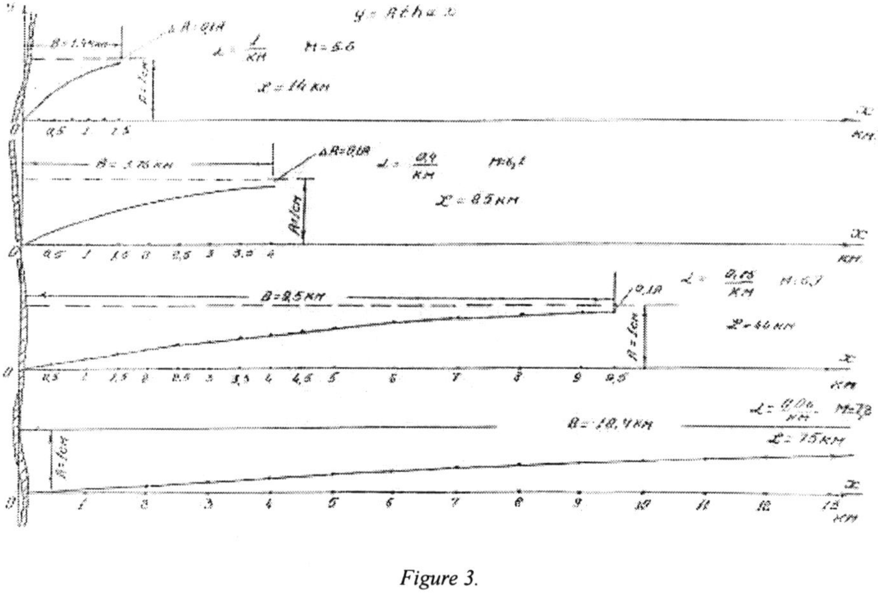

Figure 3.

3. Equation of Deformation Curve

In work (6) first were implemented calculations, in which equation of deformed curve expressed by hyperbolic tangent.

$$y = Ath\alpha x = \frac{e^{\alpha x} - e^{-\alpha x}}{e^{\alpha x} + e^{-\alpha x}} \tag{4}$$

This exponential curve, that expresses the character either right or left branch, better conform to data, received from measurements, implemented after some earthquakes. Really the character of deformed curve have to be defined more exactly at the result of the measurements of coordinates of its points in process of geodetiċ monitoring, implemented before earthquake. In equation (4) the value A equal to the half of relative dislocation of blocks during the time, passed between moments of fastening of points 1, 2, 3... in one vertical plane, and moment of measurement of coordinate of the points of deformed curve. To this value A will rush ordinates of points of curve when abscess of this points increase. The coefficient α in equation (4) characterizes the speed of increase of ordinates Y when increasing of X, its dimensions inversely to dimensions of x. The more is α, the little values of x ordinates y will near to A [6]. Figure 3 shows deformed curves, built on data, calculated for various values of α. The little is α, the farther into blocks body penetrate elastic deformations, the more will be magnitude of waited earthquake.

The value of coefficient α will depend on length α of zone of engagement of blocks, and, therefore, on value of surface of engagement S. The more will be this values, the slower will increase the deformations and the more will appear the value of B and the less α. On other side, the coefficient α will in definite degree reflect elastic-strained state of the rocks, i.e. will depend on module of medium's resilient. The more will be the resistibility of these rocks, the less will be α and the more deformation A_{max} will accumulate during the time of preparation of seismic shock. Thus, one can maintain, that there should be exist mathematical dependence between the characteristic of geodetic herald-coefficient α in the formula of deformed curve (4) and magnitude M of the prepared earthquake on the section of the engagement of two seismogenic blocks, displacing in opposite directions. As will be shown below, it is not difficult to establish a connection not only with magnitude M, but with other parameters of earthquake, too.

In order to find the function M=f(d) it is necessary to have enough data about parameters of the earthquakes, that already took place, and also theoretical preconditions to the solution of the problem. In the work of outstanding Russian scientist Y.W. Riznichenko [11] was considered the question about mutually coordinated determination of the average dependencies of the parameters of center of crust earthquake–its seismic moment M_0, power value K, magnitude M, length of surface of the break L, dislocation U on break, width W and so on (Table 1).

Seismic moment $M_0=\mu SU$ in contemporary seismology may be compared with such bearing conception as seismic energy E. M_o – is that potential work, i.e. possible energy, that would be spend on overcoming the strengths of friction on surface of fault for dislocation of its blocks at average distance U. It supposed that surfacing density of this strengths equal to the modulus of elasticity μ on displacement – $3 \cdot 10^{11}$ din/cm^2, i.e. near to theoretical density of undefecfive crystal material. Power value K presents the logarithm of energy in joules.

TABLE 1. *Dependence of earthquake's parameters on coefficient α*

Mag-nitude, M	Power analogy M logarithm of energy in joules K	Seismic moment dyn.cm M_o	Length of surface of rupture of stretch in km	Width of surface of rupture of fall in km	Value of arm's center motion in cm A=U:2	α 1/km, when $\varepsilon=4.7\cdot10^{-5}$	α, when $\varepsilon=(4.7-1.9)\cdot10^{-5}=2.8\cdot10^{-5}$	α, when $\varepsilon=(4.7+1.9)\cdot10^{-5}=6.6\cdot10^{-5}$
5	13	$0.25\cdot10^{24}$	8.3	3.6	3.5	2.69	1.60	3.77
5.6	14	$0.19\cdot10^{24}$	14	6	9.2	1.02	0.61	1.43
6.1	15	$0.15\cdot10^{26}$	25	10	24	0.39	0.23	0.55
6.7	16	$0.12\cdot10^{27}$	44	17	64	0.15	0.09	0.21
7.2	17	$0.90\cdot10^{27}$	75	28	170	0.055	0.033	0.078
7.8	18	$0.70\cdot10^{28}$	130	48	440	0.021	0.013	0.03
8.4	19	$0.54\cdot10^{29}$	230	80	1200	0.008	0.005	0.01
8.9	20	$0.42\cdot10^{30}$	410	140	3100	0.003	0.002	0.004

With m_0 are closely connected basic geometrical dimensions of the center, as dislocations. The value m_0 can be determine by not seismic data: field geological description, geodetic measurements, that sometimes have done (11). As will be shown below, the value m_0, like all other parameters of earthquake, bringing in table 1, are connected with functional dependence with coefficient α, i.e. with characteristic of deformed geodetic herald. in order to find this dependencies in the form of empiric formulas, we'll find before the dependence between coefficient α and relative deformation ε, using formula (4). By differentiation formula (4), we'll receive.

$$\frac{dy}{dx} = \frac{A\alpha}{ch^2\alpha x} = \varepsilon \qquad (5)$$

Maximum value ε_{max} when x=0 will be equal

$$\varepsilon_{max} = A\alpha \qquad (6)$$

It follows

$$\alpha = \frac{\varepsilon_{max}}{A} \qquad (7)$$

Correlation's (6) and (7) follow also from formulas (2) and (3).

As have already pointed, from data's treatment of considerable number of earthquakes was received average value of relative maximal deformation

$$\varepsilon_{max} = 4.7\cdot10^{-5} \pm 1.9\cdot10^{-5} \qquad (8)$$

By this value ε_{max} and bringing in Table 1 values U=2A of dislocations on rupture by formula (7) was calculated the value of α, bringing in column 7 of Table 1.

4. Relationships between Parameters of Earthquake and Characteristics of Deformation Curve

Analyzing received results of calculations α, it is not difficult to come to the conclusion that between them and all the parameters of earthquake, bringing in Table 1 there is steady dependencies, that express by formulas bringing below.

Dependence of the magnitude M on α express by formula

$$M = 8.9 - 0.6\frac{\lg\frac{\alpha}{0.003}}{\lg 2.6} \tag{9}$$

or

$$M=5.5-1.4\lg\alpha \tag{10}$$

By received formula are calculated magnitudes of earthquakes by values of the coefficient α, bringing in the Table 1. Similarity of the calculated data with values of magnitudes, bringing in Table 1 estimates at 0.2 units.

Dependence of the power analogy of magnitude– value K on α.

$$K = 20 - \frac{\lg\frac{\alpha}{0.003}}{\lg 2.6} \tag{11}$$

$$K= 14-2.41\lg\alpha \tag{12}$$

Dependence of seismic moment M_0 on α

$$M_0 = \frac{420000\cdot 10^{24}}{7.7^{(\frac{\lg\alpha}{0.42}+6)}} \tag{13}$$

Dependence between length of surface of rupture L and α

$$M_0 = \frac{410}{1.74^{(\frac{\lg\alpha}{0.42}+6)}} \tag{14}$$

Dependence between dislocation of rupture U and α

$$U_{cm} = \frac{3100}{2.6^{(\frac{\lg\alpha}{0.42}+6)}} \tag{15}$$

Dependence between width of rupture W at the direction, perpendicular to surface of earth and coefficient α

$$W_{km} = \frac{140}{1.7^{(\frac{\lg \alpha}{0.42}+6)}} \tag{16}$$

Thus, first is found mathematical dependence between characteristic of geodetic herald, coefficient α of deformed curve and all known parameters of earthquake. In calculations given above relative deformation $\varepsilon$was accepted equal to its average value. Naturally, in dependence on local conditions ε will vary on the average on $1.9 \cdot 10^{-5}$, the result of which also will vary the value of coefficient α. In columns 8, 9 of Table 1 are given the results are given the results of the calculations of the coefficient α, conforming to values $\varepsilon_{max}+1.9 \cdot 10^{-5}$ and $\varepsilon_{max}-1.9 \cdot 10^{-5}$

Therefore, one can say for example, that, if value of α calculated at the result of geodetic measurements will be in the interval 1.60-3.77, the parameters of earthquake will conform the data of the row 1, etc.

We can hope, that by the means of studying of geological conditions, conforming earth in zone of preparation of earthquake, it is succeeds to know exact in future the value ε_{max}, included in formula (7) for determination α. It allows exacting also the values of parameters of waiting earthquake. Nevertheless, approximate estimations are also enough for easing the seismologists the solution of the basic problem of the forecast –deformation of time of the coming of seismic shock.

5. References

1. The conception of International system of seismic researches and forecast of seismic Lender of countries participants of UDS.
2. Strakhov V.N. To the new paradigm of seismology, M. "Nature" N12, 1989, p.4-9.
3. Grigoryan S.S. About prediction of earthquakes, USSR, 1989, v306, p.1083-1087
4. Pevnev A.K. The forecast of earthquakes-geodetic aspects of problem, N12, 1988
5. Sobolev G.A. The problem of earthquakes forecast, M., "Nature", N12, 1989, p. 47-54
6. Movsessian R.A. Geodetic monitoring along active fault in order to find out possible centers of hard earthquakes and measurement of their magnitude. Lecture HAH Armenia, v96, 1996, N2-4
7. Movsessian R.A. Geodetic deformed herald of earthquakes. Lecture HAH Armenia, v98, N2
8. Reid G.F. The elastic-rebound theory of earthquakes. Bull Department Geology. Univ. Calif. Publ. 1911 v.6. N19, p.413-444
9. Rikitake T. Prediction of the earthquakes. M. 1979
10. Magnitski V. N. Internal building and physics of earth. M.1965
11. Risnichenko Y. V. The researches on physics of earthquakes. Sizes of center of crust earthquake and seismic moment. 1976, M., Science, p 9-27

SEISMIC HAZARD ZONATION IN EASTERN EUROPE

D. LUNGU[1], A.ZAICENCO[1], A.ALDEA[1], C. ARION[1] , T. CORNEA[2]
[1]*Technical University of Civil Engineering, Bucharest, Romania*
[2]Building Design, Research and Software Institute, IPCT-S.A., Bucharest, Romania

Key words: seismic hazard, Vrancea subcrustal source, macrozonation, Eastern Europe, Romania, attenuation laws, GIS, EUROCODE 8

Abstract:

The paper contents the probabilistic seismic hazard analysis (PSHA) and macrozonation to the Vrancea subcrustal source in Romania. The magnitude recurrence analysis uses modified format of Gutenberg-Richter law which takes into account the maximum credible magnitude of the source. That magnitude was estimated with Wells and Coppersmith (1994) regression equations. The severity of the ground motions was described by the peak ground acceleration *(PGA)*. The attenuation of the recorded Vrancea free field strong ground motions was determined. The GIS technology was used for seismic zonation of *PGA*. Present day hazard maps of Romania, Ukraine, Republic of Moldova and Bulgaria are compared with the new GIS map.

1. Seismicity of Vrancea Region in Romania

The Vrancea region is the source of subcrustal (60-170km) seismic activity which affects more than 2/3 of the territory of Romania, important parts of the territory of Republic of Moldova and small areas in Bulgaria and Ukraine. The Vrancea source induces a high seismic risk in the densely built zones of the South-east of Romania. On March 4, 1977, during the most severe Vrancea earthquake of this century, in the city of Bucharest more than 1570 casualties were registered, 11300 persons were injured, 35000 families lost their houses and 32 reinforced concrete multi-storey buildings completely collapsed.

Two catalogues of earthquakes occurred on the territory of Romania were compiled, more or less independently, by Radu (1974, 1980, 1995) and by Constantinescu and Marza (1980, 1995).

S. Balassanian et al. (eds.), Earthquake Hazard and Seismic Risk Reduction, 281–288.

TABLE 1: *Historical catalogue of Vrancea earthquakes ($I_0 \geq 8.0$) occurred during period 984-1900. Radu's manuscripts*[2]*, 1994.* $M = 0.56\ I_0 + 2.18$*, Radu 1974.*

Nr.	Date		Time (GMT) h:m:s	I_0 Epicentral intensity		M Gutenberg- Richter magnitude		Source[1]
				Radu	Others	Radu	Others	
1	984			8		6.7		*CM*
2	1038	15 Aug		8.5		6.9		*RT, CM*
3	1107	5 Feb	03:00	8	8/*CM*	6.7	6.2/*KS*	*R, KS*
4	1126	8 Aug	00:	7/8	8/*CM*	6.4	6.2/*KS*	*KS*
5	1170	1 Oct		8	8.5/*CM*	6.7	7.0/*KS*	*KS*
6	1196	13 Feb	07:	9	9/*CM*	(6.7)7.2	7.0/*KS*	*KS*
7	1230	10 May	07:	9^+	8.5/*CM*	(6.9)7.3	7.1/*KS*	*N, R, KS*
8	1258	7 Feb	15:	8		6.7		*RT, CM*
9	1327 ± 1			8		6.7	7.0/*KS*	*KS*
10	1446	10 Oct	04:	8	8.5/*CM*	6.7	7.3/*KS*	*KS*
11	1471	29Aug	10-11:	9+	8/9 *KS*	(6.7)7.4	7.1/*KS*	*R*
12	1516	8 Nov	12:	9	8/*KS*	(6.8)7.2	6.8/*KS*	*R, RT, KS*
13	1521			8		6.7		*RT*
14	1545	19 Jul	08-09:	8	7/*KS*	6.7	6.2/*KS*	*RT, R, KS*
15	1569	15 Aug		8	7/*KS*	6.7	6.2/*KS*	*R, RT*
16	1590	10 Aug	20:	8/9	8/*KS*	6.9	6.8/*KS*	*R, RT*
17	1595	21 Apr	11:	7	8/*CM*	5.5		*RT*
18	1601	24 Dec	15-16:	8		6.7	Main shock	*R, RT*
19	1620	8 Nov	13-14:	9	8/*KS*	(6.9)7.2	6.5/*KS*	*R, RT*
20	1628	4 Apr	01:	8		6.7		*RT*
21	1681	19 Aug	00-01:	9		(6.7)7.4	6.8/*KS*	*CM, R, RT*
22	1683			8.5		6.9		*RT*
23	1701	12 June	00-01:	7/8	8/*KS*	6.4	6.9/*KS*	*R, RT*
24	1711	11 Oct	00-01:	7	8/*KS*	6.1	6.5/*KS*	*R, RT*
25	1738	31/ May 11 June	10-11:	9+		(6.9)7.4	Main shock	*R, RT*
26	1790	06 Apr	19:29	8		6.7	6.9/*KS*	*R, RT*
27	1802	26 Oct	10:55	9+		(7.5)7.6	7.4/*KS*	*R, KS*
28	1829	26 Nov	1:40	7/8	8/9/*SKH*	6.4	6.9/*KS*	*R, KS*
29	1838	23 Jan	18:45	8	9/*CM*	6.7	6.9/*KS*	*R, KS*
30	1894	31 Aug	12:20	7	8/*KS*	6.1	6.5/*KS*	*R, KS*

[1] *R* Radu, C., 1971, 1974 catalogues
KS Kondorskaya, N.V., Shebalin, N.V., 1977 cat.
RT Radu, C., Torro, E., 1986 catalogue
N Niconov catalogue
CM Constantinescu, Marza catalogue
SKH Shebalin, N.V.,Karnic,V.,Hadzievski, D., 1974
[2] Selected and adapted by Lungu

The Radu's catalogues are more complete, even the majority of significant events are also included in the Constantinescu and Marza catalogue. The magnitude in Radu catalogue is the Gutenberg-Richter (1936) magnitude, M. The magnitude in Constantinescu & Marza catalogue was the surface magnitude, M_S. Tacitly, that magnitude was later assimilated as Gutenberg-Richter magnitude (Marza, 1995).

TABLE 2: *Catalogue of subcrustal Vrancea earthquakes ($M_{GR} \geq 6.0$) occurred during the 20th century. Radu manuscripts*[1)], *1994.*

No.	ate	ime GMT) :m:s	at. N°	ong. E°	ocus depth, km	I_o Epicentral intensity	M Gutenberg-Richter magnitude	w oment magnitude
1	1903 Sep 13	08:02:7	45.7	26.6	>60	7	6.3	
2	1908 Oct 6	21:39:8	45.7	26.5	150	8	6.8	
3	1912 May 25	18:01:7	45.7	27.2	80	7	6.0	
4	1934 Mar 29	20:06:51	45.8	26.5	90	7	6.3	
5	1940 Oct 22	06:37:00	45.8	26.4	122	7 / 8	6.5	
6	1940 Nov 10	01:39:07	45.8	26.7	140-150	9	7.4	7.7
7	1945 Sept 7	15:48:26	45.9	26.5	75	7 / 8	6.5	
8	1945 Dec 9	06:08:45	45.7	26.8	80	7	6.0	
9	1977 Mar 4	19:22:15	45.34	26.30	109	7 / 9	7.2	7.5
10	1986 Aug 30	21:28:37	45.53	26.47	133	8	7.0	7.2
11	1990 May 30	10:40:06	45.82	26.90	91	8	6.7	7.0
12	1990 May 31	00:17:49	45.83	26.89	79	7	6.1	6.4

[1)] Selected and adapted by Lungu: $M_w \cong M + 0.3$; $6.0 < M < 7.5$.

As a systematisation requirement for the *seismic hazard assessment* it is recommended the use of moment magnitude M_w. However, since the available catalogues of Vrancea events were prepared using the Gutenberg-Richter magnitude M, the recurrence-magnitude relationships were determined using that magnitude. The recurrence magnitude relationship for Vrancea source is determined from Radu's 20th century catalogue of intermediate depth magnitudes (threshold lower magnitude $M_o=6.0$). The average number per year of Vrancea earthquakes with magnitude equal to and greater than M, as resulting also from Figure 1, is (Lungu *et al.*, 1995):

$$\log n(\geq M) = 3.49 - 0.72\, M \qquad (1)$$

If the magnitude is limited by an upper bound M_{max}, the recurrence relationship can be modified in order to satisfy the property of probability distribution (Hwang *et al.* 1994) and, in the case of Vrancea source (Elnashai and Lungu, 1995):

$$n(\geq M) = e^{8.036-1.658M}\,\frac{1 - e^{-1.658\,(7.8 - M)}}{1 - e^{-1.658\,(7.8 - 6)}} \qquad (2)$$

In Eq.(2), the threshold magnitude is selected $M_0=6.0$, the maximum credible magnitude of the source is estimated $M_{max}=7.8$, and $\alpha = 3.49\,\ln 10 = 8.036$, $\beta = 0.72\,\ln 10 = 1.66$. M_0 and M_{max} calibrate the coefficients of Eq. (2).

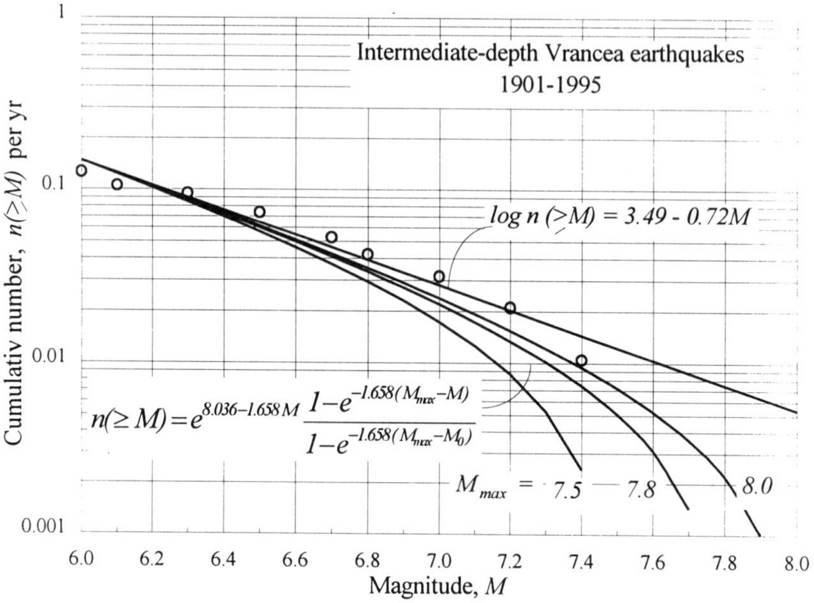

Figure 1. Magnitude recurrence relation for the subcrustal Vrancea source ($M \geq 6.0$), Lungu *et al.*, 1995.

As generally is expected, the data collected during this last century are more severe than the historical data collected over a millennium. Probably, the result is due to the inherent inaccuracies of the data from historical catalogue.

The focus depth and the position of the epicentre on the surface had a great influence on the experienced seismic intensity during the Vrancea earthquakes. The damage intensity of Vrancea earthquakes is the combined result of both magnitude and location of the focus inside the earth.

Investigating the relationship between the magnitude of a destructive earthquake ($M \geq 6.0$) and the corresponding focal depth, the following dependence was found: i.e. higher the magnitude, deeper the focus.

$$\ln h = -0.77 + 2.86 \ln M - 0.18\,P \tag{3}$$

P is a binary variable: $P=0$ for the mean relationship and $P=1.0$ for mean minus one standard deviation relationship. The mean minus one standard deviation curve in should be used in the *PSHA* as the pessimistic correlation of Vrancea magnitude with focus depth. The earthquakes with $M \leq 6.0$ display non-correlation between h and M.

The values of surface rupture area (*SRA*) and surface rupture length (*SRL*) from Wells and Coppersmith (1994) equations for "strike slip" were used to estimate maximum credible Vrancea magnitude as $M_{max} = 7.8$, Table 3. According to Romanian geologist acad. N. Sandulescu and Prof. C. Dinu for the Vrancea subduction source: $SRL \leq 150 \div 200\ km$, mean $SRA < 8000\ km^2$.

TABLE 3. *Application of Wells and Coppersmith equations to the Vrancea source (mean values).*

M	M_w	Event	Experienced	Wells & Coppersmith equations	
				$logSRA=-3.42+0.90M_w$	$logSRL=-3.55+0.74M_w$
			SRA, km^2	SRA, km^2	SRL, km
6.7	7.0	May 30,1990	1100[1)]	759	43
7.0	7.2	Aug. 30,1986	1400[1)]	1148	60
7.2	7.5	March 4, 1977	63 x 37=2331[2)]	2138	100
7.4	7.7	Nov. 10,1940	-	3236	141
7.8	8.1	Max. credible	-	7413	278

[1)]As cited in Tavera (1991)   [2)] Enescu *et al.* (1982)

The estimation of the maximum credible magnitude of the source has very important numerical consequences on the prediction of magnitude having large recurrence intervals.

TABLE 4. *Mean recurrence interval of Vrancea magnitudes, $T(\geq M)=1/\bar{n}(\geq M)$, as function of the maximum credible magnitude of the source, in years (Lungu et al., 1998).*

Data	M	M_w	From Eq. (2)	From Eq. (1)
	7.7		*475 →* *704*	114
	7.6		*323*	96
	7.4	*7.7*	135	69
Nov 10, 1940	7.3		*100→* *98*	58
March 4, 1977	7.2	*7.5*	75	50
Aug 30, 1986	7.1		*50 →* *58*	42
	7.0	*7.2*	46	36
May 30, 1990	6.7	*7.0*	24	22

2. Seismic Macrozonation In Eastern Europe Using, Gis Technology

The research devoted to the prediction of attenuation of Vrancea ground motions started in 1994. An increased accuracy in selection of the free-field accelerograms and the redigitisation (in 1997) of the accelerograms of INCERC seismic network allow the refinement of the previous predictions (Lungu *et. al*, 1995).

The database used for the present analysis of the Vrancea ground motions attenuation comprises 78 triaxial or biaxial accelerograms recorded at 49 free-field stations distributed on networks and events as in Table 5.

The free field accelerograms were obtained at the ground level or the basement of 1 - 2 storey buildings. The accelerograms recorded on instruments installed at the basement of mid-rise and tall buildings (4 ÷ 12 storeys) are not used for the attenuation analysis purposes.

TABLE 5. Distribution of the free field accelerograms used in the present attenuation analysis.

Seismic network	Romania			Republic of Moldova	Bulgaria	Total
Earthquake	INCERC[1)]	INFP[2)]	GEOTEC[3)]	IGG[4)]		
March 4, 1977	1	-	-	-	-	1
Aug. 30, 1986	22	9	3	2	-	36
May 30, 1990	22	7	3	4	5	41

[1)]INCERC, Building Research Institute, Bucharest
[2)]INFP, National Institute for Earth Physics, Bucharest-Magurele
[3)]GEOTEC, Institute for Geotechnical and Geophysical Studies, Bucharest
[4)]IGG, Institute of Geophysics and Geology, Moldavian Academy of Science, Chisinau

The following model was selected for the analysis of attenuation:

$$\ln PGA = c_1 + c_2 M_w + c_3 \ln R + c_4 h + \varepsilon \tag{4}$$

where: PGA is the peak ground acceleration at the site, M_w- the moment magnitude (for Vrancea subcrustal events $M_w \cong M+0.3$), R - the hypocentral distance to the site, h - the focal depth, c_1, c_2, c_3, c_4 - data dependent coefficients and ε - random variable with zero mean and standard deviation $\sigma_\varepsilon = \sigma_{\ln PGA}$. Taking into account the directionality of the seismic field produced by the Vrancea source, the attenuation analysis was performed on two orthogonal directions (the average direction of the rupture surface N45°E and the normal to this direction E45°S) and 3 circular sectors (of 90° each) centred on these directions: Bucharest, Moldova and Dobrogea, Table 6.

TABLE 6: *Regression coefficients for Equation 4*

	c_1	c_2	c_3	c_4	$\sigma_{\ln PGA}$
Bucharest sector	0.500	0.999	-0.471	-0.004	0.405
Complete set of data	3.598	0.971	-1.022	-0.005	0.473

Based on EUROCODE 8 and American ASCE 7-95 or UBC-1997 codes requirements, predictive GIS mapping of peak ground acceleration PGA produced by the subcrustal Vrancea earthquakes is herein illustrated for 10% probability of exceedance in 50 yr (i.e. $\overline{T} = 475\,yr$ mean recurrence interval), Figure 2. The values on the map represent $(m+1.5\sigma)$ attenuated values. That gives for Bucharest $PGA_{475yr}=0.42g$. The map containing $(m+1.0\sigma)$ values changes to $PGA_{475yr}=0.32g$ for Bucharest.

The map accounts for all available free-field ground motions recorded in Romania, Republic of Moldova and Bulgaria during the 1977, 1986 and 1990 Vrancea earthquakes. In Figure 3 is illustrated the present day seismic zonation in Romania, Ukraine and Republic of Moldova. The general pattern of the map and the content of the table describing the conversion of seismic intensity into ground acceleration are self explanatory. They clearly suggest the need for regional harmonisation of seismic macrozonation maps.

That should be achieved using probabilistic seismic hazard assessment (PSHA) procedure instead of presently used deterministic macroseismic observations procedure.

3. Conclusions

Predictive Geographic Information System (GIS) map delivered herein is consistent with ASCE 7 & EUROCODE 8 hazard format: 10% probability of exceedance in 50 yr. It represents provisions to be considered for the new Romanian code for earthquake resistance of buildings and structures.

The comparison of the maps from Figure 2 and Figure 3 proves the need for a joint zonation of the seismic hazard in the influence area of the Vrancea source.

4. Acknowledgements

The ground motion data were provided by the late Dr. C. Radu, INFP, by S.Borcia & Dr.H.Sandi, INCERC-Bucharest. We are indebted to all of them. The authors acknowledge with thanks the generous access to ArcView Spatial Analyst, ESRI Inc., provided by GEOSYSTEMS, Romania (Mr. C. Vasile).

5. References

Constantinescu, L, Marza, V., 1980, A computer-compiled and computer-oriented catalogue of Romania's earthquakes during a millennium (984-1979), *Géophysique*, Tome 24, No2, p.193-206.

Elnashai A., Lungu D., 1995. Zonation as a tool for retrofit and design of new facilities. Report of the Session A1.2 - *5th International Conference on Seismic Zonation*, Nice, France, Oct. 16-19. Proceedings Vol.3. Quest Editions, Presses Academiques, Nantes, p. 2057-2082.

Hwang H.H.M., Huo J.R., 1994. Generation of hazard-consistent fragility curves for seismic loss estimation studies. Technical Report NCEER-94-0015. National Center for Earthquake Engineering Research, State University of New York at Buffalo, Aug.

Lungu D., Cornea T., Nedelcu C., 1998. Probabilistic hazard assessment and site-dependent response for Vrancea earthquakes. In *Vrancea Earthquakes. Tectonics, Hazard and Risk Mitigation*, Kluwer Academic publishers b.v. Wenzel F., Lungu D., editors, p.251-268

Lungu D., Cornea T., Coman O., 1996 and 1995. Probabilistic hazard analysis to the Vrancea earthquakes in Romania. Part I in "Experience database of Romanian facilities subjected to the last three Vrancea earthquakes", *Research Report to the International Atomic Energy Agency*, Vienna, Contract No. 8233/EN R1 and Contract No. 8223/EN , Stevenson & Assoc. Office in Bucharest (57p+41 p. and 75 p.).

Radu C. manuscripts, 1994. Catalogues of earthquakes occurred on Romanian territory during the periods 984-1990 and 1901-1994.

Wells D.L., Coppersmith K.J., 1994. New empirical relations among magnitude, rupture length, rupture width, rupture area, and surface displacement. *Bulletin of the Seismological Society of America, Vol.84, No 4*, p. 974-1002.

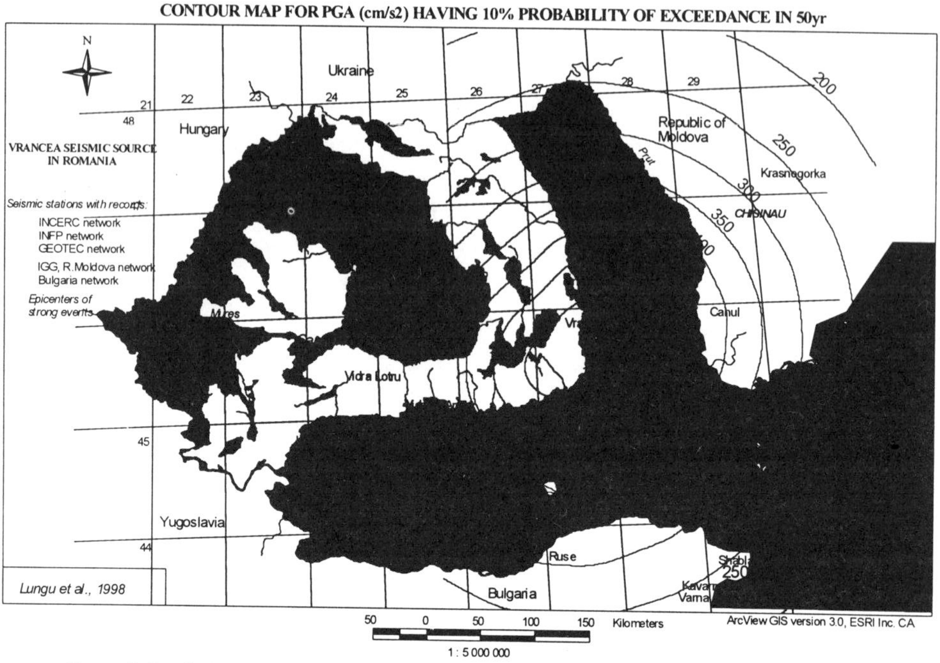

Figure 2. Predictive contour map for peak ground acceleration from subcrustal Vrancea source

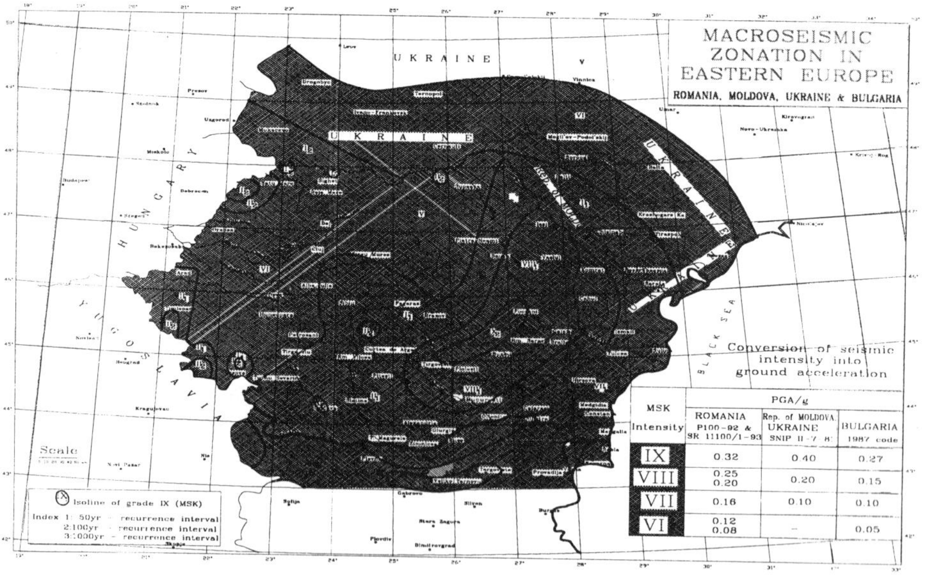

Figure 3. Seismic zonation maps for Romania, Ukraine, Republic of Moldova and Bulgaria

A NEW SEISMIC HAZARD MAP FOR THE IMPLEMENTATION IN THE NATIONAL PHYSICAL PLANNING OF IRAN

A. A. MOINFAR*, A. NADERZADEH*, E. MALEKI**
** Centre for Earthquake Studies of Tehran*
63 Padidar St., Africa Blvd., Tehran 15189, Iran
*** Mahab Ghods Consulting Engineers, Tehran, Iran*

1. Abstract

Iso-acceleration and seismic hazard maps of Iran were prepared based on the results of deterministic and probabilistic evaluation of the ground shaking hazard. The seismic source model used in this analysis was based on the results of seismotectonic and seismological studies to identify active faults and seismic source zones and to characterize them in terms of location, maximum earthquake magnitude, and earthquake recurrence. Taking into account probable ground motion attenuation effects, iso-acceleration contours of peak ground acceleration were computed for three return periods (500, 1000, and 2000 years). Finally, 1:1,000,000 scale seismic zonation map was prepared for the country. The maps are based on average site conditions.

2. Keywords

Seismic hazard; iso-acceleration lines; physical planning.

3. Introduction

During the period from 1956 through 1986, the number of cities in Iran has increased 3.3 times from 186 to 606, while during the same period, the urban population has grown 4.7 times from 6 to 28.3 million. This is to say that the urbanization has taken place both by the increase of the urban centres as well as by the growth of the existing townships. In the coming decades the country has to construct two times more new urban areas-in the form of the expansion of existing cities and/or the creation of new ones-than all that have been inherited in several thousand years of the Iranian history. The spontaneous and unplanned growth of cities has resulted in numerous disasters.

Most of the cities in Iran, are situated in the midst of prime irrigable lands. Hence, their unplanned growth, will inevitably lead to the plundering of those most valuable resources.

S. Balassanian et al. (eds.), Earthquake Hazard and Seismic Risk Reduction, 289–296.

Referring to the map of the urban centres of Iran and the earthquake hazard map of the country, one will see that the so-called "natural" growth of the urban centres has not resulted in the most rational pattern that one can imagine. On the contrary, many important cities of the country, including the capital city of Tehran with over 8 million inhabitants, are located on the high hazard zone.

Therefore, in order to achieve a wiser use of land one should deliberately take into account, as far as possible, all positive and negative factors relevant to identifying suitable locations for future urban growth. These factors are manifold. Among others are: the urgency of minimizing (and if possible eliminating) vulnerability by avoiding (or regulating/managing/proofing) high hazard zones for earthquakes, landslides, subsidence, and liquefaction.

In the case of the national physical planning in Iran, items such as the following, were studied: Topography, Slope, Landuse and Cover, Land Capability, Priority of Existing Cities, Distance to Major Highways, Decentralization, Access to Water Resources, Climatic Comfort, Job Opportunities, Other infrastructures, and Earthquakes.

The objective of this paper is, however, limited to presentation of the methodology used for preparation of the seismic hazard map part of the National Physical Planning Project of Iran.

In this study seismotectonic and seismological data were integrated into a probabilistic seismic hazard model for Iran located between 28-40°N and 44-64°E (about 2,000,000 square kilometers). The seismotectonic investigation provided information on the regional stress pattern that produces faulting, and the likely location of future large earthquakes. Seismological data provided information on the location and frequency of occurrence of earthquakes during the past several hundred years. The results of both investigations were analyzed and combined to produce a seismic source model for the region that describes the potential locations of future earthquakes within an area extending approximately 100 km beyond the limits of the study region. Because of earthquake attenuation effects, it is unlikely that sources more than 100 km away would have a significant effect on the probability of strong ground shaking in the study region. Frequency/magnitude relations and maximum earthquake magnitudes were estimated for each seismic source. Earthquake ground motion data provided information on the variation of ground motion levels with earthquake size and source- to- site distance (attenuation characteristics).

4. Characterization of Earthquake Sources

Seismotectonic and seismological investigations were carried out to identify the faults in the region that maybe potential earthquake sources; delineate source areas where smaller magnitude earthquakes that can not be associated with specific active faults are most likely to occur; assess the potential maximum magnitudes of earthquakes that can occur on the various seismic sources; and assess how frequently earthquakes occur (recurrence intervals).

Two types of earthquake sources were identified: faults or line sources, which may be sources of future earthquakes; and area sources, which are characterized by uniform levels of random background seismicity that can not be attributed to identified faults. The faults are located in parts of nine known seismotectonic provinces of Iran. 568 line sources were identified and selected in various regions of the country (Ghorashi and Barzegar, 1993). The fundamental seismogenic line sources lying along faults were classified into four categories as follows:

1. Quaternary faults with previous seismic activities
2. Quaternary faults
3. faults with previous seismic activities
4. Old faults

Many identified faults, which have been the sources of several past earthquakes, are considered critical potential for future major earthquakes. Tehran, the capital of Iran and Tabriz, the capital of Azarbaijan province, are located in the vicinity of such faults. For example, the length of the fault from Bostan Abad (southeast of Tabriz) to Sufian (northwest of Tabriz) is over 90 km. But it seems that its extension continues to Ali Khalaj in southeast and to southwest of Marand in northwest, in which case, its length will reach to 150 km. This fault has been the causative fault of many historical earthquakes in the region such as the 858 M_S 6.0; 1042, M_S 7.6; 1721, M_S 7.7; and 1960, M_b 5.1.

Not all the earthquakes that have occurred in the region can be assigned with mapped faults. Random earthquakes occur in areas away from the mapped faults. Although these earthquakes tend to have smaller magnitudes than those assigned to mapped faults, they can cause damage if they occur close to an urban area. Therefore, it is important to include the potential for random earthquakes in the seismic source model. These events are more likely to occur in the highly deformed areas than they are in the less severely deformed areas. Nine area sources have been selected based on the nine seismotectonic provinces mentioned above. Comparison of the geologically defined provinces with the pattern of historical seismicity shows a good correlation between the rate of seismicity and the relative degree of post-Neogene deformation. The nine area sources considered in this study are Azarbaijan, Alborz, Central Iran, Kopeh Dagh, Binalud, Western Zagros, Eastern Zagros, Lut, and Makran.

5. Earthquake Recurrence

Probabilistic seismic hazard analysis requires the specification of the recurrence or frequency of occurrence of earthquakes of various magnitudes-each seismic source requires its own recurrence relationship. Typically, the recurrence curve (also called a frequency-magnitude curve) has the general form described by Gutenberg and Richter (1954):

$$\text{Log } N = a - b\,M$$

Where N is the number of earthquakes of a given magnitude M or larger per unit time, a is the log of the number of earthquakes of magnitude zero or greater, and b is the slope of the curve thus characterizing the relative proportion of large earthquakes to small earthquakes. This general form which says that magnitudes are exponentially distributed, was originally developed based on observations of seismicity over large regions, including global catalogues. Variation in this magnitude distribution may be appropriate for individual source.

Determination of seismicity parameter β (the slope of the above curve on a semi-logarithmic paper) which represents the frequency of occurrence of earthquakes in an area, is very important in seismic hazard analysis. The well known empirical method of Gutenberg-Richter can be used to determin β if the earthquake data bank is complete and the level of errors in estimation of earthquake magnitudes are identical. Hence, this method can only be applied for the earthquakes of the last three decades in the study region; and this time duration is quite insufficient to estimate the seismicity parameter. Therefore, it was decided to apply a method which can enable us to use even pre-twentieth century earthquakes in this analysis. The maximum likelihood estimation for fitting the seismicity parameter β presented by Kijko and Sellevol (1990) has a good applicability correlation with the Iranian earthquake data. This is due to the fact that earthquake magnitudes have always been reported with some uncertainty.

Applying the Kijko and Sellevol method, the earthquake hazard parameters, as shown in Table 1, were estimated for the nine seismotectonic provinces of the region.

TABLE 1. *Estimated earthquake hazard parameters for the nine seismotectonic provinces.*

	β	Threshold magnitude (M_S)	Annual rate
Azarbaijan	1.69	4.2	1.85
Alborz	1.45	4.2	0.78
Central Iran	1.59	4.2	0.58
Kopehdagh	1.25	4.1	0.30
Binalud	1.60	4.1	1.18
Western Zagros	1.96	4.2	3.50
Eastern Zagros	1.71	4.5	1.47
Lut	1.48	4.5	1.16
Makran	1.03	4.5	0.22

6. Attenuation Relationships

The available data on ground motion characteristics in the region are limited' consisting of shaking intensity assessments for several historical earthquakes and some not very strong motion recording from recent earthquakes. Therefore, in this study we relied on analogy with other parts of the world having strong motion data and relatively well-constrained attenuation relationships and compared the predicted ground motions from these relationships to the limited regional data in order to select appropriate relations for Iran.

The seismotectonic regime of the region is similar to California in termsofgross crustal structure and types of earthquakes (shallow crustal events). This suggested that attenuation relationships based largely on California strong motion data would be appropriate for this region. The reasonability of using California attenuation relationships had been further assessed by comparing such relationships with ground motion data from some Iranian earthquakes with similar tectonic settings. However, one should realize that the observed intensity of attenuation for Iranian earthquakes are more compative to that of California earthquakes. Near-source ground motion predictions from the attenuation relationships were also found to be similar to approximate back-estimates of acceleration in the near-source of the study region and to the limited body of world-wide acceleration data in the near-source region for higher magnitude shallow crustale earthquakes. The reasonability of near-source predictions was particularly important because it was found in the probabilistic ground motion analysis that the hazard was dominated by nearby events.

As a result of these evaluations, the Campbell (1990) attenuation model was used in the seismic zonation study.

7. Program For Seismic Hazard Analysis

Although a variety of parameters can be used, peak ground acceleration (PGA) is usually choosen as the practical hazard indicator. Therefore, it was decided to chose PGA as the parameter to reflect the ground motion in the seismic hazard zoning maps of Iran.

In this study SHAZAP (Seismic Hazard Analysis Program) was used to calculate PGA's. This program computes the peak ground motions or response spectral ordinates for a range of specified exceedance probabilities given an assumed' seismic source model and an appropriate attenuation relationship. The method used in this analysis is a variant of the Cornell method (Cornell, 1969, 1971; McGuire, 1976, 1978) and is similar to the formulation given in Bender and Perkins (1987). The algorithms used in SHAZAP are more efficient and result in much more compact program.

8. Probabilistic Ground Motion Analysis

Probabilistic ground motion analyses (seismic hazard analyses) were made for sites located throughout the region on a 0.5°N×0.5°E grid. At each site, the results of the hazard analyses were used to obtain the levels of peak ground acceleration level at different return periods (hazard curves). The return period for a selected acceleration level is equal to the inverse of the annual frequency of exceedance. The resulting acceleration values were then contoured to provide iso-acceleration maps for the region for return periods of 500, 1000, and 2000 years. The range in computed acceleration values throughout the area are 0.14 to 0.64 g for a 500-year return period; 0.19 to 0.72 g for a 1000-year return period; and 0.24 to 0.80 g for a 2000-year return period. The iso-acceleration contours generally delineate one area of higher accelerations, running along the great north Tabriz fault and the main Zagros thrust fault. Figure 1 shows the iso-accelaration lines for 500-year return period.

9. Seismic Hazard Maps

Zonation maps were prepared in 1:1,000,000 scale to provide suitable locations for urban development plans in Iran. As shown in Figure 2, six zones were defined for zonation mapping of the region. Within each zone, a similar overall severity of hazard was judged to exist, based on subjective evaluation of the effects of the earthquake hazard within a given area. The defined zones were rated as: very high hazard, high hazard, relatively high hazard, moderate hazard, relatively low hazard, and low hazard.

Very high hazard zone comprises conditions that generally pose very significant earthquake-related ground shaking hazard to ordinary buildings; low hazard zone comprises conditions that generally do not pose significant hazard; and other zones comprise conditions that are intermediate between those in very high hazard or low hazard.

Very high hazard zone lies along the great North Tabriz fault and the main Zagros thrust fault. Therefore, any expansion of the regional capital city of Tabriz toward north, east and west where the fault passes should be restricted. The zonation map is intended to apply to ordinary building construction in urban development planning. Special investigations are required for very important or critical facilities, regardless of the zonation zone in which the building is located.

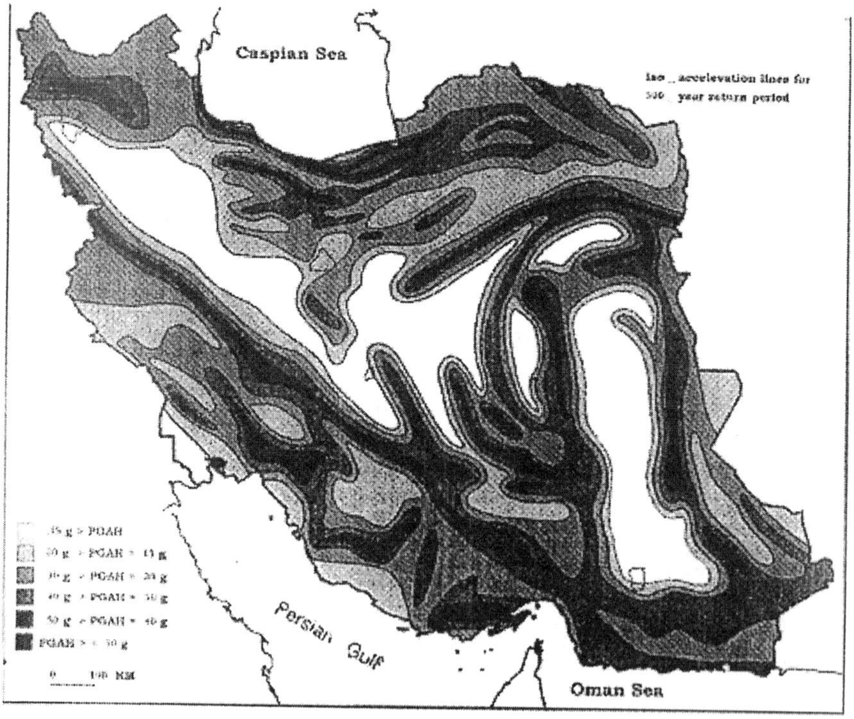

Figure 1. Iso-acceleration lines for 500-year return period for Iran

10. Conclusions

Seismic hazard study was conducted as a part of the National Physical Planning in Iran to provide the suitable locations for future urban developments. The methodology used for preparation of the most recent iso-acceleration and seismic hazard maps of the country was presented in this paper.

Using this method in future, will enable us to prepare more accurate seismic zonation maps. This will, however, require more accurate information on seismotectonic (preparation of fault maps in larger scales, and classifying them for seismic hazard and faulting) and seismicity. Then, other than giving the horizontal peak ground acceleration, suitable ground response spectra can be introduced for each zone.

11. Acknowledgment

This study was done as a part of the National Physical Planning Project of Iran sponsored by the Ministry of Housing and Urban Development. The fault map used in this study was prepared by M. Ghorashi and F. Barzegar. The authors are greatful to Dr. F. Tofigh of the National Physical Planning Project for his help.

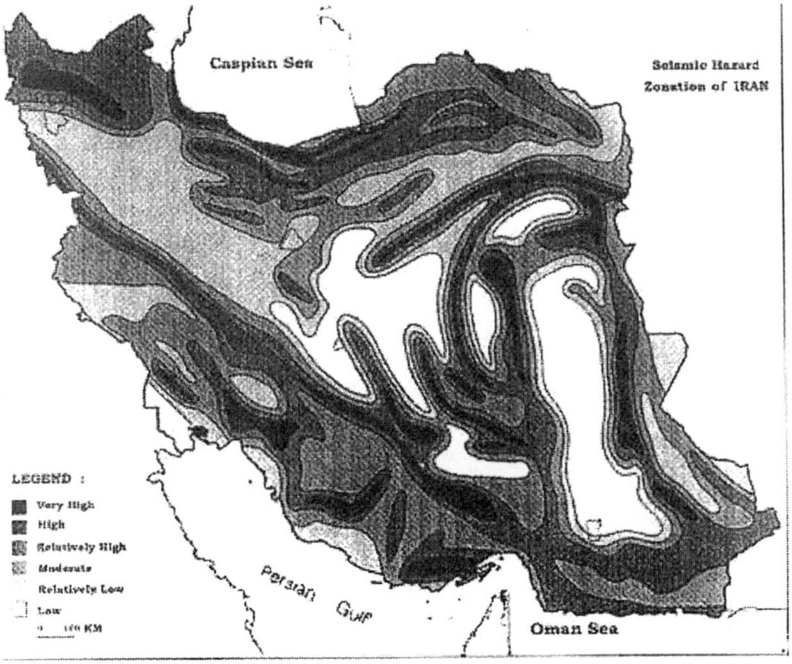

Figure 2. Seismic zonation map of Iran.

12. References

1. Ambrasys, N.N. and Melville C.P. (1982) AHistory of Persian Earthquakes. *Cambridge University Press*. 219 PP.
2. Campbell, K.V. (1990) Empirical Prediction of near source soil and soft rock ground motion for the Diablo Canyon power plant site. Report to Lawrence Livermore National Laboratory by Dames and Moore, Evergreen, Colorado.
3. Cornell, C.A. (1969) Engineering Seismic Risk Analysis, *Bull, Seism. Soc. Am.* **58**, 1583-1606.
4. ICOLD (International Commission on Large Dams) (1989) Selecting seismic parameters for large dams (Recommendations) ICOLD, *Bull.***72**.
5. Dunbar, W.S. and Charlwood R.G. (1991) Empirical suethods for the prediction of response spectra. *Earthquake Spectra*, **7**, 333-354.
6. Ghorashi, M. and Barzegar F. Seismotectonic study of the northwestern region of Iran. National Physical Planning Project of Iran. *In press.*
7. Kijko, A. and Sellevole M.A. (1990) Estimation of earthquake hazard parameters for incomplete and uncertain data files. 21th General Assembly of the European Seismological Commission, Sofia.
8. McGuire, P.K. (1976) Fortran computer program for seismic risk analysis. *US.Geol. Surv.* Open File Report **76-67**.
9. Moinfar, A.A., Mahdavian A. and Maleki E. (1994) Historical and Instrumental Earthquakes Data Collection of Iran. *Iranian Cultural Fairs Institute.*

9\. Nowroozi, A.A. (1985) Empirical relation between magnitudes and fault parameters for earthquakes in Iran. *Bull. Seism. Soc. Am.* **75,** 1327-1338.

10\. Wichert, D.H. (1980) Estimation of the earthquake recurrence parameters for unequal observation periods for different magnitudes. *Bull. Seism Soc. Am.* **70**.

TEN YEARS AFTER THE 1988 SPITAK EARTHQUAKE: RECONSTRUCTION AND NEW LINES OF EARTHQUAKE ENGINEERING DEVELOPMENT IN ARMENIA

M.G. MELKUMYAN
Earthquake Engineering Center of the National Survey for Seismic Protection under the Government of Armenia

1. Introduction

Armenia suffered a series of devastating shocks between late 1988 and the early 1990's - the earthquake, the dissolution of the Soviet Union and a trade blockade due to regional political disagreements. The December 7, 1988 Spitak earthquake struck Northern Armenia, causing extensive damage and loss of life. It has been estimated that the lives of 800,000 individuals had been disrupted as a result of the earthquake, which was considered at a Richter scale of 7.0. Between 25,000 and 30,000 people died in the earthquake and approximately 530,000 persons were left homeless, a serious blow to a country of less than 4 million people. In addition, 40 percent of the country's manufacturing capacity was destroyed just before the country achieved independence and was faced with the need to adopt market reforms and find new trade partners. Leninakan (Gumry), population 270,000, Kirovakan (Vanadazor), population 185,000, Stepanavan, population 30,000 and Spitak, population 25,000, as well as more than 80 villages were paralyzed.

According to announcement by government leaders in Moscow and Yerevan, the disaster zone should have been completely reconstructed in two years. More realistic estimates that were made as early as 1989, pointed out that this period would be ten years or longer. Ten years have elapsed already after the disastrous earthquake, but even today we still call the affected territory a "disaster zone". This name will unfortunately persist for a longer period, since only 30-35% of the collapsed have been restored up today.

2. International Efforts on Reconstruction

Efforts to recover from the earthquake were thwarted by political and economic upheaval that followed the breakup of the Soviet Union. International rescue efforts begun in December, 1988, and early 1989 were followed rapidly by reconstruction projects led by a specially created committee, which mobilized material and labor from the Soviet Republics in a concerted campaign of rebuilding. Work continued until mid -

S. Balassanian et al. (eds.), Earthquake Hazard and Seismic Risk Reduction, 297–300.

1991, but the collapse of the Soviet Union disrupted the organized reconstruction plans of the central government. The work crews were withdrawn to their newly independent home countries without fulfilling their commitments. A total of 40 countries, in addition to all the republics of the Soviet Union, participated in the reconstruction program. And yet, when 1991 came around, the volunteer builders from other countries left, because of lack of construction materials and a collapsed infrastructure. This left vast areas of partially completed apartment blocks, factories, and infrastructure projects, and even more damaged, uncleared, and unrehabilitated buildings and infrastructure. A visit to the disaster region showed an overwhelming misery, an infrastructure that was far from normal, and numerous partially built structures. The population settled as refugees in buildings that were unfit for living. The magnitude of the reconstruction effort that was needed to bring back a normal fife to the region was beyond the capacity of the local government.

In the first two years less than 10% of the residences planned for reconstruction were successfully completed. During the first year after the earthquake, one building only out of 1,195 planned for reconstruction was completed in Gumry. In Vanadzor 19 instead of the planned 366, in Spitak 12 instead of the planned 455, in Stepanavan 27 instead of the planned 145 were completed. During the second year, however, construction by volunteer units from many countries provided additional living space, as well as facilities for schools and hospitals. Some examples of the contribution by volunteer units are cited here: school for 400 pupils (United Kingdom); 90 bed hospital (France); construction materials factory (Austria); rubber processing factory (Germany); village with school (Italy); hospital (Norway); rehabilitation center (Finland); school for 480 pupils (Check Republic); 24 cottages and kindergarten (Denmark); housing building complex (USA); policlinic (Poland), etc. Those examples illustrate the volunteer efforts for reconstruction that brought hope and feeling of gratitude to the disaster region. However, they ameliorated to a limited extent the misery of the population, but the socio-economic conditions in the region remains dismal. The newly independent Republic of Armenia was unable to keep the reconstruction program on track.

3. The World Bank Project

In 1992 Armenia joined the World Bank and requested bank assistance in re-mounting the reconstruction effort. World Bank judged that International Development Association (IDA) involvement was justified to alleviate pressing economic and social problems and to provide a unique opportunity to support implementation of basic economic reforms in housing and infrastructure. It was acknowledged that the volume of reconstruction needed was well beyond the scope of a single operation. However, the project was intended to provide an opportunity for the government to reaffirm its commitment to the reconstruction effort and provide the physical basis and a coherent plan for continuing the process. Although it came some five years after the earthquake, the World Bank's Earthquake Reconstruction Project - and IDA credit - was the first significant donor activity in Armenia in the post-Soviet era.

It is well known that at the beginning, according to the decision of the Soviet leadership, all attention was concentrated on constructing new large housing districts, while no importance was placed on reconstruction of buildings directly within devastated populated areas. Within the old city limits, the plan was to reconstruct schools, hospitals and policlinics with donations from governments who had shown readiness to help. Thus, the cities of Gumry, Vanadzor and Spitak were to be relocated to new cities where new apartments were planned for construction. Villages, however, were to be reconstructed in their old locations. For the new cities, 580, 240 and 1490 hectares, respectively, were allocated. On the one hand, this decision had a positive impact. It enabled an early start to build new dwellings for the relocation of the inhabitants of the cities, while clearing the cities from the rubble of the collapsed buildings. In addition, it allowed the deployment of equipment and construction material that could not have been deployed for large-scale construction in the devastated cities. On the other hand, however, there were negative repercussions. First, the greater proportion of the new areas allocated for the new buildings were on fertile land used for agriculture. Second, the new cities required new infrastructures involving transportation lines, water and sewerage networks, and communication systems. Under the centrally controlled mass construction techniques that were well developed in the Soviet Union, this approach did not seem to present any concern to the construction agencies, but with the collapse of the Soviet Union, the responsibility to execute the plan fell on a newly emerging small republic that at best could carry the plans over decades instead of years, Third, the population was inconvenienced and depressed. They did not see dwellings being built where they had lost their ancestral homes, and they knew that the reconstruction program had slowed down considerably. The temporary houses, in which they lived, containers, tents, and shacks, had no sanitary facilities or utilities. The infrastructure of the cities was not being rebuilt with the expectation that new cities would emerge in the surrounding areas. The slams in the cities caused sanitary and hygienic hazards, and psychological problems compounded the problems. In hindsight, a continuing effort in reconstruction of the devastated area rather than building new cities would have received the approval of the population.

Taking into account the above mentioned, the Earthquake Reconstruction Project pursued three objectives: a) to provide improved housing and living conditions to residents of the earthquake zone; b) to reconstruct basic infrastructure, which will support employment creation; and c) to support a longer term sustainable program for rehabilitation in the earthquake zone. The Project comprised of four components:

1. Housing, including completion of unfinished apartment buildings; repair and strengthening of damaged buildings; and land development and construction of serviced plots and starter houses for single family ownership and self-help expansion.
2. Municipal services, including provision of selected water supply and sewerage sub-projects in Spitak, Vanadzor and Gumry; and community facilities.
3. Factory shells, including the completion of factory shells for existing profitable industries.
4. Technical Assistance, Training, Equipment and Studies.

The credit of SDR 20.1 million was disbursed locally. At appraisal, this was equal to 28.0 million USD. Disbursement followed the projected profile totaling at closing 29.75

million USD, thanks to SDR appreciation. Moreover, additional civil works valued at about 600,000 USD were achieved through the use of contractor penalties, for a total value of 30.3 million USD. The Project was implemented in about 20 cities and villages of the earthquake zone. Construction of 2,862 apartments, 10 factories, 6 schools, 2 hospitals, 11 bathhouses, a stadium, community center, kindergarten and library, as well as 40 km of water pipes has been accomplished.

4. New Lines of Earthquake Engineering Development in Armenia

The earthquake engineering approaches have changed after the Spitak earthquake principally. Construction of prefabricated multistory frame buildings in the earthquake zone was rejected. Erection of mainly monolithic up to 5-story high buildings was started in the towns, while rural areas started applying up to 2-story high stone masonry buildings in complex with reinforced concrete monolithic elements. Construction of large-panel buildings comprised a small percentage. Along with the construction of new buildings, attention has been paid to seismic protection of existing buildings that have survived Spitak earthquake. Revolutionary transformations in Armenia were pertinent just to that field.

New technologies, using seismic isolation systems for the upgrading of the earthquake resistance of existing buildings were developed in the Earthquake Engineering Center of the National Survey for Seismic Protection of Armenia, which have attracted international professional attention. These technologies are characterizing the new lines of earthquake engineering development in Armenia. They allow upgrading the earthquake resistance of existing buildings without interruption of their functioning. According to one of the technologies the earthquake resistance increases by means of erection of an additional isolated upper floor (AIUF). This technology was successfully implemented for two 9-story buildings and, in general, can be applied efficiently for the buildings, which have the first mode vibration period around 0.8-1.2sec. The AIUF represents a dynamic vibration damper and is connected to the building by means of laminated rubber bearings. It reduced stress-deformed state of the protected building in earthquakes and increases the earthquake resistance by a factor of 1.5-1.7. In order to provide reliable interaction between the building and AIUF, a rigid structure is created on the top of the building to transfer forces from AIUF to the building. The second technology is applicable for the rigid buildings with the period of vibration of up to 0.3-05 sec. This was successfully implemented for a five story building in Armenia. The idea is to cut gradually the building from its foundation, to create under the building two horizontal reinforced concrete frames, one of which is connected to the foundation and the other - to the building, and to install laminated rubber bearing between this frames simultaneously with their construction. High damping rubber bearings were used for both types of the buildings. The design of such bearings was done in collaboration with MRPRA (UK). The bearings were mainly manufactured by Min Rubber Products Sdn, Bhd., Beranang, Selangor, Malaysia. The same type of bearings were used for the construction of a new base isolated four story building in Spitak. This is the first newly constructed building in Armenia with a seismic isolation system.

THE LESSONS OF ARMENIA-SPITAK, 1988, AND OTHER RECENT EARTHQUAKES IN VIEW OF SEISMIC BUILDING CODES IMPROVEMENT

J.M.EISENBERG
Professor, Head, Earthquake Engineering Research Center, TSNIISK, Russian Government Construction Committee
4, Berezhkovsky nab., apt. 110, Moscow, 121059, Russia

1. Introduction

The recent strong earthquakes consequences have demonstrated surprisingly different seismic behaviour of different structural systems designed in accordance with Current Seismic Building Codes (SBS).

For example no sufficient damages and no harm to inhabitants were observed in prefabricated large-panel buildings during the strongest earthquakes.

On the other hand, many multi-storey reinforced concrete frame buildings have collapsed in Leninakan during Armenia earthquake, December 7, 1988, during Erzinacan-Turkey earthquake, March, 1992 and other recent earthquakes causing the Loss of dozens of thousands of human lives.

Both structural systems meet the current SBC demands.

The objective of this paper is to consider the differences of seismic behavior of different structural systems of engineered buildings and to discuss the SBC improvements.

2. Seismic behavior of different structural systems

Some common features in seismic behaviour of buildings with different bearing systems during Armenia-Spitak earthquake, 1988, Turkey-Erzinacan earthquake, 1992 and some other recent earthquakes are considered here.

2.1.ARMENIA-SPITAK, 1988, EARTHQUAKE

Seismic behaviour of buildings in Leninakan, December 7, 1992 is illustrated in Table 1, where damages of buildings with different structural bearing systems are given as well as the initial fundamental periods of the buildings.

S. Balassanian et al. (eds.), Earthquake Hazard and Seismic Risk Reduction, 301–307.

TABLE 1. *Damages of buildings with different structural systems in Leninakan*

Structural system	Number of storeys	Average fundamental periods, sec	The total number before event, 1988	Collapsed	
				Total number	Percent
Prefabricated Large-Panel	9	0,35	16	0	0%
RC frame buildings	5 to 9	0,6	65	58	89%
Masonry	4 to 5	0,3	78	48	62%
Masonry	1 to	0,1-0,2	768	264	34%

The highest ground motion amplification corresponds to the periods 0,5 to 2,5 second and it was different at different sites depending on the thickness of soft soil layer. During the earthquake after the initial cracks occurred the fundamental periods of frame buildings became longer (Table 2). That time the strength of the frame decreased as well as the stiffness.

TABLE 2. *The change of natural periods in damaged RC frame buildings in Leninakan during Armenia-88 earthquake*

	Type of structural system	Number of storeys	Fundamental period, sec	
			Natural Before earthquake	After earthquake
1	Reinforced concrete frame	9	0,6	1,8
2	RC frame plus shear walls	10	0,95	1,6
3	Lift-slab building	16	0,9	1,34

5 reasons of poor seismic behaviour of RC frame buildings.

1. High vertical loads (static as well as vertical components of seismic loads) on columns. At the ground storey level the loads on columns which have cross-section sizes 40 cm x 40 cm were in 9-storey buildings near the bearing capacity of columns. The high vertical loads led to brittle failure mechanisms and decreased the ductile properties of the columns.

2. High earthquake intensity which was unpredicted and was 2 times higher than the design intensity, it means 4 times higher forces than designed seismic forces.

3. The predominant long periods of ground acceleration due to thick layers of soft soil in Leninakan which cause amplitudes in period range 0,5-2,0 seconds.

After the initial cracking following stiffness decreasing and fundamental periods Lengthening the seismic loads became even larger because of the resonance with long period motions. The absolute horizontal displacement became very high, and due to the p-delta effects or just because of crushing the compressed concrete area and buckling of steel bars the buildings collapsed.

The same scenario was in Erzican, during the March 13, 1992, earthquake, and in Izmit area, Turkey, earthquake of August 17, 1999.

4. The relatively long duration of seismic motion and the second aftershock which occured inside the 5-minute interval after the main shock led to repeated very large horizontal amplitude swinging of the frame buildings. And this was an additional factor of the total collapses of RC frame buildings in Leninakan.

5. In many cases the bearing elements and especially the connections were performed with serious violations of technology of construction, in many cases the differences between design and actual constructed frame elements, welding in the connections of columns, beams and also vertical diaphragms were observed.

These reasons were enough for total collapses of the RC frame buildings.

5 reasons of good seismic behaviour of RC prefabricated large-panel buildings.

At the same sites, in the close neighbourhood with collapsed frame buildings prefabricated RC large-panel buildings of the same height as frame buildings behaved excellently during the December 7, 1988, event and even in the epicentral area, in Spitak, 2 five storey large-panel building have no sufficient damages and nobody of the inhabitants have been even injured in these buildings.

The principal reasons of the good seismic behaviour of large-panel buildings:

1. The vertical loads at the walls were much smaller than loads on columns of frame buildings.

2. Due to the small relative displacements between the panels during the earthquake dry friction energy dissipation occurred and it decreases the seismic loads as well as general displacements of the building.

3. The large-panel buildings have many reserve “defence lines”. If even one of several panels are damaged it does not lead to the total collapse of the building. Quite to the contrary, collapse of even one column in the ground storey could lead to the collapse of the part of a building or even to the total collapse of the building.

4. The buildings are rigid and therefore the design seismic load, in accordance with the Code Standard response spectra, was 2 times higher than the design load at the more flexible frame buildings.

5. The panels are prefabricated at factories where the quality of concrete is usually higher than at the construction site and the panels are good connected using welding of steel details in horizontal and vertical connections between the panels.

2.2. TURKEY-ERZINCAN, 1992, EARTHQUAKE

Table 3 makes an outline of the damaged residential and office buildings in numbers in Erzincan as well as in the vicinity of Erzincan /8,9/.

TABLE 3. *The number of damaged apartment buildings and offices during Erzincan earthquake, 26.III.1992*

	Area	Collapse or heavy damages		Medium damades		Weak damages	
		Apart-ment buildings	Offices ,shops	Apart-ment build-ings	Offices ,shops	Apart-ment build-ings	Offices ,shops
1	Erzincan-City	1344	825	2881	409	3832	229
2	75 villages surrounding Erzincan	1469	29	1547	24	2382	44
3	Uzumlu-town	23	0	30	0	294	13
4	25 villages surrounding Uzumlu	406	0	346	2	623	6
	Total	3242	854	4804	435	7131	292

An investigation is being carried out for having a detailed classification of the damaged structures using different indicators and parameters.

Roughly the buildings mostly damaged and collapsed during erzincan-92 earthquake could be classified in two major groups in accordance with their structural systems.

First group. To this group belong structures which are composed of rubble stone and/or adobe walls with flat and heavy earth roof. In zones where the intensity of the earthquake reached IX MSK-degree most of these buildings collapsed or they have heavy damages. For example, 25 km to the west from Erzincan in village Davarli which was crossed by visually observed rupture cracks after March 13, 1992, earthquake, 60% of the buildings collapsed. Very poor behaviour demonstrated also the unreinforced masonry walls of hollow bricks which are very popular in Turkey.

In a town Usumlu, around 20 km to the East of Erzincan, most of 3 storey buildings constructed recently collapsed generally or partly. The walls were of hollow bricks, and their thickness was 32 cm. Many buildings of the first group have been made not by specialists but by dwellers themselves and of materials available near at hand. Thus advanced and analytical approaches do not work well for estimation of the structural strength.

Second group. The second large group of heavily damaged or collapsed buildings includes RC frame buildings of different types and different quality of structural elements.

Sometimes the frames have fillings, sometimes they have not. The most popular fillings are hollow bricks. The sizes of cross-sections of the columns differed in wide range: from 20 cm to 40 cm and more. The heights of the beams were more often 55 to 70 cm.

The main response of almost total collapses of reinforced concrete buildings in Erzincan-1992 were practically the same as in Leninakan, 1988. The response spectra calculated using a strong motion accelerograms recorded in Erzincan, March 13, 1992, demonstrated the relatively long predominant periods and high displacements.

3. Some General remarks.

1. During the recent earthquakes as well as during the strong earthquakes in the past times many weak buildings with walls from adobe, mud and stones, unreinforced masonry from hollow bricks were heavily damaged or collapsed.

The society has to solve two problems related to these buildings.

One is social and economical, and it is beyond engineering profeccion. The other is an engineering problem. The problem is to develop not expensive, simple in construction and more earthquake resistant structural systems and to teach people to construct them. We will discuss this problem in more detail later here.

2. The main important conclusion common for practically all considered here earthquakes is the poor behaviour of reinforced concrete frame buildings including cases when the designes of these buildings are carried out using current Seismic Building Codes.

Reinforced concrete frame buildings collapsed in Leninakan, during the December 7, 1988, earthquake [1-5], most of reinforced concrete frame buildings collapsed also in Erzincan, during the March 13, 1992 earthquake [8, 9]. The frame buildings demonstrated poor behaviour also during the Greece earthquakes, Phillipine and other earthquakes [9].

3. The principal reasons of poor behaviour of RC frame buildings even designed with use of designed with use of current Codes were mentioned above.

4. The long periods of seismic motions occurred in most considered cases due to the soft soil amplification. It is a very usual reason because very often the cities are situated at valleys with thick alluvium soft soil layers. But some times the long period acceleration pulses were recorded at rock sites in the near-fault zone, for example, in California, during Loma Prieta, 1989, and other earthquakes [6, 7].

5. The ductility concept adopted in many Seismic Building Codes do not work in many cases. This concept is not universal and needs improvements.

The ductility concept demand is that the maximum horizontal inelastic displacements which the structure could sustain not collapsing should be 2-4, in some Codes 10 to 12 times, higher than the elastic (yield) displacements. It means the ductility factor is 2 to 4 or to 12 [10, 11].

But, if taking into account the stiffness degrading in RC frames, the actual relations between maximum horizontal displacements and yield displacements were dozen times higher, so the ductility demands should be unrealistic high.

This concept should be revised. A time-history dynamic analysis using realistic hysteretic loops affirmed in tests and using strong motion accelerograms, recorded or artificial, which take into account the uncertainty of parameters of the future earthquakes should be demanded in Codes in cases when structural systems are design sensitive to the mistakes in prediction of parameters of seismic motions and vertical loads as RC frame buildings are.

6. At sites Erzincan type, Leninakan type and others situated on thick layers of soft soils RC frame buildings should not be used especially for apartment and office buildings where thick columns and beams are undesirable from architectural point of view.

In these cases buildings with rigid high energy dissipation bearing walls are preferable.

7. There are more earthquake resistant structural systems than RC frame buildings and they have also economical advantages. These buildings are: RC prefabricated large-panel buildings, with 3-layered walls with inner from RC concrete and external layers from any available material including local materials as hollow bricks, limestone's. Hollow concrete blocks and even adobe, monolithic RC wall building.

8. The 3-laiered buildings have some advantages comparing with other types of bearing wall buildings. Some important advantages are:

* no need to construct factories for prefabricated elements as in case of large-panel buildings;

* no need in wooden or other materials for formwork during concreting as in case of RC 1-layer monolithic walls;

* the possibility of using for external layers any available material, including local materials as hollow bricks, tuff, limestones, adobe, hollow concrete blocks a.a.;

* the simplicity of construction which make it possible for use in rural areas where the majority of people were killed during the past earthquakes.

9. In case when frame systems are necessary to meet the architects demands the horizontal loads resisting shear walls use is necessary and the amount of shear walls should be established providing a comprehensive dynamic time-history analysis.

10. When the architects10. When the architects demands are to have an open ground story with columns or other thin vertical supports it is possible to use the additional rigid energy dissipating reserve elements could be provided using the methods based on low-cost seismoisolation concepts.

11. Many of the recent earthquakes demonstrated the highly important role of seismic microzonation for sound structural design. For example, the mistakes in establishing the actual soil condition in Leninakan (Armenia) was one of the causes of thousands people deaths during the 1988, Spitak event, because the design of frame structures and lift slab structures was performed using design loads for rock soils and in reality most of buildings are situated on thick soft soil covered with thin rock (tuff) layer. And the used design loads for rock soil are 2 to 4 time lower comparing with design loads for buildings on soft soil. The methods of microzonation should be improved and also attention should be paid to scrupulous carrying out the microzonation demands during design.

12. The research efforts should be continued and enlarget to decrease the many uncertainties in prediction of the future seismic motion parameters as well as to create structural systems less sensitive to these uncertainties.

4. Conclusions

1. The consequences of recent strong earthquakes in different countries (Armenia, USA - California, Greece, Turkey) were considered.
2. The recent earthquakes demonstrated poor seismic behaviour of RC frame buildings designed in accordance with the current Seismic Building Codes.
Improvements of the design concepts should be discussed revising the Codes, for example, the ductility concept.

3. For sites type Leninakan, Erzincan, situated on thick soft soils buildings are preferable with rigid and energy dissipating bearing walls instead of frame buildings.
4. 3-layered wall buildings are preferable in rural areas and in cases when some local materials are available and desirable for use. A research Program should be carried out to study optimal parameters of 3-layered wall buildings.
5. More attention should be paid to microzonation as well as to seismic motion parameters prediction and to creation structural systems less sensitive to the seismologists mistakes.

5. References

1. J.M. Eisenberg. Some Lessons of the Armenia Earthquake, December 7, 1988 (in Russian) Earthquake Engineering VNIINTPI Gosstroy USSR, vol. 2, 1991.
2. Armenia Earthquake Reconnaissance Report. Earthquake Spectra, Special Supplement, August, 1989.
3. Chalturin V.I., Shomachmadov A.M., Gedakian E.G., Sargsian N.M., MhitarianL.A. The Intensity Amplification of Seismic Ground Motion in Leninakan, (in Russian). Proc. Institute Physics of Earth, Academy of the USSR, vol. 32, "Science" editors, Moscow, (1991), 5-46.
4. Reflections on the Loma Prieta Earthquake, October 17, 1989, SEAOC. April, 1991.
5. J. Eisenberg, F. Karadogan, Short Report of the September 13, 1992. Erzincan Earthquake, Rep. ITU, 1992.
6. G.W. Housner, P.C. Jennings. Earthquake Design Criteria, EERI Monograph series, 1982.
7. R. Park, T. Paulay. Reinforced Concrete Structures. John Wiley and Sons, New York, 1975.

THE NATURE OF SPITAK EARTHQUAKE OCCURRED ON DECEMBER 7, 1988. REINFORCEMENT AND REPAIR TECHNOLOGY OF BUILDINGS AND CONSTRUCTIONS

T. G. MARKARYAN[1], S.G. SHAHINIAN[2]
[1]*Corresponding member of the Engineering Academy of Armenia*
[2]*Academician of the Engineering Academy of Armenia*

1. Introduction

"Armenia underwent and undergoes frequent and severe ground shaking" - the French geographer F. Duba de Monpere wrote nearly in 1839 after his travelling about Caucasus. Indeed, the destructive actions of disastrous and severe earthquakes are known in Armenia long ago. The 893 and 1319 earthquakes fully destroyed the capitals of ancient Armenia Dvin and Ani. The splendid architectural monument, the temple Zvartnots was collapsed during the earthquake in 973, the temple Garni was damaged resulting the disastrous earthquake in 1679. The 1926 destructive earthquake in Leninakan, where the low-rise buildings were damaged, the 1931 and 1968 earthquakes in Zangezur are also known.

On December 7,1988 at 11:41 a.m. of local time at a depth of 13-15 km near Spitak in North Armenia just another fault break - a tectonic rupture was found, where the earth was offset up. The fault spread all over 50 -kilometer area of rupture from Leninakan to Kirovakan and it came to the surface as a sloped overlap of north bort of rupture in epicentral zone. The experts assure that the slip coincides with secondary Pambak-Sevan fault which in its turn it is included into the system of biggest faults in Middle East.

According to the conclusion of State Committee of Council Ministers of the former USSR the Spitak earthquake was the biggest earthquake in North Armenia and the most severe instrumentally recorded one in Caucasus. The Spitak quake killed up to 25,000 people, the small town of Spitak was leveled by the quake, Leninakan, Kirovakan. Stepanavan and other populated areas of the republic were severely damaged. For more than 7,5 thousand dwelling houses, 140 industrial buildings were destroyed in the cities, 500 thousand houses, 83 schools, 88 kindergartens and many other objects were collapsed in rural locality.

The energy is estimated by $10^{16} - 10^{17}$ joule which corresponds to Richter magnitude M = 7.0. The seismic intensities are 10 in Spitak, 8.5-9 in Leninakan, in Kirovakan - 8 in terms of MSK-64 scale.

S. Balassanian et al. (eds.), Earthquake Hazard and Seismic Risk Reduction, 309–316.

2. The Specific Character of the Earthquake and Evaluation of its Intensity

During the earthquake the majority of recording instruments was broken down and the others had the off-scale reading of their pendulum. Nevertheless the recording was obtained by 4 points of the device SBM in Leninakan, by oscillograph SSRZ in Gukasian and seismometric station in Yerevan, therefore in most cases the earthquake intensity was evaluated by the data obtained during macro- seismic inspection of damaged buildings in accordance with the descriptive part of the scale MSK-64. For example, for more than 1600 one-storey buildings were inspected by experts of ArmNIISS and IGIS. The earthquake intensity of inspected areas had the magnitude 7,8,9 and more points. Such variety of spread is explained first by different soil conditions and by the existence of tectonic disturbance of surveyed areas.

And now few words about the features of the earthquake. The increased factor of the earthquake was the following - four minutes 20 seconds after the main shock the second quake occurred. The intensity of the second quake was near to normative. Such combination of two strong earthquakes seldom occurs and it is unfavorable factor for buildings and constructions, because after the main shock the bearing structures of buildings have been already damaged to a certain degree and their capacity to the perception of new load was either completed or reduced. So additional ground motions were observed the effect of which increased the loads for high (9 and more storey) buildings.

Some experts consider that according to the character of damage and seismogeological symptoms the Spitak earthquake was quite unusual and specific one, accounting the second severe earthquake of magnitude 5.8-6.5 that occurred in 4 min.20 sec. We will try to explain its specific features on case-study of damaged buildings and constructions where mainly 5- and 9-storey masonry and frame (including industrials buildings were destroyed, whereas the low- rise ones suffered comparatively less.

Apparently, except of high intensity of earthquake considerably exceeding (2-4 times) the design seismicity of buildings and constructions provided by Building regulations - SNIIPII-7-81, alongside with poor quality of workmanship, building materials and designs, there was the following:

- the resonance phenomenon with exceptional duration ($35 \div 40$sec.) of main shock vibration;
- the great heterogeneity of geological structure and soil conditions (engineering-geological, hydro-geological ones) of the city;
- the considerable effect of vertical component of earthquake not accounting for dwellings and civil buildings in our Codes.

The investigations of ArmNIISS experts together with the authors and other organizations showed that 9-storey buildings of series 111 with design seismicity 7-8 erected in Kirovakan where the earthquake intensity reached 8, did not suffer and their bearing structures didn't have any signs of damage, whereas_in Leninakan they were collapsed. The quality of' workmanship was hardly differed from that of Leninakan and Kirovakan. Then what is the main reason of mass damage of buildings of the series 111? The first is that in Leninakan the buildings experienced the shocks with intensity considerably exceeding their design seismicity. It is clearly seen from response spectrum. When T is equal to 0.6 sec, the actual response S_a approximately exceeds the normative one S_n twice. For the evaluation of actual horizontal load exceeding the design one, the S_a: S_n should be multiplied through by a ratio of peak actual and normative accelerations and for Leninakan it is equal to 2-4. So when T is equal to 0.6 sec., then the actual seismic load may exceed the design one 4-8 times (Eizenberg Y.M., 1989).

It should also be noted, that in the buildings of the series 111 the columns of the ground floor had sections of dimensions 40x40 cm and they were reinforced by eight rods with ∅28 mm The similar buildings are designed in Japan - the columns of 80x80 cm with 12 reinforced rods ∅28 mm. They are the constructive standards of Japanese Codes.

Secondly, the engineering analyses of damaged buildings and approximate estimates showed that the period of natural vibrations of buildings of the series 111 (it is confirmed by the measuring of Japanese experts) was fluctuated $T = 0.6 \div 0.8$ sec, and it is agreed with the period of ground shaking. According to the investigations conducted by V.I.Khalturin and V.V.Shteinberg "the thick 400-meter lens of solid clays near Leninakan have indeed the property to concentrate and sharply to increase the vibrations with the periods of 1-2 seconds". So it is quite possible the resonating of the building with seismic waves. The damage of 10-storey building with the rigid core may be explained by this fact-Further during the ground shaking with periods in the range between $0{,}5 \div 0{,}7$ sec and more and the earthquake duration of 30 sec, the buildings of this series were exposed to $30 \div 40$ cycles of vibrations entering into resonance and there is no analogy in seismology up to now. During the earthquake in Mexico in 1985 "record" number of cycling vibration reached 28. Just majority of people was killed up in 9-storey flexible frame buildings of the 111-th series.

The above mentioned is confirmed by Japanese scientists in their consolidated report. It is pointed out: "...Predominant period of ground motion was $0{,}5 \div 0{,}6$ sec, in the northern part of Leninakan and this period was close to the natural period of 9-storey frame buildings. So it was guessed that this fact might be one of the causes of serious damage to this kind of buildings. The predominant period of ground shaking in central area of the city was $0.2 \div 0.3$ sec. This period coincides with the period of low-rise ($3 \div 5$) storey stone-masonry buildings which caused serious damage. (Sh.Suyehiro and others, 1989). As to small-storey ($1 \div 2$-storey) buildings that suffered comparatively less, the long period vibrations were not dangerous for them.

The third. The high intensity of the earthquake to some degree depends on complex ground engineering-geological and hydro- geological conditions of the city (geotectonic). Decipherments of cosmic thermal (infrared) and other surveys revealed a great number of tectonic fractures in various geographical directions.

Methods of strengthening the earthquake effect for several times in connection with lake deposits are described in materials of American scientists who carried out instrumental measurements. There is a report of National Committee of USA Academy of Sciences (prof. O.Rork). The intensity of the earthquake was also increased due to significant rise of ground water level in Leninakan. This is the main cause of different intensities in various areas of the city. For example, there was no evidence of damage in the village of Azatan located in 5 km from Leninakan, whereas Akhurian located in the same distance heavily suffered. The analogues phenomenon was observed in Son-Francisco and Mexico during the 1906 and 1985 earthquakes'.

The fourth. There were also rather severe vertical motions together with horizontal ones. Their ratio reached up to 07 from horizontal one. Such phenomenon is unfavorable factor for buildings and constructions. By the evidence of experts the architectural monuments, the quality of which was beyond all manner of doubt were damaged by vertical shocks. During the main shock the doctor from city hospital 2 was uplifted with his chair and the gas-stove was turned over. The above mentioned phenomena determined the nature of Spitak earthquake.

3. Reinforcement and Repair Technology of Buildings and Constructions

For the past few years more attention is paid to reinforcement and restoration of damaged buildings in countries with high seismicity. From the point of view of building maintainability three groups of buildings should be distinguished, evaluating the degree of their damage:

1. Buildings damaged by the earthquake of the intensity which is close to design one.
2. Buildings having antiseismic equipment, but they were subjected to the seismic effect of higher intensity which caused the damage.
3. Buildings having no aseismic equipment and damaged by the earthquake.

The suggested classification allows to reveal the main causes of damaged buildings and to formulate the desired level of their restoration and reinforcement.

Following the mentioned approach and basing on post Spitak earthquake that occurred in Armenia on December 7,1988, it should be noted that majority part of damaged buildings are residential and cultural and service objects. They mainly include stone masonry dwellings of series 1A-450, 1-451 and their modifications, frame- panel buildings of series Ill, public buildings and two frame-panel buildings erected by lift-slab methods .

Returning to the adopted classification, it should be noted that on the whole the second and the third groups of buildings were damaged, because in Kirovakan and Stepanavan the earthquake intensity was somewhat higher than 8, in Leninakan - 9 and in Spitak 10 against 7,8 and 7 points respectively provided for seismic zones by Building codes. It is necessary to distinguish the main types of damaged buildings in Spitak earthquake.

3.1. IN 5-STOREY STONE-MASONRY BUILDINGS WITH ANTI-SEISMIC STRENGTHENING:

- stability loss of end walls due to unreliable connections to longitudinal walls;
- falling out of walls due to the lack of floor anchorage in bearing walls; due to the lack of reinforced concrete aseismic belts or their poor quality;
- collapsing of end sections due to their insufficient cross rigidity; horizontal displacements of slabs and disturbance of joint performance of floor disks and bearing walls due to the lack of poured-in place joints between floor slabs;
- weakening of building bearing walls by niches, openings, removal of wall parts and the erection of additional heavy walls, not provided in design and balconies supported on floors and creating additional masses;

- separation of masonry wall in longitudinal direction in some cases due to insufficient masonry strength under the effect of shearing forces.

The failure mechanism of these buildings may be as following first, the falling out of the end sections followed by the collapse of end wall, then the failure of middle sections due to loss stability of longitudinal walls having no reliable joints to floors. In the case of buildings having the "flexible" ground floor, at first, the most rigid stringer walls were collapsed, then the columns and the piers of exterior walls were damaged during the horizontal displacements.

3.2. IN 9-STOREY FRAME-PANEL BUILDINGS WITH ASEISMIC STRENGTHENING:

- the damage of stiffening diaphragm as a result of their
- weakening or poor workmanship, that caused the disturbance of joint performance of shear-walls and the building frame under horizontal loads and the collapse of the building occurred;
- failure of exterior wall suspended panels due to their excessive mass which increased the inertia loads on bearing structures of buildings;
- failure of frame joints and precast bearing reinforced concrete elements due to bad workmanship.

The failure mechanism of these buildings is the following: first, the failure of stiffening diaphragms occurred, bringing forward horizontal displacements of the upper storey up to 10 cm. At the same time the concrete was crushed in framework joints and the longitudinal reinforcement of the columns was bloated. The failure of framework joints was caused by gross violation of construction: the poor quality of concrete in columns, girders and joints, insufficient number of cross reinforcement; the poor welding of reinforcing bars; the lack of reliable joint of diaphragms to frame. Then the wall separation of wall panels and the collapse partitions occurred.

3.3. IN 9-STOREY LARGE-PANEL BUILDINGS WITH ASEISMIC STRENGTHENING

No significant damages were observed. Some deformations in joints of large-panel buildings were of usual nature in Spitak. Minor damages in large-panel buildings confirmed the justified volume-planning design with frequent arrangement of bearing walls and positive ductility of joints during seismic effects and also the more reliable and easily quality control of aseismic measure performance. Besides, they were rather rigid with the period of vibrations $0{,}3 \div 0{,}5$ sec. and they did not enter into the resonance zone of ground motion.

The analysis of structural systems and the described types of damages showed, that during the Spitak earthquake we were mainly faced with the second and in the last case with the third ground of buildings according to their maintainability.

Basing mainly on the materials of post earthquake inspection which occurred on December 7, 1988, let us state our opinions concerning to the essential and available level of building restoration.

At the first stage the restoration expediency of damaged buildings should be evaluated. For the buildings of mass purposes it should be admitted that the restoration is available, if the restoration expenditures, as a rule, do not exceed 50% of the remaining cost of the object not accounting the possible expenses for its disassembling

$$X_o \leq 0.5\ \eta\ S, \tag{1}$$

where S - is the remaining cost of the building not accounting the possible expenses of its disassembling; η - is the coefficient characterizing the objects correspondence to modern requirements of comfortability and technology and so on ($\eta \leq 1$).

The value of the coefficient 0,5 should be given for each populated area differentially in the range of deviation till 20% depending on the base of stroyindustria (building industry), the purpose of the object, precise design seismicity etc. At this stage the first serious problem is arising connected with the building maintaining, the exploitation of which was discontinued after the earthquake. The restoration of all objects immediately after the earthquake is severely increased (often 3 times) by the replacement of floors, windows, doors, built-in closets, bathroom and lavatory, electrical equipment, radio and telephone and other members, which became useless or they were lost because the object was out of service.

At the second stage two levels of building restoration should be distinguished:

- the reinforcement to the state which is close to design seismicity, often increased after the earthquake;
- the restoration to the state preceding the earthquake.

The restoration includes additional antiseismic measures, primarily not foreseen by design, bringing the building to the state satisfying the requirements of up-to-date codes of aseismic building. The restoration stipulates the elimination of damages in bearing structures, their restoration to original state. The goal of elimination of post earthquake in reference to concrete object may be formulated as follows: to ensure the spatial stiffness of the construction and its ability to resist seismic effects to the same or greater extent of that before the damage. The structural aseismic measures play an important role for building restoration and reinforcement.

Among those structural aseismic measures in the first place is the reinforced masonry, the arrangement of aseismic belts, the imposition of restriction sizes on buildings and constructions, limiting the number of floors etc.

The total expenditures of the building restoration includes the expenditures for its restoration to the state preceding the earthquake (X_o) and the expenditure value for additional earthquake resistance of buildings. As an economical criterion of the restoration. It may be assumed, that when: $X_1 > X_o$ - the building should be restored to the state proceeding the earthquake; $X_1 \leq X_o$ - the building should be restored to the state totally satisfying up-to-date requirements of seismic resistance. Of course, other evaluation techniques are quite acceptable.

It is known that the expenses for traditional systems of seismic buildings increase the cost of masonry and brick houses on the average by 4, 8, 12%, that of the large panel buildings by 3, 6, 9% and for frame ones by 5,8,11% for the seismicity of 7, 8, 9 point respectively. The earthquake inspection shows that the expenses are fully warranted.

However, after the Spitak earthquake in January 1989 the temporary scheme of seismic zoning of Armenia was adopted for the period 1989-1990 according to which 86% of its territory was referred to 9 point seismic zone having various earthquake recurrence. The seismic increasing led to the development of technical solutions for the structural reinforcement of masonry, framework and large-panel buildings admitted by Gosstroy (State Building Committee) of Armenia.

The reinforcement of buildings designed for design seismicity lower, than the subsequently corrected value, in our view, is the serious and private problem that should be solved taking into account the elements of seismic risk, reliability, the physical wear of buildings, the degree of their comfort, predicted durability and the value of their service. The engineering task of restoration or reinforcement consists in providing the desired bearing capacity of both separate members and the whole building by the effect of normative designed seismic loads.

The choice of building restoration technique depends on its purpose, structures, the degree of damage, service life, antiseismic protection measures and the frequency of earthquake recurrence in region. For the restoration of buildings damaged by Spitak earthquake, the following recommendations should be given. For masonry buildings - in case of necessity the erection of additional longitudinal and cross walls instead of partitions, through walls (Technical solutions,1989); placing of fine-grained concrete of class B12.5 with thickness of 40-50 mm on a welded mesh for spatial stiffness of floors; guniting over the welded metallic screen of damaged bearing walls and piers on both sides or the injection by cement, polymer and other mortars; strengthening of narrow piers of width less than 100 cm and stone posts by metal holders-pillars.

For frame panel buildings - the arrangement of additional metallic braces or reinforced concrete stiffening diaphragm in frame and brace directions; the strengthening of reinforced concrete columns and girders by the installation of metallic holders the strengthening of damaged stringer wallaby guniting over the metallic screen or the injection of polymer-mortar in cracks and also the installation of additional braces for the attachment of staircase elements; the arrangement of additional braces providing the attachment of exterior wall panels to bearing structures; the replacement or the strengthening of partitions and their attachment to bearing structures.

One of the most effective trends for the restoration is the use of synthetic glues mad polymer-mortars on their base having the high adhesive properties and durability prepared by TbilZNIIEP (Alexandrian and others).

4. Conclusions

The intensity of the earthquake was 2-4 times (according to the data of same authors 4-8 times higher than that of design seismicity of buildings provided by maps of seismic zoning of the republic, by SNIIP (Building Regulations), whereas in seismic codes of 1981 the intensity of Kirovakan, Spitak and other regions was reduced from 8 to 7 points.

It is necessary to begin mass construction of large-panel, cast-in-place and masonry buildings, restricting the number of their stories; widely to use metal frame, lightweight concrete structures and materials in earthquake engineering.

It is necessary to pay a special attention to engineering- geological. hydro-geological conditions of regions while making maps of seismic zoning and microseismic zoning taking into account the spectrum of earthquakes.

The designs should be based on two factors: the first - the erected buildings should resist forces provided by designs (the design level of the first order); the second - the building should not be suffered by severe earthquake that occurred once in a thousand year, i.e. safety of life should be guaranteed by design (the design level is of second order).

SOME STRUCTURAL ASPECTS OF BUILDINGS DESTROYED IN THE SPITAK EARTHQUAKE

E. LUZ
University of Stuttgart, Germany

Abstract

Paper deals with a critical evaluation of the construction scheme of residential buildings in the cities of Spitak and Leninakan, which were severely destroyed during the Spitak earthquake in 1988. Besides this a suggestion is made to protect buildings in the region which are constructed in a similar manner and which ought to be retrofitted for further earthquakes to be expected.

1. Introduction

Considering the type of the buildings destroyed in the Spitak earthquake in 1988 the question will be allowed, if the construction scheme of the up to ten stories high residential flats has influenced the disastrous collapses especially in this city. From the look on the havoc in the area in the early year 1989 the author learned to identify some elementary failures made by constructing these buildings. This look was made possible by the invitation by the former soviet society of architects and engineers, which is acknowledged in gratitude hereby.

The buildings under consideration here are 4 to 6 stories high and of masonry walls, the story floors are of prefabricated concrete plates with holes for weight reduction, and of a size of 0.8 to 4.5 meters approximately. The connection of these plates was very poor in the buildings considered in the city of Spitak. This aspect will be discussed in detail, regarding the comments by Soviet Engineers at the visit in the area. Besides this a suggestion is made for a adequate earthquake resistant construction of such plates including their implementation in the general concept of such blocks of flats. Important for such prefabricated concrete structures for the floors is a proper connection so that the single elements can act as a diaphragm or a plate and also can connect the walls.The design concept of the original buildings have covered a minimum of earthquake safety - if it would have carried out according to the plan. But as there was a lack in a proper making of the buildings and obviously a lack in supervision during erecting, disastrous collapses followed. But besides this, the original concept was not the most favourable. Therefore a more proper and earthquake resistant construction will be presented.

S. Balassanian et al. (eds.), Earthquake Hazard and Seismic Risk Reduction, 317–324.

Most of the residential flats in the Spitak and Leninakan region under consideration here which are 5 to 6 stories in height and of origin since 1950 approximately. They are constructed in a scheme where prefabricated floor plates are placed on masonry walls of tuff stones which are available in the area. The width of the flats is about 10 to 12 m. Here we have to distinguish between the design of those buildings and the actual realisation. Here first a look on the design as it was shown by soviet engineers in Moscow during a visit there and in the earthquake area in early 1989, Fig 1.

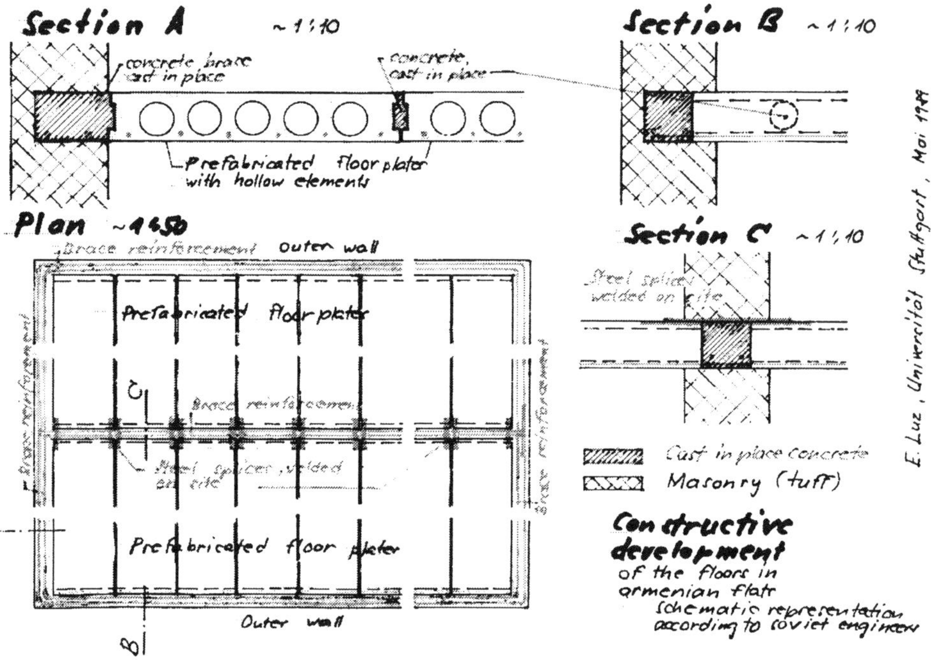

Figure 1. Construction scheme of Armenian flats, schematic representation according to the statement of soviet engineers.

These prefabricated concrete plates of approximately 0.8 to 4.5 m with holes inside for weight reduction which were placed on the supporting outer and intermediate walls from tuff stone masonry. There were designed cast in place concrete braces on the supporting walls with a small reinforcement. Also it was designed to connect the prefabricated plates over the intermediate walls by welding on site using steel lashing. The connection of the plates among each other was performed by small cast in place concrete tholes. The whole floor was to be connected by brace reinforcement over the outer walls.

This construction scheme would have avoided disastrous collapses, as the floors would have acted as diaphragms and so the floors would have not disintegrated during the quake in single plates. But nevertheless this design was not optimal, because there was only a poor connection between floors and outer walls. In such earthquake areas reinforced concrete columns within the masonry walls eventually with a frame reinforcement would have been indicated.

2. Main failures in making and inspection during erection

The main failure in performance of the buildings was the lack of the welded steel lashing between the floor plates.There were no steel lashes visible in the ruins in Spitak. This was obviously a problem of a shortage in supervision of the building site during erection. Another failure can be seen in the fact, that the floor plates are not sufficiently anchored in the concrete braces at the top of the supporting walls. So the floor plates could disintegrate from the walls. It would have been better to arrange reinforced columns in the walls, cast in place for better connection with the masonry. In one ruin was visible the reinforcement of two columns in a wall. But the concrete was fallen out there and the number of stirrups was too small. Probably there was too few cement used in the concrete. This might have been a singular view, but it seems to be a hint on systematic failures in making the buildings.

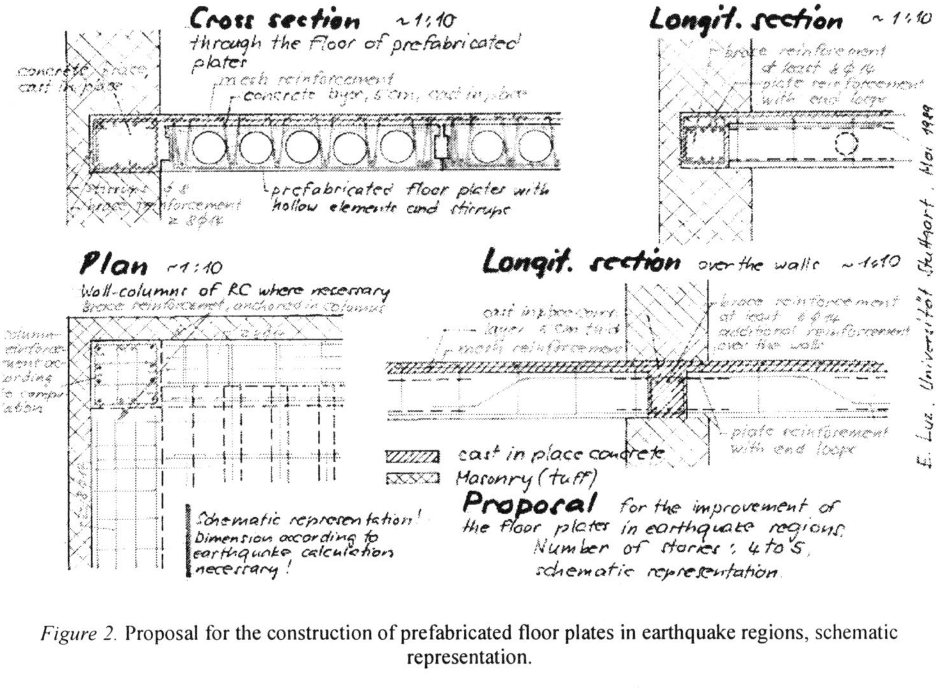

Figure 2. Proposal for the construction of prefabricated floor plates in earthquake regions, schematic representation.

In Fig. 2 a proposal is sketched for a proper construction of such floors of prefabricated plates. To receive the floor plates acting as a rigid diaphragm it is necessary to connect the prefabricated floor plates by a reinforced layer of concrete cast in place after having positioned the single elements and this together with reinforced braces on the top of the outer and partition walls. To connect prefabricated elements with braces it is necessary to have reinforcement loops coming out of these elements to be cast together with the braces. Also the concrete layer has to be connected with the plates by stirrups in the plates which are anchored in the layer which is itself reinforced by mesh reinforcement. The rigid floor diaphragm developed in this way has to be connected by wall columns at least at the corners of the building properly reinforced against the horizontal earthquake loads. The reinforced brace act together with the wall-columns and the masonry of the walls as shear walls. This proposal it's true is more complicated and more expensive than the solution chosen by the soviet engineers, but nevertheless by proper making it is earthquake safe and in this way it saves lives and money.

It should be mentioned here that the dimensions indicated in the figure 2 and the figure itself are schematic representations and should be dimensioned in actual cases according to the appearing earthquake loads.

3. Measures to improve seismic safety of existing buildings in the Yerevan region

Walking through the city of Yerevan a lot of new or elder residential buildings can be seen, which are obviously constructed in a similar manner as the collapsed ones in Spitak or in Leninakan. So the question rises if these existing buildings can be retrofitted and in which way this could be performed. Such a retrofitting would not be an easy task - neither with respect to the financial effort nor by considering the psychological problems, which could arise among the inhabitants. But in any case such measures would be the only way to protect the cities in the future.

The experience of the Spitak earthquake suggests that an improvement of existing buildings is strongly indicated as it is supposed that they are similarly poor constructed as those. To improve seismic resistance of existing buildings in the Yerevan region in the following some suggestions are made. As an example a 5 story residential building is considered and it is assumed that its construction is similar to that of Fig. 1, but without steel lashes over the partition walls.

There are two different methods of improvement for which proposals are made:

The first one is to make a corset around the building consisting of prefabricated concrete elements screwed or welded together at site. The grid made of these elements can be stretched together with the corresponding one at the opposite side of a building at story levels. The stretching elements can be arranged just under the ceilings of the flats. Such grid-elements could be also made of steel girders, but concrete elements seem to be preferable with respect to maintenance. At street level a foundation belt has to be erected, from which the grid-elements can be mounted piece by piece, Fig. 3.

The second one is to support the building against the full earthquake loads by external cast in place reinforced concrete pillars, Fig 4. As this cannot be performed by prefabricated elements - the loads are too large and the connections would become too complicated and the elements too heavy - for the lower part of the building these concrete pillars would absorb the earthquake loads completely. For the higher stories a solution like the first one - a corset similar to that of Fig. 3 - is chosen, Fig.4.

Figure 3. Prefabricated concrete elements mounted at the existing buildings piece by piece and screwed together by high tension bolts. In the center of the elements a steel bar is guided immediately under the ceiling to the corresponding element on the opposite side of the building and slightly prestressed.

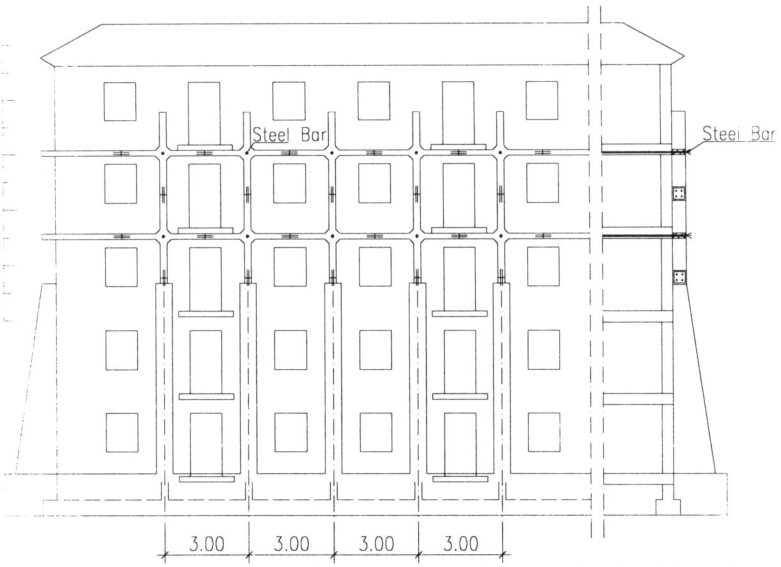

Figure 4. Concrete pillars cast in place at the lower part of the building to resist the whole earthquake load, with attached prefabricated concrete elements like in Fig. 3 in the upper stories and prestressed steel bars to connect the elements on the opposite side of the building.

The proposals presented here are two possibilities. Of course there exist different solutions according to the local conditions and the buildings under consideration.

The corset-solution consists of prefabricated concrete elements (or optionally adequate steel elements) of a grid width of approximately 3 to 3 m. The elements are made of C40/50 according to EC 2 with connected steel splices according to Fig.5.

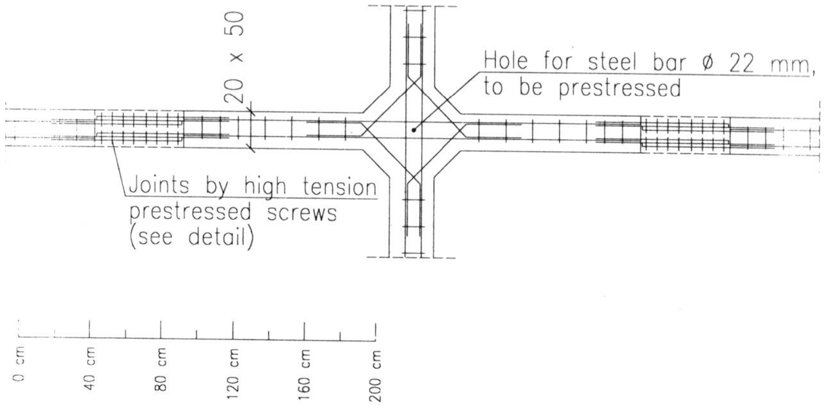

Figure 5. **Prefabricated concrete elements to build up a corset around existing buildings.**

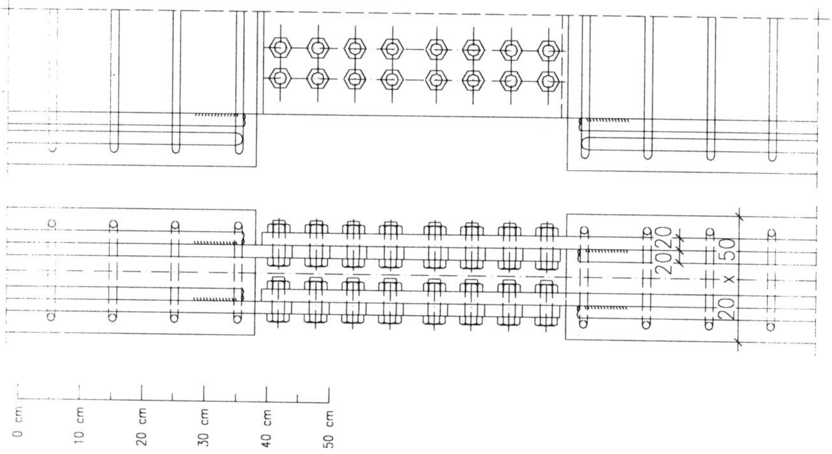

Figure 6. Connection of **the cross shaped elements by screwing together at site by high tension bolts.**

The elements are connected in place by screwing with high-strength prestressed screws, Fig 6. A connecting by welding is possible, but not recommended because this cannot easily be controlled at the building site. The elements are connected at the building by tension bars (Bst 500 S, according to EC2) directly under the ceilings connecting corresponding elements at opposite sides of the building. By the bars a slight prestressing should be applied to prevent the set of floor plates from disintegrating horizontally. By this more simple solution it is assumed that the horizontal shear forces can be carried by the masonry walls by shear resistance. The elements forming a horizontal belt around the building should have from their level branches up and down in their center. These branches could reach over the whole height of the grid and also connected in vertical direction similar to the horizontal connection.

As such a corset system is too weak to carry the whole horizontal earthquake load of the building the solution by pillars is proposed in Fig. 4. Here it is necessary, to carry the whole horizontal earthquake load of the building by external reinforced concrete pillars cast in place. The pillars are clamped at the foundation girder which can withstand the horizontal shear force as well a the overturning moment resulting from the horizontal earthquake load. It seems enough to guide these pillars up to a height of two or two and a half stories. The part of the building above this level can be protected by a corset according to the corset solution, Fig 3. It is important at this solution, that the corset part above the pillar level will be connected ring-shaped by elements described in Fig 5 and 7, to prevent the floor plates from disintegrating horizontally. At the top of the pillars therefore a connection has to be arranged for the corset elements.

As an intermediate solution not the whole corset solution according to Fig. 3 should be considered and not the heavy solution with outside pillars, Fig. 4, but a solution consisting of corset belts at story level is considered, Fig. 7.

Figure 7. Belt of prefabricated concrete elements only at story levels, connected by prestressed steel bolts with corresponding elements on the opposite site.

This solution just serves as a belt at story level to prevent the floor plates from disintegrating. The connection by steel bars and slight prestressing against the elements on the opposite site of the building seems to be enough to fix the single floor plates together as a whole plate. But for this solution it would be necessary to make full size tests to study if such a corset belt would protect the building from collapse under earthquake loads. This could be performed in a test in a sufficiently large shaking table, or even in tests at smaller scales than in natural size.

4. Concluding remarks

In the soviet code from 1970 the intensity in the Yerevan region is stated to be 8. But actually in Leninakan and in Spitak it reached intensity 10 at the quake of 1988. So the reached acceleration was about 4 times as high than expected in the code. This fact contributes to the disastrous result of 1988. So in future the code should be adapted to this fact and the intensity for the region should be raised from 8 to 10. Also the constructive details mentioned above should be noticed then.

5. References

Luz, E., Steinwachs, M., Berz, G. (1991) Bericht über das Erdbeben am 7. Dezember 1988 in Armenien. DGEB-Publikations, Vol. 4, Berlin.

SEISMIC UPGRADING OF EXISTING STONE-MASONRY BUILDINGS: LESSONS FROM THE EARTHQUAKE OF BOVEC OF 1998

M. TOMAŽEVIČ
Slovenian National Building and Civil Engineering Institute
Dimičeva 12, 1000 Ljubljana, SLOVENIA

Abstract

On the basis of extensive experimental and analytical research in seismic behaviour of existing stone and brick masonry buildings, carried out in the last decades, methods for seismic upgrading have been developed and verified. The interventions, based on the tying of the walls at floor levels, replacing the existing wooden floors with r.c. slabs, and strengthening the walls with cement-grouting, have been first used on a large scale in the Soča River Valley after the earthquakes of Friuli of 1976. When analysing the damage to buildings, caused by the earthquake of Bovec, which affected the same region in 1998, it has been found that houses, damaged in 1976, but strengthened afterwards according to recommendations issued in 1976, resisted the earthquake of 1998 without significant damage.

1. Introduction

The earthquake of April 12, 1998, which struck the area of Upper Posočje (Soča River Valley) in western Slovenia, did not represent the strongest expected earthquake in the region. The epicentral intensity of the M = 5.5 main shock was estimated to be of grade VII - VIII by the new European Macroseismic Scale (EMS). According to definitions of EMS intensity degrees, grade VII intensity earthquake may cause moderate structural damage to many and heavy structural damage to a few traditionally built buildings, whereas in the case of a grade VIII intensity earthquake, many such buildings suffer heavy damage and a few may collapse.

Although not strong by the magnitude, the earthquake of April 12, 1998, caused considerable damage to traditional buildings in a radius of about 10 km from the epicentre. In the very first days after the earthquake when the usability of buildings has been assessed by a governmental commission, 311 buildings have been found as temporarily not usable because of heavy damage to structural elements. Among them, about 1/3 have been damaged seriously, in most cases even beyond repair, whereas 2/3 suffered repairable damage (Report, 1998).

The region of Posočje has already suffered damage from a series of earthquakes with epicentres in Friuli, Italy, in 1976. At that time, after three strong earthquakes in

S. Balassanian et al. (eds.), Earthquake Hazard and Seismic Risk Reduction, 325–332.

May and September, 6175 buildings have been damaged in Posočje. Among them, 1709 had to be demolished and rebuilt, and 4476 have been repaired and/or strengthened (Ladava, 1982).

It is a rare case that the same region is affected by two relatively strong earthquakes within a short time interval of 22 years. It is even more rare that the buildings, strengthened after an earthquake, have been subjected to repeated strong ground shaking. However, it has to be mentioned that the consequences of earthquakes of 1976 were not so severe in the area which suffered in 1998. Consequently, not many buildings in the recently affected area have been severely damaged in 1976 and thoroughly strengthened by applying the technical measures recommended at that time. Nevertheless, the earthquake of 1998 provided a good opportunity for the analysis of the effectiveness of technical measures, recommended in 1976.

2. General Criteria for Seismic Upgrading

There are many criteria which should be considered when deciding upon the repair and upgrading (retrofit or rehabilitation) of a building. Besides the technical aspects, regarding the technical measures needed to improve structural deficiencies, some general criteria, related with costs of intervention and the importance of the building, availability of chosen technology and skilled workmanship, duration of the works and problems of occupancy, etc., need to be taken into serious consideration before the final decision is made. In addition to that, the principles of preservation and conservation of historical monuments should also be taken into account in the case of historical buildings.

The basic criterion for seismic upgrading is based on the results of seismic resistance verification. Although the level of design seismic loads, which should be taken into account for the redesign of existing buildings, is usually the same as in the case of a new construction, the modern seismic codes, such as Eurocode 8, Part 1-4, specify that, if the code required values of design ground acceleration immensely increase the anticipated total costs, or lead to unacceptable architectural alterations (e.g. in the case of historic monuments), the values of effective peak ground acceleration defined for a general case may be reduced for redesign purposes.

With regard to the assessment of the seismic resistance, the mechanisms of seismic behaviour and causes of damage can be determined on the basis of earthquake-damage observations and damage analysis. Mathematical models and methods should be calibrated taking into account the observed seismic behaviour and experimental data. In addition, investigations to establish the structural characteristics and material properties need to be carried out in order to evaluate the seismic resistance of existing buildings.

According to earthquake-damage, which is a result of inadequate structural layout and insufficient load-bearing capacity of the walls and/or inadequate connection between the walls, the technical measures for repair and strengthening of masonry buildings are classified into measures for providing structural integrity, such as tying the walls, and anchoring and stiffening of floors, as well for measures for strengthening of masonry wall, whenever necessary. The following technical criteria should be considered in order to achieve adequate seismic behaviour of masonry buildings:

- The walls should be adequately tied and connected. Rigid floor diaphragm action should be provided, and the floors should be well anchored into the walls to prevent out-of-plane vibration of the walls.
- The structural walls should be uniformly distributed in both orthogonal directions of the building. They should be sufficient in number and strong enough to resist the expected seismic loads.
- The foundation system must be capable of transferring the increased ultimate loads from the strengthened upper structure into the soil.

3. Buildings in Posočje and Methods for Repair and Strengthening

3.1 BUILDING TYPOLOGY

Traditional construction material in the region of Posočje is locally available lime-stone. Stone masonry walls are made of rubble stone, built in two outer layers of bigger stones, with an inner infill of smaller pieces of stone, in poor mud mortar with a little lime. Connecting stones are rare and only sometimes, stones are cut or partly cut at the corners. With the exception of some traditional houses in remote villages, which are categorised as important architectural cultural assets of the region, most buildings in towns have been rebuilt after the destruction during the World War I. It has been found that the quality of stone-masonry of larger public buildings in towns is much better than the quality of, usually smaller, residential houses. It has been also found that the quality of stone-masonry in some villages is better than the quality of masonry of houses in towns, destroyed during World War I and rebuilt afterwards.

Typically, stone-masonry houses are 2 - 3 storeys high. Floor structures and lintels are traditionally wooden, without any wall-ties provided to connect the walls. Wooden floors and lintels have been replaced with reinforced-concrete slabs in the cases where houses have been remodelled to meet the demands of the growing tourism in the region in the last decades. However, wall-ties have been installed only in the case of a small number of buildings after the earthquake of 1976. Roof structures are wooden. They are covered with ceramic tiles, sometimes laid in mortar. As a rule, the buildings are built without any foundation, and foundation walls are of lesser quality than the walls of the structure above the ground level.

Structural layout, i.e. the distribution of structural wall in plan of typical houses, is usually adequate. The distribution of walls is uniform in both orthogonal directions, and, because of the thickness of load-bearing and cross-walls, as well as relatively small rooms, the wall/floor area ratio is very large, in many cases exceeding 10 %.

A large number of voids and lack of connecting stones were the main reason of dilamination and disintegration of stone-masonry walls, whereas lack of connection between walls and anchoring of floors into the walls caused separation of the walls at vertical joints and intersection zones, as well as collapse of the walls, orthogonal to the main direction of the seismic motion.

On the basis of damage analysis, mechanisms which have caused damage can be identified and adequate measures to prevent the damage during repeated earthquakes proposed. However, any quantitative data regarding the structural resistance and effectiveness of interventions cannot be assessed unless additional experimental investigations have been carried out.

3.2 SEISMIC STRENGTHENING MEASURES

In order to ensure the integrity of the structure and common action of structural walls of stone-masonry houses during earthquakes, wooden floors are either replaced by r.c. slabs, properly anchored into supporting walls, or the walls are tied with steel ties and the existing wooden floors are anchored to the walls and/or braced with diagonal ties, especially in the case of large floor spans. Steel ties are installed symmetrically on both sides of the walls, usually just below the floor structures (Figure 1). Reinforcing steel bars, threaded at the ends so that they can be bolted on steel anchor plates at the ends of the walls, are used. As a rule, the diameter and number of bars are determined by calculation. In many cases, however, the ties are designed on the basis of experience.

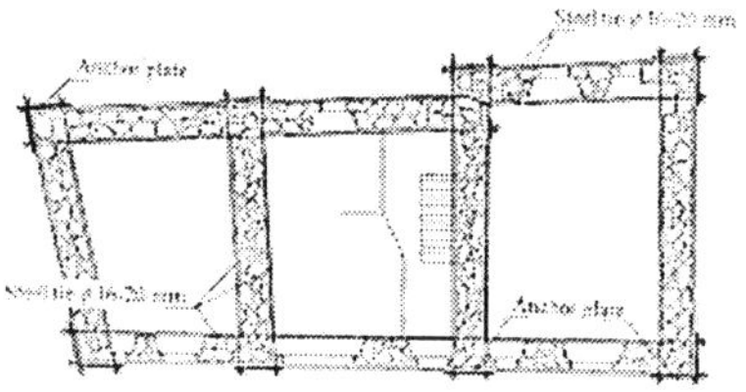

Figure 1. Distribution of steel ties in plan of a typical stone-masonry house

The basic information regarding the effectiveness of tying the walls with steel ties has been obtained already before the earthquakes of 1976 (Boštjančič et al., 1976). The effectiveness of tying has been also confirmed by recent experimental studies (Tomaževič et al., 1993, 1996). Freely supported wooden floors of buildings without ties at floor levels do not prevent the separation of the walls. Because of the separation of the walls, the upper stories of the building may partially collapse. By the tying of the walls with steel ties, however, the separation is prevented and the integrity of the structure retained until its final collapse (Figure 2). By retaining the integrity, the seismic resistance and energy dissipation capacity of the structure are significantly improved.

Figure 2. Mechanism of collapse of stone masonry model without (a) and with steel ties at floor levels (b)

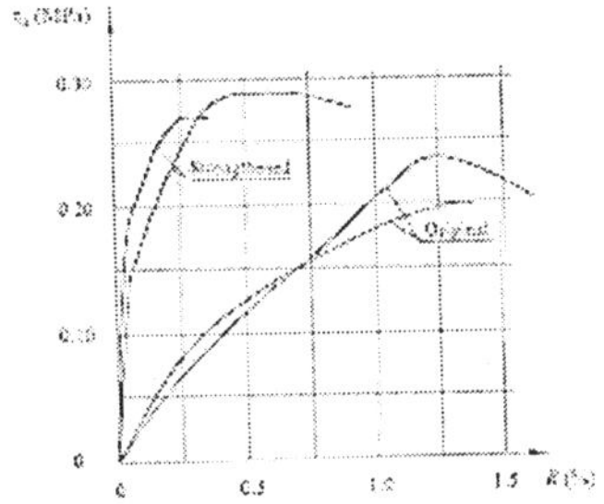

Figure 3. Effect of cement-grouting of mixed stone-and-brick-masonry walls

In the case of stone- and mixed, stone-and-brick-masonry walls, the systematic filling the voids with injected cementitious grout proved to be the most efficient method of strengthening. The simple idea that, after hardening, the injected grout will bond the loose parts of the wall together into a solid structure, has been followed. By preventing the dilamination and disintegration of the stone-masonry during the earthquake, the monolithic action of the wall is ensured and, hence, the lateral resistance significantly improved.

At the time of the Friuli earthquake of 1976, indicative laboratory tests to determine the mechanical characteristics of existing and cement-grouted stone-masonry walls have been already carried out (Turnšek et al., 1978). New data have been obtained in the years which followed, by laboratory and in-situ testing of stone-masonry walls (Tomaževič and Sheppard, 1982). Typical diagrams, showing the relationship between the average shear stress in the wall and storey rotation angle before and after cement-grouting, are shown in Figure 3.

As can be seen, by cement-grouting, the tensile strength of the stone-masonry, which determines the shear resistance of walls, is significantly increased. It can be also seen, that the values of shear modulus, which determines the stiffness of the walls in the particular case of stone-masonry, is also increased. This should taken into account in the redesign.

4. Lessons from the Earthquake of Bovec of April 12, 1998

The effectiveness of strengthening measures, adopted in 1976, has been confirmed already at that time. Namely, a building, located in Lusevera (Bardo), Italy, in the epicentral region of Friuli earthquakes, which has been severely damaged in May of 1976, and strengthened in summer, did not suffer damage during the subsequent earthquakes of the same intensity in September (Tomaževič, 1989).

Figure 4. Adequately strengthened buildings survived the earthquake of Bovec of April 12, 1998, without any damage

The effectiveness of these measures has also been verified during the recent earthquake of Bovec of April 12, 1998. The observations have clearly shown that all buildings in the region of Posočje, which have been thoroughly strengthened according to recommendations, passed in 1976, resisted the earthquake without any damage (Figure 4). In the cases, however, where these recommendations have been only partly implemented, the houses behaved as if they have not been strengthened at all.

Most frequently, only wooden floors have been replaced with r.c. slabs and the walls have not been strengthened by cement-grouting. The slabs have been supported by the inner wythe of the stone-masonry wall, and have not been properly anchored and connected with the outer wythe. In such a case, the rigid slabs sheared the outer wythe of the wall during the earthquake and, frequently, pushed the outer wythe outwards. This caused the dilamination of the wall and falling out of the outer wythe, especially at corner zones. In the rare cases, where the slabs have been adequately anchored into the walls, the resistance of existing stone-masonry walls was not adequate, what resulted into the development of shear cracks in the walls in all stories (Figure 5).

Figure 5. Typical diagonal cracks indicate that the available resistance of walls has been utilised after replacing wooden floors with adequately anchored r.c. slabs

Even in the case where stone-masonry walls have been only partly grouted, the behaviour was not adequate. Because of their increased rigidity and strength, locally grouted parts of the walls caused heavy damage to surrounding existing, but weak, masonry, during the vibration of the building subjected to earthquake. The observed local damage was even more severe than in the case of the wall, not grouted at all.

5. Conclusions

Criteria and technical measures for seismic upgrading of existing masonry buildings have been discussed. In order to achieve adequate seismic behaviour of stone-masonry houses in the zones, where earthquakes of EMS intensity VIII are expected, the integrity of masonry structural system should be provided by means of the tying the walls with steel ties and anchoring the floors into the walls. Since the quality of existing stone-masonry is poor, the resistance should be improved by cement-grouting of the walls.

The analysis of the behaviour of stone-masonry houses during the earthquake of April 12, 1998, indicated that houses, thoroughly strengthened after the earthquakes in 1976 by applying the recommended technical measures, resisted the earthquake of 1998 without, or with only minor damage. The earthquake of 1998 was strong enough to cause heavy damage to existing buildings as well as to buildings, in the case of which the strengthening measures have been only partly applied. The fact that partial implementation of structural interventions does not ensure adequate seismic behaviour was especially evident in the case where the existing wooden floors have been replaced with r.c. slabs, not properly anchored and connected to the walls, and the walls have not been strengthened by cement-grouting. In such cases, the rigid slabs pushed the outer wythe of the wall outwards, thereby causing the development of horizontal cracks just below the slabs, and subsequent dilamination and disintegration of corner zones.

The analysis of earthquake damage has shown, that the replacement of wooden floors with rigid slabs is not always necessary. In the case where the walls need not to be strengthened, the tying of the walls with steel ties will provide the integrity of the structure and, hence, utilise the available resistance. If the wooden floors are replaced with r.c. slabs, the slabs should be adequately supported by the inner wythe of the wall, as well as properly connected and anchored with the outer wythe by means of steel anchors and r.c. dowels. In addition, the integrity of the building should be ensured by installing the steel ties on the outer side of the walls. Otherwise, horizontal cracks just below the slab develop, and pushing out of the corner zones and subsequent dilamination and falling out of the outer wythe of the wall at these zones can occur.

6. References

Boštjančič, J., Sheppard, P., Terčelj, S., Turnšek, V. (1976). Use of a modeling approach in the analysis of the effects of repair to earthquake-damaged stone-masonry buildings. *Bolletino di geofisica teorica ed applicata*, Part 2, Vol. XIX. no.72, Udine: 1091-1116.

Ladava, A. (1982). Guidelines and procedures used to eliminate the impact of the earthquake in the Soča Valley. *Proc. Social and Economic Aspects of Earthquakes*, Jones, B.G., Tomaževič, M., Eds., Institute for Testing and Research in Materials and Structures, Ljubljana - Cornell University, Ithaca, pp. 413-423.

Report on mitigating the consequences of the earthquake of Bovec of April 12, 1998. (1998). Administration for Civil Protection and Disaster Relief, Ljubljana (in Slovene).

Sheppard, P., Tomaževič, M. (1986). In-situ tests of load-bearing capacity of walls of old masonry buildings. *Proc., 4th Nat. Congress on Earthquake Engineering*, Vol.2, pp. 85-92, Cavtat (in Serbian).

Tomaževič, M., Sheppard, P. (1982). The strengthening of stone-masonry buildings for revitalization in seismic regions. *Proc., 7th European Conference on Earthquake Engineering*, Vol.5, pp. 275-282, Athens.

Tomaževič, M. (1989). Some aspects of structural strengthening of historic buildings in urban and rural nuclei against earthquakes. *European Earthquake Engineering*, 3 (1), Patron, Bologna, pp. 19-28 (1989).

Tomaževič, M., Apih, V. (1993). The strengthening of stone-masonry walls by injecting the masonry friendly grouts. *European Earthquake Engineering*, 7 (2), Patron, Bologna, pp.10-20.

Tomaževič, M., Lutman, M., Weiss, P. (1993). The seismic resistance of historical urban buildings and the interventions in their floor systems: an experimental study. *The Masonry Soc. J.* 12 (1), The Masonry Soc., Boulder, pp. 77–86.

Tomaževič, M., Lutman, M., Weiss, P. (1996). Seismic upgrading of old brick-masonry urban houses: Tying of walls with steel ties. *Earthquake Spectra*, 12 (3) Earthquake Engineering Research Institute, Oakland, pp. 599–622.

Turnšek, V., Terčelj, S., Sheppard, P. and Tomaževič, M. (1978). The seismic resistance of stone-masonry walls and buildings. *Proceedings of the 6th European Conference on Earthquake Engineering*, Vol.3, Dubrovnik, pp. 275-282.

ON BASIC CONCEPTS FOR DEVELOPMENT OF UNITED INTERNATIONAL EARTHQUAKE RESISTANT CONSTRUCTION CODE

E. E. KHACHIAN
Institute of Geological Sciences of NAS RA

1. Problem Statement

Earthquake resistance of buildings and structures and hazardous consequence level significantly depend on those regulation documents based on which these buildings and structures have been designed and erected. Despite that the code establishes minimum requirements to a structure resistance, their observance results in the considerable reduction of seismic risk level. Now in various seismically active countries various national earthquake-resistive design codes are in force, though the bases for them are the same natural phenomena: ground oscillations at earthquakes and various constructions' responses to these specific disturbances. Comparison of the codes from different countries reveals a good deal of discrepancy between them both in quantitative features and in preconditions served as a basis. We consider such a situation as an unnatural one. In our opinion it has been caused by lack of exchange of valuable information among experts from different countries, especially during 1950 to 1970, when extensive scientific studies on development of national earthquake resistant construction codes had been started. These discrepancies are also a consequence of conservative thinking and national prestige trend of national codes' developers. Economic aspects of earthquake-resistant constructions and social system structure of a given country played a non-negligible role. Of course, earthquake-resistant construction codes shall take account of local geological and tectonic conditions and actual statistical data of weak and strong earthquakes occurred in the territory of a given country. However even these specific features being accounted for in national codes shall not result in significant diversity, which is observed in codes of 40 seismically active countries in the world [1, 2]. Recently new proposals and drafts of earthquake-resistant construction codes for different regions have appeared: Eurocode, and Codes of Balkanian Countries. All this evidences on necessity to coordinate efforts of scientists and engineers and aim them at development of the United Earthquake-Resistant Construction Code. We believe that statistical sets of strong and disastrous earthquakes' records, which have been collected in all the seismically active regions of the world during the last 50 years, engineering analysis of these earthquakes, and theoretical and experimental studies of seismic impact on buildings and structures, as well as the world experience of earthquake-resistant constructions allows now to commence developing of such international earthquake-resistant construction code. Below the main concepts, which, in our opinion,

S. Balassanian et al. (eds.), Earthquake Hazard and Seismic Risk Reduction, 333–343.

may be accepted as a basis for development of the United International Earthquake-Resistant Construction Code, are given.

2. Seismic Zoning

It is natural that the basis for earthquake-resistant construction code shall be a general seismic zoning map for a given country. In principle, the codes of all the countries in the world shall differ from each other only by seismic zoning maps. Of course, these maps shall be updated periodically due to accumulation of data and taking account of new earthquakes. The main question is: what principles and basic quantitative parameters shall be used in drawing up of national maps of seismic zoning. Seismic hazard level is assessed by the rate of strong earthquake occurrence and according to geodynamic and tectonic conditions of the region. The longer the time period, for which data on earthquakes occurred in the given area is available, is, then, naturally more reliable the seismic hazard level prediction will be. Such data for some regions are available for a period of 2000 years, for others only for less than 100 years. On the other hand, service life of ordinary buildings and structures is not more than 50 to 100 years. Therefore national maps of seismic zoning shall be drawn up taking account of intervals of strong earthquake repetition in the territory of a given country and structures' service life. If the average repetition period for earthquakes of a given intensity is about 500 years, then the probabilities that earthquake intensity would not exceed this intensity within 50 years and 100 years are approximately 90% and 80 %, respectively. These figures could be considered as acceptable from the viewpoint of safety and economic efficiency of large-scale civil and industrial construction. In our opinion, zoning maps of countries shall have a similar structure: the country's territory shall be divided into a small number (2 to 5) seismically hazardous zone being attributed with specific parameters, which characterize the seismic hazard level most probable for a given time interval, and which are specified by a seismic regime of a given region and adjacent (from 20 to 30 km) areas. It is clear that a number of such zones depends on the area and geological- and tectonic complexity of a given country.

Since additional forces (horizontal, vertical and torsional ones) are inertial ones, then the parameters characterizing seismic hazard level of the given zone shall be determined, in the first place, by ground particles' motion acceleration magnitude. The most valuable information on these parameters is contained in three-component accelerogrammes of strong earthquakes. Their substantiated choosing for any construction site in a given zone requires a large number of parameter's values recorded for a long period. Therefore to facilitate the problem, the seismic hazard level for a given zone may be characterized by a maximum value of horizontal acceleration **A** of ground motion. Valuable data classified by Ambraseys and Bommer for Europe, Joyrner and Boore for America and Fukushima and Tanaka for Japan, based on dependence of ground acceleration on earthquake magnitude and epicentral distance [3, 4, 5, 6] may be used for these purposes. As for the frequency spectrum of ground acceleration it is caused mainly by ground conditions of a construction site and is accounted for by the response spectrum shape, which will be described in the next section. It is expedient to find out what minimum ground acceleration value could be accepted as a threshold one with the lower magnitudes being not dangerous for buildings. This threshold value of ground acceleration may be the same for all the national earthquake-resistant construction codes. In our opinion this threshold magnitude of ground acceleration should not be less than 0.1g.

Determination of the maximum of ground acceleration **A** for the territory of a given country is of no less importance. In the first place historical information on consequences of past earthquakes, corresponding data bases for the past earthquakes in the given region and national surveys of ground's strong motion recording will be very helpful for this task. It would be also expedient to show active faults locations in the seismic zoning maps, their widths, as well as recommendations on seismic hazard level along the faults and when walking away from them. Development of seismic zoning maps for the country is one of the basic and important sections of earthquake-resistant construction codes. Not only leading seismologists, geophysicists and geologists of the country but also leading experts of the world should be engaged in these works. In principle, seismic zoning maps may be drawn up by a united international group of leading experts of the world under the aegis of the International Association on Earthquake-Resistant Construction.

3. Accounting of the Ground Conditions Impact

Ground conditions significantly impact the seismic resistance of buildings and structures and therefore codes shall contain recommendations on their accounting. And this procedure shall not be of national nature, because we are dealing with grounds of the same planet, the Earth.

Earthquakes records analyses show that magnitudes of horizontal, vertical and rotational displacements, ground velocities and accelerations, with other conditions being equal, depend significantly on geological selection of recording place, physical-and-mechanical, strength, deformation and acoustic properties of formations. In addition, earthquakes consequence results analyses show that depending on ground conditions and structural concept of structure their damages were caused either by settling, tilting and turning-over or formation of fractures in the whole volume of structure. All these circumstances should be accounted for in codes as both quantitative coefficients and designing principles. Aimed at this objective it is expedient to divide grounds of the given zone into 4 categories; rock, medium (semi-rock), loose and soft. Here the main problem is: what are quantitative parameters to comprise the basis for such division? The methods now existing for such division of grounds are restricted mainly to their description without specifying their thickness, strength indices, and longitudinal and transverse wave propagation velocities. The task of ground category determination became more difficult especially for heterogeneous bases comprising several layers with thickness of $\mathbf{H}_k$, density of ρ_k and shear modules of $\mathbf{G}_k$, from firmest to softest. A principal role in seismic wave energy transfer to a structure is played the upper surface layer of ground, which is overlain by the structure base. On the other hand, kinematic parameters of an upper layer, in its turn, are specified by physical-and-mechanical and acoustic features of overlain layers. A natural question arises here: what shall be the depth **H** of overlain layers, which affect tangibly the acceleration amplitude and frequency of the upper layer. The value **H** for rocky grounds may be 30 to 50 meters, for alluvial grounds down to the bedrock level – 10 to 300 meters and more. That is why it is expedient to determine the ground category, as its integral feature, according to the value of dominant oscillation period $\mathbf{T_0}$ for the whole heterogeneous thickness **H** because the ground category formation is affected by both density and strength features and layer thickness. $\mathbf{T_0}$ value may be determined according to the formula [10]:

$$T_0 = \frac{4H}{\overline{V}_S}; \quad \overline{V}_S = \sqrt{\frac{\sum_{k=1}^{n} \rho_k v_{sk}^2 \left[H_k - \frac{H}{\pi}\left(\sin\frac{\pi h_k}{H} - \sin\frac{\pi h_{k-1}}{H} \right) \right]}{\sum_{k-1}^{n} \rho_k \left[H_k + \frac{H}{\pi}\left(\sin\frac{\pi h_k}{H} - \sin\frac{\pi h_{k-1}}{H} \right) \right]}}, \tag{1}$$

where $G_k = \rho_k v_{sk}^2;\ h_k = \sum_{i=1}^{k} H_i;\ h_0 = 0;\ h_n = H$, and n is a number of layers.

$\mathbf{T_0}$ value could be found out experimentally as well, by means of the spectral analysis of ground microoscillations, which were recorded by a highly sensitive measuring equipment located in the construction site.

The next step for revising the ground conditions impact is to find out quantitative coefficients $\mathbf{k_0}$, which correct maximum ground accelerations **A** for a given zone, depending on a ground category. For this purpose it is necessary to generalize strong earthquake accelerogrammes recorded at equal epicentral distances but in different grounds. Such data has been collected in sufficient amount during the earthquakes of San Fernando (1971), Lomo-Prieta (1989), Kashiro-Oki (1993), Nortridge (1994) and Kobe (1995).

One of the options for dividing grounds into categories according to $\mathbf{T_0}$ magnitude and values of ground condition coefficient $\mathbf{k_0}$ is given in Table 1 [10]:

TABLE 1. *Grounds Categories According to* $\boldsymbol{T_0}$

Magnitude of dominant period $\mathbf{T_0}$ of the base oscillation	Category of the ground-base	Value of the dimensionless coefficient $\mathbf{k_0}$
$T_0 \le 0.25$	I	0.8
$0.25 < T_0 < 0.5$	II	1.0
$0.5 < T_0 \le 0.7$	III	1.2
$T_0 > 0.7$	IV	1.4 ÷ 1.6

To reduce a probability of resonance occurrence it is expedient to observe one of the following conditions at the designing stage:

$$T_1 > 1.5\,T_0 \ \text{ or } \ 1.5\,T_1 < T_0, \tag{2}$$

where **T** is a basic period of free oscillations of the onland structure.

Ground conditions significantly affect the duration and spectral composition of ground motion. They shall be accounted for in the sections of codes concerning determination of inertial force magnitudes. Thus the longer the ground vibration time is the higher is the probability of resonant phenomena occurrence and damage accumulation.

4. Determination of Internal (Seismic) Force Values

The next basic section of codes shall be devoted to the issues of determination of the level of seismic inertial effects (horizontal, vertical and torsional) on buildings and structures, which are planned to be erected in the given seismic zone with particular ground conditions. The section of a code shall also be international because it is based on generally known laws of mechanics and oscillation theory. However, even here big discrepancies are found in the earthquake-resistant construction codes in force in most of the countries. We believe that the reason is that the code developers trying to facilitate the designers' work as much as possible have simplified the real problem too much. In the result we obtain various theoretical models for structures, which do not correspond to the reality. Thus the red thread of this code's section shall be choosing of the actual theoretical model of the structure. Of course, ordinary buildings and structures are rather complicated spatial mechanical systems due to complicated geometry in the plan and vertical sections, and presence of a large number of window and door openings, partitions, staircases and internal equipment. But nevertheless they are mechanical systems with distributed and concentrated masses connected to each other by certain ties, which have definite elastic and plastic properties. Displacements and accelerations of ground are also of complicated spatial character. Thus, such general problem statement could result in complicated mathematical formulae for calculation of inertial seismic force values and significant reduction of calculation accuracy, because it is impossible to take all factors into account. Aiming at eliminating of such situation, codes should impose certain restrictions on the geometry and shape of structures. Buildings and structures of complicated geometrical shapes should be reduced to simpler ones by structural concept recommendations.

The next aspect of the problem is determining what oscillations of the structure should be chosen - either elastic ones or elastic-plastic ones with damages - when designing cross-sections of its structural elements. It is clear that oscillations of structures built of traditional construction materials are always of nonlinear character. Frequently they are accompanied by individual local fractures and damages, which are not dangerous for intactness and preservation of the structure proper and safety of people in it. Since such damages are not only unavoidable, but also expedient to some extent from the economic viewpoint, codes shall contain corresponding recommendations on these aspects. Proceeding from this, it is expedient to calculate design inertial loading for buildings and structures assuming their elastic regime and rigid constraining in the ground, and to take account of nonlinear oscillations, allowable damages and interaction between ground and structure indirectly when choosing cross-sections of structural elements by means of increasing or decreasing of elastic force values through changing of corresponding coefficients. Having assumed elastic regime of the structure, it is possible to consider horizontal, vertical and torsional oscillations of ground and structure separately. The effects related to their mutual influence are also taken into account indirectly: by introduction of substantiated correction coefficients.

Determination of inertial forces values for linear (elastic) theoretical models with an arbitrary accelegramme **A(t)** of an earthquake (horizontal, vertical and rotational) at the level **k** along the structure height is expressed by the product of mass $\mathbf{m_k}$, free oscillation mode shape η_k and the generally known response spectrum **a(T, n).** Since of these parameters only the response spectrum **a(T, n)** is related to factors of seismic zones and grounds, in our opinion, it shall be divided into three separate components: 1) characterizing a maximum ground acceleration **A** for this zone; 2) correcting this maximum acceleration by dimensionless coefficients $\mathbf{k_0}$ in dependence on a ground category; and 3) dimensionless dependence **β(T)** of ratio of maximum acceleration of the structure with the period of free oscillations **T** and a coefficient of critical attenuation **n** to maximum ground acceleration: **β(T, n) = a(T, n)/A**. In other words, seismic inertial load on **i**-th oscillation mode shape in the point **k,** where the mass $\mathbf{m_k}$ is concentrated, is represented by the general formula [9]:

$$S_{ki} = m_k \eta_{ki} A k_0 \beta(T_i), \quad \eta_{ki} = \frac{C_{ki} \sum_{j=1}^{n} m_j C_{ij}}{\sum_{j=1}^{n} m_j C_{ij}^2},$$

$$\beta(T) = \frac{\frac{2\pi}{T} \int_0^t e^{-n(t-u)} A(u) \sin \frac{2\pi}{T}(t-u)\, du}{A}, \tag{3}$$

where $\mathbf{C_{ik}}$ is an amplitude of the structure's free oscillation in i-th oscillation mode shape at the point **k**.

The principles related to determination of **A** and $\mathbf{k_0}$ values were described in Sections 2 and 3. As for the dependencies **β(T, n)** then, in our opinion, at present the records of strong earthquakes accelegrammes are collected in a sufficient number and response spectra for choosing of substantiated reference dependencies **β(T)** are constructed based on the strong earthquakes accelerogrammes. Based on the analysis of the response spectra, which were constructed on various earthquake accelerogrammes recorded in different ground conditions in different seismic regions, it is possible to represent generally a dependence of dynamic coefficient **β(T)** (for **n** being equal to 3 to 8% of the critical one) as three segments: a straight line ascending from 1, horizontal straight line and descending less steep line:

$\beta(T) = 1 + T/T_a$	at	$0 < T < T_a$;
$\beta(T) = \beta_{max}$	at	$T_a < T < T_b$;
$\beta(T) = (T_b/T)^p\, \beta_{max}$	at	$T > T_b$.

Attenuation levels and their distinctions related to material type or structural design are not included into Eq. (4) directly. For materials with low attenuation (steel construction) and lower scattering of strength indices, and for materials with high attenuation (reinforced concrete and stone constructions) and higher scattering of strength indices it could be assumed that impacts of these factors are mutually balanced. In addition, for flexible systems with a period of free oscillations of **T** > 1.0 sec, and I and II category grounds the response spectra' ordinates are nearly independent on attenuation. The question is choosing of the united values for parameters β_{max}, **p**, T_a and T_b and k_0 in dependence of ground categories. Now the values of these parameters in national earthquake-resistant construction codes are nearly stabilized and are changing within narrow ranges. For example, their values for II category grounds are:

$$\begin{aligned} &\beta_{max} \sim 2.5 \div 3.0 \\ &P \sim 2/3, 1 \\ &T_a \sim 0.2 \div 0.4,\ T_b \sim 0.6 \div 0.7 \\ &k_0 \sim 1.0 \div 1.2 \end{aligned} \qquad (4)$$

In our opinion these parameters do not depend on national conditions of a country and shall be the same for all the seismic regions of the world. Such dependencies shall be determined not only for horizontal components of ground oscillations, but also for vertical and rotational ones. However the problem of seismic inertial loads' values is not completed by this. The matter is that maximum values of seismic inertial force for elastic oscillations at the given level of the structure are the result of superposition of similar forces for first several oscillation mode shapes. The task complexity is determining of both minimum number of oscillation mode shapes for different structures and their maximum values' summation methods. Recently a significant success was achieved in this issue. Due to researches by N. Newmark, E. Rosenbluet, G. Pezenen, M. Vatabe, A. Der Kyureghian, and E. Wilson it have been agreed that the most acceptable method for summation of seismic loads according to oscillation shapes is the CQC method and its varieties [8].

Interaction between a ground and a structure plays an important role in seismic effects. Besides the condition [2] mentioned above, which indicates a possible prevention of resonance, this interaction may be taken into account by introduction of a special coefficient $k_1 < 1$, which depends on both ground category and free oscillation period of the structure. Seismic inertial forces calculated according to Eq. (3) are obtained assuming oscillation of the structure with one component of ground motion only. In fact, motions of ground and, hence, of the structure have spatial character. In addition due to big length of structure (bridges, dams, etc.) its individual supports located on a non-rocky base may oscillate differently, which could produce additional torsional oscillations of the structure and change horizontal inertial forces values considerably [3]. These effects have been studied in detail by many experts, and in our opinion their results allow introducing of the corresponding coefficient k_2 into Eq. (3); this coefficient value could be both more than unity and less than unity.

The analysis of large number of records of three-component accelerogrammes of strong earthquakes shows that a vertical component of ground motion is on average 75 % of a horizontal component value. When such an acceleration is taken into account then it results in small increase of structural elements' cross-section up to 10 %. The character of a dynamic coefficient (T) for a vertical component of ground motion differs from that of horizontal one only by a small shifting of maximum value towards smaller values of **T**. Vertical component shall be taken into account when designing those elements for which the combination of vertical and horizontal forces is definite (strength of bridge and span structure supports, stability of dams and slopes, etc.).

It is known that the strength properties of materials for the effect applied once are increased depending on a loading rate. On the other hand, the load carrying capacity is significantly reduced depending on a number of alternating loading cycles (this is especially true for reinforced concrete). These indices for materials and structural elements made of these materials are significantly differing from each other. For earthquakes it is necessary to proceed from the operation character of structural elements and not from that of materials. Studying of dynamic models with using of strong earthquakes accelerogrammes shows that structural elements in low buildings with a relatively low value of free oscillation period are subject to a large number of oscillations with high loading rate, and on the contrary, in multi-storeyed flexible structures they are subject to a small number of oscillation with a relatively low loading rate. It allows considering that these two factors – high loading rate (positive) and low-cycle fatigue (negative) – compensate each other for both rigid and flexible structures during earthquakes.

A disastrous earthquake is rather a rare phenomenon, and the requirement of completely undamaged state and integrity of structure is not economically justified, to say nothing about the fact that in some cases it is nearly impossible to realize technically. It is very important to prevent such damages, which can cause casualties and require exceedingly high cost of restoration. On the other hand, granting of damages at earthquake occurrence is related to a certain risk. Thus, the allowable damage level shall be specified within reasonable limits, which will assure people safety and further reliable exploitation of a building after spending certain minimum means for its restoration. Naturally, different coefficients (levels) $\mathbf{k}_3 < 1$ of allowable damages for buildings erected of metal, reinforced concrete, stone and wood shall be specified because of different plastic and adaptation properties of these materials. Different levels of damage could be specified for various structural elements of a structure (wall, columns, girders, structural connections, joints and ties) in dependence on their role in a general vulnerability of constructions. In our opinion, a level of allowable damages for buildings and structures of the same structural concept and functional purpose may be the same in all the countries.

However not all the structures shall have the same level of protection at an earthquake. Destruction of individual structures such as nuclear power plants, dams, bridges, airports, tunnels, buildings and structures of a life support system, as well as structures where large gatherings of people are possible, causes big social-and-economic shocks in the society as a whole. Therefore their reliability during earthquakes shall be much higher than that, let us say, of buildings and structures of repair shops, storehouses and agricultural premises, separate transport and irrigation structures, which can be designed sometimes even without taking account of seismic effects. For these issues it is also possible to recommend united coefficients of structure's responsibility $\mathbf{k}_4$ (from 0 to 1.5) for all national earthquake-resistant construction codes.

From the above said it follows that choosing of cross-section of structural elements, their assemblies and joints shall be done during the elastic stage of their operation (according to load carrying capacity) assuming a static effect at the level **k** of full inertial seismic characteristic forces according to the following formula:

$$S_k = k_1 k_2 k_3 k_4 \sqrt{\sum_{i=1}^{n} \sum_{j=1}^{n} \rho_{ij} S_{ki} S_{kj}}\,, \tag{5}$$

where the coefficients ρ_{ij}, which depend on the periods $\mathbf{T}_i$ and attenuation coefficients, are determined by the CQC method in accordance with [8]. A technical and economic efficiency of earthquake-resistant construction and the level of seismic protection of the society in general will be conditioned by a reasonable determination of the coefficients $\boldsymbol{k_0}$, $\boldsymbol{k_1}$, $\boldsymbol{k_2}$, $\boldsymbol{k_3}$ and $\boldsymbol{k_4}$.

5. Basic Principles of Designing and Structural Requirements

Naturally a code shall also contain some known principles of designing and structural requirements, which shall be observed when designing buildings and structures, independent on the results of calculations. As the examinations of disastrous earthquake consequences have shown, their observance significantly increases the level of seismic safety for people and structures. They include, in the first place, observance of the principles of symmetry and uniform distribution of stiffness and masses, definite ratio of structure's geometrical dimensions, restriction of minimum values of strength indices of applied materials, regulation of reinforcement methods for reinforced concrete and stone constructions, location and dimensions of window and door openings and their reinforcement, reinforcement of non-bearing walls, partitions, staircases, chimneys, parapets, water supply and sewage pipes, and household and other equipment. Such recommendations are included in many national codes. It is necessary to analyze them comprehensively and to represent as united structural requirements for all the seismically active countries.

6. Restoration, Reinforcement and Certification of Buildings and Structures

Reduction of damages caused to the society by earthquakes is also closely related to the regular preparation (stabilization) of old buildings to earthquakes, urgent determination of the damage level for this or that object immediately after the earthquake and works on their quick restoration or stabilization after the earthquake. After disastrous earthquakes uncertain situations with respect to further availability of this or that damaged object for exploitation often occur. Codes shall contain recommendations on these issues as either appendices to the code or a separate section of the code. It would be expedient to restore buildings and structures, which were damaged insignificantly at designed earthquakes, to their initial conditions. Aiming at this it is expedient to specify a standard scale for damages of buildings erected of various materials and of various structural concepts. It would be desirable that this scale be based on the description of this or that element of a structure but also on changes of its dynamic characteristics due to damages caused by an earthquake. To this end it is expedient to draw up a standard technical passport for each new object after its construction being completed, which will contain brief structural specific features and characteristic loads, ground conditions of the site, as well as a magnitude of free oscillation's basic mode shape period **T** of the object, which is found out by recording of its microoscillations. Comparison of the ratio of periods for damaged and undamaged object may serve as an additional quantitative parameter describing its damage level.

In the conclusion we would like to emphasize once more that the main objective of this report is to establish a necessity to develop united international codes for earthquake-resistant construction. The arguments, considerations and proposals presented above are just pointing out a possibility to commence such a procedure and do not pretend for their universal character or completeness.

7. References

1. Earthquake Resistant Regulations - A World List, 1984. Compiled by the International Association for Earthquake Engineering. July 1984, Tokyo, Japan, 904 pp.
2. International Handbook of Earthquake Engineering Codes. Programs and Examples. Edited by Mario Paz. Chapman and Hall, An International Publishing Company, 1994.
3. N. N. Ambraseys and J. J. Bommer. Database of European Earthquakes Associated with Strong-Motion Records. European Earthquake Eng., v. N2, 1991, pp. 18 - 37.
4. W. B. Joyrner and D. M. Boore. Peak Horizontal Acceleration and Velocity from Strong-Motion Records, Including Records from the 1979 Imperial Valley (California) Earthquake. Bull. Seis. Soc. Am. 71, 1981, pp. 2011-2038.
5. Y. Fukushima and Tanaka, A New Attenuation for Peak Ground Acceleration, Velocity and Displacement Based on Multiple Regression Analysis of Japanese Strong Motions. Bull. Seis. Soc. Am. 71, 1981, pp. 757-783.
6. F. Yamazaki, Comparative Study of Attenuation Characteristics of Ground Acceleration in Europe, North America and Japan. Bulletin of Earthquake Resistant Structure Research Center, N26, March 1993. Inst. of Ind. Science Univ. of Tokyo, pp. 39-56.
7. N. M. Newmark and Posenblueth. Fundamentals of Earthquake Engineering. Moscow, 1980, 344 pp. (in Russian).
8. A. Der Kiureghian. A Response Spectrum Method for Random Vibration Analysis of MDF Systems. Earthquake Engineering and Structural Dynamics, N9, 1981, pp. 419-435.
9. E. E Khachyan. The Seismic Influence o the Multi-Storeyed Buildings and Constructions. Hayastan, Yerevan, 1973, 327 pp. (in Russian).
10. E. E. Khachyan. Study of Dynamic Characteristics of Heterogeneous Ground Bases. Izv. NAN RA, Ser. Earth Sciences, 1995, XLVIII, N2-3, pp. 112-119.

11. Basic Concepts of Seismic Codes. v. 1. The International Association for Earthquake Engineering, Tokyo, 1980, 103 pp.
12. S. I. Poltavtsev, Ya. M. Eisenberg, G. L. Koff, A. M. Melentyev, V. I. Ulomov. Seismic Zoning and Seismic Resistant Construction. Moscow, 1998, 259 pp.

ANALYSIS OF NEW SEISMIC BUILDING CODES OF CIS COUNTRIES

T.MUKHADZE, I.TIMCHENKO
Institute of Srtuctural Mechanics and Earthquake Engineering, Georgian Academy of Sciences
8, M.Aleksidze str., Tbilisi 380093, Georgia

1. Abstract

The level and character of distribution of seismic load in new codes of Armenia, Russia, Uzbekistan is analyzed in the paper. The various approach to the account of the factors influencing on seismic resistance of buildings and structures is discussed. The values of different design factors and values of seismic load for various type of buildings are compared.

2. Introduction

The most effective way of seismic risk reduction is realized with the help of aseismic normalization of building. The seismic building codes of the former Soviet Union reissued some times. At each stage they corresponded to the level of world standards. However, as the last strong earthquakes have shown on the territory of the former Soviet Union (Spitak, 1988; Zaysan, 1990; Racha, 1991; Yakutia, 1991; Soosamir, 1992; Neftegorsk, 1995), the seismic hazard of separate regions has appeared essentially high. The next revision of codes practically has coincided with disintegration of the USSR.

After that the development of national codes has taken a special significance in the independent states located in seismic regions.

The new codes are already developed in Armenia, Uzbekistan, Russia and Kazakhstan. The development of new codes is also carried out in Georgia.

3. Main Results

The main principles of the former Soviet Union seismic building codes are in many respects in a basis of these codes. Local features of building, experience of the most developed countries generalized in Eurocode 8 was taken into account.

In particular, the preference is given to those constructions, which seismic resistance is poorly depends on consrtuction work quality.

S. Balassanian et al. (eds.), Earthquake Hazard and Seismic Risk Reduction, 345–349.

The analysis of seismic load calculated for rigid and flexible buildings of various height is given below. Shear load values distribution is compared for codes of Armenia, former USSR, Uzbekistan, Russia (I and II editions) and for Eurocode 8.

The considered codes contain practically all sections, which account, under the judgement of international association on aseismic building, is necessary for development and for coordination of national codes at an international level.

Among all factors, methods of determination of seismic load play the main role in ensuring of seismic resistance of buildings and structures. And they mainly determine level and degree of antiseismic measures of buildings and structures and their vulnerability.

New codes are on a way of adjustment of seismic load on buildings and structures by adjustment and introduction of new design coefficients. The level of seismic load depends on dynamic properties of building, of it type and materials, ground conditions, seismicity of building site etc.

Factor which takes into account seismicity of building site was not changed in new codes of Russia and Armenia (A = 0.1, 0.2, 0.4 for seismicity of 7, 8, 9 degrees accordingly). In code of Uzbekistan this factor is equal to 0.25, 0.5, 1 accordingly.

Dynamic coefficient β varies depending on dynamic characteristics of a building. In code of Uzbekistan the regional curves (for 4 regions) are given for different types of soils.

K_f – coefficient, depending from building height (number of floors), has remained in code of Uzbekistan. In codes of Russia and Armenia this coefficient is absent.

In all considered codes the factors which are taking into account reduction of seismic load because of nonlinear behavior of construction (reduction factor) are introduced. In code of Uzbekistan this reduction factor is taken into account at the stage of calculation of separate elements and their sections. It depends on the value of limiting relative elastic strain μ of an element and depends on a period of natural oscillations of a building. In codes of Armenia and Russia reduction factor is introduced in a basic formula of seismic load. For brick buildings K_1 = 0.7 (Armenia); Kz = 0.48 (Russia, I edition), Kz = 0.35 (Russia, II edition). For reinforced concrete frame buildings with diaphragms K_1=0.45 (Armenia); Kz=0.36 (Russia, I edition) and Kz = 0.25 (Russia, II edition).

The influence of soil-structure interaction is taken into account with coefficient K3 (Armenia) which depends on the period of fundamental form of natural oscillations of a building. In codes of Russia Ks factor takes into account nonlinear properties of soils and is equal to 0.7 for I category; 1.0 for II category and 1.4 for III category of ground (I edition). In the second edition Ks varies from 1 (rock) up to 0.7 (soft soils). As a whole the values of Ks and Kz factors in I and II editions of Russian codes were significally changed.

For purposes of practical design the understanding of modifications in new codes and their comparison with preceding code of the former USSR acting long time and which act now, particularly, in Georgia, is rather important.

Let's evaluate how seismic load is formed for flexible and rigid buildings by new codes and how it corresponds to the old codes of former Soviet Union and to Eurocode 8.

Initial information for the comparative digital analysis is following. The grounds are of average category by seismic properties; design seismicity of building site is 8 degrees.

Reinforced concrete frame buildings with diaphragms (1, 2, 3, 4, 5, 7, 9, 12 - story) and brick buildings (1, 2, 3, 4 - story) are compared by character and values of seismic load.

The periods of fundamental form of natural oscillations of buildings are determined by the empirical formulas recommended in codes of Armenia: $T_1 = 0.055n$ for brick buildings; $T_1 = 0.06n$ for frame reinforced-concrete buildings; n- number of floors. The values of dynamic coefficient are determined by curves and formulas of appropriate norms for average ground conditions. The relative values of a seismic load S_{ik}/Q_kh_{ik} are used in calculations. In Eurocode8 η coefficient is defined from supposition, that fundamental mode shape is approximated by a straight line along the height: $S_iW_i/\Sigma S_iW_i$. The difference between methods of determination of η is taken into account by introduction of a correction factor equal to 1.17÷1.25.

All buildings are supposed regular in plan and height.

The relative floor seismic load in reinforced-concrete buildings and in brick buildings (calculated by different codes is shown below. The corresponded values of seismic load are given in Tab. 1 and Tab. 2.

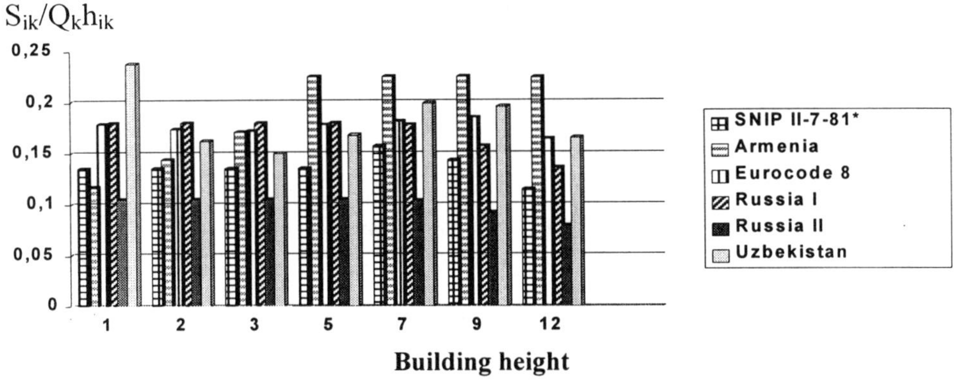

Figure 1. Relative floor seismic load in reinforced-concrete buildings

TABLE 1. *Values of relative floor seismic load in reinforced-concrete buildings*

Codes / Building height :	**1**	**2**	**3**	**5**	**7**	**9**	**12**
SNIP II-7-81* (USSR)	0,135	0,135	0,135	0,135	0,157	0,143	0,115
Armenia	0,117	0,144	0,171	0,225	0,225	0,225	0,224
Eurocode 8	0,179	0,174	0,172	0,178	0,181	0,185	0,164
Russia I edition	0,18	0,18	0,18	0,18	0,178	0,157	0,136
Russia II edition	0,105	0,105	0,105	0,105	0,104	0,092	0,079
Uzbekistan	0,238	0,162	0,15	0,168	0,199	0,196	0,165

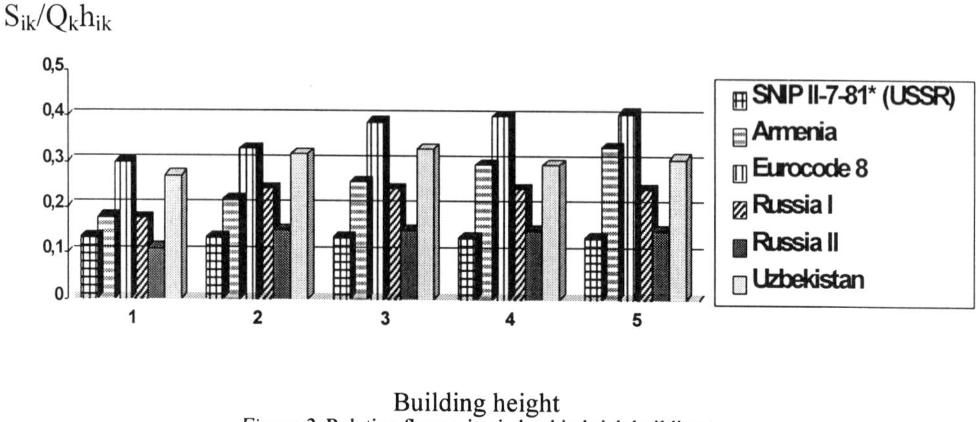

Figure 2. Relative floor seismic load in brick buildings

TABLE 2. Values of relative floor seismic load in brick buildings

Codes / Building height:	**1**	**2**	**3**	**4**	**5**
СНиП II-7-81* (USSR)	0,135	0,135	0,135	0,135	0,135
Armenia	0,178	0,217	0,256	0,294	0,332
Eurocode 8	0,298	0,327	0,387	0,4	0,406
Russia I edition	0,175	0,24	0,24	0,150	0,150
Russia II edition	0,109	0,150	0,150	0,144	0,144
Uzbekistan	0,305	0,415	0,509	0,555	0,576

Relative seismic loads calculated by different codes for 1-12 story frame buildings with diaphragms are shown in Fig.1. One can see in the diagrams that seismic load calculated by different codes vary significally. The significant difference of obtained values (up to two times) is observed. The maximum values of seismic forces correspond to codes of Armenia and Uzbekistan. A level of load calculated by codes of the former USSR takes in this comparison an average position for one-story buildings. It is interesting to note, that seismic loads determined by the first edition of Russian codes, exceed seismic loads determined by old Soviet code. In the second edition of Russian codes seismic loads on the contrary are much lower. Thus, the minimum value of seismic load corresponds to the code of the former USSR and to the second edition of Russian code. The significant difference in a level of loads is observed in the first and second edition of Russian codes. It is explained by difference of coefficients Ks and Kz in these editions of codes. It is necessary to note, that for frame and brick buildings a reduction factor is taken into account in the main formula of seismic load of codes of Uzbekistan. It is made for facilitation of the analysis of results. Otherwise, in case of disregarding of a reduction factor, the values of seismic load determined by code of Uzbekistan, will be significantly higher.

In Fig. 2 relative story seismic loads for brick buildings of various height (1-5 story) are shown. In Tab. 2 the appropriate values of seismic load are given. Significant differences of relative values of seismic load is also indicated by different codes.

The minimum values correspond to codes of former USSR and Russian second edition. The maximum values are calculated by Eurocode 8 for 3-5 story buildings.

In new codes of Russia and Uzbekistan the additional coefficients of the working conditions are taken into account at strength and stability calculations of separate elements and their sections. In seismic code of Armenia these factors are absent.

4. Conclusions

Seismic loads determined by new codes or by projects of new codes of CIS countries, have a significant dispersedness. The executed comparison of a level of seismic load by different codes is useful for practical strength calculations of buildings and structures and for evaluation of a level of seismic load. The obtained information is taken into account at developing of national codes of Georgia.

5. References

1. Building Codes of Armenia. Earthquake Engineering. Design Codes SNRA II-2.02-94. Erevan, 1995.
2. SNIP II-7-81*. Design Codes. Construction in seismic regions. Moscow, Stroiizdat, 1982.
3. SNIP of Russian Federation. Construction codes and rules. Construction in seismic regions. (I edition). Moscow, Gosstroy of Russia, 1997.
4. SNIP of Russian Federation. Construction codes and rules. Construction in seismic regions. (II edition). Moscow, Gosstroy of Russia, 1997.
5. Construction Codes of Uzbekistan. Design Codes. Construction in seismic regions. Строительство в сейсмических районах. Tashkent, 1993.

RECENT TECHNIQUES AND REGULATIONS ON SEISMIC RETROFIT AND DIAGNOSIS FOR BUILDINGS IN JAPAN

T. SAITO
Building Research Institute, Ministry of Construction
1 Tatehara, Tsukuba, Ibaraki 305-0802, JAPAN

1. Introduction

On the 17th of January 1995, a disastrous earthquake hit the west part of Japan. This earthquake is called "the Hyogoken Nanbu Earthquake" and its disaster is called "the Great Hanshin-Awaji Earthquake Disaster." During these three years, structures such as buildings, roads and bridges which were severely damaged, were demolished and their quick restoration work was carried out.

In Japan, the building design code is controlled by the Building Guidance Division, Housing Bureau, Ministry of Construction. The Building Research Institute (BRI), supports the back up of the code's maintenance from the viewpoint of research. After the Hyogoken Nanbu Earthquake, the BRI responded immediately to investigate building damage and prepare possible countermeasures. Based on the report prepared by the BRI, the Building Standards Law was revised and a new law was enacted for promoting seismic improvement of existing buildings. Also many research projects were prompted in the BRI on the issues of seismic performance evaluation and seismic upgrading of existing vulnerable buildings.

This paper introduces the outline of building damage at the Great Hanshin-Awaji Earthquake Disaster, and explains recent techniques and regulations on seismic retrofit and diagnosis for buildings in Japan. Also, the current situation of the revision work of the Building Standards Law is briefly explained.

2. BRI Response To The Earthquake Disaster

The responses of the BRI to the Great Hanshin-Awaji Earthquake Disaster is summarized in the followings:

1) On the same day of the earthquake, the headquarters for the countermeasure to the earthquake disaster were established in the BRI.

2) From the day after the earthquake occurred, BRI staffs participated on the spot to give technical guidance for the emergency risk assessment being done by the voluntary building engineers who came from all over the country. The main purpose of the risk assessment is to prevent disaster at possible aftershocks of the earthquake. Finally they assessed about 46,000 buildings in total.

S. Balassanian et al. (eds.), Earthquake Hazard and Seismic Risk Reduction, 351–358.

3) The investigation project team was organized to investigate structural damage and possible countermeasures. Based on the investigation data, "The damage survey report of the Hyogoken Nanbu Earthquake (Prompt Report)" was published in February, 1995.

4) The Survey Committee of Earthquake Damaged Buildings was established on January 31st. The committee concluded the interim report on July 28th and the final report on Dec.27th. Based on these reports, a new law, "the law for promotion of seismic retrofit of buildings," was enacted. Also, a notice for "revision of the Building Standards Law and related notification" were issued.

3. Summery of Building Damage at the Earthquake Disaster

From the data in the final report of the survey committee [1], building damage at the Great Hansin-Awaji Earthquake Disaster is summarized below:

1) *General situation of building damage* (Fig. 1): Over 442 thousands buildings were damaged at the earthquake, and 11% of them got severe damage.

2) *Relation between structural type and damage level* (Fig.2): The ratio of fatal or damaged case is less than 10% for reinforced concrete buildings, about 30% for steel buildings, and about 70% for wooden buildings.

3) *Relation between construction year and damage level* (Fig.3): Japanese design regulations have been revised two times. First revision was made in 1971, and the next one was made in 1981. The ratio of severe damage or collapsed buildings is considerably decreased for the buildings constructed after each code revision.

4) *Damage patterns of collapsed buildings* (Fig.4): For reinforced concrete buildings and steel reinforced concrete (SRC) buildings, more than 50% of the collapsed buildings were those with story collapses at the first or the middle story.

5) *Damage patterns of collapsed RC&SRC buildings* (Fig.5): In the case of the buildings designed before 1971, nearly 30% of the collapsed buildings were the buildings with pilotis (=soft first story.) Even the number of collapsed buildings is decreasing after each code revision, still a large ratio of collapsed buildings with pilotis is observed. To prevent such type of collapse, the Building Standards Law was revised to avoid significant vertical irregularity of story stiffness,

6) *Damage patterns of steel buildings* (Fig.6): In the case of collapsed steel buildings, the damage patterns are classified into two groups, one is damage at the column foot and the other is damage at the joint. Fig. 7 shows the detail of joint damage and column foot damage. The large number of joint damage occurred at the welded joints, and in the case of column foot damage, damage at the anchor bolts is the main reason of damage.

4. Regulations Responding to the Earthquake Disaster

4.1 LAW FOR PROMOTION OF SEISMIC RETROFIT OF BUILDINGS

As mentioned in the above, most of the buildings which got serious damage were those designed and constructed before 1981, especially before 1971. In order to promote the seismic retrofitting of existing vulnerable buildings, a new law, " the law for promotion of seismic retrofit of buildings," was enacted in December 1995. The objective of this law is to enforce the owners of the large occupant buildings, such as schools, hospitals, theaters, hotels and so on, to make seismic retrofit. If the building officials approve the retrofit plans of such buildings, the owners are eligible to apply lower interest loan, tax exemption, and exemption from regulations for land use and fire protection codes.

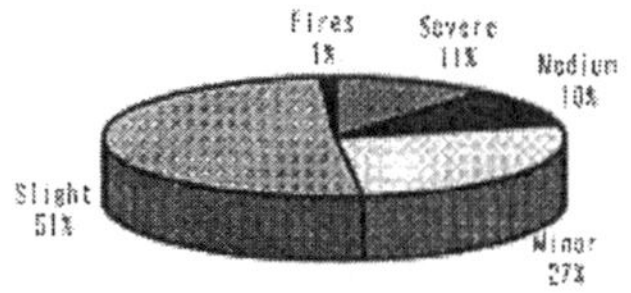

Figure 1. Ratio of building damage

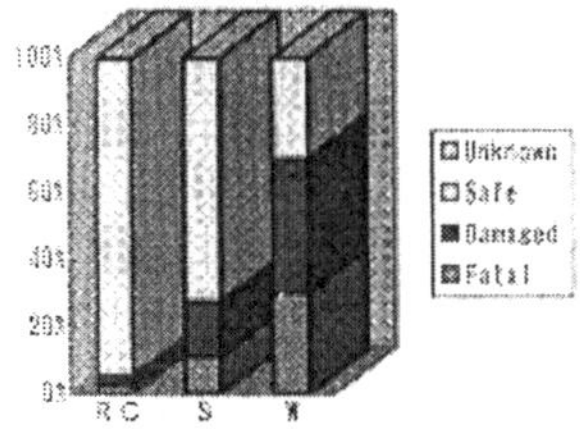

Figure 2. Structural type and damage level

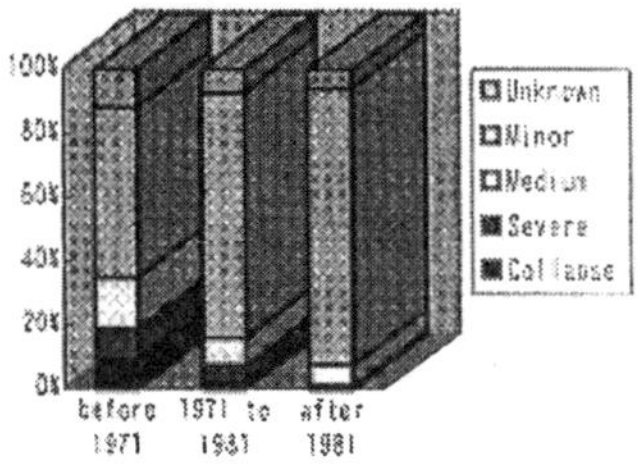

Figure 3. Construction year and damage level

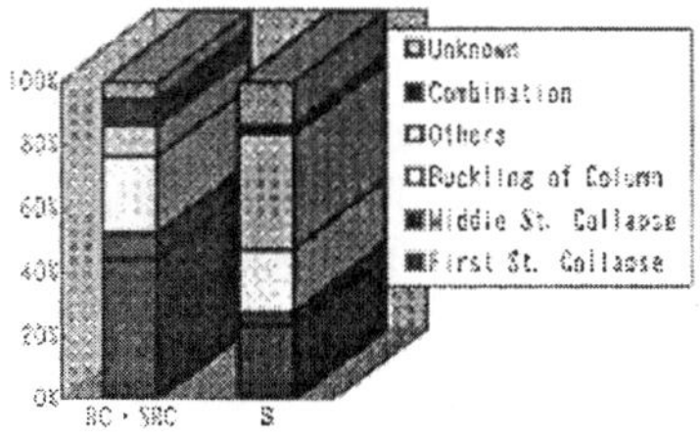

Figure 4. Damage pattern of collapsed buildings

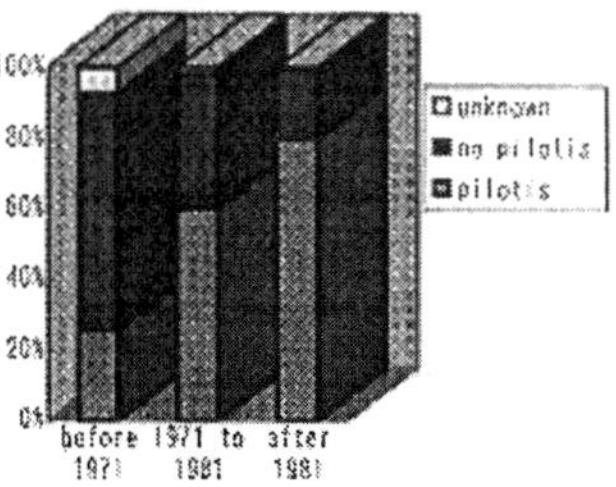

Figure 5. Damage pattern of RC&SRC buildings

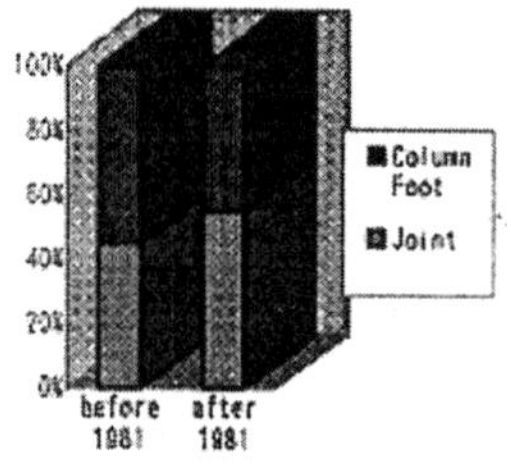

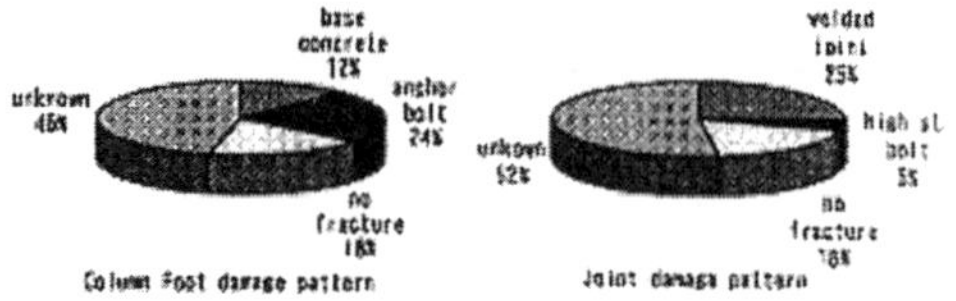

Figure 6. Damage pattern of steel buildings *Figure 7.* Detail of damage pattern of steel buildings

4.2 REVISION OF THE BUILDING STANDARDS LAW AND RELATED NOTIFICATION

At the Great Hanshin-Awaji Earthquake Disaster, the buildings designed according to the present seismic design code showed fairly good performance in general even to such severe ground motions. However, as mentioned in the above, still some of the buildings took serious damage. The notification of the Building Standards Law was revised in December 1995 to reinforce the present code in the following points:

1) In order to avoid the story failure at the first story or middle story of RC&SRC buildings, the revised code requires designers to consider the weak-beam strong column mechanism. Also, the strength modification factor for the vertical stiffness irregularity was increased.

2) For Steel buildings, many of the buildings which were damaged in the first story had pin joints of columns with the base plate and anchor bolts. The revised code added the notification which requires designers to confirm the design of anchor bolts at column foot if they can carry the shear force safely.

5. Evaluation of Seismic Capacity of Existing Buildings

Since the Great Hanshin-Awaji Earthquake Disaster, a large number of retrofit works for existing buildings have been carried out around the country. Seismic diagnosis, or evaluation of seismic capacity of buildings, must be done before implementing seismic retrofit. As the guideline of seismic diagnosis, "Standards for evaluation of seismic capacity and Guidelines for retrofit of existing reinforced concrete buildings," published in 1977 has been widely used or adopted in the regulations such as "the law for promotion of seismic retrofit of buildings." In these evaluation procedures, the seismic capacity of a building is measured by the seismic index, Is, which is defined as,

$$I_S = E_0 \times S_D \times T \qquad (1)$$

where, I_S: seismic index, E_0: basic seismic index, S_D: structural design index, T: time index. Basic seismic factor E_0 is evaluated as a function of strength index, C, and ductility index, F.

$$E_0 = C \times F \tag{2}$$

For example, if a story is idealized as a series of vertical members such as in Fig.8, the load deflection relationship for the story under a monotonic loading could be represented by the curves in Fig.9. The idea of the seismic index, Is, is to evaluate the capability of energy consumption of the story. There are three levels of procedure to evaluate the index, E_0, depending on its simplicity. The first level procedure is a simple but useful for buildings with considerable amounts of walls, where the strength index C is calculated based on section areas and unit strengths of columns and walls. The second level procedure requires the calculation of the ultimate strengths of columns and walls based on their reinforcements. In the third level procedure, it is required to calculate the ultimate strength and ductility of beams as well as columns and walls.

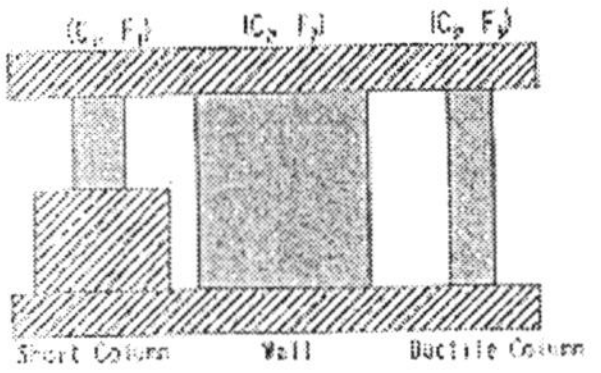

Figure 8. Idealized Story of a building

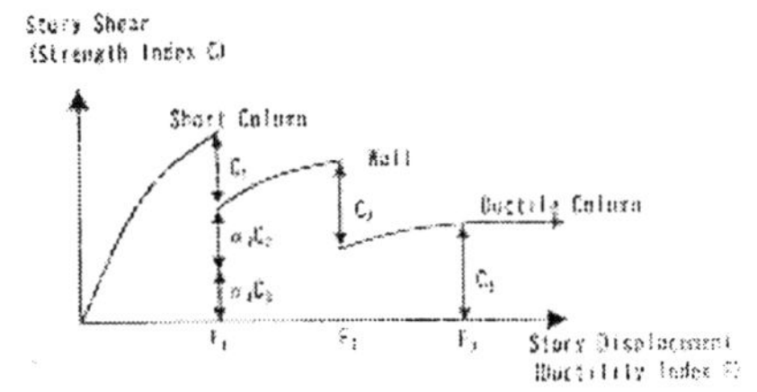

Figure 9. Story Shear-Displacement Relationship

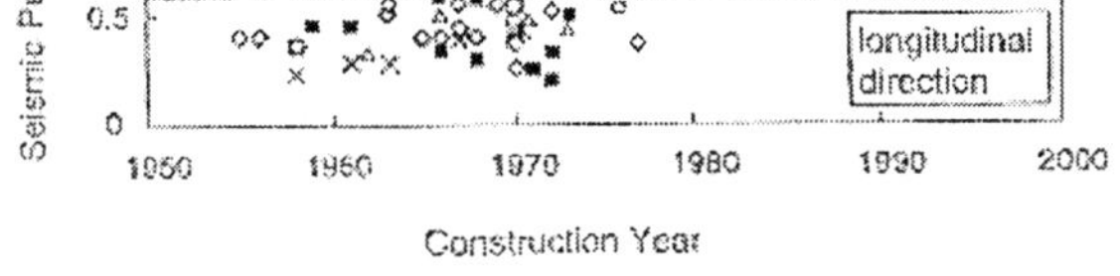

Figure 10. Is indices for the school buildings (adopted from the reference [4])

Fig.10 shows Is indices for the school buildings which were suffered at the Great Hanshin-Awaji Earthquake Disaster. Similar study was carried out for other earthquakes. and it was found out that the Is index of 0.6 is almost border between severe damage and moderate damage. Considering these studies, the Is index of 0.6 is adopted as a standard criterion for the retrofit of reinforced concrete buildings in Japan.

6. Implementation of Seismic Retrofit of Buildings

The purpose of seismic retrofit is to recover and upgrade original seismic performance of a building providing 1) increased strength, 2) increased ductility, and 3) a proper combination of these two features (see Fig.11.) Typical retrofit techniques available in Japan are summarized in Fig.12 and Fig.13. Generally, new elements are added to the original frames to increase strength, and new materials are applied to the original elements to improve ductility. Dowel connections are usually used for infilled walls and side walls. Steel elements may be simply attached to the existing concrete elements through mortar fill. The techniques of column jacketing may be used to improve column ductility. In the jacketing techniques, it is important to provide a narrow gap at the end of steel or concrete encasement to avoid undesired increase of shear forces resulting from the increase of flexural capacity.

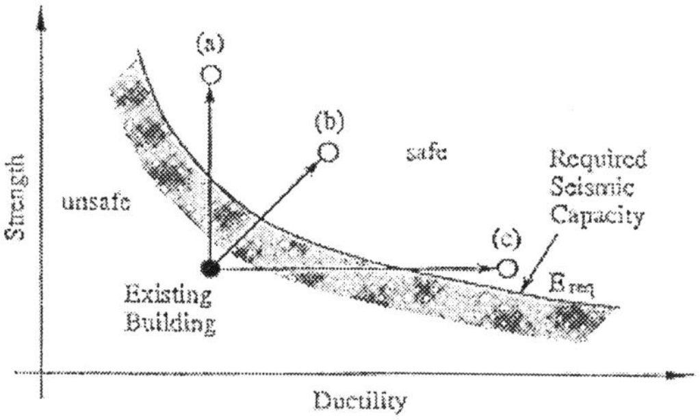

Figure 11. Construction year and seismic performance index Is (adopted from the reference [4])

A large amount of public buildings, such as schools and hospitals have been retrofitted based on "the law for promotion of seismic retrofit of buildings" after the Great Hanshin-Awaji Earthquake Disaster, however, retrofit works for private office buildings and apartment buildings seem not to be progressed well. The reasons could be summarized as follows:

1) Usually residents should move during the work of retrofitting.
2) Sometimes retrofit requires unreasonable amount of strengthening of members.
3) Construction works may produce dust, chemical smell, noise etc.
4) Law for promotion of seismic retrofit mainly target the public buildings.

To solve these problems, the BRI promoted a research project on applying new techniques on retrofit works. The research topics are described below:

1) Application of fiber sheet reinforcement techniques as easy, fast and clean methods for seismic retrofit.

2) Application of seismic isolation system to retrofit low performance buildings for the purpose of reducing input ground motions.

3) Application of seismic control devises such as viscous dampers for seismic retrofit.

4) Development of performance evaluation guideline for seismic retrofit using advanced devises.

7. Towards the Performance-Based Design Code

Currently, the Building Standards Law is under revision work to be performance-based code from prescriptive code. This revision work started in responding the need to reduce the international trade friction in the building industry. Furthermore, at the Great Hanshin-Awaji Earthquake Disaster, the performance of buildings widely ranged from no damage to collapse even they were designed according to the same current building code, and this fact also suggests the need to clarify the level of performance of a building in the design code.

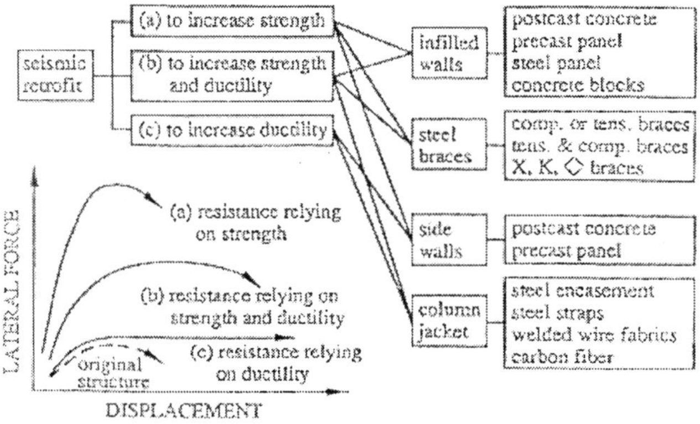

Figure 12. Typical retrofit methods in Japan (adopted from the reference [4])

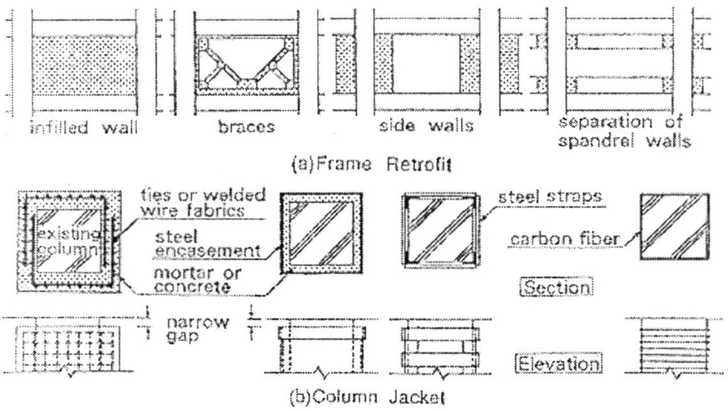

Figure 13. Typical retrofit techniques used in Japan (adopted from the reference [4])

The conceptual framework of the performance-based structural code proposed by the BRI is shown in Fig.14. It should be noted that the presented framework is up-to-now level of development and the final official version will be ready in year 2000. In the framework, following the principles of structural safety, there are several different routes are prepared for verification procedures to examine the conformity of structure with the required performance level; 1) proposed route, 2) conventional route, 3) small building route, and 4) other route. The proposed route represents a new verification procedure proposed by the BRI to be used instead of the current one. Alternative verification procedures such as those developed by private institutions are also accepted in the fourth route. In the procedures proposed by the BRI for major seismic, a building is modeled as an single-degree-of-freedom model with equivalent natural period and equivalent damping, and its maximum structural responses are estimated with the site-dependent response spectrum. The estimated response values are then compared with the limit values to confirm whether they satisfy the required performance levels.

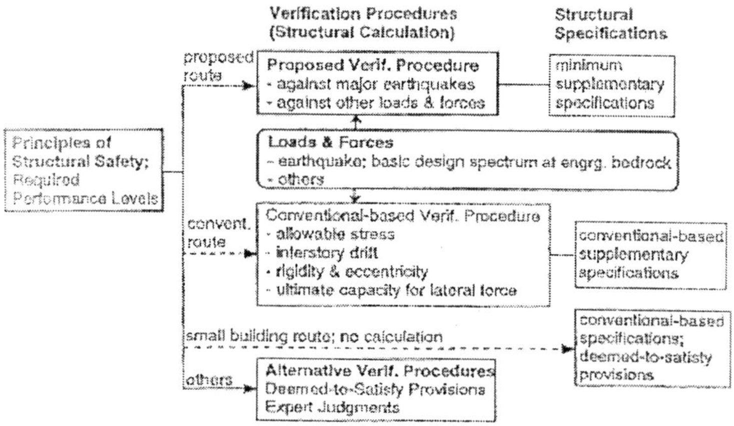

Figure 14. Conceptual Framework of Proposed Performance-Based Structural Provisions (adopted from the reference [5])

References

1. Committee for Building Damage Investigation of Hanshin-Awaji Earthquake Disaster (1995) Report on the Damage due to the Hyogoken Nanbu Earthquake, Ministry of Construction, Japanese Government.
2. Okada, T. (1998) Promotion of Seismic Retrofit of Buildings, Lecture Notes of IISEE, Building Research Institute, Ministry of Construction.
3. Japan Building Disaster Prevention Association (1977) Standards for evaluation of seismic capacity and Guidelines for retrofit of existing reinforced concrete buildings.
4. Sugano, S. (1998) Seismic Performance Evaluation and Strengthening of RC Buildings (1)-(3), Lecture Notes of IISEE, Building Research Institute, Ministry of Construction.
5. Hiraishi, H., Midorikawa, M., Teshigawara M. and Gojo, W. (1998) Development of Performance-based Building Code in Japan, the 30th Joint Meeting, UJNR.

SOME MAIN PRINCIPLES OF NEW REPUBLIC CODES ON DESIGN AND CONSTRUCTION OF SEISMIC- RESISTANT MASONRY STRUCTURES

T.G. MARKARYAN[1], L.A. DAVIDIAN[2]
[1]Associate Professor
Ministry of Town Planning of the Republic of Armenia
[2]Engineer
Armenian Earthquake Engineering Research Institute

After destructive Spitak earthquake of December 7, 1988 important scientific and engineering problems and questions appeared in the sphere of construction, in the first place referring to overestimation of the seismicity of the territory of the republic and providing required level of seismic resistance of newly erected and also damaged buildings and constructions, including masonry bearing walls.

It should be noted, that masonry buildings of various number of storeys and structural decisions are the basis of existing urban and rural building in the republic. According to predicted data during many years natural stone (mainly tuff) will be one of the main wall materials for the mass construction.

Because of increasing of normative seismicity of the territory of the republic for 1-2 points, new conceptional approaches for seismic zoning, defectiveness of Building Codes functioned before /1-3/ for the buildings with bearing walls of local atone materials and others there was a need in the development of new Republic Codes on Earthquake Engineering and masonry structures taking into account the results of analyses of the consequences of the earthquake.

SNFA II-2.02-94 "Earthquake Engineering. Standards of Design" and SNPA IY-13.01-96 "Masonry and armo-masonry structures" were developed and used at present, in which material changes and additions have been done in comparison with earlier current standards.

The following vies and circumstances ware taken into account for the development of Codes:

- the main properties and features of all types of local stone materials and masonry systems were not reflected in Building Codes functioned before in the Republic having no analogues outside the Republic;
- in projects of buildings and constructions it is necessary to note special requirements and technology of the performance of some kind of works being shown in independent Codes
- after Spitak earthquake of December '7, 1988 in Armenia the restoration (strengthening) of damaged buildings at disaster zone and also increasing seismic resistance of the buildings outside that zone became the actual problem - because of general increasing of design seismicity of the territory of the republic. Meanwhile, the

S. Balassanian et al. (eds.), Earthquake Hazard and Seismic Risk Reduction, 359–362.

chapter on restoration and strengthening of masonry structures was lack in Union Codes functioned before;

- for the past few years after the earthquake, separate republic normative and instructive documents were developed.

Let's in brief pay our attention to main principles of new Republic Building Codes on Earthquake Engineering. Specifically, in SNPA 11-2.02-94 the conception of the category of the masonry for the resistance to seismic effects is lack and depending on applied method of increasing their seismic resistance the classification of masonry is given by its structural decisions (types), namely:

1. The masonry of complex structure with vertical reinforced concrete cores by compulsory joint with anti-seismic belts in all storeys and foundation. In addition for the provision of joint work under the load of the masonry and reinforced concrete core, the last should be bined with-the wall masonry in 60 cm (two rows of masonry) by height with reinforced nets passed through the body of cores and appeared on both sides at a distance of 60 cm. The walls of complex structure can be designed both with cores forming a distinct frame (I a) and also with strengthening walls and partitions, that is to say, forming indistinct frame (1b).
2. Masonry of vertical and horizontal reinforcement applied by design. Vertical reinforcement should be anchored in antiseismic belts and foundation. Horizontal reinforcement (as a mesh) should be fixed at the pace not more than 60 cm by height;
3. The masonry being reinforced by horizontal steel meshes at the pace not more than 60 cm by height according to its design.
4. The masonry in which structural reinforcement of conjugations and intersections of walls is performed by steel grids. Besides the grids are fixed in the masonry of buildings at a distance of 90 cm arranged at the first, 60 cm - at the second and 30 cm - at the third seismic zones.

In all cases for the masonry of all types minimum value of temporary masonry resistance to axial tension along unbinded sections (normal adhesion of the mortar and, stone) is fixed which is equal to 120 KPa (1,2 kgs/cm^2);

- the requirement for the arrangement of concrete layer of the thickness 8 cm on the top of the wall masonry of each storey being reinforced by grids (for the floors of precast reinforced concrete panels) is suggested;
- structural decisions for the floors poured in place by precast reinforced concrete slabs (panels) are given and depending on applied method of pouring the acceptable pace between reinforced concrete frames is shown, which substitute cross and longitudinal walls;
- appropriate structural and other requirements are given for the applied wall materials and their volume mass, arrangement and rigidity of walls and others.

Specifically the following requirements are introduced:

- the type of the masonry and the material of walls within the building (section), as a rule, should be the same. The use of different structures or materials is accepted for the performance of the following conditions:
- horizontal rigidity of walls of each storey shouldn't differ from the same of adjacent storey for more than 20% ;
- wall materials within one storey should be similar;

- volume mass of the masonry of walls for the higher storey shouldn't exceed the same for lower storey.

Limitations of number of storeys of masonry buildings, maximum distances between the wall axes, minimum sizes of structural elements and other requirements depending on the type of applied masonry and seismic zone are introduced in Codes.

The Chapter on restoration and strengthening of buildings and constructions is shown.

Taking into account the changes and additions in SPNA II-2.02-94 the following is given in SNPA IY-13.01-96:

- design characteristics of all types of masonry which are used for building practice in Armenia including various types of finishing;
- in comparison with republic Codes functioned before normative aspects for design of masonry and armo-masonry structures on limiting states of the first Mid the second groups are given for different force effects which most actually are found during the design of masonry buildings (axial and eccentric compression, crushing, bending, axial tension and so on);
- the chapter on design of masonry buildings and its elements is added and extended;
- the chapter on armo-masonry structures is added by main aspects of design of elements with longitudinal reinforcement, masonry elements strengthened by reinforced concrete (complex structures) and also elements being strengthened by metal rings, reinforced concrete and mortar;
- the chapter on restoration and strengthening of masonry structures is introduced;
- some compulsory and reference supplements (for design characteristics of applied materials, structural decisions of masonry and compositions of applied mortar);
- the chapter on the performance of works for all types of local masonry in accordance with rules is introduced, including the works which are performed in winter and during the restoration and strengthening of structures;
- the main requirements and rules for quality control are given during the erection of masonry structures and also their acceptance.

In chapters referring to restoration (strengthening) of masonry structures the main methods of the achievement of required level of seismic resistance of buildings are listed which are accepted for the conditions of the republic.

Taking into account the achieved experience in production during the last years much attention is paid to new effective methods of restoration and increasing of monolithity, bearing capacity and seismic resistance of masonry by using polymer solutions. Besides minor amount of materials is used, lateral sizes and the weight of structures do not change and the achievement of bearing capacity of structures is provided during one day.

The main value of these methods is also the possibility of strengthening of buildings face walls preserving their appearance which is indispensable condition for the restoration and reconstruction of buildings having historical and cultural values.

It should be noted that during the development of SNPA IY-13.01-96 "Masonry and Armo-Masonry Structures" the materials of interstate and republic normative and instructive documents were used including those which were developed after Spitak Earthquake, specifically.

So, during the period after Spitak earthquake the basis of republic normative documents where actually all normative data are reflected for the design, erection and acceptance of newly constructed and restored buildings with bearing walls of local masonry under seismic conditions of the republic.

Observance of normative requirements of SNPA II-2.02-94 and SNPA IY-13.01-96 will provide the desired level of seismic resistance of erected and operated buildings with bearing masonry walls in Armenia.

INTERNET AVAILABILITY OF THE VULNERABILITY DATABASE

F. CASCIATI, L. FARAVELLI, S. PODESTA
Department of Structural Mechanics, University of Pavia
Via Ferrata 1 - Pavia, Italy

Abstract

The availability of the National Seismic Vulnerability Database in Internet is the central topic of this paper. From the different formats adopted for the vulnerability forms in different successive structural campaigns, the data were homogenised and the resulting database was made available to restricted access. The use of these data is able to supply a preliminary answer to most of the questions a building hit once a seismic event occurs.

1. Introduction

In planning seismic mitigation actions, vulnerability analysis is addressed to collect a knowledge of the buildings in the region of interest, with particular reference to their proneness to be damaged by earthquake. The purpose is the derivation of strategies for the prioritisation of seismic retrofitting. Indeed, the study of the seismic response of every single building through a mechanical model is not practicable at a territorial level. The appraisal models, therefore, must be simple and adequate for classes of buildings, and must be based on a limited number of typological and dimensional parameters.

The methodologies till now proposed in Italy are mainly conceived for buildings in masonry and in reinforced concrete. In particular, in the framework of researches supported by the National Group for Seismic Mitigation (GNDT) of the National Research Council (CNR) (GNDT, 1989), forms of I and II level have been processed. Level I forms allow one to classify every building in a single class of vulnerability, on the basis of its typological and structural characteristics. Level II forms conduct one to give each building a score (vulnerability index) through the combination of various items of heuristic nature or based on very simple mechanical models.

In both cases the evaluation of the expected damage due to a seismic event of given intensity is possible: through the damage probabilistic matrices (DPM), which supply a probabilistic distribution for all thc buildings of a class of vulnerability (Braga et al. 1982), or through the curves of fragility (Casciati & Faravelli, 1991a), which provides a correlation between the index of vulnerability of every single building and a suitable damage index, (Benedetti & Petrini, 1984; Angeletti et al., 1998; Bernardini et al., 1998; Corsanego, 1997).

The survey campaigns, carried out in Italy by GNDT, have produced an enormous amount of data. Approximately, 100000 forms of first level were filled (50000 cover the regions of the centre Italy, and nearly 40000 the area of Irpinia, shaken by the

S. Balassanian et al. (eds.), Earthquake Hazard and Seismic Risk Reduction, 363–370.

earthquake of 1980). Moreover 1745 forms of second level are available (even if those effectively readable and complete are about 1600) (Casciati & Faravelli, 1991b, 1992, 1996; Casciati et al., 1994; Baratta et al., 1996). Moreover the seismic vulnerability forms have been modified year after year, making them more and more complete and precise with the goal of supplying new informative requirements.

It is worth noting that for masonry buildings, which cannot be regarded as mechanically controlled, the knowledge of the vulnerability is just the starting point on which a detailed study of the structural response can be based.

Consequently, the acquisition and the management of such data assume a special character. The only way is to conceive an automatic management, that allows several fields to be homogenised and the information to be accessed in a fast and practical way.

2. Building A User-Friendly Database

The available data, (GNDT '84; GNDT 86/89 "A" - I and II level -; GNDT 86/89 "B" - I and II level -; Irpinia), were read by a Fortran code that manages the processing phase of the data in an automatic way: the pieces of information are made consistent

The adoption of an efficient and flexible Data Base Management System (DBMS) is then required together with techniques of accessing multi-source data. For this purpose one developed methodologies of access to the data, based on techniques of OBDC (Open Data Connectivity Base). This allows the exchange of information between different DBMS and specific applications.

In addition to the adoption of tools able to support the most common database management system, the database integrators became a need. They provide both the access to data of various origin and their storage following a procedure defined by the customer.

In order to allow the dissemination of the information within the scientific community, the resulting Seismic Vulnerability Database was made available in Internet. The access and the consultation through a browser are realised by a GUI (Graphics User Interface) which allows the selection of a subset of data answering the characteristics assigned by the operator. The procedure was created using a database package for platforms UNIX (m-SQL2.0) operating according to a standard SQL ANSI (Mini SQL, 1997).

This tool was realised through two fundamental phases:

- Design of the *tables* according to the m-SQL structure and import of the data into the tables.
- Creation of a WEB interface and successive drawing up of the programs necessary to construct the query (extraction of data and visualisation of the results).

In m-SQL language, the data were organised in tables: one must specify the number, the dimension and the typology of the fields associated with each table.

For a corrected consultation of the Database, logical-mathematical operations can be carried out between integer or real numbers, but not between strings of characters which can only be compared by analogy.

m-SQL offers the possibility of joining tables. It was used for the database 'GNDT 86/89 "B"' (made of 2 files) through the property of the single index. m-SQL does not allow to consider tables too much extended, therefore every survey has been subdivided in two sections; the first one of a descriptive nature, while the second one is based on the structural properties of the building. Each table is identified by a field which contains a progressive integer number. In this way the query causes that the research motor reads the index and automatically associates it the adjacent table, which contains the second section. Once all the tables are organised, the WEB pages have been created by programs in HTML (Hyper Text Markup Language) and LITE (Mays, 1998; Nielsen, 1996).

Database di Vulnerabilità Sismica GNDT
GNDT Seismic Vulnerability Database

Consultazione:

db1 (formato Abruzzo '84): schede di primo livello per le provincie di AQ (Aquila), FR (Frosinone), CH (Chieti), PE (Pescara), IS (Isernia), CE (Caserta).

db2 (formato GNDT '86-'89): schede di primo livello per le provincie di PZ (Potenza) e SR (Siracusa).

db3 (formato GNDT '86-'89): schede di secondo livello per la provincia di PZ (Potenza).

db4 (formato GNDT '86-'89): schede di primo livello per le provincie di PR (Parma), AR (Arezzo), CT (Catania), LU (Lucca), RG (Ragusa), PZ (Potenza).

db5 (formato GNDT '86-'89): schede di primo livello per le provincie di PR (Parma), AR (Arezzo), CT (Catania), LU (Lucca), RG (Ragusa), PZ (Potenza).

db6 (formato GNDT '86-'89): schede di secondo livello (MURATURA) per le provincie di PR (Parma), AR (Arezzo), PZ (Potenza), LU (Lucca).

db7 (formato GNDT '86-'89): schede di secondo livello (CEMENTO ARMATO) per le provincie di PR (Parma), AR (Arezzo), PZ (Potenza), LU (Lucca).

db8 (formato Irpinia): schede di primo e secondo livello riguardanti il terremoto del 198 in Irpinia, le provincie interessate sono: NA (Napoli), AV (Avellino), CE (Caserta), BN (Benevento), SA (Salerno), MT (Matera), PZ (Potenza).

Figure 1 – Initial applet of the database

HTML is a language used to construct a hypertext and represents the tool to describe the different components of an hypertext. In a single file of text, markers (TAG) are placed. When, read and interpreted from the browser, they produce the corresponding visualisations. The WEB pages are not capable of processing the data; this is why it was necessary to compile another program: in LITE language. It is based on the semantic of the C language and contains a bookcase of commands capable of directly communicating with the WEB interface.

After all that, the visualisation (figure 1) of the WEB pages results. An introductory page lists the eight implemented Databases. For the selected Database, it is possible to approach the single Province of interest.

Moreover, the customer will be able to visualise, as a result, all the fields available or only some (with the limit of eight) of them. The eight fields cover both information of descriptive type and data of engineering interest.

3. Consulting the Database

The way of using the Seismic Vulnerability Database, changes depending on the purposes one is pursuing. This example of consultation compares the indices of vulnerability, deduced from the database, with the hazard values, resulting from the response spectrum of the site. Two and three storeys buildings at Monchio delle Corti and Palanzana, two villages in Parma province, are considered (Picciocchi, 1998).

The vulnerability indices obtained from the consultation of the database are reported in Table 1.

The further data, extracted from the vulnerability forms are: β_1, β_2, H (the buildings height) and A_t (the total area). The coefficients β_1, β_2 are defined as the ratios between the main plan dimensions of the building. After that, the natural periods of the buildings were computed, according to the Italian law "D.M. 16/01/1996 : Norme tecniche per le costruzioni in zona sismica" (*Technical code for the constructions in seismic zone*).

$$T = 0.1\frac{H}{\sqrt{B}}$$

where:

$$B = \sqrt{\frac{A_t \times 100}{\beta_1 + \beta_2}} \; ; \quad H = \text{total height of the buiding}$$

TABLE 1

ISTAT Code	n° storeys	At (m^2)	%b1	%b2	Vulnerab. index
22 (1)	2	122	48	21	**61.25**
22	2	145	83	0	**155**
22	2	484	61	8	**86.25**
22	3	182	29	29	**116.25**
22	3	110	70	20	**80**
26 (2)	2	100	50	18	**120**
26	2	114	62	0	**41.25**
26	2	235	26	28	**61.25**
26	3	415	22	24	**145**
26	3	144	87	22	**117.5**
26	3	463	52	0	**112.5**
26	3	72	56	0	**67.5**

(1) Palanzana

(2) Monchio delle Corti

The response spectrum at the two sites, can be used to derive an estimate of the single building risk (figure 2), provided it was accurately built.

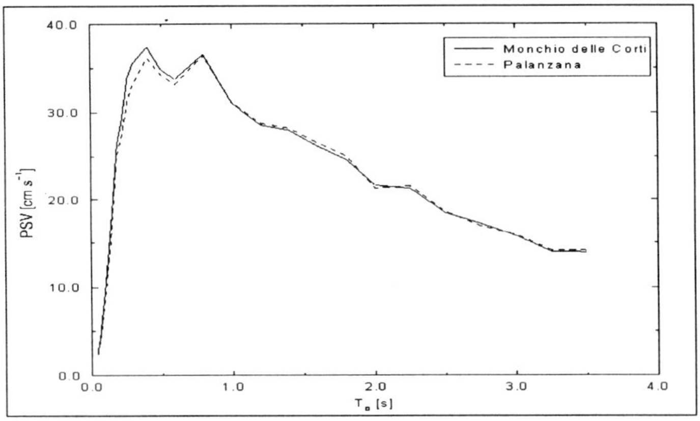

Figure 2 - Response spectrum at the two sites

In the village of Palanzana, the data for five buildings (of three and two storeys) were extracted. The vulnerability indices (Table 1) vary in the range (61.25, 155) for the two-storey buildings and (80, 116.25) for those with three storeys. The natural periods cluster around the values of 0,15 sec for two storey buildings, and 0.22 sec for three storeys ones: indeed they are rather uniform. For the frequency value of 4.55 to which a period of 0,15 sec corresponds, the spectrum gives, a value of pseudo - velocity (PSV) of approximately 26 cm/sec, while for a value of 6.67 (T = 0.22 sec) the pseudo - velocity (PSV) of 22 cm/sec (figure 3).

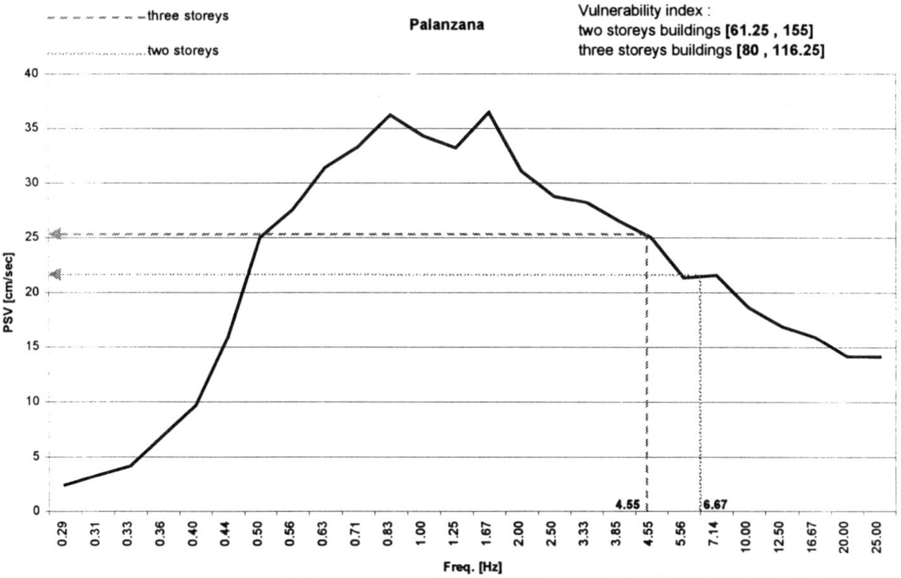

Figure 3 - Response spectrum of Palanzana – It is possible to observe the PSV value for the frequency of two and three storey buildings

In the village of Monchio delle Corti, the data relative to seven buildings were extracted. Also in this case, like in the village of Palanzana, the natural periods of two storey buildings are all around the value of 0,13 sec; while, for the buildings with three storeys, the natural periods change from values of 0,16 sec until 0,24 sec.

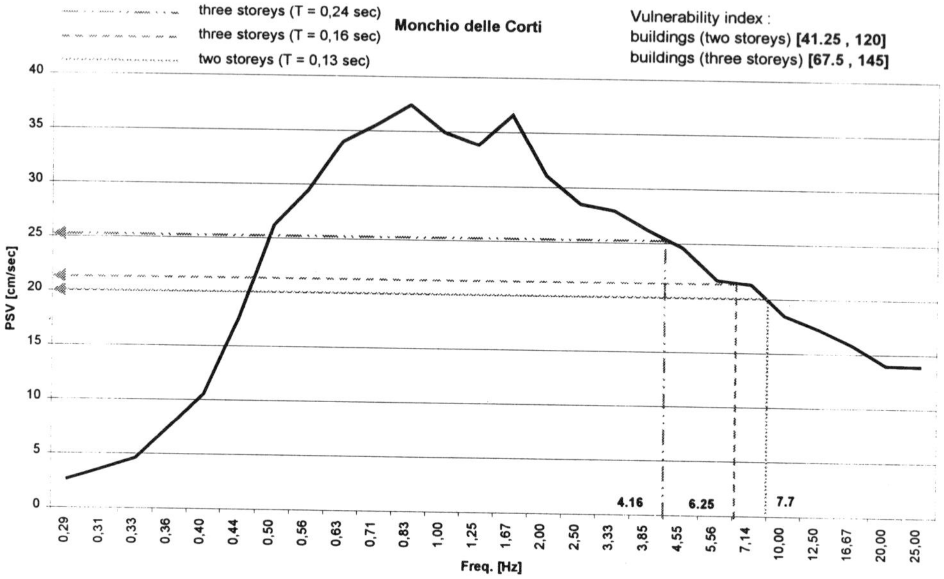

Figure 4 - Response spectrum of Monchio delle Corti - It is possible to observe the PSV value for the frequency of two and three storey buildings

In this case three values of pseudo velocity are determined (figure 4): one for two storey buildings, according to the natural period of 0.13, and two for three storey buildings according to the maximum and minimum value of natural period (0.16 – 0.24 sec).

From the response spectrum at the site it is possible to observe:

- *three storey buildings*: for periods equal to 0,16 sec (→ natural frequency equal to 6.25 sec^{-1}), the pseudo - velocity is approximately 21 cm/sec, while for values around to 0,24 sec (→ natural frequency equal to 4.16 sec^{-1}), it achieves 26 cm/sec.
- *two storey buildings:* for the average value of the period equal to 0,13 sec (→ natural frequency equal to 7.7 sec^{-1}) the pseudo velocity is 20 cm/sec.

From the analysis of the results, it is possible to observe that:

- the dissemination of the vulnerability index does not allow one to deduce a classification of risk degree on the basis of the number of building storeys;
- the appraisal of the natural period, also with the approximations of the adopted formulation (cfr. DM. 16/01/96), allows one to obtain values with negligible standard deviation;
- from the shape of response spectrum it is possible to underline that the more stressed buildings (three storey buildings) are the less vulnerable.

4. Conclusions

The possibility of having a database of easy consultation, represents a valid tool for the planning of areas with high seismic risk. To learn from past earthquakes, in fact, means to exploit the base of knowledge in order to understand the seismic behaviour of the structures. In particular, ancient buildings offer an intrinsic vulnerability to the seismic actions, because the horizontal inertial forces can cause the loss of static equilibrium, mainly for poorly connected elements.

Moreover the masonry strength to traction is low. The design trend was, therefore, to arrange the elements in such a way that the weaker plans (mortar joints) are stressed by action of pure compression. But the seismic action induces tangential stresses, with consequent cracks for sliding or separation between the elements.

Counteractions rely on the good quality of the connections, the use of adequate stiffners, the realisation of a box behaviour, the insertion of adequate elements to contrast the overturning mechanisms. The rules of thumb, nearly never written, arising from the constructor's experience, are built observing the damage mechanisms of collapse, and translating these observations into a series of structural expedients. Such rules are applied for one or two generations and then are gradually lost.

Presently the information from vulnerability analysis represents the base of indispensable knowledge for a seismic design improvement. Only through a friendly user database, it is possible to make the collected experience effectively applicable. The database discussed in this paper must now be enlarged, one class of buildings of present interest are the churches.

The churches are probably, within the monumental heritage, the mainly vulnerable typology, also for earthquakes of a low-medium intensity. The observation of the damages after earthquakes in Italy and the successive researches have already produced a broad and consolidated base of knowledge of the more frequent damage mechanisms (Doglioni et al. 1996, Lagomarsino 1997). The next step is to make them available.

5. Acknowledgement

The research activity reported in this paper was supported by GNDT (National Group of Seismic Mitigation) of CNR (National Research Council). The present paper is supported by a grant from the University of Pavia (FAR) to the second author. Moreover a particular thanks goes to Dr. Marco Pagani for his precious help in building the database.

6. References

Angeletti P., Bellina A., Guagenti E., Moretti A., Petrini V., 1988, "Comparison between Vulnerability Assessment and Damage Index, Some Results", *Proc. 9th World Conference on Earthquake Engineering*, 7, Tokyo-Kyoto. pp. 181-186.

Baratta, A., Colletta, T. & Zuccaro, G. 1996. *Seismic risk of historic centres. A preliminary approach to the Naples' case*. Napoli: La Città del Sole.

Benedetti D., Petrini V., 1984, "On Seismic Vulnerability of Masonry Buildings: Proposal of an Evaluation Procedure", (in Italian) *L'Industria delle Costruzioni*. Vol. 18, pp. 66-78.

Bernardini A., Gori R., Modena C., 1988, "Valutazioni di Resistenza di Edifici in Muratura per Analisi di Vulnerabilità Sismica", (in Italian) Istituto di Scienza e Tecnica delle Costruzioni, Università di Padova, R.I. 2/88.

Braga, F., Dolce, M. & Liberatore, D. 1982. "A statistical study on damaged buildings and ensuing review of MSK-76 Scale" (in Italian), in: *Terremoto in Irpinia del 23 Novembre 1980* (Capitolo 5), *Progetto Nazionale Geodinamica (CNR)*, Roma: Ed. Scientifiche Associate.

Casciati F., Faravelli L., 1991a, *Fragility Analysis of Complex Structural Systems*, Research Studies Press, Taunton, UK.

Casciati F., Faravelli L.,1991b, "A Knowledge-based System for Seismic Vulnerability Assessment of Masonry Buildings", *Microcomput. Civil Engineering*. Vol. 6, pp. 291-301.

Casciati F., Faravelli L., 1992, "Exploiting expertise in Monumental Buildings Diagnosis", *Proc. 10th WCEE*, Madrid, Spagna, pp. 6283-6289.

Casciati F., Faravelli L., Liu X.D., 1994, Lessons of Rehabilitation Design Learned from Statistical Analyses of Masonry Seismic- Vulnerability Data. The vulnerability Shell", *Structural Safety* Vol. 16, pp. 73-89.

Casciati F., Faravelli L., 1996, "Managing a seismic vulnerability database", *Post-earthquake rehabilitation and recostruction,* Elsevian NY, pp. 241-249.

Corsanego A., 1997, *Valutazione della Vulnerabilità. Rischio Sismico di Edifici Pubblici.* (in Italian) Gruppo Nazionale per la Difesa dai Terremoti (GNDT), Consiglio Nazionale per le Ricerche.

Doglioni, F., Moretti, A. & Petrini, V. 1994. *Le Chiese e il terremoto* (in Italian). Trieste, LINT.

GNDT – CNR, 1989, Istruzioni per la Compilazione della Scheda di Rilevamento Esposizione e Vulnerabilità Sismica degli Edifici in Muratura.(in Italian)

Lagomarsino, S., 1999. "Damage survey of ancient churches: the Umbria-Marche experience", in A. Bernardini (ed.), *Seismic Damage to Masonry Buildings, Proceeding of the International Workshop on "Measure of seismic damage to masonry building* Monselice, Padova, Italy, June 25-26, 1998, Rotterdam: Balkema.

Mini-SQL, 1997, *User Guide*, Hughes Technologies.

Mayr D., 1998, *The History of the Internet and the WWW.*

Nielsen F.H., 1996, *The Original Hypertext Transfert Protocol. A Beginner's Guide to HTML*, NCSA

Picciocchi S 1998, "La vulnerabilità sismica: Elaborazione della banca dati italiana" (in Italian), Master Thesis, Department of Structural Mechanics, University of Pavia

ASSESSMENT OF DAMAGE AND USABILITY OF EARTHQUAKE-DAMAGED BUILDINGS: EXPERIENCE OF SLOVENIA

M. TOMAŽEVIČ
Slovenian National Building and Civil Engineering Institute
Dimičeva 12, 1000 Ljubljana, SLOVENIA

Abstract

Within the seismic preparedness plan of Slovenia, prepared by the Administration for Civil Protection and Disaster Relief, a methodology for the assessment of damage and usability of earthquake damaged buildings has been recently developed, based on the experiences obtained in the country and internationally. On the basis of damage evaluation, the buildings are classified into three categories of usability: usable, temporarily unusable, and unusable buildings. In Slovenia, a two level system of damage and usability assessment has been adopted. In order to provide information regarding the safety of occupancy of buildings and to propose immediate safety measures, rapid evaluation is carried out immediately after the earthquake. In the case of detailed evaluation, however, the earthquake-damaged buildings are inspected for a second time, and damage to structural elements and structure as a whole is inspected in detail.

1. Introduction

Slovenia is an earthquake-prone country, where earthquakes of intensity VII or more by European Macroseismic Scale (EMS - European, 1993) are expected within 475-years return period. In the most densely populated areas, the expected intensity of earthquakes is VIII and more by EMS scale. Buildings, designed and constructed before the adoption of the first seismic code in 1963, still represent more than 50 % of the existing building stock. Many of them are traditional stone and brick masonry buildings of historical value, in the case of which substantial damage can occur even when subjected to moderate earthquakes. Seismic vulnerability of these and other buildings, constructed before the adoption of the first seismic code, has been demonstrated by earthquakes, as well as by recent seismic vulnerability studies. In some cases, even the buildings, designed and constructed after the adoption of the first seismic code of 1963, proved to be vulnerable.

Since rapid and reliable evaluation of damage to buildings and assessment of usability represent the basis for subsequent earthquake disaster management activities, the preparation of an adequate methodology has been brought to the fore within the

S. Balassanian et al. (eds.), Earthquake Hazard and Seismic Risk Reduction, 371–378.

national earthquake disaster mitigation program. In order to be successfully applied, methodology should be prepared and experts should be trained to become familiar with the basic criteria, well before the earthquake.

2. Previous Experience

In the period from 1963, when the first regulations for seismic resistant design and construction have been passed just before the earthquake of Skopje, Macedonia, which represented the milestone in seismic resistant construction of buildings in former Yugoslavia, to 1998, Slovenia suffered from three damaging earthquakes. Namely, the earthquake of Kozjansko in 1974, the series of earthquakes with epicentres in Friuli, Italy, which damaged Posočje, i.e. the area of Soča River Valley, the western-most part of Slovenia, in 1976, and the earthquake of Bovec, which struck the Upper Soča River Valley in 1998. There were other earthquakes on the territory of former Yugoslavia, where also Slovenian experts took part in earthquake mitigation programs, such as Banja Luka earthquake in 1969, and the disastrous Montenegro earthquake in 1979, where an important part of experience has been gained.

In 1974, a governmental committee for damage evaluation has been formed and a questionnaire for damage evaluation of buildings has been prepared. At that time, usability of buildings has been defined by the category of damage. Since the earthquake of 1974 affected a rural area, damage to only masonry buildings has been considered. According to the observed damage, the buildings have been classified into four categories: category I buildings with no or only minor damage, category II buildings with damage to non-structural elements, category III buildings with damage to structural walls, and category IV buildings where structural damage beyond repair or partial collapse has been observed. In the proposed questionnaires, the necessary repair works have been specified for each category of damage, and the amount of works had to be defined during the inspection of each particular building. In this way, the costs of repair and reconstruction of earthquake-damaged buildings have been evaluated as a basis for the estimation of direct economic property losses.

After the subsequent earthquake of Friuli in 1976, the questionnaire has been modified to make possible automatic data analysis. Again, since very few modern r.c. structures have been affected by the earthquake in Posočje, only masonry buildings have been considered. During the inspection, inspectors answered the questions and estimated the amount of works needed for repair and strengthening. On the basis of the approved unit costs and the estimated amount of damage, the costs of repair and strengthening of each building have been automatically evaluated.

In addition to the estimation of the amount of damage, inspectors also assessed the usability of inspected buildings, classified the buildings into one of the three predefined usability classes, and marked the buildings with visible marks according to their decision. In this regard, a building usability classification scheme, shown in Table 1, has been proposed.

TABLE 1: *Classification of usability and damage of earthquake-damaged buildings (Posočje, 1976)*

Category	Colour	Usability	Damage
1	-	continuing occupancy	undamaged structural system, minor damage to non-structural elements
2	yellow	temporarily unusable buildings; reoccupied after repair and/or strengthening	minor damage to structural elements, damage to non-structural elements, repair and strengthening economically justified
3	red	to be demolished	collapse, partial collapse or heavy damage to structural system; repair and strengthening not economical

Whereas only a few hundreds of buildings have been damaged in the region of Kozjansko in 1974, the number of inspected buildings exceeded 6500 in Posočje in 1976 (Ladava, 1982). Nevertheless, the assessment of both damage and usability of buildings has been carried out simultaneously in a relatively short period of time in both cases. However, in the case of Montenegro earthquake of 1979, where the number of damaged buildings exceeded 57000, the usability of buildings had to be assessed in the first place (Pavičević, 1980). Consequently, in the questionnaire, proposed in 1979 by a Federal Co-ordination Committee, the basic identification data about the inspected buildings have been collected (location, occupancy, and structural system), and the answers to questions regarding the overall damage and usability of buildings had to be given. No questions, regarding the type and amount of damage to individual structural and non-structural elements, are specified in the questionnaire. The costs of damage, or economic losses, have been estimated on the basis of a detailed damage evaluation of a random sample of 150 buildings of all characteristic typologies existing in the region. In the years to follow after the earthquake of Montenegro, a new questionnaire has been proposed for the assessment of damage and usability of earthquake-damaged buildings (Methodology, 1984).

The damage and usability of building classification scheme, proposed after the earthquake of Montenegro in 1979, is presented in Table 2.

TABLE 2: *Classification of usability and damage of earthquake-damaged buildings (Montenegro, 1979)*

Category	Colour	Usability	Damage
I	green	usable	1. undamaged buildings 2. damage to buildings, with no damage to structural elements 3. minor damage to structural elements
II	yellow	temporarily unusable	1. damage to structural system 2. severe damage to structural system
III	red	unusable	1. heavy damage to structural system 2. partial collapse 3. total collapse

3. Slovenian Methodology

Within the earthquake preparedness plan of Slovenia, recently set by the Administration for Civil Protection and Disaster Relief (ACPDR), a methodology for the assessment of damage and usability of earthquake-damaged buildings has been prepared, based on the experiences obtained in the country and internationally. In order to train the professionals, not familiar with the effects of earthquakes on buildings, to be ready for an efficient and timely action in the case of the expected damaging seismic events, a manual for the assessment of damage and usability of earthquake-damaged buildings of all types has been recently published (Tomaževič, 1998). In the manual, the proposed methodology is described in details, and the basic lessons are given regarding the effects of earthquakes on building and civil engineering structures, including the principles of post-earthquake repair and strengthening. As an appendix to the manual, a CD-ROM has been prepared with more than 500 slides showing, in a systematic way, typical damage to buildings, and explaining the causes of damage.

3.1. CLASSIFICATION OF DAMAGE AND USABILITY

Since the safety of a structure depends on the degree of damage to structural elements, structural damage is the decisive parameter to be taken into account in the assessment of usability of earthquake-damaged buildings. In this regard, the possibility of occurrence of strong aftershocks should also be carefully considered. However, the usability of buildings after an earthquake also depends on the degree of damage to non-structural elements. Therefore, in the assessment of damage and usability, non-structural elements should be inspected, especially with regard to falling.

For different purposes and building typologies, different systems of classification of damage to buildings have already been proposed. International Association for Earthquake Engineering (IAEE) proposed a five category damage classification (not including undamaged buildings) for non-engineered construction (Guidelines, 1986). Similarly, damage to buildings is classified in the new European Macroseismic Scale. To be compatible with definitions given in EMS scale, and having in mind the idea that the same questionnaires, used for the assessment of damage and usability of buildings, could be used to estimate the intensity of the earthquake in the particular location, as well as for various post-earthquake studies, Slovenian proposal also adopted five categories (classes) of damage as follows:

- damage category 1: negligible to slight damage;
- damage category 2: moderate damage,
- damage category 3: substantial to heavy damage,
- damage category 4: very heavy damage, and
- damage category 5: partial or total collapse.

On the basis of classification of damage to structural and non-structural elements, the usability of buildings is assessed and classified into three categories:

- *Usable buildings* - continuing occupancy: damage category 1 and 2, marked green;
- *Temporarily unusable buildings* - restricted occupancy, to be resettled after repair and/or strengthening: damage category 3 and 4, marked yellow;

- *Unusable buildings* - unsafe buildings, to be demolished: damage category 5, marked red.

In the case of usable buildings, building occupancy may continue without limitation. In the case of temporarily unusable buildings, occupancy may continue if the safety can be ensured by emergency safety measures. In most cases, however, the occupancy is not possible until the building is completely repaired and strengthened. Unusable buildings are demolished and the debris removed.

3.2 ASSESSMENT PROCEDURE AND QUESTIONNAIRE

Assessment of usability of buildings should be carried out in as short period of time as possible. On the basis of the number of temporarily unusable and collapsed buildings, the number of residents which need additional assistance in terms of temporary shelters, such as tents and mobile homes, or should be otherwise moved to a safe place, can be determined. This is, besides rescue and immediate relief, one of the basic tasks of the earthquake emergency management body. Also, the function of important public buildings, such as hospitals and fire stations, as well as basic infrastructure and industry, should be assessed as rapidly as possible. The reconstruction planning and the estimation of direct cost of losses, however, which is based upon the detailed assessment of damage, can be postponed for some time.

Therefore, the procedure of assessment of damage and usability is divided in two phases. In the first phase, which, depending on the number of earthquake-damaged buildings, can be completed in a few days after the earthquake, the usability of buildings is assessed by rapid inspection of buildings, without entering into details. In the phase of detailed evaluation, however, the earthquake damaged buildings are inspected for a second time, and damage to structural elements and structure as a whole is inspected in detail. The questionnaires needed for the assessment of damage and usability have been conceived accordingly.

In the first part of the questionnaire, the building identification data are collected, such as location, owner's name and address, building's usage, number of residents and/or employees, age, size and orientation of buildings. Data about structural system and foundation soil, as well as condition of the building before the earthquake, are also obtained. Since these data will be used for various kinds of analyses, resulting in various kinds of maps, the location is defined with geographical co-ordinates of building's centroid. In this way, all data can be presented and analysed in a Geographic Information System (GIS).

In the second part of the questionnaire, data regarding the observed damage and the assessed usability of the inspected building, obtained during the rapid evaluation, are presented. Category of damage and resulting usability is estimated, data regarding the observed soil instability and possible casualties are presented, and recommendation are given for necessary emergency measures. If needed, different relevant remarks and observations regarding the safety and usability of the inspected buildings are written in the provided space (Figure 1).

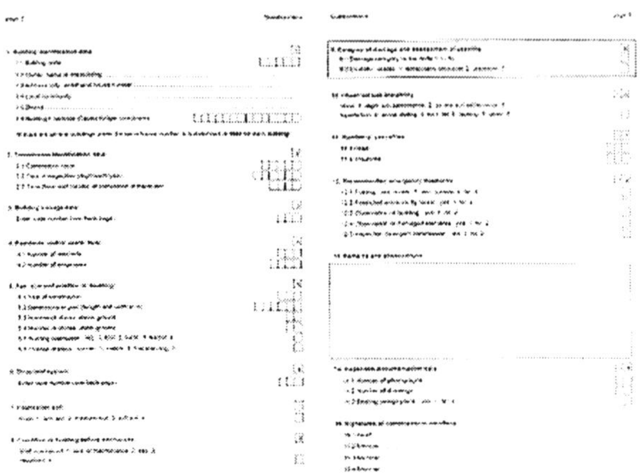

Figure 1. Pages 2 and 3 of the Questionnaire with building identification data (page 2) and assessed damage and usability during rapid evaluation (page 3)

In the Appendix to the Questionnaire, which is filled in during the detailed evaluation of buildings, structural material data are collected (materials of structural and non-structural elements), and damage to structural and non-structural elements is assessed in detail. A sketch of the structural system, in both plan and elevation, with relevant dimensions is plotted in the appropriate space. The data are collected in such a way that costs of damage, or economic losses can be estimated on the basis of the known unit prices for strengthening and repair works.

4. The Case of the Earthquake of Bovec of April 12, 1998 and Conclusions

On Easter Sunday, April 12, 1998, an earthquake of estimated epicentral EMS intensity VII-VIII damaged the area of Upper Posočje (Soča River Valley), a mountainous region in the western part of Slovenia. The epicentre of the main shock was located at 46.31 degrees north and 13.64 degrees west, focal depth was somewhere between 15 and 18 km, and the local magnitude was estimated to M = 5.5 by Richter scale (Report, 1998).

Following the proposed methodology for the assessment of damage and usability of earthquake-damaged building, and based on information received from local staff of Civil Defence regarding the approximate amount of damaged buildings, ACPDR immediately mobilised 25 professionals, who inspected the damaged buildings and assessed their usability according to the proposed methodology. Out of 952 inspected buildings, where the damage has been reported immediately after the earthquake, 311 have been found as temporarily not usable. Among them, 1/3 have been damaged seriously, in some cases even beyond repair, and 2/3 have suffered repairable damage. More than 710 people have been temporarily moved to mobile homes, located close to their homes. Four days have been needed to complete the rapid evaluation of the earthquake-damaged area. It has to be mentioned that only buildings with permanent residents and some cultural monuments have been included in the rapid evaluation of damage and usability.

According to the time-plan of activities within the earthquake preparedness program of Slovenia, prepared by ACPDR, a continuos training of experts, members of committees for the assessment of damage and usability of earthquake damage buildings has been set forth. On the basis of the action plan, following the strategy of earthquake disaster mitigation, the proposed methodology for the assessment of damage and usability has been first exercised during a simulation of an earthquake disaster, carried out in Ljubljana in 1996. To facilitate the training of professionals, a manual has been prepared (Tomaževič, 1998) and the official training has been planned to start in 1998/99. Unfortunately, the manual has not yet been printed, and the training has not yet started at the time of the earthquake of April 12. Consequently, professionals, mobilised by ACDPR, had to be briefly educated by experts on the site. In order to prevent any misjudgement in the estimated degree of damage and assessed usability, in the four days of the assessment activities, the inspected buildings, found unusable by one committee, have been re-inspected by the other committee and marked accordingly.

Despite the timely and efficient assessment of usability of earthquake-damaged buildings, an important lesson can be learned on the basis of experience of the earthquake of Bovec of 1998, regarding the assessment of damage. In 1998, the rapid evaluation of buildings has been successfully carried out under the authority of ACDPR, which is responsible for disaster relief. The proposed methodology, described in this contribution, has been used.

However, the detailed evaluation of buildings and assessment of damage has been carried out under the authority of the Ministry of environment and physical planning, which is responsible for the construction buildings in the Republic of Slovenia, and, hence, for the reconstruction of the earthquake-damaged area. For damage evaluation, a methodology, not consistent with the methodology proposed by ACDPR has been used, and professionals, not included in the rapid evaluation of earthquake-damaged buildings immediately after the earthquake, have carried out the damaged assessment.

Although the number of all damaged buildings in the area, including buildings where new damage caused by the aftershocks, and only minor non-structural damage has been reported by the owners, by far exceeded the number of buildings, inspected immediately after the earthquake, the non-consistency of both methodologies was one of the reasons for the late beginning of the reconstruction program.

Hopefully, the activities between the responsible authorities will be co-ordinated in the future and unique methodology for the assessment of damage and usability will be used during future seismic events. The experiences obtained also confirmed fact that timely and constant training of professionals, who will be mobilised in the case of an earthquake disaster, for both usability assessment and damage evaluation, is needed. Otherwise, significant delays and unnecessary complications in earthquake disaster mitigation activities can be expected in the case where an expected severe earthquake would affect the most densely populated earthquake-prone areas of Slovenia.

5. References

European Macroseismic Scale - 1992 (up-dated MSK scale) (1993). G. Grünthal, (ed.), European seismological Commission. Luxembourg.

Guidelines for Earthquake Resistant Non-Engineered Construction (1986). International Association for Earthquake Engineering, Central Office, Tokyo.

Ladava, A. (1982). Guidelines and procedures used to eliminate the impact of the earthquake in the Soča Valley. *Proc. Social and Economic Aspects of Earthquakes,* Jones, B.G., Tomaževič, M., Eds., Institute for Testing and Research in Materials and Structures, Ljubljana - Cornell University, Ithaca, pp. 413-423.

Methodology and Procedure for Earthquake Damage Assessment (1984). Institute for Earthquake Engineering and Engineering seismology, Skopje, Macedonia.

Pavičević, B. (1980). Activities and technical, design and planning measures for the mitigation of consequences of catastrophic earthquake of Montenegro. *Proc., Experience and conditions for design and construction of buildings after the earthquake of Montenegro.* Njivice, Montenegro, pp. 23-44 (in Serbian).

Report on Mitigating the Consequences of the Earthquake of Bovec of April 12, 1998. (1998). Administration for Civil Protection and Disaster Relief, Ljubljana (in Slovene).

Tomaževič, M. (1998). *Assessment of Usability of Earthquake-Damaged Buildings.* Administration for Civil Protection and Disaster Relief, Ljubljana (in Slovene).

DESIGN AND TESTING OF HIGH DAMPING RUBBER EARTHQUAKE BEARINGS FOR RETROFIT PROJECT IN ARMENIA

K.N.G. FULLER[1], C.L. LIM[2], S. LOO[3], M. MELKUMYAN[4], K. MUNIANDY[5]
[1]*TARRC, Brickendonbury, Hertford, UK*
[2]*RRIM Technology Centre, Sungai Buloh, Selangor, Malaysia*
[3]*Min Industries Sdn. Bhd., Lot 4-9 Beranang Industrial Estate, Selangor, Malaysia*
[4]*Earthquake Engineering Center, National Survey for Seismic Protection, Davidashen-IV Massiv, Yerevan, Armenia*
[5]*Sime Engineered Rubber Products Sdn Bhd., 2 Jalan Tandang, Petaling Jaya, Selangor, Malaysia*

1. Introduction

Isolation of structures from horizontal ground motions is gradually becoming a more common method of providing protection from earthquake damage. By reducing the seismic forces transmitted, isolation protects the contents and secondary structural features as well as the main structure; the safety of occupants and passers-by is thus also enhanced. High damping rubber (HDR) bearings are a simple and economical means of providing isolation. They have the low horizontal stiffness required to give a long vibration period (typically 2s) to the structure when mounted on the bearings. Their vertical stiffness is high, thus minimising any rocking of the structure during an earthquake. The damping needed to limit the displacement of the structure and reduce the response at the isolation frequency is incorporated into the rubber compound so that there is generally no need for auxiliary dissipation devices. The bearings can be designed to withstand safely the large horizontal displacements imposed during an earthquake. The service life of the bearings is expected to be several decades (Fuller & Roberts, 1997), and they should require no maintenance. Many projects throughout the world have installed seismic isolation systems based on HDR bearings. During the 1994 Northridge earthquake in California and the Kobe earthquake of 1995 seismically-isolated buildings performed well.

There has been much emphasis on the suitability of base-isolation for critically important structures such as hospitals and emergency centres. The protection of both structure and contents, however, would be advantageous for ordinary civil structures such as apartment blocks. Projects in Indonesia and China have confirmed the technical and economic feasibility (Fuller et al., 1995). Isolation using HDR bearings has potential for the rehabilitation of ordinary civil structures such as apartment blocks. The first retrofit of such a building has been carried out in Armenia. This paper describes the

S. Balassanian et al. (eds.), Earthquake Hazard and Seismic Risk Reduction, 379–385.

design of the isolation system and the individual bearings, and presents data on the properties of the bearings and the high damping rubber developed for their manufacture. The severe winter weather at the site meant that particular attention had to be paid to the low temperature crystallization resistance of the rubber compound.

2. Design Of Isolation System And Bearings

The rehabilitated building is located in Vanadzor (formerly Kirovakan) about 70km north of Yerevan. It is a five-storey masonry apartment block (plan dimensions 52x15metres), symmetrical about the centre of the long side. The rehabilitation involves cutting the masonry walls at foundation level. Under the load-bearing walls two sets of reinforced concrete beams are constructed one directly above the other (Melkumian, 1994). One set rests on the foundations and the other directly supports the load-bearing walls. The upper beams rest upon the high damping rubber isolators. The range of the loads supported is quite large so it was decided to have two types of bearing, but differing only in the shear modulus of the rubber. One type (designated Hard) would support loads in the range 74 - 90t and the other (Soft) loads in the range 50 - 68t. This strategy would lead to a somewhat more uniform distribution of horizontal seismic forces, and help ensure that the centre of stiffness of the isolation system is close to the projection of the centre of mass of the building on to the ground plan. The latter minimises the excitation of torsional vibration of the building on the isolators during an earthquake.

An isolation frequency of 0.5Hz was chosen. For the design earthquake input and a damping in the isolation system of 10%, this implies a design horizontal displacement of 130mm. The bearings were to operate safely up to 1.5x the design displacement. The design requirements including the horizontal stiffness for the two types of bearing and the performance details of the design chosen are summarized in Table 1. The rubber shear strain at the design displacement is chosen to be 100%; thus the shear moduli in Table 1 refer to that strain. The overall diameter of the bearings is 380mm and their overall height is 202.5mm.

Because of the desire to simplify the manufacture and installation on site of the bearings, the simple recess connection detail, as used for the previous projects in Indonesia and Shantou, China (Fuller et al., (1995)) was chosen. Such an option necessitates a check that the bearings are safe against roll-out at the maximum horizontal displacement; the check has to take account of the reduction in vertical load on some of the bearings due to the overturning of the building at large displacements. Using a simple dynamic analysis (Muhr, 1994) the maximum and minimum loads for each type of bearing, after accounting for the effect of overturning of the building at the maximum horizontal displacement, are listed in Table 1. None of the bearings is close to being totally unloaded at the extreme displacement. Regarding the maximum loads estimated, there is a substantial margin against instability (at zero displacement) and yielding of the internal reinforcing plates (at zero displacement) The displacement for the onset of roll-out is included in Table 1. All the bearings are safe against the onset of roll-out at the maximum displacement; for all but one there is a margin of at least 20%.

TABLE 1: *Requirements And Performance Details Of Bearings*

Bearing type	Hard	Soft
Horizontal stiffness, kN/mm	0.81	0.58
Design vertical load, t	82	59
Design horizontal displacement, mm	130	130
Max. horizontal displacement, mm	195	195
Max. static vertical load, t	90	68
Static compressive stress (max), MPa	8.7	6.5
Vertical loads (t) at max horizontal displacement:		
Max.	124	98
Min.	44	14
Rubber shear modulus, MPa	0.97	0.69
Critical load (t)	326	236
Load for internal plate yield (t)	480	480
Horizontal displacement at onset of roll-out, mm		
for design vertical load	300	300
for min. vertical load	260	196
Nominal vertical stiffness, kN/mm	400	310

Because the bearings are located in a recess rather than bolted the end-plates can be covered with rubber to afford corrosion protection.

Hard and Soft rubber compounds were developed to satisfy the requirements of the two types of bearing. The dependence of the compound dynamic properties upon strain amplitude and temperature are summarised in Figures 1 and 2. For both compounds the modulus increases by a factor of 3 to 3.5 between 100% and 5% strain.

At the site of the building prolonged spells of low temperature may be anticipated. It is required therefore that the high damping rubber should not stiffen excessively due to low temperature crystallization after a period of 7 days at -20°C or 30 days at -10°C. In order to ensure this, special compounds were developed. Their performance at -20°C under a static shear strain of 35% (a strain equivalent to that imposed on the rubber by the design static vertical load acting on the bearings) was checked by intermittent measurement of the shear modulus for 2% incremental shear strain. The Soft compound showed no significant stiffening after 30 days at even -20 C. In contrast the Hard compound stiffened much more rapidly, the modulus rising at -20°C by about 40% over 7 days. The increase in stiffness after 30 days at -10 C was found to be about 45% for the Hard compounds, a figure regarded as just acceptable.

3. Bearing Tests

The shear dynamic stiffness of the Soft and Hard bearings was measured over a range of rubber strains at a frequency of 0.1Hz under a constant vertical load of nominally 50 tonnes. (For the Soft bearing test at 130mm displacement, the vertical load was 38t.)

These tests were carried out on a single bearing facility using a biaxial load-cell directly under the bearing. The magnitude of the vertical load applied was limited by the capacity of the biaxial load-cell, and thus the bearings could not be tested under the design vertical load; this should not have influenced the results significantly for the Soft bearing, but the Hard bearings would be expected to be about 3% stiffer under 50t instead of the design vertical load. The horizontal capacity of the dynamic test facility limited the rubber strain amplitudes to 100% for the Soft-type bearing and 50% for the Hard type.

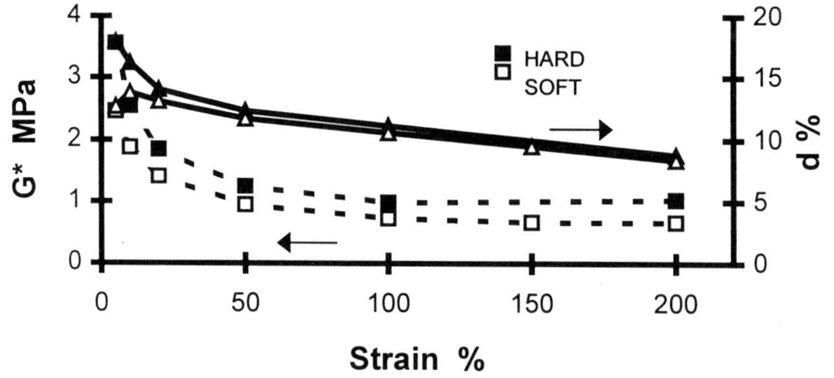

Figure 1: Dependence of shear modulus, G* (, ~) and damping, d (, Δ) on strain amplitude. Filled points are Hard compound and open points Soft compound. Frequency 0.5Hz; 6th cycle analysed

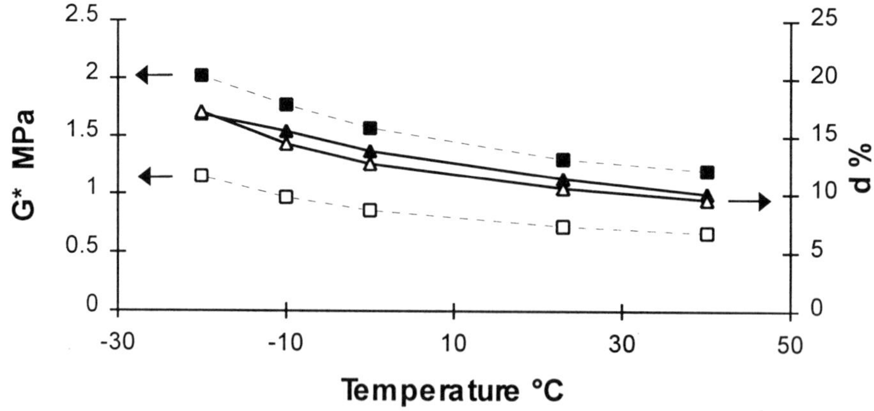

Figure 2: Dependence of shear modulus and damping on temperature. Strain 100%

The sixth cycle force-displacement loops are given in Figure 3 for the Soft bearing types. The stiffness and damping values are listed in Table 2 as a function of the displacement amplitude for tests on both types of bearing. The increased stiffness at small displacements reduces the movement of the building under wind loading without the need for additional wind restraint devices. The magnitude of the non-linearity in stiffness is very close to that expected from the variation of the rubber modulus with strain amplitude (Figure 1).

TABLE 2: *Shear Dynamic Stiffness (K_s) And Damping (D) Of Bearings*

Displ. mm	Soft		Hard	
	K_s,kN/mm	d%	K_s,kN/mm	d%
6.5	1.8	15	2.2	19
13	1.3	14	1.6	17
26	1.04	12	1.3	15
65	0.76	11	0.87	14
130	0.63	9		-

In comparing the stiffnesses and damping with the original design values some allowance must be made for the test frequency of 0.1Hz (cf. isolation frequency of 0.5Hz). The small effect of frequency (2 to 4% increase between 0.1 and 0.5Hz) can be obtained from tests on rubber specimens. Allowing for this, the figure for dynamic stiffness of the Soft bearing becomes 0.66kN/mm - just within the ±15% target band. The dynamic modulus of the particular batch of rubber used to fabricate the test bearing was 7% above the target value (Table 1); this fact explains much of the discrepancy. The observed dynamic stiffness of the Hard bearings appears somewhat lower than expected. Using the test data (Figure 1) for the variation of rubber shear modulus with strain to estimate a bearing stiffness at 100% rubber strain from the observed value at 50% strain, and making an allowance for the effect of test frequency and the reduced vertical load of 50t gives a dynamic stiffness of 0.73kN/mm -12% below the design value. It therefore appears that calculation of the bearing horizontal stiffness from the rubber shear modulus and using the standard design equations worked less well for the Hard bearings than for the Soft bearings. The damping of the bearings (after allowing for the effect of frequency, and (in the case of the Hard type, the reduction between 50 and 100% strain amplitude indicated by the measurements on rubber samples (Figure 1)) is, at 9% for the Soft and 12% for the Hard, slightly greater than that obtained directly from tests at 100% strain on rubber samples.

The bearings were tested quasistatically in shear under the design vertical load up to the maximum horizontal displacement of 195mm. A force-deflection plot (Figure 4) shows a slight stiffening at large deflections; there is no sign of an approach to the displacement capacity of the isolator.

4. Conclusions

The economic and technical feasibility of upgrading the seismic resistance of low-cost housing by the installation of high damping rubber isolators at the foundation level has been demonstrated. A rubber compound suited to sites with severe winter temperatures has been developed. Dynamic tests on the bearings showed the performance of the design to be satisfactory. The bearing test results confirmed that their stiffness and damping is predicted reasonably well from the design equations and rubber properties as measured on small testpieces.

5. References

FULLER, K.N.G., MUHR, A.H. AND ZHOU, F.L. (1995) The design of natural rubber base-isolation system for apartment blocks in Indonesia and China. *Proceedings of 10th European Conference on Earthquake Engineering*, Duma, G., ed., Balkema, Rotterdam, Vol.1. pp.739-744

FULLER, K.N.G. and ROBERTS, A.D. (1997) Longevity of Natural Rubber Structural Bearings. *Proceedings IRC 97, Kuala Lumpur*, pp.777-87

MELKUMIAN, M. (1994) Base Isolation Retrofit Project in Armenia. *Proc. of Workshop on the Use of Natural Rubber based Bearings for Earthquake Protection of Small Buildings*, Jakarta, TARRC

MUHR, A.H. (1994) Design of Isolation Systems using rubber bearings. *Proceedings of International Workshop on use of rubber based bearing for earthquake protection of building*, Shantou, China

6. Acknowledgements

We should like to thank UNIDO (Project US/GLO/91/160) for funding part of this work. The assistance of Hamid Ahmadi, Ian Goodchild, Julia Gough and Richard Ward with the manufacture and testing at TARRC is gratefully acknowledged.

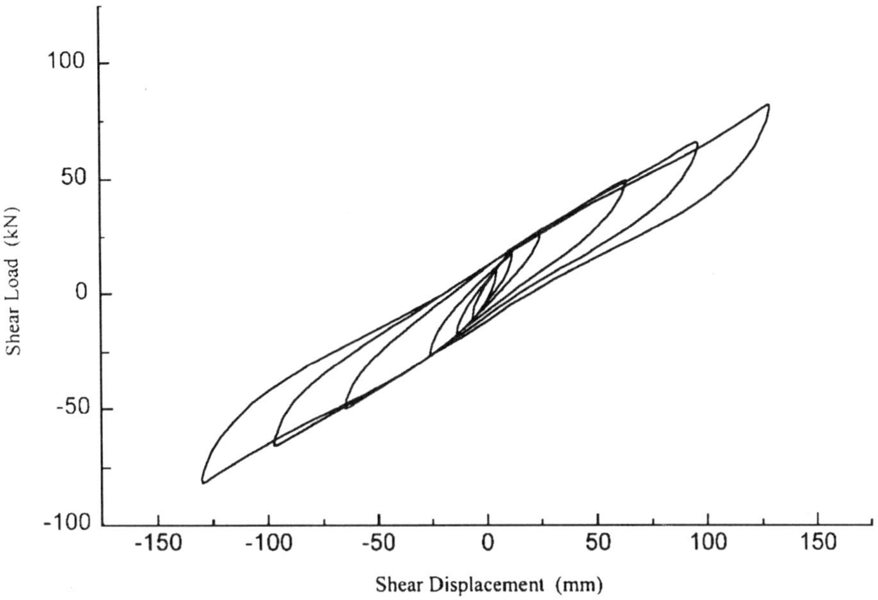

Figure 3: Shear force-deflection loops (6th cycle) for Soft-type bearing. Test frequency 0.1Hz. Nominal vertical load 50t

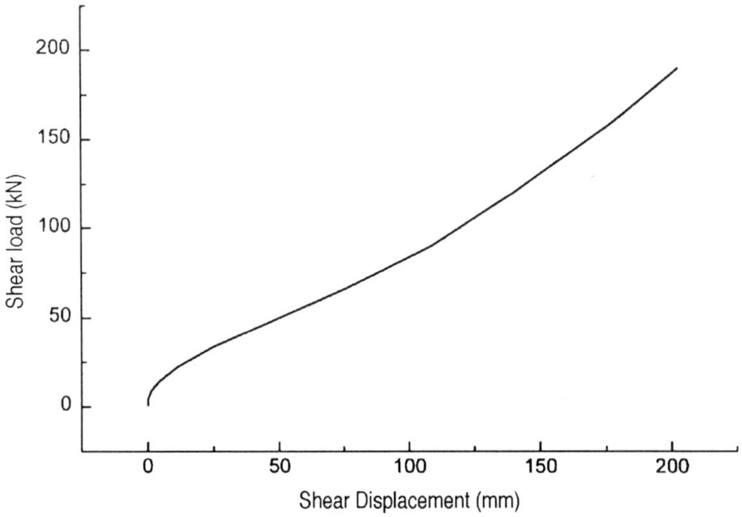

Figure 4: Quasistatic shear force-deflection curve for Hard type bearing under constant vertical load of 82t

THE RESEARCH, APPLICATION AND DESIGN CODE OF SEISMIC ISOLATION AND ENERGY DISSIPATION IN CHINA

FU LIN ZHOU
South China Construction University
No.248 Guang Yuan Zhong Road, Guangzhou 510405, P. R. CHINA

Abstract

This paper describes the most recent development and the design code of seismic isolation and passive energy dissipation In China. Using seismic isolation and passive energy dissipation, which are more reasonable, safe also saving, is becoming a strong tendency in China.

Seismic isolation is mostly used for 2-30 stories civil buildings, bridges, subway and tank of liquid or gas in China. There are over 170 buildings with seismic isolation constructed, include new designing buildings and existed buildings, RC frame structures, shear wall structures and multi-stories masonry structures. The isolation buildings, the acceleration response of which can be reduce to 1/4-1/12 of the traditional anti-seismic buildings, and the building cost of which can be saved 3-10% comparing the traditional anti-seismic buildings. More and more buildings with seismic isolation are constructed, the era of strong earthquake proof buildings is coming in China !

There are three kinds of locations putting isolators in China: put on the base of building if the building is with basement, or put on the top of the first story of buildings if the building is without basement, or put on the top of existed building -- like TMD if the building can be added 1-2 stories to get more floor area also reduce the seismic response of existed structure.

The rubber bearings are as isolators in most cases in China. A very large factory, Vibro Tech" Company, produces high quality also low price rubber bearings, to satisfy the great demand of rubber bearings in China and other countries in the world.

The design rule-- "Design code of Seismic Isolation with Rubber Bearings" of China will be published at the end part of 1998. This paper will briefly introduce the main principle of this design code.

Energy dissipation is mostly used for 20-100 stories high rise buildings or long span bridges in China, including new designing structures or existed structures. The energy dissipation bracing is the main type of energy dissipater which dissipate energy by yielding of steel plate, hybrid friction or some elastic-damping material in China. The structure with energy dissipating bracing, the displacement response of which can be reduce to 60% of the structure without energy dissipating bracing, and the building cost can be saved about 5-10%. Some high rise buildings used energy dissipating bracing to reduce the response simultaneously from earthquake and wind load.

1. Development of Seismic Control of Structures in China

1.1 EVOLUTION OF TRADITIONAL ANTI-SEISMIC STRUCTURE

During earthquake attacks, the traditional building structure which is fixed on the ground will respond gradually increasing from the building's bottom (ground) to the building's top, liking an "Amplifier" (*Figure 1a*). This will result in damage or collapse of structure or damage of nonstructural components, decoration and equipment's

S. Balassanian et al. (eds.), Earthquake Hazard and Seismic Risk Reduction, 387–394.

facilities in building. In order to reduce the response and avoid the damage of structure or internal facilities, some methods have been developed:

(1) Rigid Structure System: Increasing the structural stiffness to be very large, form a "Rigid Structure System"(*Figure 1b*) whose structural response may be nearly as same as the ground motion. But this is very expensive and very difficult to be realized..

(2) Flexible Structure System: Decreasing the structural stiffness to be very small, form a " Flexible Structure System " (*Figure 1c*) whose structural response should be also very small. But this kind of structure system is not suitable for normal use because it is too flexible under wind load or minor earthquakes.

(3) Soft First Story Structure System: Making the first story columns very soft and allowing it to deform into the inelastic range to dissipate the energy for reducing the response of the upper structure, form a "Soft First Story Structure System "(*Figure 1d*). This system can reduce the upper structure response, but the building may collapse in earthquake because of large inelastic deformation of first story columns (Zhou, 1990).

(4) Inelastic Structure System: Increasing the structural ductility and allowing the structural elements or joints to work in inelastic range to dissipate the energy of structure under earthquakes (*Figure 1e*). This is the general structural system for earthquake resistance in many countries including China at present. This system is called "traditional anti-seismic structure system". But its usefulness is limited or it is not very safe in some cases. There are many problems existed in this system (Zhou,1991).

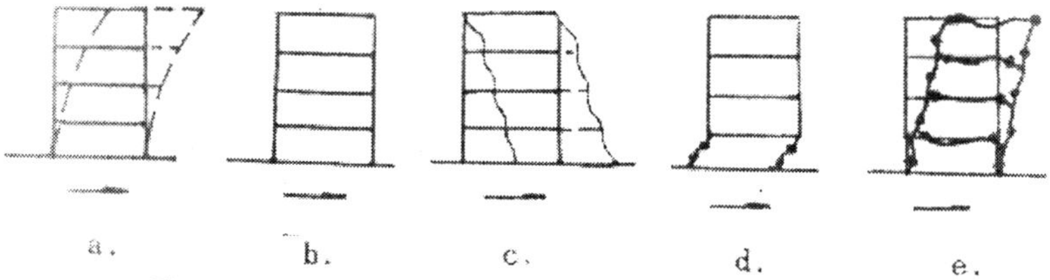

Figure 1. ***Evolution of traditional anti-seismic system of structures***

1.2 PROBLEMS IN TRADITIONAL ANTI-SEISMIC STRUCTURE

(1) It is not very safe: It is difficult to control the structural damage level due to the inelastic deformation, and it may be dangerous in severe unpredicted earthquake.

(2) It is limited to be used: It is not able to be used in some important structures whose elements are not allowed to work in inelastic range, such as some buildings whose decoration is very expensive, nuclear power plant, museum building, and the buildings with precise instruments in it.

(3) It is more expensive: The anti-seismic way of traditional technique is to strengthen the structures: making the columns, beams, shear walls or other structural elements bigger or stronger, increasing the stiffness of the structures. It will cause to increase the seismic load (by increasing the stiffness), then, cause again to strengthen the structures. Finally, it will rise the cost of structure, but the safety level of anti-seismic is still unsure. The cost of building using traditional anti-seismic technique will

rise as bellow comparing with the cost of no-anti-seismic building depending the present design code in China (Zhou,1995):

For Intensity I-7 (moderate earthquake), rise the cost 3- 10 %.

For Intensity I-8 (major earthquake), rise the cost 10-20 %

For Intensity I-9 (strong earthquake), rise the cost 15-30 %

For Intensity I>9 (worst earthquake), rise the cost 30-60 %, also the design is difficult.

Because the traditional anti-seismic system is not very safe in some cases, also is more expensive, the people do not like to pay more money to "strengthen" the structures for "future unknown" earthquake. It causes the dangerous state of many structures at present in China, including existed buildings, new designing buildings, and the infrastructure structures.

1.3 DEVELOPMENT OF SEISMIC CONTROL OF STRUCTURE

The urgent task for earthquake engineering is to find out a new system which needs to satisfy following requirements:

(1) Safe: ensure safety for structures in any strong earthquake.

(2) Inexpensive: do not rise the cost of structures comparing with the traditional anti-seismic technique.

(3) Wide range of application: is able to be used in, both new design structures and existed structures, both important buildings and civil buildings or house buildings, both for protecting the structures and for protecting the facilities inside the buildings.

This new system seems to be found out -- The Seismic Control of Structure, which consists of Seismic isolation, Energy dissipation, Passive control, Active control and Hybrid control.

The range of application and technical maturity of Seismic Control of Structures in China are shown in TABLE 1 (Zhou,1997).

TABLE.1. *The range of application and maturity seismic control of structure*

Name of control	Ranges of application	Technical maturity
Seismic isolation	2- 30 stories buildings (new design or existed) bridges, subway equipment or facilities	Mature technique Rich theoretical and testing results Widely application Successfully experienced the strong earthquake in China,
Energy dissipation	20-150 stories high rise building Long span bridges High tower or skeleton	Mature technique Rich theoretical and testing results Application for seismic or winds
Passive control (TMD,TLD.etc)	High rise buildings Long span bridges High tower or skeleton	Basically mature technique Some research and testing results Application for seismic or winds
Active control	High rise buildings Long span bridges High tower or skeleton	Not become a mature technique Some research and testing results Application for seismic or winds

2. Seismic Isolation in China

2.1 GENERAL STATE OF DEVELOPMENT

The history of application on seismic isolation in China has proved that the isolation system is more safe, economic and reasonable than the traditional structure system, specially suitable for using in the cases where the earthquakes are unpredictable. In mainland China, About 140 existed ancient buildings, towers and temples built thousands years ago with isolation concept have successfully withstood several strong earthquakes.

In recent 15 years, there are 170 buildings and 21 bridges being built with isolation which isolators mostly were rubber bearings, only a few were sand sliding layer or graphite layer. In order to solve some problems for more wide application of isolation , a great numbers of tests for rubber bearings and shaking table tests for building models with rubber bearings were carried out (Zhou,1994,1997). Based on the testing results, analysis and application, a set of optimal design methods, computer programs were raised in China (Zhou,1988,1992,1997). The "Chinese design code for seismic isolating structures with rubber bearings" will be published (Zhou,1996).

2.2. PROGRESS IN TECHNIQUE AND ECONOMY USING SEISMIC ISOLATION IN CHINA

Many significant advantages of structures using rubber bearing isolators in China (TABLE 2):

(1) Safe in strong earthquake: It is very effective to reduce the response of structures in earthquake and is able to prevent the structure from damage or collapse in earthquake. So it can ensure the structures with rubber bearings to be safe in any strong earthquakes.

Comparing the seismic isolation structures with the traditional anti-seismic structures, the response of isolating structures can be reduce to 1/4-1/12 of the response of traditional structures, according the testing results and the records in real earthquake also (Zhou,1997).

(2) Save the building cost in many cases: Comparing the seismic isolation structures with the traditional anti-seismic structures, the building cost of isolation structures can be saved 5-20 % of the building cost, because re-designing the supper structure which seismic response is very small, according the final statistics results of 30 buildings with rubber bearings completed in southern, western and northern China.

(3) Wide ranges of application: The technique of seismic isolation of structure can be used in, both new design structures and existed structures, both important buildings and civil buildings or house buildings, both for protecting the structures and for protecting the facilities inside the buildings.

(4) Long working life of rubber bearings isolators: According the permanent testing and investigation, the safely working life of rubber bearings are over 70-100 years which is longer than the working life of structure itself (Zhou,1997).

TABLE 2 *Technical and economical results of comparing different structures*

	Traditional anti-seismic structure	isolation structure
Acceleration response	1.00	1/4 - 1/12
Working state of structure during earthquake	inelastic	elastic
Building cost	1.00	0.80 - 0.95

2.3. THREE KINDS OF LOCATION OF ISOLATION BEARINGS

There are four kinds of locations putting isolating rubber bearings in China:

(1) Put on the base of building if the building is with basement (*Figure 2*).

(2) Put on the top of the first story of buildings if the building is without basement (*Figure 3*)

(3) Put on the top of existed building -- like TMD if the building can be added 1-2 stories to get more floor area also reduce the seismic response of existed structure (*Figure 4*)

(4) Put on the top of the pier of bridges (*Figure 5*)

2.4. PRODUCTS OF RUBBER BEARINGS

The rubber bearing is as isolator in most cases in China. A very large factory, "Vibro-Tech" Company, produces high quality also low price rubber bearings, to satisfy the great demands of rubber bearings in China and other countries in the world. The "Vibro-Tech" produces many different types of Rubber Bearings, which main characteristics are following:

Diameter (mm)	∅200 - ∅1200
Design compression loads (kN)	400 - 16000
Design maximum displacement (mm)	130 - 800
Horizontal stiffness (kN/mm)	0.30 - 2.30
Vertical stiffness (kN/mm)	200 - 4300
Equivalent damping ratio (%)	8 - 26

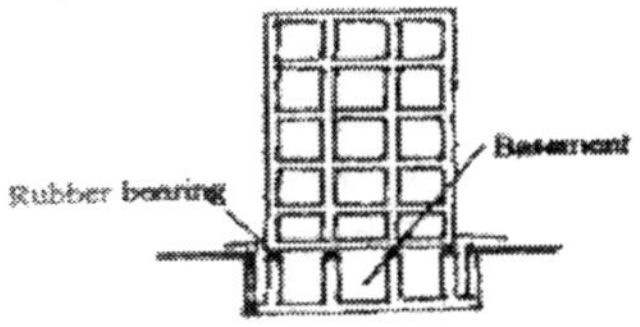

Figure 2. Rubber bearings on the base of building

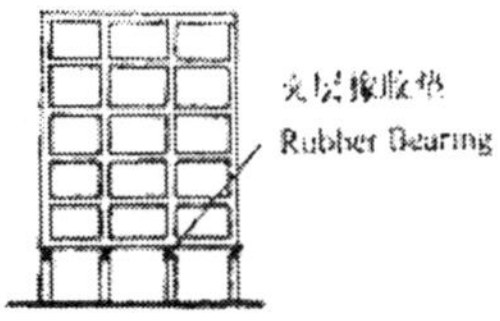

Figure 3. Rubber bearings on the top of the first story

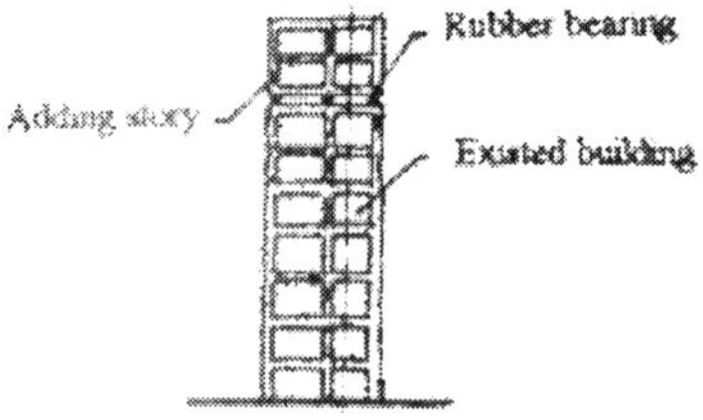

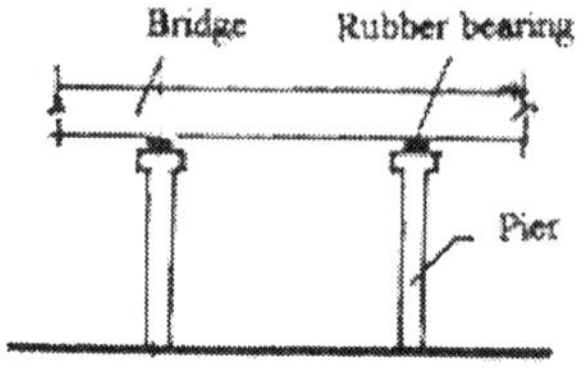

Figure 4 Rubber bearings on the top of building *Figure 5* Rubber bearings on the top of piers for bridge

3. Energy Dissipation in China

3.1. DESIGN, TESTS AND APPLICATION

Two kinds of energy dissipation bracing system to be used:

(1) Composite elastic-friction energy dissipation bracing (Zhou,1991), some high rise buildings with this energy dissipating bracing were built in China (*Figure 6*).

(2) Plastic steel energy dissipating bracing (Zhou,1982), some high rise building or industrial structures with this energy dissipating bracing were built in China .

In order to prove the behavior of this new bracing system, many sets of full-scale models of this bracing and shaking table tests have been carried out in China. Base on the testing results, two sets of computation programs for designing have been compiled (Zhou 1992).

3.2. PRINCIPLE CALCULATION OF ENERGY DISSIPATION

For high rise building structure with energy dissipating bracing, the energy balance equation in any instant of time during earthquake is:

$$Ein = Ep + Ek + Ep + Eb \qquad (1)$$

where Ein -- energy input to the structure. Ep -- potential energy in structural vibration. Ek-- kinetic energy in structural vibration. Ed--energy dissipated by viscous damping of structure. Eb--energy dissipated by bracing system (energy dissipater). Researches indicate that the energy dissipater can dissipate about 90% of the total energy input at the end of earthquake. So, some items whose effect are relatively small can be neglected in Eq.(1), then the energy dissipating design for earthquake resistance need to be satisfied with:

$$Ein < Eb \qquad (2)$$

For calculating energy input Ein, the system can be considered as multi-degree of freedom system. The energy dissipated by bracing system Eb depends on the area enclosed by Load -Displacement loop curve (Zhou et al. 1988).

Figure 6 Composite elastic-friction energy dissipation bracing system

4. The Future developing Tendency of Seismic Isolation and Energy Dissipation

The traditional earthquake resistant structure system and its design method have resulted in many problems. Since the Intensity of Earthquake is always uncertain, the ground motion and structural response is too complicated, and the requirement of structures with modern contents are more and more strict. The traditional system and its design method for earthquake resistant structures are not satisfied for using. The seismic isolation and energy dissipation system of structure will give a new way to resist earthquake for structures in seismic zone. The new system is more safe, effective, inexpensive and simple. Many engineers and scientists in China predict: " The coming decade will be a speed developing decade in Base Isolation, Energy Dissipation and Controlling Structure System in China and in the world !"

Reference

Zhou Fu Lin, Yu Gong Hua (1982), "Testing and Design for Energy Dissipation Bracing." *Report on Earthquake Engineering*, Vol.1. 1982.(in Chinese)

Zhou Fu Lin, S.F.Stiemer, S.Cherry (1988), "Design Method of Isolating And Energy Dissipating System for Earthquake Resistant Structures." *Proc. of 9th World Conference on Earthquake Engineering*. Tokyo-Kyoto. Aug.1988.

Zhou Fu Lin, (1989), " Development and Application for Technique of Isolation and Energy Dissipation and Control of Structural Response (1st Part) " *Journal of International Earthquake Engineering*. VoL.4. 1989 (in Chinese).

Zhou Fu Lin (1990a), "Development and Application for Technique of Isolation and Energy Dissipation and Control of Structure Response (2nd Part)" *Journal of International Earthquake Engineering*. Vol.1, 1990 (in Chinese).

Zhou Fu Lin, Stiemer S.F. and Cherry S. (1990b), "A New Isolation and Energy Dissipating System for Earthquake Resistant Structures". *Proc. of 9th European Conference on Earthquake Engineering*. Moscow. Sept. 1990.

Zhou Fu Lin (1991) " New System of Earthquake Resistant Structures in Seismic Zone. " *Recent Development and Future Trends of Computational Mechanics in Structural Engineering*. Published by ELSEVIER APPLIED SCIENCE PUBLISHERS LTD, London and New York.

Zhou Fu Lin. et al. (1992), " Recent Research , Application, and development of Seismic Control for Structures ".*Proc. of China National Conference on Seismic Control for Structures*. 1992, Wuhan, China.

Zhou Fu Lin, Kelly J M, Fuller K N G, Pan T C, et al, (1994) " Recent Research Development and Application on Seismic Isolation of Buildings in P R China". *Proc. of International Workshop IWADBI*, Shantou, China May 1994.
Zhou Fu Lin (1995),"Technical Report on Mission to Santiago, Chile" as Consultant to attend the *International Post-SMiRT conference Seminar on Seismic Isolation, Passive Energy Dissipation and Active Control of Vibrations and Structures*, August 1995.
Zhou Fu Lin (1996), " Discussion on Compiling Chinese Design Code of Seismic Isolation", *Proc. of The PRC-USA Bilateral Workshop on Seismic Codes*, Dec. 1996, Guangzhou, China.
Zhou Fu Lin (1997), " Seismic Control of Structures" (Chinese), *Chinese Seismic Publishing House*, 1997.

ON ONE APPROACH OF STUDYING THE FREE AND FORCED VIBRATIONS OF BASES AND FUNDAMENTS OF STRUCTURES

L.A. AGHALOVIAN
Institute of Mechanics of National Academy of Sciences RA,
24b M.Baghramian Ave, Yerevan,375019, Republic Armenia

Abstract

One of the widespread models of bases and fundaments of structures is the model of compressible layer. For the calculation of bases and fundaments on dynamic, and particularly, seismic actions under a strict approach, it is necessary to find solutions of the equations of dynamics of the elasticity theory. It is shown that for these domains the above mentioned equations are small parameter singularly perturbed. An efficient asymptotic method of solution of the equations is offered.

Questions of determination of frequencies and forms of free vibrations of one-layered and two-layered beam-strips and plates, laid on a rigid foundation are considered. Forced vibrations under dynamic (seismic) actions are studied. Conditions causing a resonance are established.

1. Free Vibrations of Beam-Strips and Plates on a Rigid Foundation

Consider a problem of finding frequencies of free vibrations of a strip-base, i.e. in general case of an orthotrope layer $D=\{(x,y): x\in[0,l],\ |y|\leq h,\ h<<l\}$, when its lower edge is rigidly fixed.

It is required to find non-zero solutions of homogeneous equations of dynamics of elasticity theory:

$$\frac{\partial\sigma_{xx}}{\partial x}+\frac{\partial\sigma_{xy}}{\partial y}=\rho\frac{\partial^2 u}{\partial t^2}$$

$$\frac{\partial\sigma_{xy}}{\partial x}+\frac{\partial\sigma_{yy}}{\partial y}=\rho\frac{\partial^2 \mathrm{v}}{\partial t^2}$$

$$\frac{\partial u}{\partial x}=a_{11}\sigma_{xx}+a_{12}\sigma_{yy} \qquad (1)$$

S. Balassanian et al. (eds.), Earthquake Hazard and Seismic Risk Reduction, 395–402.

$$\frac{\partial v}{\partial y} = a_{12}\sigma_{xx} + a_{22}\sigma_{yy}$$

$$\frac{\partial u}{\partial y} + \frac{\partial v}{\partial x} = a_{66}\sigma_{xy}$$

under the following two variants of boundary conditions modeling the behavior of the base or the fundament more realistically:

$$a)\quad \sigma_{xy}(h) = 0, \quad \sigma_{yy}(h) = 0 \qquad (2)$$
$$u(-h) = 0, \quad v(-h) = 0$$

$$b)\quad u(h) = 0, \quad v(h) = 0 \qquad (3)$$
$$u(-h) = 0, \quad v(-h) = 0$$

where σ_{ik} are the components of the stress tensor, u and v are the components of the displacement vector, a_{ik} are the elasticity constants and ρ is the density of the medium.

In order to find the solution of system (1), let's switch to the dimensionless variables $\xi = x/l$, $\zeta = y/h$ and the dimensionless components of the displacement vector $u_1 = u/l$ and $v_1 = v/l$. The solution of the transformed system (1) is sought in the form:

$$\sigma_{xx} = \sigma_{11}(\xi,\zeta)\exp(i\omega t), \quad \sigma_{xy} = \sigma_{12}(\xi,\zeta)\exp(i\omega t)$$
$$\sigma_{yy} = \sigma_{22}(\xi,\zeta)\exp(i\omega t) \qquad (4)$$
$$u_1 = U(\xi,\zeta)\exp(i\omega t), \quad v_1 = V(\xi,\zeta)\exp(i\omega t)$$

For determination of coefficients in (4) we shall get the following system singularly perturbed by the small parameter $\varepsilon = h/l$:

$$\frac{\partial \sigma_{11}}{\partial \xi} + \varepsilon^{-1}\frac{\partial \sigma_{12}}{\partial \zeta} + \varepsilon^{-2}\omega_*^2 U = 0$$

$$\frac{\partial \sigma_{12}}{\partial \xi} + \varepsilon^{-1}\frac{\partial \sigma_{22}}{\partial \zeta} + \varepsilon^{-2}\omega_*^2 V = 0 \qquad (5)$$

$$\frac{\partial U}{\partial \xi} = a_{11}\sigma_{11} + a_{12}\sigma_{22}, \quad \varepsilon^{-1}\frac{\partial V}{\partial \zeta} = a_{12}\sigma_{11} + a_{22}\sigma_{22}$$

$$\varepsilon^{-1}\frac{\partial U}{\partial \zeta} + \frac{\partial V}{\partial \xi} = a_{66}\sigma_{12}, \qquad \omega_*^2 = \rho\omega^2 h^2$$

The solution of singularly perturbed system (5) is sought in the form of an asymptotic expansion [1].

$$\sigma_{ik} = \varepsilon^{-1+s}\sigma_{ik,s}, \quad U,V = \varepsilon^{s}(U_s, V_s), \qquad s = \overline{0,S} \qquad (6)$$

Substituting (6) into (5), we shall find a closed system for the determination of coefficients of expansion (6). All the variables can be expressed through U_s and V_s which, in their turn, are determined from two common differential equations of the second order [2]. Solving these equations, later determining the stresses and satisfying the boundary conditions (2), we shall find two independent systems of algebraic equations relative to the constants of integration. From the solvability of the system (its determinant is equal to 0) the equations of frequencies of free vibrations follow. Conditions $\sigma_{xy}(h)=0$ and $u(-h)=0$ generate free shear vibrations with frequencies:

$$\omega_{*n}^{\mathrm{I}} = \frac{\pi}{4\sqrt{a_{66}}}(2n+1) \quad \text{or} \quad \omega_{n}^{\mathrm{I}} = \frac{\pi}{4h}(2n+1)\frac{1}{\sqrt{\rho a_{66}}} = \tag{7}$$

$$= \frac{\pi}{4h}(2n+1)\sqrt{\frac{G_{12}}{\rho}} = \frac{\pi}{4h}(2n+1)\mathrm{v}_s, \qquad n \in N$$

where $\mathrm{v}_s = \sqrt{\dfrac{G_{12}}{\rho}}$ is the well-known velocity of seismic shear waves (S-waves).

And conditions $\sigma_{yy}(h)=0$ and $\mathrm{v}(-h)=0$ generate free longitudinal vibrations with frequencies

$$\omega_{*n}^{\mathrm{II}} = \frac{\pi}{4\sqrt{A_{11}}}(2n+1), \qquad A_{11} = \left(a_{11}a_{22} - a_{12}^2\right)/a_{11}$$

or (8)

$$\omega_{n}^{\mathrm{II}} = \frac{\pi}{4h\sqrt{\rho A_{11}}}(2n+1) = \frac{\pi}{4h}(2n+1)\mathrm{v}_p, \qquad n \in N$$

where $\mathrm{v}_p = \dfrac{1}{\sqrt{\rho A_{11}}} = \sqrt{\dfrac{E_2}{\rho(1-\nu_{12}\nu_{21})}}$ is the well-known in seismology and elasticity theory velocity of propagation of longitudinal waves (P-waves) in a plate. Conditions (3) again correspond to the free shear and longitudinal vibrations, but with frequencies:

$$\omega_n^{\mathrm{I}} = \frac{\pi n}{2h}\mathrm{v}_s, \quad \omega_n^{\mathrm{II}} = \frac{\pi n}{2h}\mathrm{v}_p, \qquad n \in N \tag{9}$$

It is natural to find out the question– how the medium's being layered, influences the free vibrations. We have established that the free shear and longitudinal vibrations can be generated in a layered medium as well. But the frequencies cannot be determined by simple formulae like (7)-(9). For example, for a two-layered strip $D = \{(x,y) : x \in [0,l],\ -h_2 \le y \le h_1\}$, where h_1 and h_2 are the thicknesses of the layers, the frequencies of the free vibrations corresponding to the boundary conditions (2) on the front surfaces, are determined from the equation [3]

$$a\cos b\omega_* = c\cos d\omega_* \tag{10}$$

where for the shear vibrations

$$a = 1 + \sqrt{\frac{\rho^{II} G^{II}}{\rho^{I} G^{I}}}, \qquad b = \sqrt{\frac{\rho^{I}}{G^{I}}}\zeta_1 + \sqrt{\frac{\rho^{II}}{G^{II}}}\zeta_2$$
$$c = 1 - \sqrt{\frac{\rho^{II} G^{II}}{\rho^{I} G^{I}}}, \qquad d = \sqrt{\frac{\rho^{I}}{G^{I}}}\zeta_1 - \sqrt{\frac{\rho^{II}}{G^{II}}}\zeta_2 \tag{11}$$

$$\zeta_1 = h_1/h, \ \zeta_2 = h_2/h, \ h = \max(h_1, h_2), \ \omega_*^2 = h^2\omega^2$$

and for the longitudinal vibrations

$$a = 1 + \sqrt{\frac{\rho^{II} \overline{E}_2^{II}}{\rho^{I} \overline{E}_2^{I}}}, \qquad b = \sqrt{\frac{\rho^{I}}{\overline{E}_2^{I}}}\zeta_1 + \sqrt{\frac{\rho^{II}}{\overline{E}_2^{II}}}\zeta_2$$
$$c = 1 - \sqrt{\frac{\rho^{II} \overline{E}_2^{II}}{\rho^{I} \overline{E}_2^{I}}}, \qquad d = \sqrt{\frac{\rho^{I}}{\overline{E}_2^{I}}}\zeta_1 - \sqrt{\frac{\rho^{II}}{\overline{E}_2^{II}}}\zeta_2 \tag{12}$$

$$\overline{E}_2^{i} = \frac{E_2^{i}}{1 - \nu_{12}^{i}\nu_{21}^{i}}, \qquad i = \mathrm{I}, \mathrm{II}$$

If the base is a medium with general anisotropy, again two types of free vibrations exist, but they are not purely shear and longitudinal vibrations [4].

According to the model of a compressible layer the fundament or the base of the capacity (thickness) $2h$ lies on a rigid foundation [5]. The offered asymptotic method makes also possible the solution of the corresponding equations of dynamics of a spatial problem of elasticity theory (equations for plates). If one of the front edges of a plate is free and the other one lies on a rigid foundation, then two types of the free shear vibrations are generated in the plate, taking place in the planes (xz) and (yz) which have frequencies:

$$\omega_n^{xz} = \frac{\pi}{4h}(2n+1)\sqrt{\frac{G_{13}}{\rho}}$$
$$\omega_n^{yz} = \frac{\pi}{4h}(2n+1)\sqrt{\frac{G_{23}}{\rho}} \qquad n \in N \tag{13}$$

where G_{13} and G_{23} are the shear modules.

Longitudinal vibrations with frequencies are generated as well:

$$\omega_n^{p} = \frac{\pi}{4h}(2n+1)\mathrm{v}_p \tag{14}$$

$$\mathrm{v}_p = \sqrt{\frac{E_3}{\rho}\frac{1-\nu_{12}\nu_{21}}{1-\nu_{12}\nu_{21} - \nu_{31}(\nu_{12}\nu_{23} + \nu_{13}) - \nu_{32}(\nu_{21}\nu_{13} + \nu_{23})}}$$

where E_3 is Young's module in the direction of the axis Oz, and ν_{ik} are Poisson's coefficients.

2. Forced Vibrations of Bases and Fundaments

Asymptotic method offered above can also be used to study forced vibrations of bases and fundaments generated by dynamic, and particularly, by seismic actions.

As an illustration of the words cited above, let's consider forced vibrations of beam-strips caused by harmonically varied over the time displacements on the lower edge of the beam as a result of a seismic action.

It is required to find the solution of the dynamic equations (1) under the boundary conditions

$$u(-h) = u^{-}(\xi)\exp(i\Omega t), \quad \mathrm{v}(-h) = \mathrm{v}^{-}(\xi)\exp(i\Omega t) \tag{15}$$

where $u^{-}(\xi)$ and $\mathrm{v}^{-}(\xi)$ characterize the amplitude, and Ω – the frequency of the forcing action. Different variants of boundary conditions for $y = h$, which model the behavior of the fundament or the base, can be discussed. We limit our consideration to the cases of free: $\sigma_{xy}(h) = \sigma_{yy}(h) = 0$ and rigidly fixed: $u(h) = \mathrm{v}(h) = 0$ edges. The solution is again sought in the forms of (4) and (6) with the only difference that Ω and $\Omega_* = \rho\Omega^2 h^2$ appear instead of ω and ω_*. For the coefficients in expansion (6), it is possible to get

$$\begin{aligned}
\sigma_{12,s} &= \frac{1}{a_{66}}\left(\frac{\partial U_s}{\partial \zeta} + \frac{\partial V_{(s-1)}}{\partial \xi}\right)\\
\sigma_{22,s} &= \frac{1}{A_{11}}\left(\frac{\partial V_s}{\partial \zeta} - \frac{a_{12}}{a_{11}}\frac{\partial U_{(s-1)}}{\partial \xi}\right)\\
\sigma_{11,s} &= -\frac{1}{A_{11}}\left(\frac{a_{12}}{a_{11}}\frac{\partial V_s}{\partial \zeta} - \frac{a_{22}}{a_{11}}\frac{\partial U_{(s-1)}}{\partial \xi}\right)
\end{aligned} \tag{16}$$

where U_s and V_s are determined from the equations

$$\begin{aligned}
&\frac{\partial^2 U_s}{\partial \zeta^2} + a_{66}\Omega_*^2 U_s = R_u^{(s-1)}\\
&\frac{\partial^2 V_s}{\partial \zeta^2} + A_{11}\Omega_*^2 V_s = R_{\mathrm{v}}^{(s-1)}
\end{aligned} \tag{17}$$

$$R_u^{(s-1)} = \frac{1}{A_{11}}\left(A_{26}\frac{\partial^2 V_{(s-1)}}{\partial \xi \partial \zeta} + \frac{a_{66}a_{22}}{a_{11}}\frac{\partial^2 U_{(s-2)}}{\partial \xi^2}\right)$$

$$R_{v}^{(s-1)} = \frac{1}{a_{66}}\left(A_{26}\frac{\partial^2 U_{(s-1)}}{\partial\xi\partial\zeta} + A_{11}\frac{\partial^2 V_{(s-2)}}{\partial\xi^2} \right)$$

$A_{26} = (a_{11}A_{11} - a_{12}a_{66})/a_{11}$

Having solved the equations (17), calculated stresses by formulae (16) and having satisfied the boundary conditions for $y = h$, we shall find the final solution. The solution corresponding to the free edge $\sigma_{xy}(h) = \sigma_{yy}(h) = 0$ has the form:

$$U_s = \frac{1}{\cos 2\Omega_* \sqrt{a_{66}}}\left[\left(u_0^{(s)} - U_\tau^{(s)}(\xi,-1)\right)\cos\Omega_*\sqrt{a_{66}}(1-\zeta) - \right.$$
$$\left. - \frac{\sqrt{a_{66}}}{\Omega_*} f_\tau^{(s)}(\xi,1)\sin\Omega_*\sqrt{a_{66}}(1+\zeta) \right] + U_\tau^{(s)}(\xi,\zeta)$$
$$V_s = \frac{1}{\cos 2\Omega_* \sqrt{A_{11}}}\left[\left(\mathrm{v}_0^{(s)} - V_\tau^{(s)}(\xi,-1)\right)\cos\Omega_*\sqrt{A_{11}}(1-\zeta) - \right. \tag{18}$$
$$\left. - \frac{\sqrt{A_{11}}}{\Omega_*} \psi_\sigma^{(s)}(\xi,1)\sin\Omega_*\sqrt{A_{11}}(1+\zeta) \right] + V_\tau^{(s)}(\xi,\zeta)$$

where $U_\tau^{(s)}(\xi,\zeta)$ and $V_\tau^{(s)}(\xi,\zeta)$ are particle solutions of the non-homogeneous equations (17), and

$$f_\tau^{(s)}(\xi,\zeta) = \frac{1}{a_{66}}\frac{\partial U_\tau^{(s)}}{\partial\zeta} + \frac{1}{a_{66}}\frac{\partial V_{(s-1)}}{\partial\xi}$$
$$\psi_\sigma^{(s)}(\xi,\zeta) = \frac{1}{A_{11}}\left[\frac{\partial V_\tau^{(s)}}{\partial\zeta} - \frac{a_{12}}{a_{11}}\frac{\partial U_{(s-1)}}{\partial\xi} \right] \tag{19}$$
$$u_0^{(0)} = \frac{u^-(\xi)}{l}, \quad \mathrm{v}_0^{(0)} = \frac{\mathrm{v}^-(\xi)}{l}, \quad u_0^{(s)} = \mathrm{v}_0^{(s)} = 0, \ s > 0$$

Particularly, for $u^- = \mathrm{const}, \mathrm{v}^- = \mathrm{const}$, the solution of the problem is:

$$u = \frac{u^-}{\cos 2\Omega_* \sqrt{a_{66}}}\cos\Omega_*\sqrt{a_{66}}(1-\zeta)\exp(i\Omega t)$$
$$\mathrm{v} = \frac{\mathrm{v}^-}{\cos 2\Omega_* \sqrt{A_{11}}}\cos\Omega_*\sqrt{A_{11}}(1-\zeta)\exp(i\Omega t) \tag{20}$$

Solutions (18) and (20) are finite, if

$$\cos 2\Omega_* \sqrt{a_{66}} \neq 0, \qquad \cos 2\Omega_* \sqrt{A_{11}} \neq 0$$

i.e. $$\Omega \neq \frac{\pi}{4h}(2n+1)\sqrt{\frac{G_{12}}{\rho}} = \frac{\pi}{4h}(2n+1)\mathrm{v}_s, \qquad \Omega \neq \frac{\pi}{4h}(2n+1)\mathrm{v}_p \tag{21}$$

Because of (7) and (8), the right-hand sides of relations (21) contain frequencies of free vibrations. Therefore, coincidence of frequency Ω with the frequencies of free vibrations will cause a resonance. As far as the range of seismic action frequency changes during powerful earthquakes is known [6], it is necessary to choose parameters of the fundament and the base (thickness, density and elasticity modules) so that to avoid the resonance. Relations (7)-(11) and (21) may serve as a starting point. The solution corresponding to the rigidly fixed edge $(u(+h) = \mathrm{v}(+h) = 0)$ is

$$\begin{aligned} U_s &= \frac{1}{\sin 2\Omega_* \sqrt{a_{66}}}\Big[\big(u_0^{(s)} - U_\tau^{(s)}(\xi,-1)\big)\sin \Omega_* \sqrt{a_{66}}(1-\zeta) - \\ &\quad - U_\tau^{(s)}(\xi,1)\sin \Omega_* \sqrt{a_{66}}(1+\zeta)\Big] + U_\tau^{(s)}(\xi,\zeta) \\ V_s &= \frac{1}{\sin 2\Omega_* \sqrt{A_{11}}}\Big[\big(\mathrm{v}_0^{(s)} - V_\tau^{(s)}(\xi,-1)\big)\sin \Omega_* \sqrt{A_{11}}(1-\zeta) - \\ &\quad - V_\tau^{(s)}(\xi,1)\sin \Omega_* \sqrt{A_{11}}(1+\zeta)\Big] + V_\tau^{(s)}(\xi,\zeta) \end{aligned} \tag{22}$$

For $u^- = \mathrm{const}, \mathrm{v}^- = \mathrm{const}$ we find the precise solution:

$$\begin{aligned} u &= \frac{u^-}{\sin 2\Omega_* \sqrt{a_{66}}}\sin \Omega_* \sqrt{a_{66}}(1-\zeta)\exp(i\Omega t) \\ \mathrm{v} &= \frac{\mathrm{v}^-}{\sin 2\Omega_* \sqrt{A_{11}}}\sin \Omega_* \sqrt{A_{11}}(1-\zeta)\exp(i\Omega t) \end{aligned} \tag{23}$$

Having calculated the stresses by formulae (19), it is possible to determine the time of the fracture and to solve the reverse problem of determining the displacements $u^-(\xi), \mathrm{v}^-(\xi)$, causing the fracture.

As a conclusion, note that the mentioned method can be used to study the influence of the seismic waves on the layered structures (propagation of waves inside the crust from the earthquake center) as well as other problems of seismology and seismosteady construction.

References

1. Aghalovian, L.A. (1997) *Asymptotic Theory of Anisotrope Plates and Shells.* Nauka, Moscow.
2. Aghalovian, L.A. (1994) On Frequencies of Free Vibrations of an Anisotrope Strip. In *Proceedings of Jubilee Scientific Conference to the 60-th Anniversary of Gjumry State Pedagogiical Institute of Armenia*, Publishing House "Vishaja Shkola", Gjumry, pp.23-26.
3. Aghalovian, L.A. and Sargsian, L.S. (1997) On Free Vibrations of a Two-Layered Strip. In *Transaction of the XVIIIth International Conference on Shells and Plates Theory.* Saratov State Technical University,. Saratov, Russia, v.1, pp.30-38.
4. Aghalovian, M.L. (1997) On Eigenvalues and Eigenfunctions of One Differential Operator. *Scientific Papers of Yerevan State University* **2(187),** 8-14
5. Vlasov, V.Z. and Leontyev, N.N. (1960) *Beams, Plates and Shells on an Elastic Base.* Fizmatgiz, Moscow.
6. Kasahara, K. (1981). *Earthquake Mechanics*. Cambridge University press.

ACTIVITIES OF THE EUROPEAN LABORATORY FOR STRUCTURAL ASSESSMENT FOR SEISMIC RISK REDUCTION IN EUROPE

V. RENDA, G. VERZELETTI, G. MAGONETTE, J. MOLINA, D. TIRELLI, F. BONO
European Commission, Joint Research Centre, I - 21020 Ispra (VA)

Abstract

The European Commission contributes to Seismic Risk Reduction (SRR) in Europe through research actions performed by the Joint Research Centre (JRC) that is in charge of the European Laboratory for Structural Assessment (ELSA). The laboratory, sited at Ispra (VA) in Northern Italy, is a unique facility in Europe for dimensions and characteristics conceived and built to contribute to the advance in earthquake engineering.

ELSA contributes to SRR in Europe through its own institutional research programme and competitive actions performed in networks including international partners. ELSA is also made available to the construction industries for the validation of projects of relevant interest that need for experimental tests to be approved. This is mainly the case of innovative projects for which no reference can be done to well established design standards.

ELSA is also engaged in research for innovation techniques for the preservation of historical buildings and monuments that is one of the most challenging tasks in Europe because of the large number of structures having an inestimable cultural value. The contribution of ELSA is relevant for the attempt to transfer the recent advances in earthquake engineering from civil buildings to cultural heritage structures taking into account the limitations imposed by the durability demand for the materials and the architectural constraints. Many of these are carried out in close collaboration with the Authorities of European Regions and Member States.

1. Introduction

Earthquakes are the result of a sudden liberation of energy accumulated at the geological faults (tectonic earthquakes). The consequences of earthquake events are well known to the public: thousands of persons killed or injured each year, thousands of homeless, heavy damage to the building stock, complete disruption of the infrastructure, irreversible damage to the cultural heritage, very large indirect costs resulting from business interruption, loss of revenues, interruption of industrial production.

1.1 REASONS TO PERFORM TESTS ON LARGE-SCALE MOCK-UPS

The assessment of the seismic vulnerability of structures is a very complex issue due to the non-deterministic characteristics of the seismic action and the need for an accurate prediction of the seismic responses for levels beyond conventional linear behavior. Satisfactory numerical modeling (based on the finite element method) depends largely upon the availability of a complete characterization of material properties and

S. Balassanian et al. (eds.), Earthquake Hazard and Seismic Risk Reduction, 403–411.

appropriate calibration of mathematical models able to represent all structural components, including joints. Seismic vulnerability assessment and design of retrofitting solutions for existing structures can be subsequently carried out using calibrated numerical models.

The main objectives of large-scale tests on experimental models are thus:

- to calibrate analytical models suitable for different classes of structures;
- to validate rules of norms and standards;
- to validate specific designs of relevant constructions, in particular where a lack of standards occurs.

1.2. THE EUROPEAN LABORATORY FOR STRUCTURAL ASSESSMENT

ELSA, sited at JRC-Ispra, is equipped with one of the three large reaction walls in the world and the second one in capacity. The largest one is located in Tsukuba (Japan), and the third one is located in San Diego, California. Its load characteristics are given by Figure 1.

Characteristics of the ELSA Reaction Wall

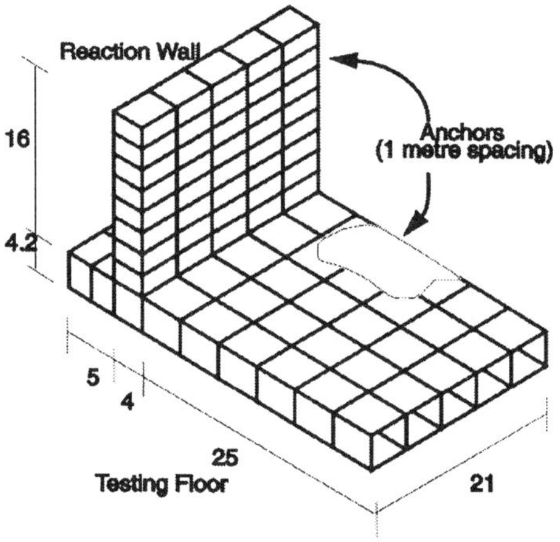

LOAD CAPACITY	**REACTION WALL**	**Bending Moment**	**200 MNm**
		Base Shear	**20 MN**
	REACTION FLOOR	**Bending Moment**	**240 MNm**
	ANCHOR LOAD	**Axial Force**	**500 kN**
HYDRAULIC CHARACTERISTICS	**FLOW**		**1500 l/min.**
	PRESSURE		**210 bar**
	ACTUATORS	**Load (MN)** **Stroke (m)**	**(0.5 - 1.0))** **(0.25) - (1.0)**

Figure 1: Characteristics of the ELSA Reaction Wall

2. Protection Of Civil Structures

The seismic risk reduction in Europe through the protection of civil engineering structures is one of the major items of JRC and the reason leading to the construction of ELSA.

2.1. IMPROVEMENT AND VALIDATION OF EUROCODE-8

The Eurocodes (the European provisional standards for constructions) are being prepared by the European Committee for Standardisation (CEN, Technical Committee No 250) under mandate of the European Commission. The procedure involves a first release as provisional norms. After a period of three years, during which comments are expected, the Eurocodes are intended to be accepted as European Norms. The process of approval of each Eurocode has reached different stages. Since its establishment, the ELSA laboratory has been used in close co-operation with the European experts on Earthquake Engineering to provide the necessary validation experiments for the final drafting of Eurocode 8 [1].

The work is expected to support the verification and possible improvement of the code. In particular, the work carried out for the DG-III will allow the specific clauses for infilled frames and the rules for composite structures to be assessed. The work carried out under the different shared cost actions will be used to study the effects of particular geological conditions in the seismic action, as well as to enforce the adoption of new materials and technologies in both the construction of new structures and the retrofit of existing ones.

In addition to the pan-European partnership in the work carried out at ELSA, the worldwide dimension of the problem represented by the earthquake safety of structures has been recognised, and collaborations with foreign countries have been established and will be enlarged in the near future.

2.2. INNOVATION TECHNOLOGY IN EARTHQUAKE ENGINEERING

Techniques for mitigating the effects of dynamic loading such as strong earthquakes and high winds can be grouped into three broad areas, including base isolation of structures, passive vibration dampers and active control of structural vibration.

2.2.1. Buildings

A base-isolated system is characterised by a very low natural frequency compared to those occurring during a strong earthquake. The superstructure moves like a rigid body over its isolation system, deformations and energy dissipation being mostly concentrated in the isolators [2] & [3].

Passive energy dissipation systems encompass a range of materials and devices for enhancing damping, stiffness and strength, and can be used both for natural hazard mitigation and for rehabilitation of ageing or deficient structures.

The objective of using energy dissipation devices is to preferentially dissipate earthquake-induced energy in devices designed especially for this purpose, and to minimise the energy dissipation demand on primary structural members such as beams, columns or walls.

Active control systems are force delivery devices integrated with real-time processing controllers and sensors within the structure. They must react simultaneously with the hazardous excitation to provide enhanced structural behaviour for improved service and safety. In this field remarkable progress has been made over the last years and among the current activities, the damping of cable-stayed bridges has become a major issue.

The contribution of ELSA, through its capability of testing large scale models of protected systems (Figure 2) and the performance of its control systems, will be important for the validation of designs which include these innovative technologies.

Figure 2: PsD tests performed on a 4-storey model of base isolated structure

2.2.2. Bridges

A leading motivation behind the research on the earthquake behaviour of bridges is the evidence brought about by recent seismic events all over the world that bridge structures are particularly sensitive to the effects of earthquakes.

A significant example in Europe is the severe bridge damage observed after the 1976 Friuli earthquake. Furthermore, the recent evidence from earthquakes in California (1994) and Japan (1995) definitely points to long bridges and highway and railway viaducts as the potentially weakest elements in transportation systems.

The failure of such critical elements having lifeline character seems to be, in most cases, the result of deficient design, as much as of the severity of ground motion causing large displacements and non-linear behaviour of the foundation soil.

Insufficient consideration for seismic risk in bridge design has resulted in inadequate detailing of confining steel and insufficient shear reinforcement in the bridge piers, insufficient seat length of bearings, and inadequate design of the bridge abutments. Furthermore, there are many open questions concerning the ductile behaviour of large bridge piers, in particular those with rectangular hollow cross-section.

3. Preservation of Cultural Heritage Structures

The main objective is to contribute to the definition of intervention methodologies and design guidelines for the structural diagnostics and retrofitting/repair techniques for structures of the Cultural Heritage.

To reach the objective, an effort is done to extend the recent advances in optical technologies and earthquake engineering to the structures of the Cultural Heritage. This attempt to make compliance between the engineering and the architectural approach for the retrofitting/repair techniques is a first step to set-up a unified theory for the restoration of Cultural Heritage structures.

The purpose of the activity is the development of innovative techniques for the structural diagnostics and for the improvement of the overall stability, in particular seismic protection, applicable to the rehabilitation of cultural heritage.

- The technologies used by JRC for the structural diagnostics are related to photogrammetry and laser-vibrometry.
- The research and validation tests for strengthening and retrofitting the structures of cultural heritage is performed at the European Laboratory for Structural Assessment (ELSA) that is a unique facility in Europe for dimensions and characteristics [4] & [5].

3.1. TYPOLOGY OF INTERVENTION

The most relevant activities are related to:

- Application of Photogrammetry for the tridimensional digital description of the geometry.
- Dynamic characterisation of structures by laser-vibrometry.
- Characterisation of new materials (Composite, Shape Memory Alloys and others) for static and seismic loads and experimental validation of devices.
- Non-linear numerical analyses of masonry buildings
- Pseudo-Dynamic (PsD) seismic testing of large-scale models of structures.
- Development of guidelines for the retrofitting of masonry and Cultural Heritage structures.

3.2. ACTIVITIES PERFORMED

3.2.1. Palazzo Geraci in Palermo (http://tintin.jrc.it./htdocs/)

Activity executed in close collaboration with the Sicilian Universities and related to the front wall of "Palazzo Geraci" and to "Fontana Pretoria" in Palermo. Photogrammetry (Figure 3) and Laser Vibrometry have been used for the geometric and mechanic characterisation of the monuments. The effectiveness of retrofitting techniques for "Palazzo Geraci" have been analysed through numerical analyses and seismic tests on a large-scale model tested at the ELSA laboratory (Figure 4).

Figure 3: Palazzo Geraci in Palermo and photogrammetric reconstruction

Figure 4: Numerical analyses and large-scale Model at ELSA

3.2.2. Sao Vicente de Fora Monastery in Lisbon

The activities, run in close collaboration with the national research centre of Portugal and other partners, focused on the behaviour of the cloister of Sao Vicente de Fora Monastery in case of earthquake.

3D numerical models able to represent the structural components including joints have been developed. These models will be calibrated and used, together with linear models, to assess seismic vulnerability and to help in the design of retrofitting systems for monumental structures in earthquake prone regions (Figure 5).

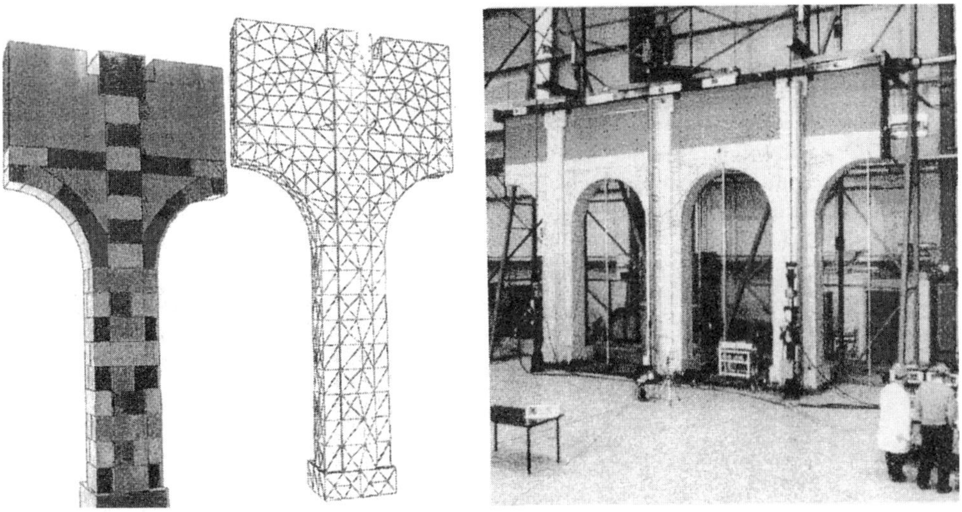

Figure 5: Numerical analyses and full-scale Model at ELSA

The testing campaign will be completed with tests on the retrofitted structure. The utilisation of *iron tie bars* at the floor and at the arch-base levels and their practical application will be investigated.

3.2.3. Innovative retrofitting techniques

The work is carried out in close collaboration with partners from different countries of the European Union and is funded by the General Directorate XII through the Environment and Climate Programme. The project, named ISTECH (Innovative Stability Techniques for the European Cultural Heritage), investigates the possibility to use the Shape Memory Alloys (SMAs) for the realisation of intrinsically energy dissipation devices to be applied for the retrofitting of cultural heritage structures.

The protection system based on SMAs devices couples strengthening due to compression on the masonry wall with energy dissipation (Figure 6) that can occur for strong earthquakes. The strengthening with constant force will be use for the seismic upgrading of the structure, while the energy dissipation will provide stability improvement to avoid the loss of the structure. A demonstrative application is foreseen to retrofit the bell tower of Trignano (RE) in Italy damaged by a recent earthquake.

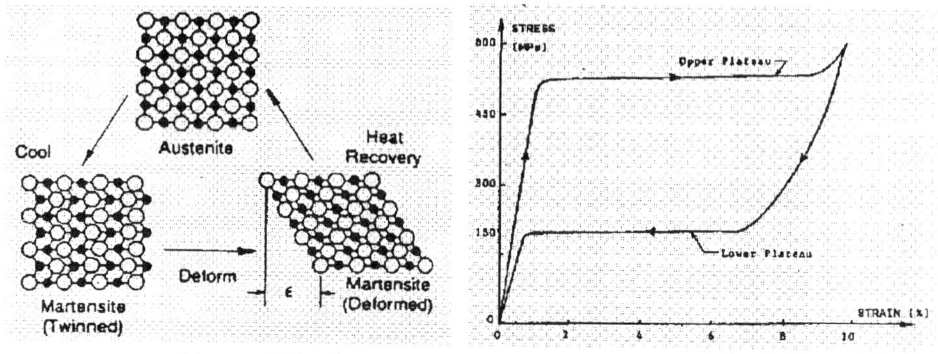

Figure 6: SMAs change of phase and super-elastic dissipation cycle

JRC is involved mainly in the characterisation tests of SMAs samples for engineering applications foresees in the project and in testing of large scale models of masonry walls (Figure 7) and comparing cases without and with the application of the protection system.

Figure 7: Full-scale masonry wall under construction at ELSA

Also in this case the design and the preparation of the experiments is supported with numerical analyses. The analyses are of relevant interest also for the interpretation of the results of the tests and for the validation of the numerical models. This is of basic importance for the vulnerability assessment of historical buildings and monuments, the classification in terms of seismic risk and the preliminary design of interventions for the retrofitting and upgrading of the structures.

References

[1] Pinto, A.V. and. Calvi, G.M. (1996), Acapulco Ongoing and Future Research in Support of Eurocode 8. Proceedings of the 11th World Conference on Earthquake Engineering,

[2] Renda, V.; Verzeletti, G.; Magonette, G.; Tirelli, D.; Papa, L.; Molina, F.J. Validation of the pseudo-dynamic method to test base-isolated structures. ASME PVP & ICPV-8 Conference, July 21-26, (1996) Montreal, Canada.

[3] Magonette, G.; Molina, F.J.; Taucer, F.; Verzeletti, G.; Tognoli, P.; Renda, V. Contribution of the JRC ISPRA to the Intercomparison of Analysis Methods for Seismically Isolated Nuclear Structures. Proceeding of Post-SmiRT-14 International Conference, August 25-27, (1997) Taormina.

[4] Renda, V.; Verzeletti, G.; Anthoine, A.; Tirelli, D.; Molina, F.J.; Bono, F.; Laudicina, M. Pseudodynamic tests of large scale isolated structures: cases of a steel frame and a historical building Proceeding of Post-SmiRT-14 International Conference, August 25-27, (1997) Taormina.

[5] Pegon, P. and Pinto, A.V. Seismic Study of Monumental Structures: Structural Analysis, Modelling and Definition of Experimental Model. Report EUR 16387 EN, EC-JRC-ISIS-SMU (1996) Ispra, Italy

EFFECTS OF INCOHERENCE, WAVE PASSAGE AND SPATIALLY VARYING SITE CONDITIONS ON BRIDGE RESPONSE

P. KESHISHIAN, A. DER KIUREGHIAN
Department of Civil & Environmental Engineering
University of California, Berkeley, USA

1. Introduction

Recent earthquakes in California and Japan have clearly demonstrated that seismic ground motions can vary significantly over distances that are of the same order of magnitude as the span lengths of extended structures, such as bridges. Such incoherent support motions may induce forces in multiply-supported structures, commonly known as pseudo-static forces, that are absent in structures uniformly excited at their base. At the same time, the forces generated by dynamic inertia effects tend to be smaller in the case of incoherent support motions due to the effect of random cancellations. The total effect, i.e., the sum of dynamic and pseudo-static forces, can be larger or smaller than the forces generated by uniform base excitation, depending on the characteristics of the structure and the ground motion field.

Spatial variability in ground motions may arise from four sources: (1) loss of coherency of seismic waves due to scattering in the heterogeneous medium of the ground, as well as due to the differential super-positioning of the waves arriving from an extended source, collectively denoted the "incoherence effect"; (2) difference in the arrival times of waves at separate stations, commonly known as the "wave-passage effect"; (3) gradual decay of wave amplitudes with distance due to geometric spreading and energy dissipation in the ground medium, denoted the "attenuation effect"; and (4) spatially varying local soil profiles and the manner in which they influence the amplitude and frequency content of the bedrock motion underneath each station as it propagates upward, denoted the "site-response effect". These effects are well characterized by the coherency function, which is the normalized cross-power spectral density of the motions at two stations.

Past studies have shed considerable light on the effects of incoherence, wave passage and attenuation (e.g., Bolt *et al.* 1982, Harichandran and Vanmarcke 1986, Abrahamson *et al.* 1991). Recently, Der Kiureghian (1996) derived a theoretical model for the coherency function accounting for all four effects mentioned above

The objective of this paper is to investigate the relative importance of the effects of incoherence, wave passage, and site response on a typical bridge. Use is made of the Multiple Support Response Spectrum (MSRS) combination rule (Der Kiureghian and Neuenhofer, 1992; Der Kiureghian *et. al* 1997). Based on the principles of random vibration theory, this rule provides a simple and accurate method for computing the

S. Balassanian et al. (eds.), Earthquake Hazard and Seismic Risk Reduction, 413–424.

bridge response directly in terms of the support response spectra and peak displacements, and a coherency function. Based on these findings, recommendations and guidelines for the seismic analysis and design of bridges are given.

2. The MSRS Method

The MSRS rule for the expected maximum of a generic response quantity $z(t)$ is expressed as follows (Der Kiureghian and Neuenhofer 1992):

$$\begin{aligned} \mathrm{E}\big[\max|z(t)|\big] = \Bigg[& \sum_{k=1}^{m}\sum_{l=1}^{m} a_k a_l \rho_{u_k u_l} u_k u_l \\ & + 2\sum_{k=1}^{m}\sum_{l=1}^{m}\sum_{j=1}^{n} a_k b_{lj} \rho_{u_k s_{lj}} u_k D_l(\omega_j,\zeta_j) \\ & + \sum_{k=1}^{m}\sum_{l=1}^{m}\sum_{i=1}^{n}\sum_{j=1}^{n} b_{ki} b_{lj} \rho_{s_{ki} s_{lj}} D_k(\omega_i,\zeta_i) D_l(\omega_j,\zeta_j) \Bigg]^{1/2} \end{aligned} \tag{1}$$

where $\mathrm{E}\big[\max|z(t)|\big]$ is the mean peak of the absolute value of the response; m is the number of support degrees of freedom; n is the number of modes used in the analysis; a_k is an influence coefficient associated with support degree of freedom k; b_{ki} is an effective participation factor associated with support degree of freedom k and mode i; u_k is the mean peak ground displacement at support degree of freedom k; $D_k(\omega_i,\zeta_i)$ is the ordinate of the displacement response spectrum associated with support degree of freedom k and the frequency and damping of mode i; $\rho_{u_k u_l}$ is the cross-correlation coefficient between ground displacements at support degrees of freedom k and l; $\rho_{u_k s_{lj}}$ is the cross-correlation coefficient between the ground displacement at support degree of freedom k and the response of an oscillator having the frequency and damping of mode i and subjected to the motion at the support degree of freedom l; and $\rho_{s_{ki} s_{lj}}$ is the cross-correlation coefficient between the responses of two oscillators having the frequencies and damping ratios of modes i and j and respectively subjected to the motions at support degrees of freedom k and l.

All the above terms, except the three cross-correlation coefficients, are available in an ordinary dynamic analysis procedure that handles multiple support excitations. The three cross-correlation coefficients are given in terms of the modal frequencies and damping ratios, peak ground displacements and response spectra, the coherency function, and the bridge geometry (Der Kiureghian and Neuenhofer 1992). A computer code named *MSRS* is available for computing these coefficients and applying the MSRS rule for arbitrary bridge structures (Der Kiureghian *et. al* 1997).

Der Kiureghian (1996) has shown that, in general, the coherency function for the motions at two support degrees of freedom k and l can be written as the contribution of three separate components in the form

$$\gamma_{kl}(\omega) = |\gamma_{kl}(\omega)|^{\text{incoherence}} \exp\left\{ i\left[\theta_{kl}(\omega)^{\text{wave passage}} + \theta_{kl}(\omega)^{\text{site response}}\right]\right\}, \tag{2}$$

where ω denotes the circular frequency and $i = \sqrt{-1}$. The incoherence component, representing the modulus of the function, must be developed by empirical means using recorded data. In the present study we have used the form

$$|\gamma_{kl}(\omega)|^{\text{incoherence}} = \exp\left[-(\alpha \omega d_{kl} / v_s)^2\right] \tag{3}$$

where d_{kl} is the distance between the two supports, v_s is the shear wave velocity, and α is a parameter. The phase angle due to the wave passage effect is given by

$$\theta_{kl}(\omega)^{\text{wave passage}} = -\frac{\omega d_{kl}^L}{v_{\text{app}}} \tag{4}$$

where d_{kl}^L is the projection of d_{kl} on the ground surface along the direction of propagation of seismic waves and v_{app} is the apparent wave velocity. The phase angle due to the site response effect is given by

$$\theta_{kl}(\omega)^{\text{site response}} = \tan^{-1}\frac{\text{Im}[H_k(\omega)H_l(-\omega)]}{\text{Re}[H_k(\omega)H_l(-\omega)]} \tag{5}$$

where $H_k(\omega)$ denotes the frequency response function of the site associated with the support degree of freedom k. This function characterizes the dynamic behavior of the soil site in the frequency domain. Methods for determining this function for a given site, while approximately accounting for nonlinear soil behavior, are described by Der Kiureghian *et. al* (1997).

3. Example Bridge

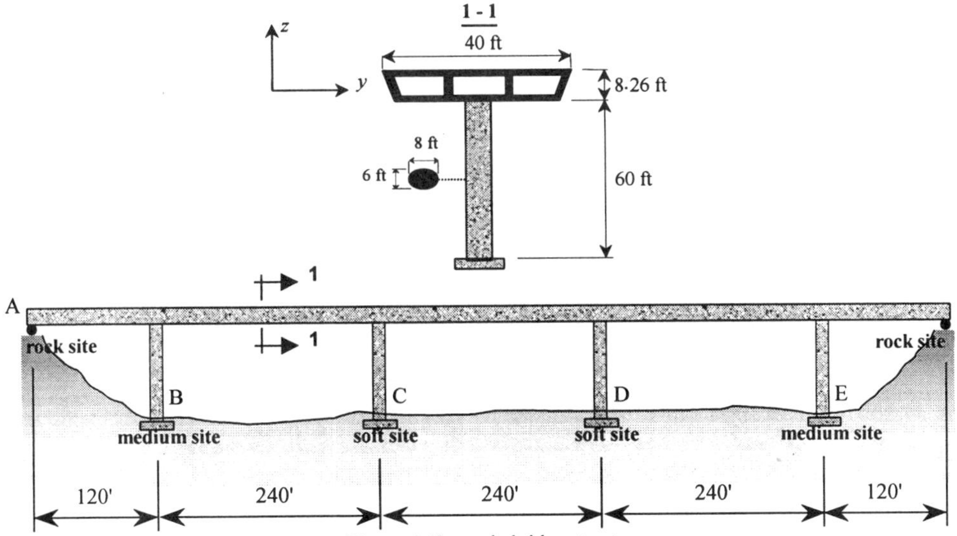

Figure 1. Example bridge structure.

We examine the individual and combined effects of incoherence, wave passage and site response on selected responses of the bridge shown in Figure 1. The bridge has the following support conditions: support A and F are situated on "rock" sites, supports B and E are situated on "medium" sites, and supports C and D are situated on "soft" sites. The soil profiles for the "medium" and "soft" sites are shown in Figure 2. All supports are fixed, except support A, which is free to rotate around the y axis and support F, which is free to translate in the x direction and rotate around the y axis. The first period of free vibration of the bridge is 1.49s. Acceleration response spectra for the three sites, shown in Figure 3, are obtained from the ATC-3 design response spectra by scaling in accordance to the peak ground accelerations listed in Table 1. The waves are assumed to propagate in the direction from support A to support F. For the incoherence effect $v_s/\alpha = 5000$ ft/s and for the wave passage effect $v_{app} = 1500$ ft/s are assumed. Below, we examine selected responses of the bridge.

Table 1. Peak ground accelerations and displacements for each site.

	"rock site"			"medium site"			"soft site"		
	x-dir	y-dir	z-dir	x-dir	y-dir	z-dir	x-dir	y-dir	z-dir
peak ground acceleration, g	0.15	0.15	0.10	0.35	0.35	0.10	0.25	0.25	0.10
peak ground displacement, ft	0.20	0.20	0.13	0.25	0.25	0.13	0.40	0.40	0.13

The results of the analysis are given in Figures 4 through 12. Each figure compares the response estimates for several cases of spatial variability (incoherence effect alone,

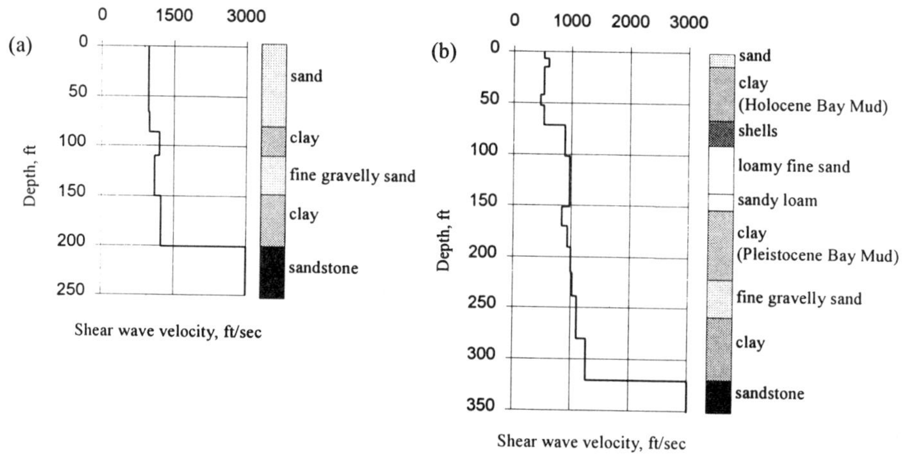

Figure 2. Soil profiles: (a) "medium" site; (b) "soft" site

wave passage effect alone, site response effect alone, and all three effects combined) with a benchmark case. For the benchmark case, the support motions are uniform and identical to the motions at the medium site. The difference between each pair of response estimates is highlighted in gray when the benchmark case over-estimates the response, i.e., when spatial variability effect tends to de-amplify the response, and in black when the benchmark case under-estimates the response, i.e., when the spatial variability effect tends to amplify the response. Comparing the highlighted areas within each figure, one can determine the relative importance of the three spatial variability effects, as well as determine the magnitude of the response amplification or de-amplification due to the spatial variability effect.

Figures 4, 5 and 6 show that the spatial variability effect has moderate influence on the nodal absolute displacements. In most cases the spatial variability effect de-amplifies the girder displacements in the x and y directions and amplifies the girder displacements in the z direction.

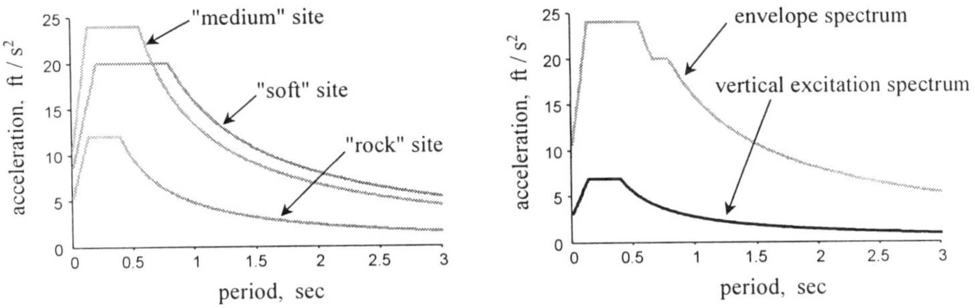

Figure 3. Acceleration response spectra.

The displacements at columns C and D in the x and y directions are amplified, primarily due to the larger displacements at the soft sites. Comparing the combined effect with the individual contributions, it is clear that the combined effect is not a simple superposition of the individual effects and that complex interactions between the various spatial variability effects occur.

Figure 6 shows large amplification of the axial forces in the columns and de-amplification of the axial force in the girder, primarily due to the site-response effect. The asymmetry in the response is due to the fact that support A is constrained in the x direction, whereas support F is not.

Figure 7 shows the shear forces in the y direction. Large de-amplification is observed due to the site-response effect, while modest amplifications occur due to incoherence and wave passage. Figure 8 shows the shear forces in the girder in the z direction and the shear forces in the columns in the x direction. Large amplifications, by factors as large as 5.6, are observed in both girder and column forces due to the site-response and wave passage effects.

For the girder torque response, shown in Figure 10, the spatial variability effect tends to de-amplify the response within spans BC and DE, and amplify the response within span CD. It should be noted that due to the symmetry of the bridge and the site conditions, the torque within span CD is zero in the cases of uniform support motions and for site response alone. Hence, the amplification within this span is solely due to the incoherence and wave passage effects.

Figures 11 shows the results for the bending moments around the y axis. Large amplifications, by factors as large as 5.8, are observed in both the columns and the girder due to the effects of site response and wave passage. Figure 10 shows the bending moments in the girder around the z axis and in the columns around the x axis. Small amplification of the girder moments due to the incoherence effect is observed.

However, the site-response and wave-passage effects tend to de-amply the moments in most places.

In the above comparisons the benchmark was a case with uniform support motions, all identical to the motions on the medium site. In practice it is common to perform uniform support response spectrum analysis using the envelope of the support response spectra at all sites. The conservatism inherent in the envelope spectrum is thought to adequately account for any amplification of the response due to the site response effect.

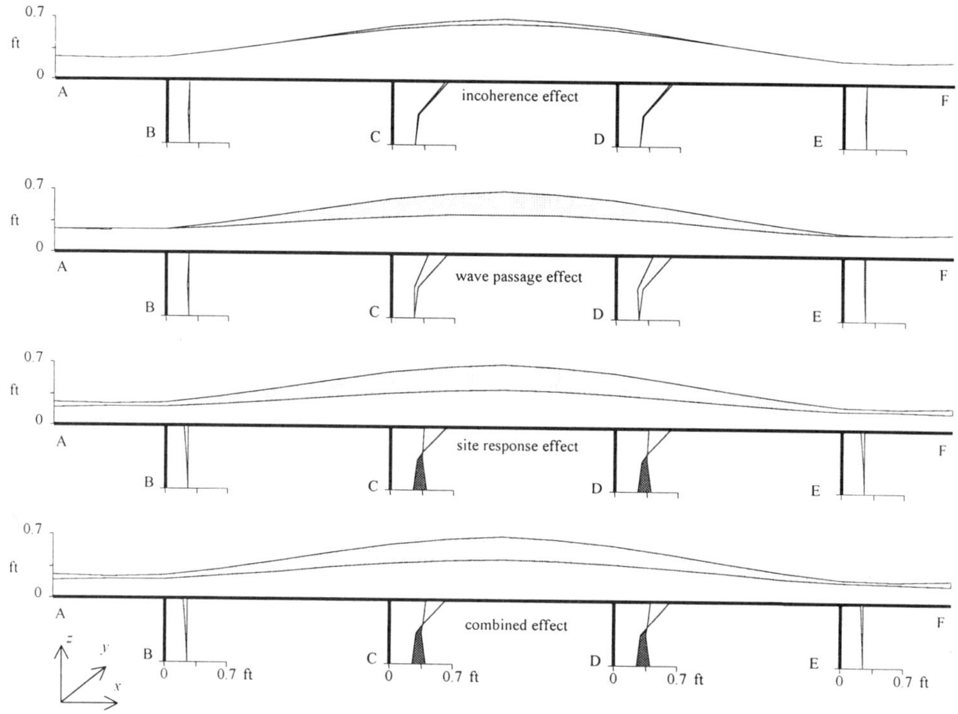

Figure 4. Amplification (▇) and de-amplification (▭) of the response due to spatial variability effects: absolute displacements in direction *y*.

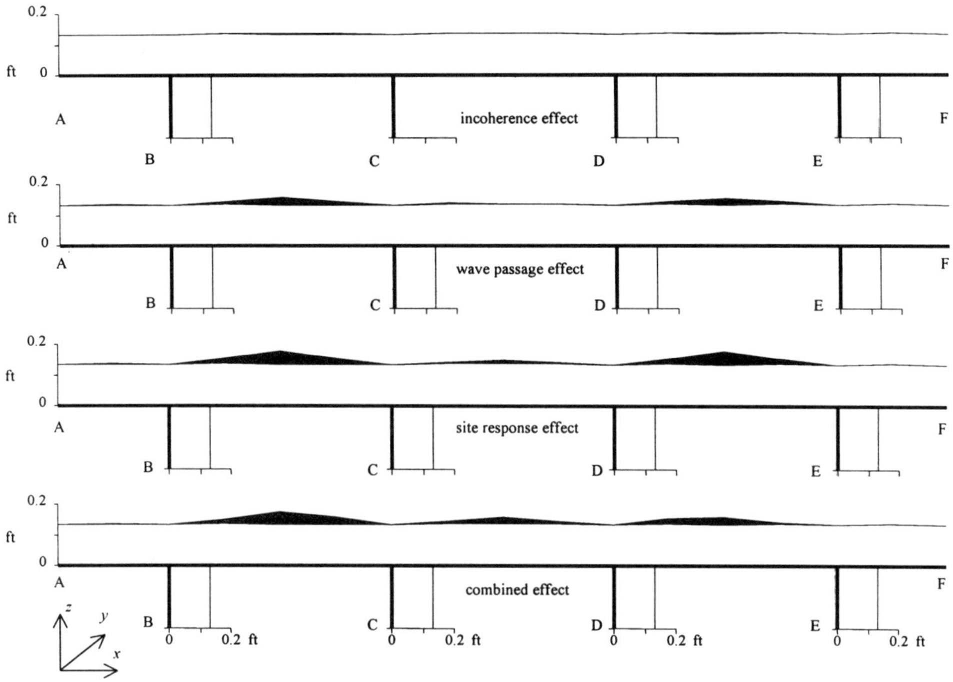

Figure 5. Amplification (■) and de-amplification (□) of the response due to spatial variability effects: absolute displacements in direction *z*.

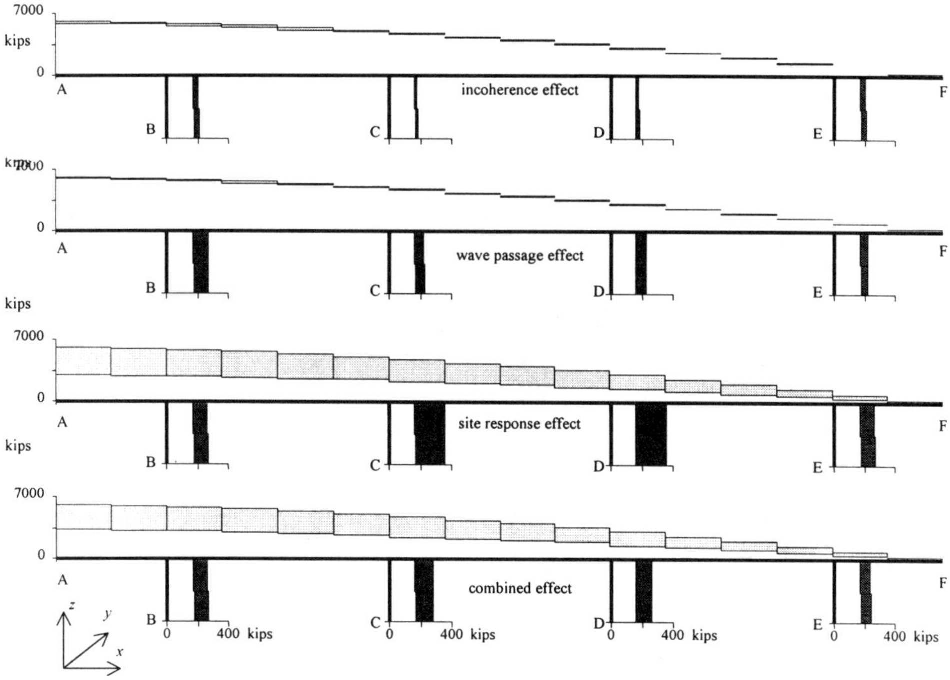

Figure 6. Amplification (■) and de-amplification (□) of the response due to spatial variability effects: axial forces.

Figure 7. Amplification (■) and de-amplification (□) of the response due to spatial variability effects: shear forces in direction *y*.

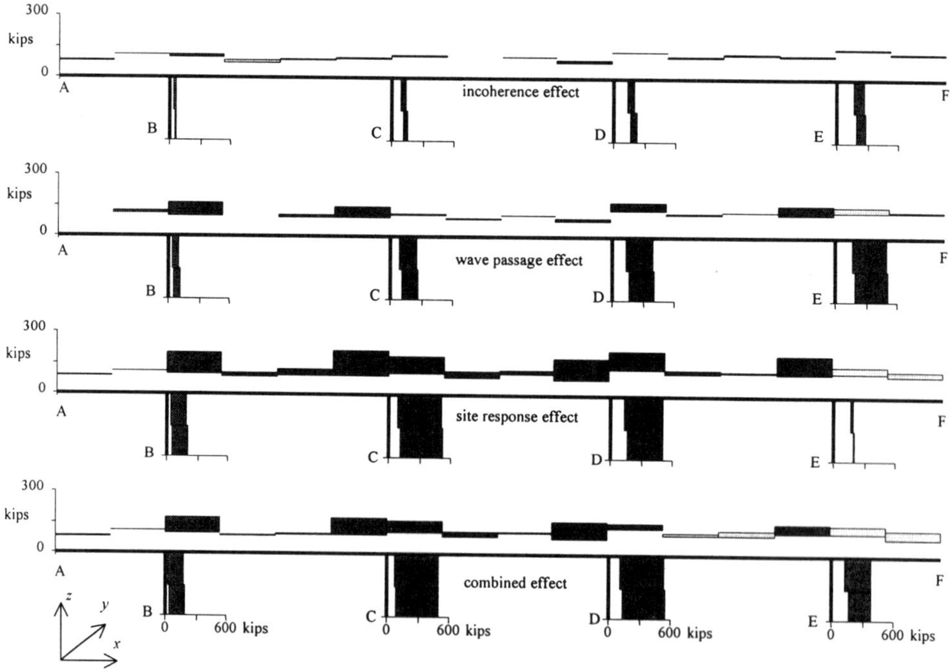

Figure 8. Amplification (■) and de-amplification (□) of the response due to spatial variability effects: girder shear forces in direction *z* and column shear forces in direction *x*.

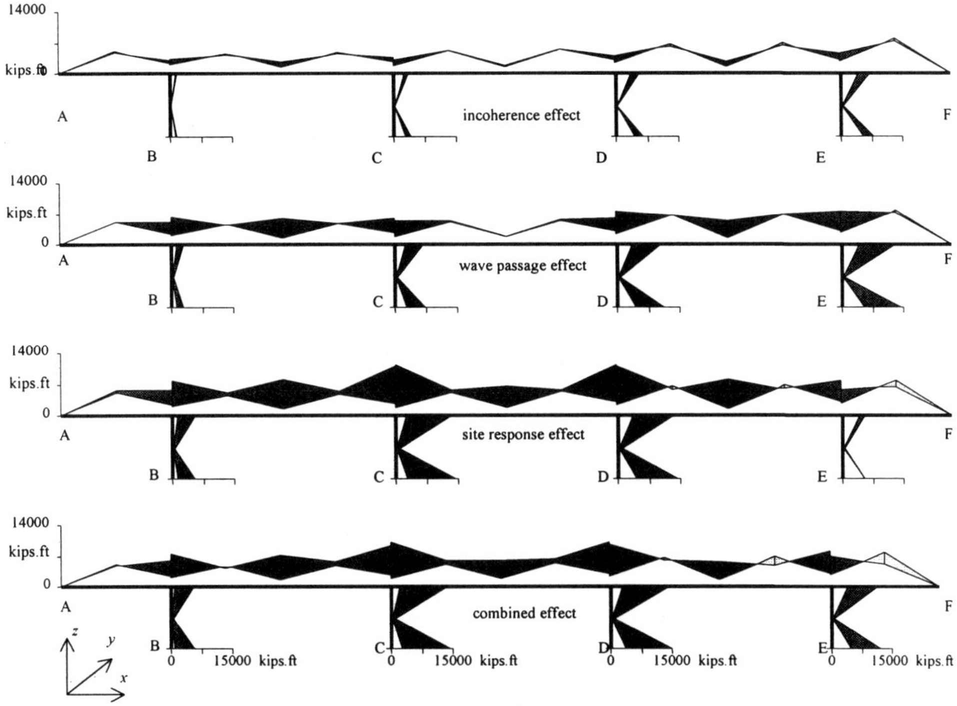

Figure 9. Amplification (■) and de-amplification (□) of the response due to spatial variability effects: bending moments around *y* axis.

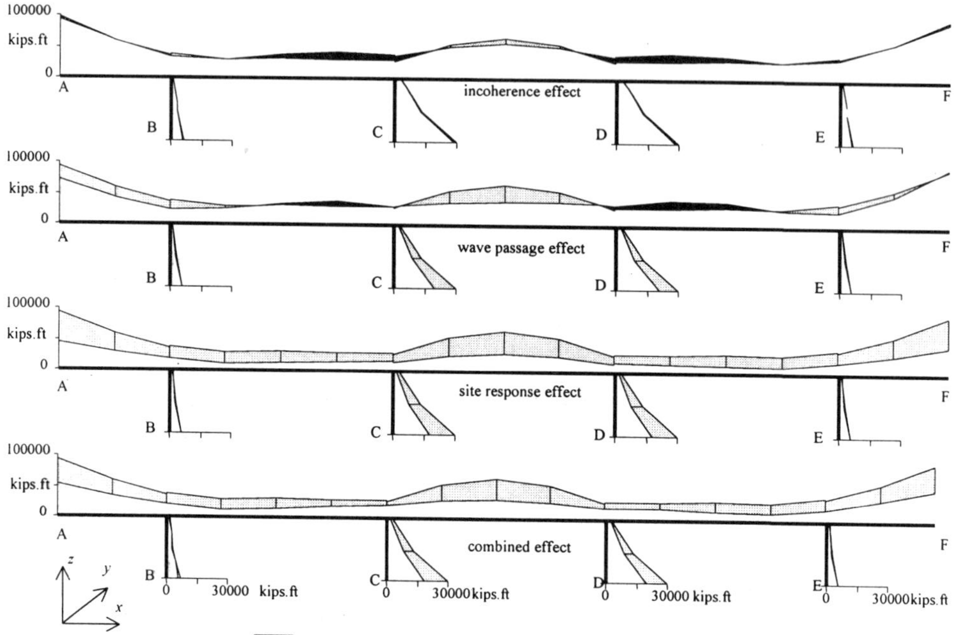

Figure 10. Amplification (■) and de-amplification (□) of the response due to spatial variability effects: girder bending moments around *z* axis and column bending moments around *x* axis.

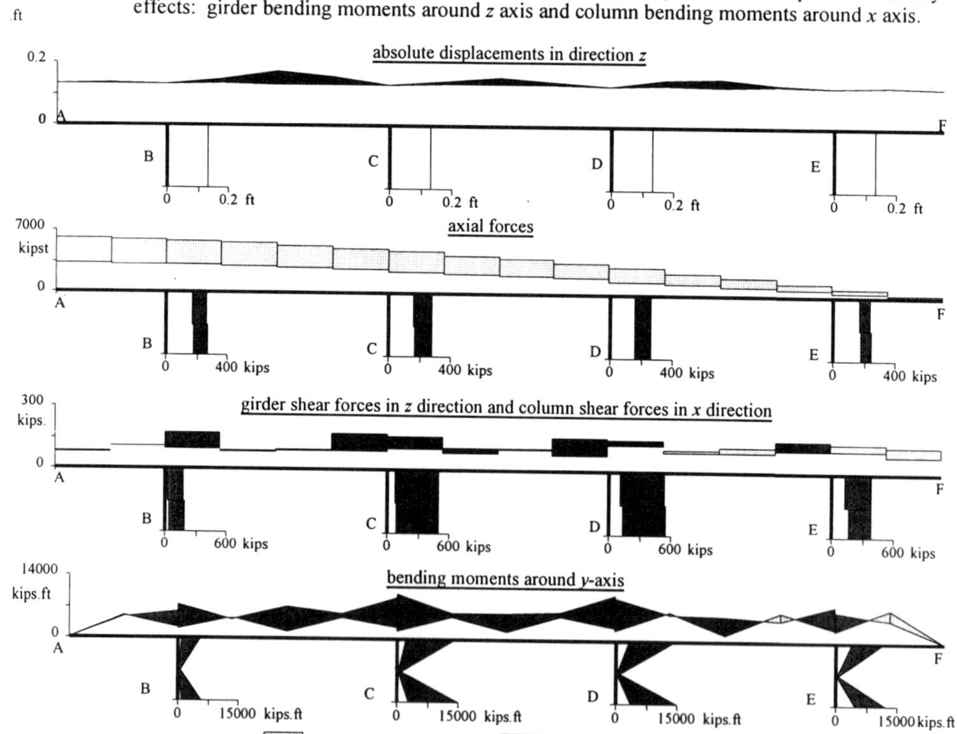

Figure 11. Overestimates (□) and underestimates (■) by the envelope spectrum method.

To examine the validity of this approach, we compare selected responses of the bridge computed based on the envelope spectrum in Figure 3 with the responses computed by the MSRS rule for the combined case of the spatial variability effects. The results are shown in Figure 11. It is clear that the envelope spectrum method severely underestimates the response at many locations. This serves to demonstrate that the envelope spectrum method is unreliable and could lead to gross errors on both the conservative and unconservative sides.

4. Main Conclusions

On the basis of the results shown in this paper, as well as more comprehensive results reported in Der Kiureghian *et. al* (1997), the following conclusions are made:

1. The spatial variability effect, consisting of the incoherence effect, the wave passage effect, and the site response effect, can strongly influence the response of bridge structures. These effects could amplify or de-amplify the bridge response by as much as an order of magnitude or more.
2. For the bridge example considered, the site response effect had the largest influence, followed by the wave passage effect and then the incoherence effect. This trend is expected to be true for bridges having short or moderate span lengths and situated on supports with varying soil conditions.
3. In the presence of the site-response effect, the direction of propagation of seismic waves could influence the responses of the bridge by as much as 40% or more. Interaction between the wave passage and site response effects is believed to be the reason for such enhanced sensitivity to the direction of propagation of waves.
4. The currently practiced method of using the envelope of support response spectra uniformly at all supports is inaccurate and potentially unconservative. This method does not account for the spatial variability of ground motions and could lead to unsafe designs.
5. The MSRS rule and the *MSRS* code are effective tools for multiple-support response spectrum analysis with full account of the spatial variability effects. In particular, this method provides maximum flexibility for parametric investigations at minimal cost.

5. Acknowledgment

This study was supported by the California Department of Transportation Contract No. 59U429, which is gratefully acknowledged.

6. References

Abrahamson, N.A., J.F. Schneider, and J.C. Stepp (1991). Empirical spatial coherency functions for application to soil-structure interaction analysis. *Earthq. Spectra*, 7, 1-28.

Bolt, B.A., C.H. Loh, J. Penzien, Y.B. Tsai, and Y.T. Yeh, (1982). Earthquake strong motions recorded by a large near-source array of digital seismographs. *Earthq. Eng. Struct. Dyn.*, **10**, 561-573.

Der Kiureghian, A. and A. Neuenhofer (1992). Response spectrum method for multiple-support seismic excitation. *Earthq. Eng. Struc. Dyn.*, **21**, 713-740.

Der Kiureghian, A. (1996). A coherency model for spatially varying ground motions. *Earthq. Eng. Struct. Dyn.,* **25**, 99-111.

Der Kiureghian, A., P. Keshishian and A. Hakobian (1997). Multiple-support response spectrum analysis of the bridges including the site response effect & the MSRS code. *Report No. UCB/EERC-97/02*, Earthquake Engineering Research Center, University of California, Berkeley, CA.

Harichandran, R.S., and E. Vanmarcke (1986). Stochastic variation of earthquake ground motion in space and time. *J. Eng. Mech.*, ASCE, **114**, 1526-1541.

Idriss, I.M., J. I. Sun and P.B. Schnabel. (1991). User's manual for SHAKE91: a computer program for conducting equivalent linear seismic response analyses of horizontally layered soil deposits. *Report of Center for Geotechnical Modeling,,* Department of Civil and Environmental Engineering, University of California at Davis, CA.

THE IMPACT OF THE SYSTEM OF TIES UPON THE LEVEL OF SEISMIC EFFECT FOR ROAD BEAM BRIDGES

R.S.AZOYAN, A.N.SARGSYAN, T.G.PETROSYAN
Yerevan Institute of Architecture and Construction,
Yerevan, Republic of Armenia

1. Introduction

The effective seismic protection of bridges is very important problem.

In the areas with high seismic hazard it is also necessary to take into consideration the possibility of recurring of the earthquakes.

Table 1 shoes data for some places on Armenian Highland with higher recurring of earthquakes. Table 2 shoes return period of historic earthquakes within 100km radius of Spitak [1].

TABLE 1.

	Place		
	Kars	Erzrum	Erznka
Numbers of earthquakes per 1000 years / Frequency	27 / 0.027	33 / 0.033	31 / 0.031
Maximal numbers of earthquakes per 50 years / Frequency	10 / 0.2	14 / 0.28	5 / 0.1
Maximal numbers of earthquakes per 25 years / Frequency	7 / 0.28	12 / 0.48	4 / 0.16
Earthquake with magnitude M>6	1 (M_{max}=6.7)	5 (M_{max}=6.4)	5 (M_{max}=7.8)

The analysis of these data demonstrates that recurring of earthquakes in this region is high, and if we will consider that average life period for the road bridges is 50-75 years, then problems of the effective seismic protection are becoming more important, because of the bridge can be exposed to earthquake influence during exploitation a several times.

TABLE 2.

Earthquake magnitude, M	Return periods (years)
M=6.0	25
M=6.5	70
M=7.0	110

S. Balassanian et al. (eds.), Earthquake Hazard and Seismic Risk Reduction, 425–432.

The structure can accumulate some small damages such as thin cracers all insignificant deformation after first earthquake impact and will destroy during-next impact, which may be lower than first.

Therefore, to our opinion the isolation of structure is more preferable and effective method of structure protection in such areas. Besides, the isolation of structure allows to endure not only main shock but also to endure the following after shocks, which may have rather high value.

However, the seismic isolation of structures brings to rising of costs of their constructions. Therefore during the design with seismic isolation, it is necessary to follow to adequate principles. For bridges these principles can be the following:

- Bridge's span length, height of piers, total length of span structure;
- Bridge's strategic importance.

2. Proposed Isolation System for Span Structure of Road Bridges.

On the figures 1 is shown proposed multi level seismic isolation system for span structures of road bridges.

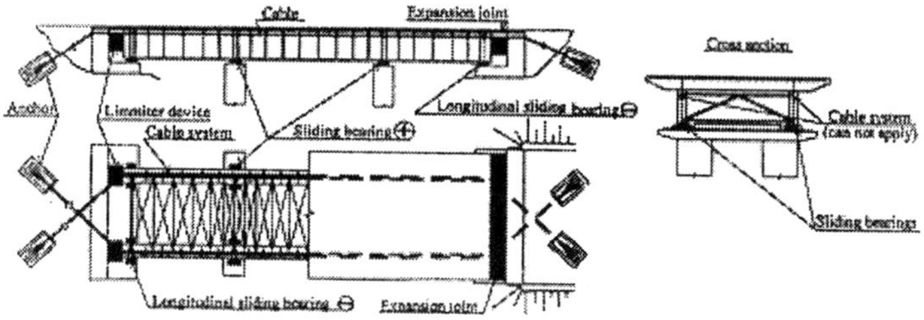

Figure 1. Proposed multi-level ties system

The basic idea of this system is obligatory providing the double-sided ties with definite stiffness between the all members of the bridge, which joints together.

The basic principles of the system are:

1. Placing of span structure through sliding bearings, which exclude structure movements in vertical direction;
2. Connection the span structure with abutments through double-sided ties;
3. Application of expansive joints with double-sided ties;
4. Arrangement along the span structure the special cable system. Cable systems can be lengthy, short and absent at all;
5. Using of special structures for expansion joints, limiters and sliding bearing.

3. Sliding Bearing Structure

The proposed sliding bearing structure for long span girder (fig.2) allows free longitudinal and transversal displacements of the span structure through leaning of low balance beams on antyfriction pad and excludes vertical movements due to ties between top and low balance beams. The span structure vertical movements exclusion is very significant because in case of considerable value of the vertical component of ground acceleration the span movement may occur in this direction as a result of this movement of the span relatively bearings.

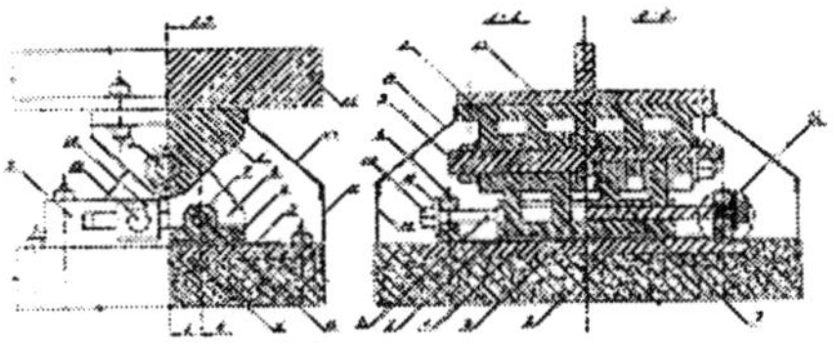

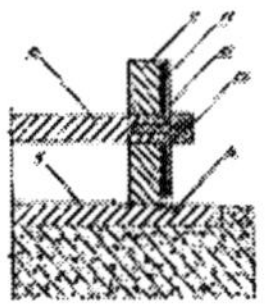

Figure 2. Proposed sliding bearing structure.

That was also corroborated by inspection of the bridges after Spitak Earthquake, where according to some evaluations the value of the vertical component of the ground acceleration was higher 1.0g in the epicentrum zone.

4. Rubber-Steel deformable module

The structure of the proposed rubber-steel deformable module consists of the package steel, wavy in the plan sheets, which are covered by rubber rug (fig.3). The structure of deformable module allowing has free transversal movements for span and dissipate the vibration energy in longitudinal direction.

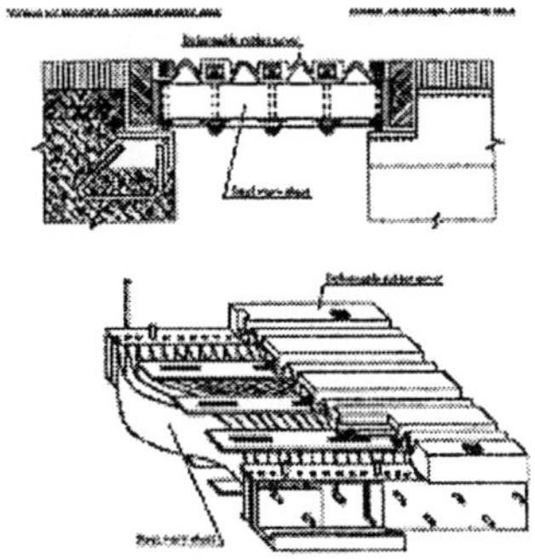

Figure 3. Proposed structure of rubber-steel deformable module for expansion joints of road bridges.

The continuously of the bridge's roadway is achieved due to wavy steel sheet and rubber rug, which are jointed with both coupled members of the bridge.

Energy's dissipation of the longitudinal vibration in the module is achieved according to deformation properties of wavy sheet. The behavior of the sheet of the joint steel-rubber deformable module has been tested. Figure 4 illustrates the dependencies of the absorption coefficient (y) versus the value relation of initial amplitude (a) of sheet wave and wavelength (L).

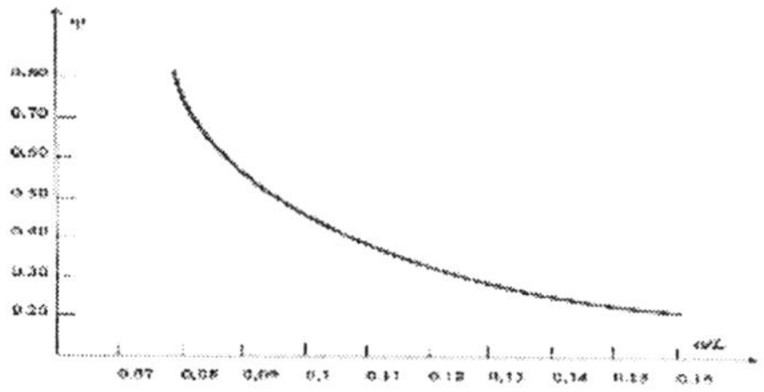

Figure 4. Dependence of absorption coefficient from the relation of amplitude and wavelength.

5. Limiter Structure

The proposed construction of limiter is represented on fig.5. This principle idea is that the steel hinge tetralinkige has a shape of square in cross section and includes rubber pad. In the process of operation the square section is transforming in to diamond-shape one.

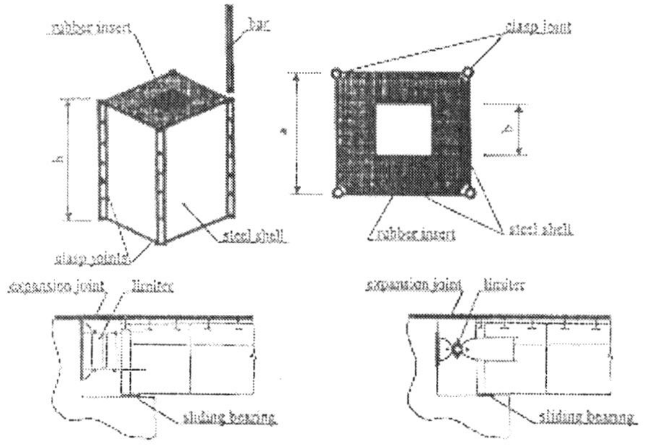

Figure 5. Proposed structure of limiter

Thus the rubber insert experiences the deformation of pure shear. Utilization of the frame of limiters package allows to compensate large value of potential displacement and to absorb energy of structure vibrations. The dissipation coefficient y [2] of rubber insert is close to 0.1 in the case when the largest diagonal displacement is equal to 20% of side of the square.

The stiffness of the rubber pad could be regulated by a hole of different size in it (fig.6).

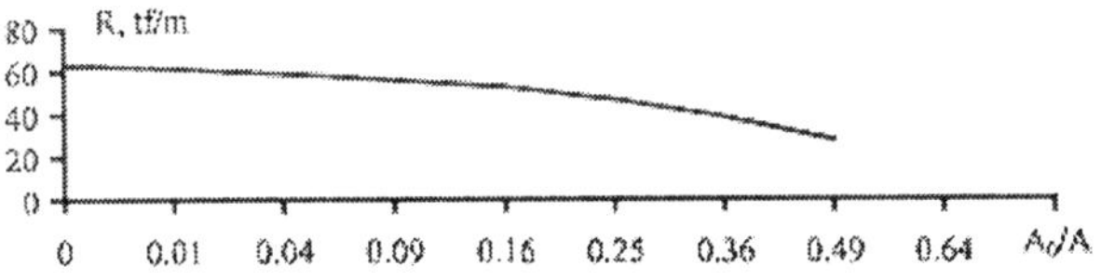

Figure 6. Dependence of the limiter 's stiffness versus the relation of hole area (A_0) and total cross section area (A) of the rubber pad.

6. Cable System

The cable system attached to the major beams of the span through the guides tubs and follows the trace of mass centers of the sections on which the span is conditionally broken.

7. Analysis of proposed isolation system and structures

Analysis had shown that for the 15-30m spans length with girder system it is enough to use only rubber-steel deformable module as expansion joints and rubber bearing, that is the minimal isolation level. The calculations for one, two and three span girder systems with 18m length beams when they are leaned through rubber bearing and when as expansion joints are used proposed deformable module, the longitudinal seismic force value is reduced on 15% compared with the case when beams are leaned through rubber bearing and used common expansion joints structures.

Besides, it is necessary to mention that application of the rubber bearings instead of widely used for these spans length cast steel roller bearings if in the same time there are proposed expansion joints in the both edges of span allows to significant reduce the seismic impact on span structure about 2.5 times.

For the span length more than 40m and continuos span beams it is necessary to use higher level seismic isolation, which includes deformable modules, limiters and rubber or sliding bearings, and for the continuos span system with span longer than 63m seismic isolation system proposed to install sliding bearing, limiters, and if it is necessary to arrange the cable system.

The diagram on the figure 7 shows the beneficial application of seismic isolation through sliding bearing and limiters for the typic continuos steel beams with reinforced concrete roadway deck. The anchoring to the one of abutment is widespread seismic protection method for big bridges in Armenia. For comparison the calculations were provided under the Armenian Seismic Construction Code (CCRA)[3] and EuroCode [4, 5].

The big difference between the research results, conducted in accordance with both (CCRA) and EuroCode is explained by different consideration the plastic hinges in the isolated span structures. The EuroCode excludes the formation of plastic hinges in the span structures under the vibrations, whereas the CCRA allows their formation (coefficient k>=0.7). Besides it, CCRA has coefficient that depends on the length of span. For the long span bridges as continuos beams (with length more than 100m) the value of this coefficient strives to 0.7. It is necessary to point that their is no analogous one in the in the EuroCode. So the difference in those two coefficients is the reason of reducing seismic loads according to CCRA in 2 times compared with seismic force value by EuroCode. The coefficients of importance for both codes were adopted equal 1, the ground acceleration value 0.4g and ground category is II-nd.

1. Dependence of value of the longitudinal force on the main girder of span structure virsuse the value of total stiffness of limiters (k).

a) according to CCRA II-2.02.94.

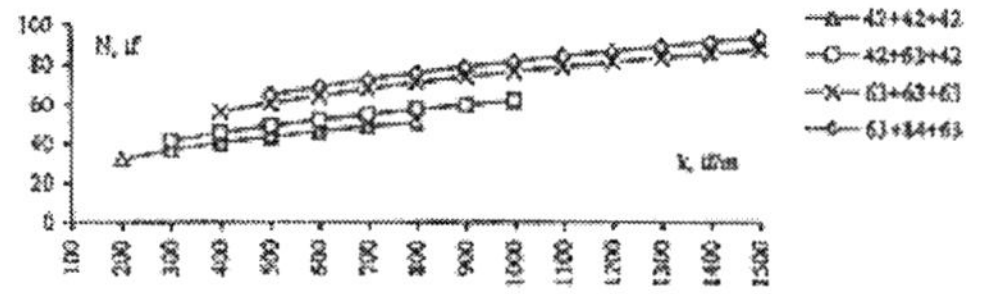

b) according to EC8, part 1.1 and 2.

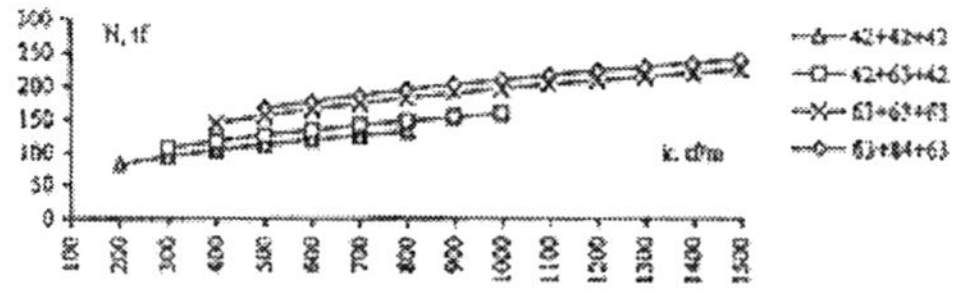

2. Dependence of periods of vibration of span structures with different length versus of the total stiffness of limiters (k).

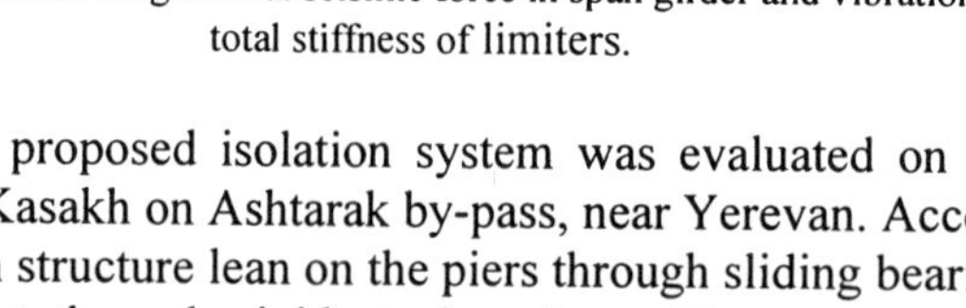

Figure 7. Dependence of value of longitudinal seismic force in span girder and vibration period of span porn total stiffness of limiters.

Efficiency of the proposed isolation system was evaluated on the span of new bridges through River Kasakh on Ashtarak by-pass, near Yerevan. According to. design, the steel continuos span structure lean on the piers through sliding bearing and jointed to the right side abutment through rigid steel anchors. The span structure scheme is 63+84+63m, and it has steel ortotropic top deck. The total weight of the span is 1640t. Two models were tested under the El-Centro, San Fernando and Spitak earthquakes (on ground horizontal acceleration record in Gukasyan). The first model was simulated the really condition of the span and next model simulated proposed isolation system. For the model with isolation was considerate that value of the limiter on the each edge equal 1100tf/m. For the calculations SAP90 program was used.

Figure 8 illustrates the diagrams of the longitudinal seismic loads at the edge of beam from dependence of time for both models under the Spitak earthquake.

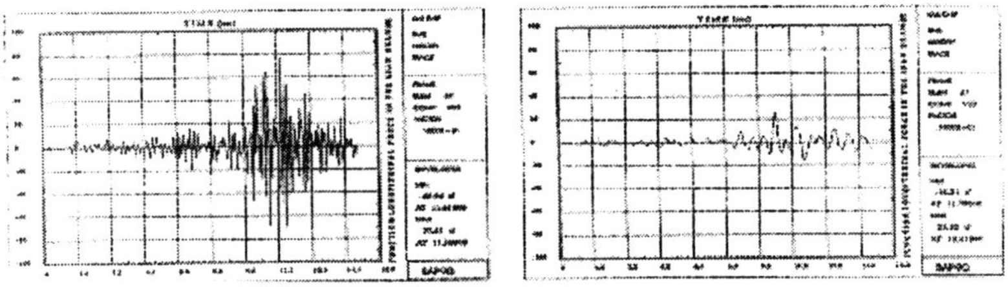

a) one edge of span anchored

b) both edges of span jointed to abutments by limiters

Figure 8. diagrams of the longitudinal seismic loads at the edge of beam porn dependence of time for both models under the Spick earthquake.

As you see the value of the seismic force reduces about 4 times, when isolation system was used. The frequency of span vibration also reduces in this case. For this models dissipation coefficient [4] was calculated, for the model with rigid fixing the value of this coefficient practical equal 0 and for model with isolation through limiters it is equal 0.11. Figure 9 illustrates dependence of value of structure reaction for both models of span, which were tested versus the frequency value of disturbing force. The isolation of the structure allows to transfer the resonance range to the low natural vibration frequency range. However, it is necessary to take into consideration the value of natural vibration frequency of ground base.

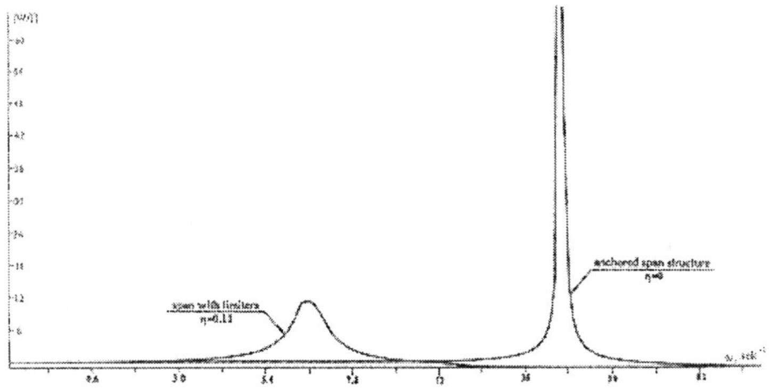

Figure 9. Dependence of dynamis reaction of structure |W/F| from the value of disturbing force frequency.

Thus, correctly selected stiffness dimensions of the double-sided ties of isolation system allow to significant reduce the seismic impact to the structure.

8. Conclusion

In conclusion it is worth noting that:

- Seismic isolation in combination with double-side connections between the joined bridge members allows having reliable seismic response behavior of structures.
- It's necessary to eliminate an uplift of the span.
- Suggested constructions are rather simple and allowing for seismic movement.
- Utilization of limiters is more effective if they are fixed at the ends of spans.
- The anticipated stronger seismic stress the more valid for the utilization of seismic isolation.

This fact becomes more effective it take into account the recurrence of earthquake.

References

1. Mishac.K.Yegian, Vahe G.Ghahraman. *The Armenia Earthquake of December 1988*. Northeastern University, Boston, Massachusetts, October, 1992.
2. E.S.Sorokin. *Dynamic calculation of carrying structures of building*. Moscow. Gosstroyizdat, 1956.
3. CCRA II-2.02-94. Seismic Construction Design Code. Vol.6, Transport Structure. Yerevan, 1997.
4. *EuroCode 8. – Design Provisions for Earthquake Resistance of Structures. – Part 1-1: General rules – Seismic actions and general requirements for structures.* European Committee for Standardization. Bruxelles, 1994.
5. *EuroCode 8. – Design Provisions for Earthquake Resistance of Structures. – Part 2:* Bridges. European Committee for Standardization. Bruxelles, 1994.

THE ASYMPTOTIC SOLUTION OF SOME MODEL PROBLEMS OF SEISMOSTABILITY CONSTRUCTION AND SEISMOLOGY

L. A. AGHALOVIAN, R. S. GEVORGYAN
Institute of Mechanics of Armenian National Academy of Sciences, Republic of Armenia

Introduction

In order to reduce negative effects of strong earthquakes on constructions, elastomeric seismoisolators are being used in construction practice. The existing methods of their calculation are based on those or other hypotheses [1]. In this paper corresponding model problems in an exact statement – on the base of the equations of elasticity theory for compressible and incompressible bodies are suggested to solve.

By means of asymptotic integration of these equations recurrent formulae for determining stress-strain state of a layer made of incompressible material (elastomeric layer) under static and dynamic mixed boundary conditions are derived.

A two-layered strip, consisting of compressible and incompressible layers, is considered, closed solutions, which can be used in the calculations of elastomeric seismoisolators, are obtained. The suggested approach can also be used for the structures from arbitrary number of layers and in spatial statement of the problems.

1. Stress-Strain State of the Incompressible Layer-Elastomer

Let an isotropic incompressible layer occupy the region $\Omega = \{x, z : |x| \le l, |z| \le h, h << l\}$. It is required to determine the stress-strain state of the layer, when on its longitudinal edges kinematic conditions are given:

$$u_x(z = \pm h) = u_x^{\pm}(x), \quad u_z(z = \pm h) = u_z^{\pm}(x) \tag{1}$$

The conditions, given on the end-walls, will be concretely defined later.

For solving the stated boundary problem in the equations of a plane strain for an incompressible body

$$\begin{aligned} &\frac{\partial \sigma_{xx}}{\partial x} + \frac{\partial \sigma_{xz}}{\partial z} = 0, \quad \sigma_{xx} = \sigma_{zz} + 4G\frac{\partial u_x}{\partial x} \quad (x, z) \\ &\frac{\partial u_x}{\partial z} + \frac{\partial u_z}{\partial x} = \frac{1}{G}\sigma_{xz}, \quad \frac{\partial u_x}{\partial x} + \frac{\partial u_z}{\partial z} = 0 \end{aligned} \tag{2}$$

we shall pass to dimensionless coordinates and dimensionless displacements according to the formulae

$$x = l\xi, \ z = h\zeta = \varepsilon l\zeta, \quad u_x = lu, \ u_z = lw, \quad \varepsilon = h/l \tag{3}$$

As a result a singular perturbed by a small parameter ε system will be obtained:

S. Balassanian et al. (eds.), Earthquake Hazard and Seismic Risk Reduction, 433–446.

$$\frac{\partial\sigma_{xx}}{\partial\xi}+\varepsilon^{-1}\frac{\partial\sigma_{xz}}{\partial\zeta}=0,\qquad \frac{\partial\sigma_{xz}}{\partial\xi}+\varepsilon^{-1}\frac{\partial\sigma_{zz}}{\partial\zeta}=0,$$

$$\sigma_{xx}=\sigma_{zz}+4G\frac{\partial u}{\partial\xi},\qquad \varepsilon^{-1}\frac{\partial u}{\partial\zeta}=\frac{1}{G}\sigma_{xz}-\frac{\partial w}{\partial\xi},\tag{4}$$

$$\sigma_{zz}=\sigma_{xx}+4G\frac{\partial w}{\partial\zeta}\varepsilon^{-1},\qquad \varepsilon^{-1}\frac{\partial w}{\partial\zeta}=-\frac{\partial u}{\partial\xi}\ \text{(condition of incompressibility)}$$

where G is the shear module.

It's easy to verify, that the correlation $\sigma_{zz}=\sigma_{xx}+4G\varepsilon^{-1}\frac{\partial w}{\partial\zeta}$ is identically satisfied, if the rest of the equations of system (4) are satisfied, that's why it won't be considered later on. In fact, it is replaced by the condition of incompressibility.

The solution of system (4) is sought in the form of [2]:

$$Q=\sum_{s=0}^{S}\varepsilon^{\chi_Q+s}Q^{(s)}(\xi,\zeta)\tag{5}$$

where Q is any of stresses and displacements, χ_Q characterizes the asymptotic order of the corresponding value and is defined in such way in order to get a noncontradictory system relative to the coefficients $Q^{(s)}(\xi,\zeta)$ of the expansion after substituting (5) into (4). This aim is reached when

$$\chi_{\sigma_{xx}}=\chi_{\sigma_{zz}}=-3,\quad \chi_{\sigma_{xz}}=-2,\quad \chi_u=-1,\quad \chi_w=0\tag{6}$$

Asymptotics (6) principally differs from the asymptotics of the values of the analogous boundary problem for a compressible layer [2]

$$\chi_{\sigma_{xx}}=\chi_{\sigma_{zz}}=\chi_{\sigma_{xz}}=-1,\quad \chi_u=\chi_w=0\tag{7}$$

i.e. the incompressibility of the material fully changes the picture of the stress-strain state.

Substituting (5) into (4), taking into account (6), a noncontradictory system relative to $Q^{(s)}(\xi,\zeta)$ will be obtained, the solution of which is

$$\sigma_{zz}^{(s)}=\sigma_{zz0}^{(s)}(\xi)+\sigma_{zz*}^{(s)}(\xi,\zeta)$$

$$\sigma_{xx}^{(s)}=\sigma_{zz}^{(s)}(\xi,\zeta)+4G\frac{\partial u^{(s-2)}}{\partial\xi}$$

$$\sigma_{xz}^{(s)}=\sigma_{xz0}^{(s)}(\xi)-\zeta\frac{d\sigma_{zz0}^{(s)}}{d\xi}+\sigma_{xz*}^{(s)}(\xi,\zeta)\tag{8}$$

$$u^{(s)}=u_0^{(s)}(\xi)-\frac{1}{G}\zeta\sigma_{xz0}^{(s)}-\frac{1}{2G}\zeta^2\frac{d\sigma_{zz0}^{(s)}}{d\xi}+u_*^{(s)}(\xi,\zeta)$$

$$w^{(s)} = w_0^{(s)}(\xi) - \frac{1}{2G}\zeta^2 \frac{d\sigma_{xz0}^{(s)}}{d\xi} + \frac{1}{6G}\zeta^3 \frac{d^2\sigma_{zz0}^{(s)}}{d\xi^2} - \frac{du_0^{(s)}}{d\xi} + w_*^{(s)}(\xi,\zeta)$$

where the values

$$\sigma_{zz*}^{(s)} = -\int_0^\zeta \frac{\partial\sigma_{xz}^{(s-2)}}{\partial\xi} d\zeta, \qquad \sigma_{xz*}^{(s)} = -\int_0^\zeta \left(\frac{\partial\sigma_{zz*}^{(s)}}{\partial\xi} + 4G\frac{\partial^2 u^{(s-2)}}{\partial\xi^2} \right) d\zeta$$

$$u_*^{(s)} = \int_0^\zeta \left(\frac{1}{G}\sigma_{xz*}^{(s)} - \frac{\partial w^{(s-2)}}{\partial\xi} \right) d\zeta, \qquad w_*^{(s)} = -\int_0^\zeta \frac{\partial u_*^{(s)}}{\partial\xi} d\zeta \tag{9}$$

for every s are known, if the previous approximations are built. In (8) until unknown functions of integration $\sigma_{zz0}^{(s)}, \sigma_{xz0}^{(s)}, u_0^{(s)}, w_0^{(s)}$ must be defined from boundary conditions.

Satisfying the boundary conditions (1), we shall get

$$u_0^{(s)}(\xi) = u_x^{-(s)} + \frac{1}{2}U_x^{(2)} + \frac{1}{2G}\frac{d\sigma_{zz0}^{(s)}}{d\xi} - u_*^{(s)}(\zeta=-1)$$

$$w_0^{(s)}(\xi) = u_z^{-(s)} - \frac{du_x^{-(s)}}{d\xi} - \frac{1}{4}\frac{dU_x^{(s)}}{d\xi} - \frac{1}{3G}\frac{d^2\sigma_{zz0}^{(s)}}{d\xi^2} + \frac{du_*^{(s)}(\zeta=-1)}{d\xi} - w_*^{(s)}(\zeta=-1)$$

$$\sigma_{xz0}^{(s)} = \frac{1}{2}GU_x^{(s)} \tag{10}$$

$$U_x^{(s)} = u_x^{+(s)} - u_x^{-(s)} + u_*^{(s)}(\zeta=-1) - u_*^{(s)}(\zeta=1) \qquad (x,z;\, u_*, w_*)$$

$$u_x^{\pm(0)} = \varepsilon u_x^{\pm}/l, \quad u_z^{\pm(0)} = u_z^{\pm}/l, \quad u_x^{\pm(s)} = 0,\ s \neq 0 \ \ (x,z)$$

where $\sigma_{zz0}^{(s)}$ is the solution of the differential equation

$$\frac{d^2\sigma_{zz0}^{(s)}}{d\xi^2} = -\frac{3}{2}G\left(\frac{d\overline{U}_x^{(s)}}{d\xi} + U_z^{(s)} \right) \tag{11}$$

$$\overline{U}_x^{(s)} = u_x^{+(s)} + u_x^{-(s)} - u_*^{(s)}(\zeta=-1) - u_*^{(s)}(\zeta=1)$$

i.e.
$$\sigma_{zz0}^{(s)} = A^{(s)}\xi + B^{(s)} - \frac{3}{2}G\int_0^\xi \left(\overline{U}_x^{(s)} + \int_0^\xi U_z^{(s)} d\xi \right) d\xi \tag{12}$$

So the solution of the boundary problem (1), (2) is determined by formulae (5)-(10), (12). This solution contains integration constants $A^{(s)}$, $B^{(s)}$ which are defined from the conditions, given on the end-walls $x = \pm l$. These conditions can be satisfied integrally. Particularly, if the both end-walls are free, then

$$\int_{-h}^{h} \sigma_{xx}(x=\pm l, z)dz = 0 \tag{13}$$

must be fulfilled.

In order to satisfy the boundary conditions at every point of the end-wall, it is necessary to use the solution of the boundary layer [2]. So a solution, which allows with arbitrary selected asymptotic exactness to solve all the components of the stress tensor and displacement vector in inner points of the layer, is obtained.

As an application of obtained recurrent formulae, the solutions of some particular applied problems will be brought.

a) Let uniform displacements be imparted to the longitudinal edges of the layer:

$$u_x^{\pm} = \text{const}, \quad u_z^{\pm} = \text{const} \tag{14}$$

Using the formulae (5)-(10), (12), (13), after three steps of iteration, the following exact solution of the inner problem is obtained:

$$\begin{aligned}
\sigma_{xx} &= \frac{3G}{4h^3}\left(l^2 - x^2 + 3z^2 - h^2\right)\left(u_z^+ - u_z^-\right) \\
\sigma_{zz} &= \frac{3G}{4h^3}\left(l^2 - x^2 + 3h^2 - z^2\right)\left(u_z^+ - u_z^-\right) \\
\sigma_{xz} &= \frac{G}{2h}\left(u_x^+ - u_x^-\right) + \frac{3G}{2h^3}xz\left(u_z^+ - u_z^-\right) \\
u_x &= u_x^- + \frac{z+h}{2h}\left(u_x^+ - u_x^-\right) + \frac{3}{4h^3}x\left(z^2 - h^2\right)\left(u_z^+ - u_z^-\right) \\
u_z &= u_z^- + \frac{1}{4h^3}(z+h)^2(2h - z)\left(u_z^+ - u_z^-\right)
\end{aligned} \tag{15}$$

From (15) follows, that normal stresses σ_{xx}, σ_{zz} depending on coordinate x are changing parabolically and receive their maximal values in the geometrical center of the layer. And in compressible layer, under the same boundary conditions (14) the stresses field is homogeneous, and the displacements are changing linearly along the height of the layer [2].

b) Let linear normal displacements be reported to face bounds of the layer (inclination to the angle $\pm\alpha$) (Fig.1).

$$u_x(z = \pm h) = 0, \quad u_z(z = \pm h) = \pm\alpha x \tag{16}$$

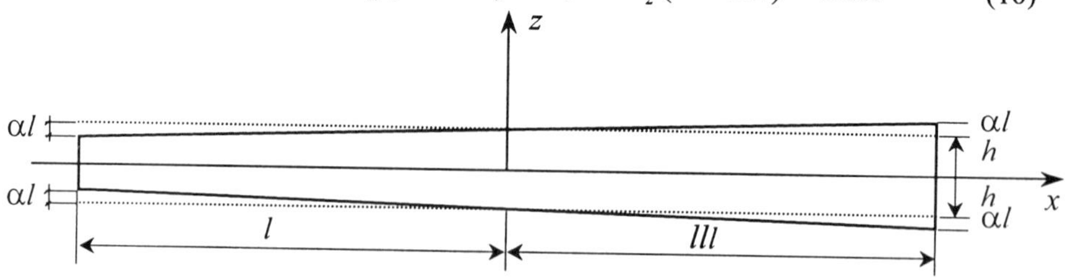

Figure 1.

After three steps the iteration stops and a mathematically exact solution of the inner problem is obtained:

$$\sigma_{xx} = \frac{\alpha G}{2h^3} x\left(l^2 - x^2\right) + \frac{3\alpha G}{2h^3} x\left(3z^2 - h^2\right)$$
$$\sigma_{zz} = \frac{\alpha G}{2h^3} x\left(l^2 - x^2\right) + \frac{3\alpha G}{2h^3} x\left(3h^2 - z^2\right)$$
$$\sigma_{xz} = \frac{\alpha G}{2h^3} z\left(2x^2 - l^2\right) + \frac{3\alpha G}{2h^3} z\left(h^2 - z^2\right) \qquad (17)$$
$$u_x = \frac{\alpha}{4h^3}\left(z^2 - h^2\right)\left(3x^2 - l^2 - z^2 - h^2\right)$$
$$u_z = \frac{\alpha}{2h^3} xz\left(3h^2 - z^2\right)$$

Note that the problems, the exact solutions of which are got in this paragraph, are basic in the calculations of the seismoisolators, and they are also discussed in [1], where approximate solutions, which for some values differ with solutions (15), (17), are obtained.

In the conclusion we shall note that recurrent formulae (5), (8) allow us to calculate stress-strain state of the layer under other conditions, given on the longitudinal edges, too.

2. Asymptotic Solution of a Boundary Problem for a Two-Layered Strip Composed of Compressible and Incompressible Layers

In section 1, we determined the principal difference between asymptotics of stress-strain states of compressible and incompressible beam-strips. Elastomeric seismoizolators, containing a two-layer beam-strip composed of compressible and incompressible layers as a basic element, have applications in contemporary construction practice. Below solutions of applied problems for two-layer seismoizolators are presented.

Assume a two-layered strip in the plain with Cartesian system of coordinates Oxz: $\Omega = \{x, z : -h_e \le z \le h,\ -l \le x \le l,\ h + h_e << 2l\}$, where layer $0 \le z \le h$ is of a compressible material and layer $-h_e \le z \le 0$ is of an incompressible material. Let displacements be given on the longitudinal edges of the strip

$$u_x(x, z = -h_e) = u_x^-(x), \quad u_z(x, z = -h_e) = u_z^-(x)$$
$$u_x(x, z = h) = u_x^+(x), \quad u_z(x, z = h) = u_z^+(x) \qquad (18)$$

and there is full contact between the layers

$$\sigma_{zz}(x, z = 0) = \sigma_{zz}^e(x, z = 0), \quad \sigma_{xz}(x, z = 0) = \sigma_{xz}^e(x, z = 0)$$
$$u_x(x, z = 0) = u_x^e(x, z = 0), \quad u_z(x, z = 0) = u_z^e(x, z = 0) \qquad (19)$$

From now on, all variables of incompressible layer will have an index e (after the word "elastic").

In general case, it is assumed that a thermal field also influences on the two-layered strip.

For the solution we use recurrent formulae obtained in [2] for the compressible layer:

$$Q=\varepsilon^{\chi}\sum_{s=0}^{S}\varepsilon^{s}Q^{(s)}(\xi,\zeta)\qquad 0\le\zeta\le\zeta_1$$

$$\chi_\sigma=-1,\quad \chi_u=0$$

$$\sigma_{zz}^{(s)}=\sigma_{zz0}^{(s)}+\sigma_{zz*}^{(s)},\qquad \sigma_{zz*}^{(s)}=-\int_0^{\zeta}\frac{\partial\sigma_{xz}^{(s-1)}}{\partial\xi}d\zeta$$

$$\sigma_{xz}^{(s)}=\sigma_{xz0}^{(s)}+\sigma_{xz*}^{(s)},\qquad \sigma_{xz*}^{(s)}=-\int_0^{\zeta}\frac{\partial\sigma_{xx}^{(s-1)}}{\partial\xi}d\zeta$$

$$\sigma_{xx}^{(s)}=\frac{\nu}{1-\nu}\sigma_{zz}^{(s)}+\frac{2G}{1-\nu}\frac{\partial\bar{u}_x^{(s-1)}}{\partial\xi}-\frac{2(1+\nu)}{1-\nu}G\alpha\theta^{(s)}\tag{20}$$

$$\bar{u}_x^{(s)}=\bar{u}_{x0}^{(s)}+\frac{1}{G}\zeta\sigma_{xz0}^{(s)}+\bar{u}_{x*}^{(s)}$$

$$\bar{u}_{x*}^{(s)}=\int_0^{\zeta}\left(\frac{1}{G}\sigma_{xz*}^{(s)}-\frac{\partial\bar{u}_z^{(s-1)}}{\partial\xi}\right)d\zeta$$

$$\bar{u}_z^{(s)}=\bar{u}_{z0}^{(s)}+\frac{1-2\nu}{2G(1-\nu)}\zeta\sigma_{zz0}^{(s)}+\bar{u}_{z*}^{(s)}$$

$$\bar{u}_{z*}^{(s)}=\int_0^{\zeta}\left(\frac{1-2\nu}{2G(1-\nu)}\sigma_{zz*}^{(s)}-\frac{\nu}{1-\nu}\frac{\partial\bar{u}_x^{(s-1)}}{\partial\xi}+\frac{1+\nu}{1-\nu}\alpha\theta^{(s)}\right)d\zeta$$

$$\xi=\frac{x}{l},\ \zeta=\frac{z}{h+h_e},\ \varepsilon=\frac{h+h_e}{2l},\ \bar{u}_x=u_x/l,\ \bar{u}_z=u_z/l$$

$$\zeta_1=\frac{h}{h+h_e},\qquad \theta=\varepsilon^{-1}\sum_{s=0}^{S}\varepsilon^{s}\theta^{(s)}(\xi,\zeta)$$

For the incompressible layer we use recurrent formulae from section 1, completed with terms representing influence of the thermal field according to Djuhamel-Neiman's model

$$Q^e=\varepsilon^{\chi}\sum_{s=0}^{S}\varepsilon^{s}Q^{e(s)}(\xi,\zeta)\qquad -\zeta_e\le\zeta\le 0$$

$$\chi_{\sigma_{xx}^e}=\chi_{\sigma_{zz}^e}=-3,\ \chi_{\sigma_{xz}^e}=-2,\ \chi_{u_x^e}=-1,\ \chi_{u_z^e}=0$$

$$\sigma_{zz}^{e(s)} = \sigma_{zz0}^{e(s)} + \sigma_{zz*}^{e(s)}, \qquad \sigma_{zz*}^{e(s)} = -\int_0^{\zeta} \frac{\partial \sigma_{xz}^{e(s-2)}}{\partial \xi} d\zeta$$

$$\sigma_{xx}^{e(s)} = \sigma_{zz}^{e(s)} + 4G_e \frac{\partial \overline{u}_x^{e(s-2)}}{\partial \xi} - 6G_e \alpha_e \theta^{(s-2)}$$

$$\sigma_{xz}^{e(s)} = \sigma_{xz0}^{e(s)} - \zeta \dot{\sigma}_{zz0}^{e(s)} + \sigma_{xz*}^{e(s)}$$

$$\sigma_{xz*}^{e(s)} = -\int_0^{\zeta} \left(\frac{\partial \sigma_{zz*}^{e(s)}}{\partial \xi} + 4G_e \frac{\partial^2 \overline{u}_x^{e(s-2)}}{\partial \xi^2} - 6G_e \alpha_e \frac{\partial \theta^{(s-2)}}{\partial \xi} \right) d\zeta \tag{21}$$

$$\overline{u}_x^{e(s)} = \overline{u}_{x0}^{e(s)} + \frac{1}{G_e} \zeta \sigma_{xz0}^{e(s)} - \frac{1}{2G_e} \zeta^2 \dot{\sigma}_{zz0}^{e(s)} + \overline{u}_{x*}^{e(s)}$$

$$\overline{u}_{x*}^{e(s)} = \int_0^{\zeta} \left(\frac{1}{G_e} \sigma_{xz*}^{e(s)} - \frac{\partial \overline{u}_z^{e(s-2)}}{\partial \xi} \right) d\zeta$$

$$\overline{u}_z^{e(s)} = \overline{u}_{z0}^{e(s)} - \zeta \dot{\overline{u}}_{x0}^{e(s)} - \frac{1}{2G_e} \zeta^2 \dot{\sigma}_{xz0}^{e(s)} + \frac{1}{6G_e} \zeta^3 \ddot{\sigma}_{zz0}^{e(s)} + \overline{u}_{z*}^{e(s)}$$

$$\overline{u}_{z*}^{e(s)} = -\int_0^{\zeta} \left(\frac{\partial \overline{u}_x^{e(s-2)}}{\partial \xi} - 3\alpha_e \theta^{(s)} \right) d\zeta$$

(dots on the top of letters mean derivatives of corresponding order with respect to x).

Note that the same-name components of stresses and displacements calculated by recurrent formulae (20) and (21) respectively for compressible and incompressible layers have different asymptotic orders.

Recurrent formulas (20) and (21) and contact conditions (19) allow us to calculate components of stresses and displacements for a two-layered strip composed of compressible and incompressible layers when on the front surface kinematic conditions (18) are given.

Solutions of Some Particular Applied Problems

a) Assume that on the longitudinal edges of the strip normal displacements are given and variation of the thermal field is constant

$$u_z^{\pm} = \text{const}, \quad u_x^{\pm} = 0, \quad \theta = \text{const} \tag{22}$$

Substituting (22) into (20) and (21), we find for the compressible layer $0 \le z \le h$

$$\sigma_{zz} = \sigma, \quad \sigma_{xx} = \frac{\nu}{1-\nu}\sigma - \frac{2(1+\nu)}{1-\nu}\frac{h\alpha}{M}\theta$$

$$\sigma_{xz} = \frac{1}{P+M}\left[\left(2h_e + h\frac{4\nu-1}{1-\nu}\right)\frac{M}{4}\dot{\sigma} - \frac{1}{6}Ph_e^3\dddot{\sigma}\right] - \left(\frac{h_e}{2} + z\frac{\nu}{1-\nu}\right)\dot{\sigma} \tag{23}$$

$$u_z = u_z^+ + (z-h)\frac{1-2\nu}{2h(1-\nu)}M\sigma + (z-h)\frac{1+\nu}{1-\nu}\alpha\theta, \qquad M = \frac{h}{G}, \quad P = \frac{h_e}{G_e}$$

for the incompressible layer $-h_e \le z \le 0$

$$\sigma_{zz}^e = \sigma + \frac{z}{2}(z + h_e)\ddot{\sigma}$$

$$\sigma_{xx}^e = \sigma - \frac{3}{2}(z^2 + zh_e)\ddot{\sigma} - 6\frac{h_e\alpha_e}{P}\theta \tag{24}$$

$$u_z^e = u_z^- + P\left(\frac{z^2 - h_e^2}{4} + \frac{z^3 + h_e^3}{6h_e}\right)\ddot{\sigma} + 3(z + h_e)\alpha_e\theta$$

where σ in the case $u_z^+ - u_z^- = ax + b$ is determined by formula

$$\sigma = A\frac{\mathrm{ch}kx}{\mathrm{ch}kl} + B\frac{\mathrm{sh}kx}{\mathrm{sh}kl} + W$$

$$W = \frac{2(1-\nu)}{1-2\nu}\frac{1}{M}\left[u_z^+ - u_z^- - \left(3\alpha_e h_e + \alpha h\frac{1+\nu}{1-\nu}\right)\theta\right]$$

$$A = C\left(T_\theta - H\frac{b - \left(3h_e\alpha_e + h\alpha\frac{1+\nu}{1-\nu}\theta\right)}{M}\right) \tag{25}$$

$$B = C\left(T_\theta - H\frac{al}{M}\right)$$

$$C = \frac{1}{\frac{\nu}{1-\nu}h + h_e + k^2\frac{h_e^3}{4}}, \qquad H = \frac{2(1-\nu)}{1-2\nu}\left(\frac{\nu}{1-\nu}h + h_e\right)$$

$$T_\theta = \frac{2\alpha(1+\nu)h^2\theta}{M(1-\nu)} + \frac{6h_e^2\alpha_e\theta}{P}$$

The strain state corresponding to (23), (24) is illustrated in Fig.2.

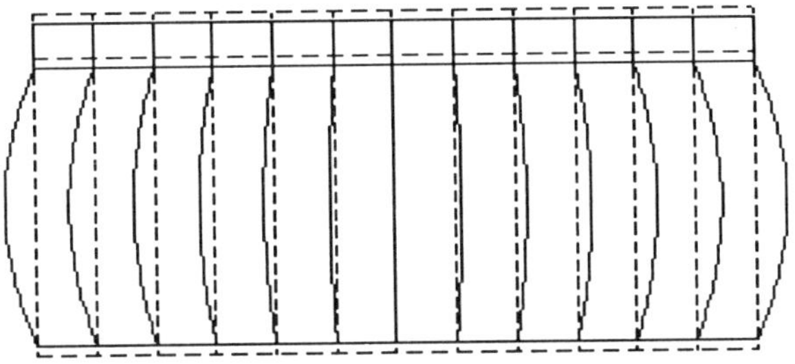

Figure 2. Dash lines - before deformation, solid lines - after deformation

b) On the longitudinal edges of the strip normal displacements variating linearly along the beam are given and variation of the thermal field is constant

$$u_z^- = 0, \quad u_z^+ = ax + b, \quad u_x^\pm = 0, \quad \theta = \text{const} \tag{26}$$

Components σ_{zz}, σ_{xx}, u_z, σ_{zz}^e, σ_{xx}^e, u_z^e are determined from formulae (23) and (24), σ is determined by formula (25) and σ_{xz}, u_x^e are determined as follows:

$$\sigma_{xz} = \frac{1}{P+M}\left[ha + \left(2h_e + h\frac{4\nu - 1}{1-\nu}\right)\frac{M}{4}\dot{\sigma} - \frac{1}{6}Ph_e^3\dddot{\sigma}\right] - \left(\frac{h_e}{2} + z\frac{\nu}{1-\nu}\right)\dot{\sigma}$$

$$u_x^e = (z+h_e)\frac{ahP}{h_e(P+M)} + \frac{(z+h_e)P}{h_e(P+M)}\left[\left(2h_e + h\frac{4\nu-1}{1-\nu}\right)\frac{M}{4}\dot{\sigma} - \frac{P}{6}h_e^3\dddot{\sigma}\right] +$$
$$+\frac{P}{12h_e}(z+2h_e)(z^3+h_e^3)\ddot{\sigma} - \frac{P}{2h_e}z(z+h_e)\dot{\sigma} \tag{27}$$

Strain state is given in Fig.3.

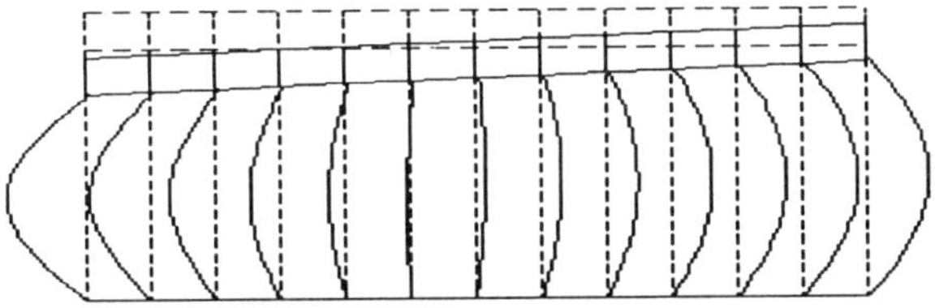

Figure 3. Dash lines - before deformation, solid lines - after deformation

On Influence of the Type of Boundary Conditions

In previous sections we made clear that there is a principal difference between asymptotics corresponding to compressible and incompressible materials if components of displacement vector are given on the front surfaces of a strip or a plate.

Let us show that the asymptotics becomes the same for compressible and incompressible materials when on one of the front surfaces components of stress tensor and on the other one components of displacement vector are given.

For example, for the following boundary conditions:

$$u_x^- = \text{const},\quad u_z^- = \text{const},\quad \sigma^+ = \text{const},\quad \tau^+ = \text{const} \tag{28}$$

we obtain

for the compressible layer $0 \le z \le h$

$$\sigma_{zz} = \sigma^+,\quad \sigma_{xx} = \frac{\nu}{1-\nu}\sigma^+,\quad \sigma_{xz} = \tau^+$$
$$u_x = u_x^- + \left(\frac{z}{G} + \frac{h_e}{G_e}\right)\tau^+,\quad u_z = u_z^- + z\frac{1-2\nu}{2G(1-\nu)}\sigma^+ \tag{29}$$

and for the incompressible layer $-h_e \le z \le 0$

$$\sigma_{zz}^e = \sigma^+,\quad \sigma_{xx}^e = \sigma^+,\quad \sigma_{xz}^e = \tau^+$$
$$u_x^e = u_x^- + \frac{z+h_e}{G_e}\tau^+,\quad u_z^e = u_z^- \tag{30}$$

3. Forced Vibrations of Incompressible Layer-Seismoisolator

Consider the behaviour of the beam-layer made of an incompressible material subjected to the action of periodically changing in time vector of displacement. The problem, particularly, models the seismic action on the base of seismoisolator.

It is required to determine the solution of dynamic equations of elasticity theory for an incompressible body in the domain $D = \{(x,z): x \in [-l,l], |z| \le h\}$

$$\frac{\partial \sigma_{xx}}{\partial x} + \frac{\partial \sigma_{xz}}{\partial z} = \rho\frac{\partial^2 u_x}{\partial t^2} \qquad (x,z)$$
$$\sigma_{xx} = \sigma_{zz} + 4G\frac{\partial u_x}{\partial x} \tag{31}$$
$$\frac{\partial u_x}{\partial z} + \frac{\partial u_z}{\partial x} = \frac{1}{G}\sigma_{xz}$$
$$\frac{\partial u_x}{\partial x} + \frac{\partial u_z}{\partial z} = 0 \qquad \text{(condition of incompressibility)}$$

under boundary conditions

$$u_x(-h) = u^-(\xi)\exp(i\Omega t)$$
$$u_z(-h) = v^-(\xi)\exp(i\Omega t), \qquad \xi = x/l \tag{32}$$

as well as under the conditions on the opposite edge

$$\sigma_{xz}(h) = 0, \quad \sigma_{zz}(h) = 0 \tag{33}$$

If on the edge $z = h$ act periodically changing loadings, then on the right parts of conditions (33) corresponding the items, which yet won't influence on the suggested way of getting the solutions of the problem, will figure.

In order to get the solution of the problem, introduce dimensionless coordinates $\xi = x/l$, $\zeta = z/h$ and dimensionless displacements $U = u_x/l, V = u_z/l$. The solution of transformed system (31) we'll seek in the form

$$\sigma_{xx} = \sigma_{11}(\xi,\zeta)\exp(i\Omega t), \quad \sigma_{zz} = \sigma_{22}(\xi,\zeta)\exp(i\Omega t)$$
$$\sigma_{xz} = \sigma_{12}(\xi,\zeta)\exp(i\Omega t) \tag{34}$$
$$U = u(\xi,\zeta)\exp(i\Omega t), \quad V = \mathrm{v}(\xi,\zeta)\exp(i\Omega t)$$

Substituting (34) into the equations of movement and Hooke's correlations, we'll have the system

$$\frac{\partial\sigma_{11}}{\partial\xi} + \varepsilon^{-1}\frac{\partial\sigma_{12}}{\partial\zeta} + \varepsilon^{-2}\Omega_*^2 u(\xi,\zeta) = 0$$
$$\frac{\partial\sigma_{12}}{\partial\xi} + \varepsilon^{-1}\frac{\partial\sigma_{22}}{\partial\zeta} + \varepsilon^{-2}\Omega_*^2 \mathrm{v}(\xi,\zeta) = 0$$
$$\sigma_{11} = \sigma_{22} + 4G\frac{\partial u}{\partial\xi} \tag{35}$$
$$\varepsilon^{-1}\frac{\partial u}{\partial\zeta} = \frac{1}{G}\sigma_{12} - \frac{\partial \mathrm{v}}{\partial\xi}$$
$$\varepsilon^{-1}\frac{\partial \mathrm{v}}{\partial\zeta} = -\frac{\partial u}{\partial\xi}$$
$$\Omega_*^2 = \rho h^2\Omega^2, \quad \varepsilon = h/l$$

As in cases of static problems and free vibrations, the correlation of elasticity $\sigma_{22} = \sigma_{11} + 4G\varepsilon^{-1}\dfrac{\partial \mathrm{v}}{\partial\zeta}$ is the consequence of the rest.

The solution of system (35) will be sought in the form of (5), where $\chi_\sigma = -1$, $\chi_u = 0$. Substituting these values into (35), for determining the coefficients of the asymptotic expansion we'll have a system and solving it will get

$$\sigma_{12}^{(s)} = G\frac{\partial u^{(s)}}{\partial\zeta} + \sigma_{12}^{*(s)}$$
$$\mathrm{v}^{(s)} = \mathrm{v}_0^{(s)}(\xi) + \mathrm{v}_*^{(s)} \tag{36}$$
$$\sigma_{22}^{(s)} = \sigma_{220}^{(s)}(\xi) - \Omega_*^2 \mathrm{v}_0^{(s)}\zeta + \sigma_{22}^{*(s)}$$
$$\sigma_{11}^{(s)} = \sigma_{22}^{(s)} + 4G\frac{\partial u^{(s-1)}}{\partial\xi}$$

where $u^{(s)}$ is determined from the equation

$$G\frac{\partial^2 u^{(s)}}{\partial\zeta^2}+\Omega_*^2 u^{(s)}=R_u^{(s-1)}$$

$$R_u^{(s-1)}=-\frac{\partial\sigma_{11}^{(s-1)}}{\partial\xi}-\frac{\partial\sigma_{12}^{*(s)}}{\partial\zeta} \tag{37}$$

and $v_0^{(s)}(\xi)$, $\sigma_{220}^{(s)}(\xi)$ are until unknown functions. The values with asterisks are calculated by formulae

$$\sigma_{12}^{*(s)}=G\frac{\partial v^{(s-1)}}{\partial\xi},\qquad v_*^{(s)}(\xi,\zeta)=-\int_0^{\zeta}\frac{\partial u^{(s-1)}}{\partial\xi}d\zeta$$

$$\sigma_{22}^{*(s)}(\xi,\zeta)=-\int_0^{\zeta}\left[\frac{\partial\sigma_{12}^{(s-1)}}{\partial\xi}+\Omega_*^2 v_*^{(s)}\right]d\zeta \tag{38}$$

The solution of equation (37) is

$$u^{(s)}=C_1^{(s)}(\xi)\sin\frac{\Omega_*}{\sqrt{G}}\zeta+C_2^{(s)}(\xi)\cos\frac{\Omega_*}{\sqrt{G}}\zeta+\bar{u}^{(s)}(\xi,\zeta) \tag{39}$$

where $\bar{u}^{(s)}(\xi,\zeta)$ is the partial solution of nonhomogeneous equation (37). Substituting (39) into (36), we'll find

$$\sigma_{12}^{(s)}=\sqrt{G}\Omega_*\left[C_1^{(s)}(\xi)\cos\frac{\Omega_*}{\sqrt{G}}\zeta-C_2^{(s)}(\xi)\sin\frac{\Omega_*}{\sqrt{G}}\zeta\right]+G\frac{\partial\bar{u}^{(s)}(\xi,\zeta)}{\partial\zeta}+\sigma_{12}^{*(s)}(\xi,\zeta) \tag{40}$$

Satisfying boundary conditions (32), (33) relative to u_x, σ_{xz} we'll find

$$C_1^{(s)}(\xi)=\frac{1}{\cos\frac{2\Omega_*}{\sqrt{G}}}\left[-f_1^{(s)}\sin\frac{\Omega_*}{\sqrt{G}}+f_2^{(s)}\cos\frac{\Omega_*}{\sqrt{G}}\right]$$

$$C_2^{(s)}(\xi)=\frac{1}{\cos\frac{2\Omega_*}{\sqrt{G}}}\left[-f_1^{(s)}\cos\frac{\Omega_*}{\sqrt{G}}+f_2^{(s)}\sin\frac{\Omega_*}{\sqrt{G}}\right] \tag{41}$$

$$f_1^{(s)}=\bar{u}^{(s)}(\xi,-1)-u^{-(s)}(\xi),\qquad u^{-(0)}=u^-/l,\quad u^{-(s)}=0,\quad s\neq 0$$

$$f_2^{(s)}=-\frac{\sqrt{G}}{\Omega_*}\left(\frac{\partial\bar{u}^{(s)}}{\partial\zeta}\right)_{\zeta=1}-\frac{1}{\sqrt{G}\Omega_*}\sigma_{12}^{*(s)}(\xi,1)$$

Using (36) and satisfying boundary conditions (32), (33) relative to u_z, σ_{zz} we'll have

$$v_0^{(s)}(\xi)=v^{-(s)}-v_*^{(s)}(\zeta=-1)$$

$$\sigma_{220}^{(s)}=\Omega_*^2 v_0^{(s)}-\sigma_{22}^{*(s)}(\zeta=1) \tag{42}$$

$$\mathrm{v}^{-(0)} = \mathrm{v}^{-}/l, \quad \mathrm{v}^{-(s)} = 0, \; s \neq 0$$

Substituting the obtained values of the integration functions into (36), (39), (40) we'll define the coefficients of the asymptotic representation and, therefore, the final solution

$$u^{(s)} = \frac{1}{2\cos\frac{\Omega_*}{\sqrt{G}}}\left[-f_1^{(s)}\cos\frac{\Omega_*}{\sqrt{G}}(1-\zeta) + f_2^{(s)}\sin\frac{\Omega_*}{\sqrt{G}}(1+\zeta)\right] + \bar{u}^{(s)}(\xi,\zeta)$$

$$\sigma_{12}^{(s)} = \frac{\Omega_*\sqrt{G}}{2\cos\frac{\Omega_*}{\sqrt{G}}}\left[-f_1^{(s)}\sin\frac{\Omega_*}{\sqrt{G}}(1-\zeta) + f_2^{(s)}\cos\frac{\Omega_*}{\sqrt{G}}(1+\zeta)\right] + G\frac{\partial\bar{u}^{(s)}(\xi,\zeta)}{\partial\zeta} + \sigma_{12}^{*(s)}(\xi,\zeta)$$

$$\mathrm{v}^{(s)} = \mathrm{v}^{-(s)} + \mathrm{v}_*^{(s)} - \mathrm{v}_*^{(s)}(\zeta=-1) \tag{43}$$

$$\sigma_{22}^{(s)} = (1-\zeta)\Omega_*^2\left(\mathrm{v}^{-(s)} - \mathrm{v}_*^{(s)}(\zeta=-1)\right) + \sigma_{22}^{*(s)} - \sigma_{22}^{*(s)}(\zeta=1)$$

$$\sigma_{11}^{(s)} = \sigma_{22}^{(s)} + 4G\frac{\partial u^{(s-1)}}{\partial\xi}$$

The solution (34), (36), (43) will be finite, if

$$\cos 2\frac{\Omega_*}{\sqrt{G}} \neq 0, \quad \text{that is} \quad \Omega_* \neq \frac{\pi}{4}\sqrt{G}(2n+1)$$

or, taking into account, that $\Omega_*^2 = \rho\Omega^2 h^2$, if

$$\Omega \neq \frac{\pi}{4h}(2n+1)\sqrt{\frac{G}{\rho}} \tag{44}$$

On the right part of (44) frequencies of free vibrations of the layer from an incompressible material occur. So, when the frequency of the outer action coincides with the frequency of free vibrations, a resonance will take place, therefore it is necessary to select the parameters of the seismoisolator so, that to avoid the resonance.

In (43) if the initial approximation is particularly restricted, we shall have

$$u^{(0)} = \frac{u^{-}}{l\cos 2\frac{\Omega_*}{\sqrt{G}}}\cos\frac{\Omega_*}{\sqrt{G}}(1-\zeta)$$

$$\sigma_{12}^{(0)} = \frac{\Omega_*\sqrt{G}u^{-}}{l\cos 2\frac{\Omega_*}{\sqrt{G}}}\sin\frac{\Omega_*}{\sqrt{G}}(1-\zeta) \tag{45}$$

$$\mathrm{v}^{(0)} = \frac{\mathrm{v}^{-}}{l}$$

$$\sigma_{11}^{(0)} = \sigma_{22}^{(0)} = (1-\zeta)\Omega_*^2 \frac{\mathrm{v}^-}{l}$$

The determination of values $u(\zeta=1)$, $\mathrm{v}(\zeta=1)$, $\sigma_{11}(\zeta=\pm 1)$, $\sigma_{12}(\zeta=-1)$ represents a certain practical interest. These values are not difficult to obtain from formulae (43), (45).

In conclusion note, that it is possible to study the influence of the seismic action on multilayered structures as well.

The paper is fulfilled to assist INTAS Grant.

References

1. Kelly J.M. (1994) The influence of plate flexibility on the buckling load of elastomeric isolators. Report No UCB/EERC-94/03, March 1994. 59p.
2. Aghalovyan L.A. (1997) Asymptotic theory of anisotrope plates and shells. M.:Nauka, Fizmatgiz. 414p.

Advances in Natural and Technological Hazards Research

Series Editor: Prof. Dr. Mohammed I. El-Sabh, *Département d'Océanographie, Université du Québec à Rimouski, 310 Allée des Ursulines, Rimouski, Québec, Canada G5L 3A1*

Publications

1. S. Tinti (ed.): *Tsunamis in the World.* Fifteenth International Tsunami Symposium (1991). 1993 ISBN 0-7923-2316-5
2. J. Nemec, J.M. Nigg and F. Siccardi (eds.): *Prediction and Perception of Natural Hazards.* Symposium Perugia, Italy (1990). 1993 ISBN 0-7923-2355-6
3. M.I. El-Sabh, T.S. Murty, S. Venkatesh, F. Siccardi and K. Andah (eds.): *Recent Studies in Geophysical Hazards.* 1994 ISBN 0-7923-2972-4
4. Y. Tsuchiya and N. Shuto (eds.): *Tsunami: Progress in Prediction, Disaster Prevention and Warning.* 1995 ISBN 0-7923-3483-3
5. A. Carrara and F. Guzzetti (eds.): *Geographical Information Systems in Assessing Natural Hazards.* 1995 ISBN 0-7923-3502-3
6. V. Schenk (ed.): *Earthquake Hazard and Risk.* 1996 ISBN 0-7923-4008-6
7. M.I. El-Sabh, S. Venkatesh, H. Denis and T.S. Murty (eds.): *Land-based and Marine Hazards.* Scientific and Management Issues. 1996 ISBN 0-7923-4064-7
8. J.M. Gutteling and O. Wiegman: *Exploring Risk Communication.* 1996 ISBN 0-7923-4065-5
9. G. Hebenstreit (ed.): *Perspectives on Tsunami Hazard Reduction.* Observations, Theory and Planning. 1997 ISBN 0-7923-4811-7
10. C. Emdad Haque: *Hazards in a Fickle Environment: Bangladesh.* 1998 ISBN 0-7923-4869-9
11. F. Wenzel, D. Lungu and O. Novak (eds.): *Vrancea Earthquakes: Tectonics, Hazard and Risk Mitigation.* 1999 ISBN 0-7923-5283-1
12. S. Balassanian, A. Cisternas and M. Melkumyan (eds.): *Earthquake Hazard and Seismic Risk Reduction.* 2000 ISBN 0-7923-6390-6

Kluwer Academic Publishers – Dordrecht / Boston / London